Modular Design
for Machine Tools

ABOUT THE AUTHOR

Yoshimi Ito, Dr.-Eng., C.Eng., FIET, is professor emeritus at the Tokyo Institute of Technology and past president of the Japan Society of Mechanical Engineers. The author of numerous engineering research papers and books, he is currently vice president of the Engineering Academy of Japan and a visiting professor at the Kanagawa Institute of Technology.

Modular Design for Machine Tools

Yoshimi Ito, Dr.-Eng., C.Eng., FIET

Professor Emeritus
Tokyo Institute of Technology

Mc
Graw
Hill

New York   Chicago   San Francisco   Lisbon   London   Madrid
Mexico City   Milan   New Delhi   San Juan   Seoul
Singapore   Sydney   Toronto

The McGraw·Hill Companies

Library of Congress Cataloging-in-Publication Data

Ito, Yoshimi.
   Modular design for machine tools / Yoshimi Ito.
      p.   cm.
   ISBN 978-0-07-149660-5 (alk. paper)
   1. Engineering design.   2. Computer-aided design.   3. Modularity
(Engineering)   4. Machine-tools.   I. Title.
   TA174.I86   2008
   621.9'02—dc22   2007041214

1 2 3 4 5 6 7 8 9 0   DOC/DOC   0 1 4 3 2 1 0 9 8

ISBN 978-0-07-149660-5
MHID 0-07-149660-2

This book is printed on acid-free paper.

McGraw-Hill books are available at special quantity discounts to use as premiums and sales promotions, or for use in corporate training programs. For more information, please write to the Director of Special Sales, McGraw-Hill Professional, Two Penn Plaza, New York, NY 10121-2298. Or contact your local bookstore.

Sponsoring Editor	**Proofreader**
Stephen S. Chapman	Terese Platten
Production Supervisor	**Indexer**
Richard C. Ruzycka	WordCo Indexing Services, Inc.
Editing Supervisor	**Art Director, Cover**
Stephen M. Smith	Jeff Weeks
Project Manager	**Composition**
Arushi Chawla	International Typesetting and
Copy Editor	Composition
Patti Scott	

Contents

Part 2 Engineering Design for Machine Tool Joints—Interfacial Structural Configuration in Modular Design

Preface

Not only in the old days, but also at present, wider availability of the machine tool is at crucial issue to enhance and rationalize the production ability of the nation, which is capable of creating wealth. With the advance of human society, the machine tool must have differing dimensional and performance specifications to a various extent, and thus the modular design has been duly employed across the whole world. The concept of modular design is, in principle, one of the most strategic ordnance in designing the machine tool, and the greater flexibility of the machine tool must be realized from the aspect of the structural design. Even in the era of NC (*numerical control* or *numerically controlled*) technology, this concept is thoroughly applicable, although the NC can provide the machine with the flexibility to a large extent by only exchanging the NC information. In fact, the structural configuration should be regenerated when the required flexibility is far beyond from that capable of being provided by the NC information only. In general, modular design sounds like *standardization*, i.e., less expandability for the structural configuration; however, the modular design is, in fact, a synergy of flexible configuration and standardization. This is derived from the modular design of hierarchical type, which was proposed by Brankamp of Aachen and Herrmann of Langen, Germany, in 1969.

Importantly, the modular design for machine tools has a long history since the 1930s, and its representative terminology has changed. As a result, there are now a handful of variants of modular design simultaneously arising out of the confusion in the related terminologies and technologies to a certain extent. More specifically, it is recommended to consider the *four design principles*, i.e., *principles for separation, unification, connection, and adaptation*, when studying on, conducting the research into, and developing the leading-edge technology of the modular design. These principles were first proposed by Doi of Toyoda Iron Works in 1963 and are applicable even in the year 2000 and beyond. Within an engineering context, these principles can be converted into the threefold cores, i.e., (1) concept and engineering guide of modular design, (2) design methodology for modular design, and (3) engineering design for machine tool joints.

Consequently, this book consists of the two parts, i.e., Part 1, Engineering Guides of Modular Design and Description Methodology of Machine Tools, and Part 2, Engineering Design for Machine Tool Joints—Interfacial Structural Configuration in Modular Design. In addition, another kernel technology in the modular design is how to measure the interface pressure between both modules by the nondestructive method, and thus the ultrasonic waves method is furthermore stated in App. 1. Obviously, the book is available for the individual use of each part and for the synthetic use, depending on the reader's requirements.

In retrospect, the history of modular design can be classified into three phases in full consideration of the epoch-making proposals of concept and idea, innovation, contrivance, and marked applications.

In the first phase up to the 1970s, the modular design was, in wider scope, applied to the structural body component of the conventional machine tool to rationalize the design and manufacture. Geminately, the modular concept was applied to the system design of the TL (*transfer line*) to reduce primarily the renewal cost of the line. The fundamental engineering technique and methodology of the modular design were duly established in this phase through the vigorous activities during the 1960s.

With the advent and growing importance of the NC machine tool in the late 1950s and 1960s, the modular design was launched out to development of the second phase. In this phase, the modular design was characterized by its capability for reinforcing further flexibility of the machining method from the aspect of the structural configuration. As widely recognized, the NC technology is an eminent innovation, which can be considered equal to that of Wilkinson's cylinder boring machine in the industrial revolution era. Consequently, some new machine tools and production systems, i.e., MC (*machining center*), TC (*turning center*), and flexible manufacturing, were contrived positively by using the NC and computer technologies, and the modular design has been applied to these machines and production systems, depending on their necessities. Nowadays, the MC, TC, GC (*grinding center*), flexible manufacturing, and their variants are dominant in the machine tool sector.

Following the second phase up to the mid-1980s, the modular design has been in the third phase, although it is difficult to observe obvious differences from those in the second phase. Two representative applications are for the FMC (*flexible manufacturing cell*) and the system machine, i.e., machine tool compatible with flexible manufacturing. With the advance of system machine, at issue is the machining or processing complex.

In the beginning of the twenty-first century, there have been some symptoms for new modular design concepts with the growing importance of localized globalization in the production in full consideration of the compatibility of the production facilities with the natural environment. More specifically, the modular design is being requested to incorporate the modules in closer relation to the multiple-function integrated type, culture- and mindset-harmonized type, and to the environment-harmonized types, e.g., indigenous available, performance up-to-date, and LCA (*life cycle assessment*) modules.

Meanwhile, a tedious work in the modular design is to determine a group of modules (principles of separation and unification) and a suitable combination of modules from it in accordance with the design specifications (principles of connection and adaptation). Up to the second phase, the designer managed these works with the trial-and-error method, especially based on her or his long-standing experience and flair. A new innovation in the third phase has thus been the machine tool description to assist the modular design of software aspect by the computer. In fact, the machine tool can be represented by, e.g., the directed graph, that is, structural pattern, where each vertex has its own property. Thus, the combination of modules, e.g., generation of structural configuration (pattern), can be carried out without any difficulties, with the aid of graph theory, where the structural pattern is converted to the adjacency matrix suitable for the computation. In contrast, the static and dynamic stiffness of the jointed surfaces was vigorously investigated at the second phase. The joint stiffness is one of the leading factors in the application of the principle of connection to the practical design, and a sphere called *engineering design for machine tool joints* was duly established, although there remains something to be seen. As a result, at present we can calculate or compute the static and dynamic stiffness and the thermal deformation of the machine tool as a whole in consideration of the joint to some extent. In fact, there are significant differences between both the structures with and without joints. In addition, nearly all industrial machines are designed on the *allowable stress principle* whereas the machine tool is, as widely known, designed on the *allowable deflection principle*, and thus the joint deflection is very dominant in the hardware aspect of the machine tool design.

In summary, the modular design is very popular now; however, the machine tool engineer often faces difficulties to get some reference books for the modular design together with touching on its long-standing history. This book can systematically provide the reader with necessary and valuable knowledge about the modular design, ranging from the basic idea and engineering guides, through the machine tool description, to the engineering design of the machine tool joint. In addition, the book touches on a valuable experimental technique, i.e., measurement of the interface pressure by means of ultrasonic waves. The author is proud of it and wishes furthermore that the book is somewhat helpful for not only machine tool engineers, but also engineers in other spheres, where the modular design is thriving such as the automobile industry. In short, the book is suitable for the CPD (*continuing professional development*) for the mature engineer.

Finally, the author would like to express his sincerest thanks to the friends listed below, who supported him vigorously by collecting valuable information and materials to various extents:

Dr.-Ing. H. Hammer of Fritz Werner, Dr.-Ing. E. Moritz of Sportskreativwerkstatt of Technische Universität München, Mr. T. Nojiri of Dunlop Co., Dr.-Ing. G. Seliger of Technische Universität Berlin, Dr.-Eng. S. Shimizu of Sophia University, and Dr.-Eng. H. Shinno of Tokyo Institute of Technology

Within a book publication context, Dr. Ruth of the Universität Bremen cooperated with the author in getting the copyright permissions, and Mr. Bok of the Singapore office and Mr. Chapman of the U.S. office of McGraw-Hill assisted the author in publishing the book to larger and various extents. In addition, Mr. Hori, President of Japan Kistler, devoted himself to refining the illustrations from the financial resources aspect. The author would like simultaneously to express his sincerest thanks to them.

Yoshimi Ito, Dr.-Eng., C.Eng., FIET

Terminology and Abbreviations

Within an engineering context, so far there have been a handful of cases in which a technological system changed the terminology to represent properly its sphere in accordance with the due development and evolution, although its principles and essential features have remained in the original states. In retrospect, the modular design of the machine tool has been innovated, modernized, and developed to a various extent since the 1930s, simultaneously changing its representative terminology, i.e., unit construction, building block system, modular design, holonic design, reconfigurable construction, and platform. In contrast, there are various abbreviated terms and jargon to represent core technologies, production systems, organizational structures, and so on within the production sphere, often resulting in some confusion. For the ease of understanding and to avoid unnecessary confusion, the key term *modular design* or *modular principle* will be commonly used throughout this book, and there is a reference table for abbreviated terms. Furthermore, the reader is requested to refer to the developing history of modular design within this book, when finding something uncertain in relation to the terminology.

In this context, the author would like to touch on something definite regarding the key term *modular design* here. The concept and method of the modular design for structural configuration are credited to Drs. G. Schlesinger and F. Koenigsberger in the 1930s. The former proposed the concept by exemplifying it through the design for the headstock of the radial drilling machine, and the latter applied the method to the design of the milling machine of Wanderer make. Actually, Prof. Schlesinger, the eminent leading engineer in the machine tool sphere, was the supervisor of Prof. Koenigsberger [1]. On that occasion, the modular design was, in general, called *BBS*, which is the acronym of the *building block system* (das Baukasten System). During the evolution and deployment over 70 years afterward, various terms were used to represent the modular design, although the essential features within the concept and method have remained the same. For example, U.S. engineers used the term *unit construction principle* before World War II. Importantly, the term *BBS* was duly replaced with *modular design* by Prof. Koenigsberger himself around 1967, when he served at the University of Manchester Institute of Science and Technology. On that occasion, the NC machine tool became accidentally thrived. We may guess that he proposed this new term to emphasize the user-oriented aspect of the BBS. Up to today, the modular design has encompassed its sphere to a various extent, ranging from the production system and machine tool, through the software and cutting tool, to other products, although remaining its major application area within the structural design of machine tools. In due course, machine tool engineers and related people have been familiar with this terminology. In contrast, there is now some confusion in the terminology with the advance of the modular design into some new spheres. Intuitively, this confusion is caused by the uncertain definition of each key term. In addition, confusion in terminology becomes often more intricate by a new proposal, which has not been reviewed the past and present perspectives of the related subject in five fathom deep.

Professors Koren and Ulsoy have, e.g., asserted the importance of reconfigurable manufacturing systems in their keynote paper of CIRP [2]. However, it appears that they must refer to the effective application of the modular design principle to the TL in the 1960s, in which the user was able to replace some modules within the user's factory. Regarding flexible manufacturing in the 1980s and beyond, e.g., expandable FMC of Hitachi Seiki make and FML *(flexible machining line)* of Fritz Werner make, the system and line have, furthermore, enough flexibility that the user has no need to replace the module within the user's factory. In fact, there is obviously a difficulty of distinguishing the technological difference between the *reconfigurability* and the *flexibility of hierarchical type* according to Heisel and Michaelis [3]. In addition, Koren and Ulsoy have asserted that a variant of the reconfigurable manufacturing systems is that of combining the flexibility of the FMS with the high throughput and low-cost dedicated manufacturing lines. Such a system has, however, already been established as an FTL *(flexible transfer line)*-FMC complex of Toyoda Iron Works make, which can also be regarded as a variant of modular-configured FMS of hierarchical type. Moreover, it appears that reconfigurable manufacturing is one of the variants of agile manufacturing with on-the-spot replacement function of the module, modular complex, and units of a machine within a system. In other words, a variant in the proposal of Koren and Ulsoy can be interpreted as a synergy for the modular principle of FTL and conventional machine tool from both the hardware and the software.

Following Koren and Ulsoy, Metternich and Würsching proposed a concept represented with the term *platform*. This concept is substantially new and can be interpreted as a mix concept based on mono- and plural-layered type in the modular design [4]. In 2004, furthermore, Abele and Wörn proposed an idea by combining the platform concept with reconfigurability; however, this idea can be regarded as a variant of the modular design of hierarchical type for producing the different kinds, when changing the viewpoint [5]. In due course, at burning issue is how to incorporate the definition of the platform within the term *modular design*.

A basic necessity is thus to authentically define the modular design with wider acceptance. Within a system configuration context, furthermore, the modular design has been and is being widely accepted in close relation to the system flexibility. In retrospect, the TL was once classified into a kind of machine tool, and thus in this book, the modular design in flexible manufacturing will quickly be shown in Figs. 1-6 and 1-9. In such cases, the author will use the key term *flexibility*, although we have other key terms, e.g., *versatility, expandability, agility,* and *reconfigurability*. In this case, the author asserts the following.

Flexibility is a definition of space domain, i.e., the flexibility of system configuration including those provided by the NC technology, whereas *agility* is the flexibility of space and time domains, i.e., time series-like flexibility of system configuration. Thus, the agility is worth proposing in the next phase of production systems.

Furthermore, we have employed the term *holonic manufacturing system* since the mid-1980s. In short, holonic concept correctly means the fusion effects in functionality and performance of the two entities into one entity having those effects more than the two entities. Marshall and Leaney [6] suggested that the concept of holon be credited to Koestler in 1967, and they stated the definition of the holon as follows.

Holons are autonomous self-reliant units which have a degree of independence and handle contingencies without asking higher authorities for instructions; simultaneously, holons are subject to occasional control from higher authorities.

Importantly, the concept of holon can be considered available for the manufacturing system; however, some machine tool manufacturers have recently characterized their products from the commercial-based viewpoint by the term *holonic machine tool*. This trend induces further terminology confusion nowadays.

AGV: Automated guided vehicle

ATC: Automatic tool changer

BBS: Building block system (das Baukastensystem)

CAD: Computer aided design

CAE: Computer aided engineering

CAPP: Computer aided process planning

CIM: Computer integrated manufacturing

CNC: Computerized numerical control

FEM: Finite element method

FMC: Flexible manufacturing cell

FML: Flexible machining line

FMS: Flexible manufacturing system

FOF: Flow of force (der Kraftfluß)

FTL: Flexible transfer line

GC: Grinding center

GT: Group technology

HSK: Hollow shank tool

ISO: International Standards Organization

LCA: Life cycle assessment

MC: Machining center

MTIRA: Machine Tool Industry Research Association

NC: Numerically controlled, numerical control

NCTL: Numerically controlled transfer line

QFD: Quality function deployment

SME: Small- and medium-size enterprise

TC: Turning center

TL: Transfer line

UMIST: University of Manchester Institute of Science and Technology

VDI: Vereinigte Deutscher Ingonier

References

1. Spur, G., *Produktionstechnik in Wandel*, Carl Hanser Verlag, München/Wien, 1979.
2. Koren, Y., et al., "Reconfigurable Manufacturing Systems," *Annals of CIRP*, 1999, 48(2): 1–14.
3. Heisel, U., and M. Michaelis, "Reconfigurable Manufacturing Systems," *Production Engg.*, 2001, 8(1): 129–132.
4. Metternich, J., and B. Würsching, "Plattformkonzepte im Werkzeugmaschinenbau," *Werkstatt und Betrieb*, 2000, 133(6): 22–29.
5. Abele, E., and A. Wörn, "Chamäleon im Werkzeugmaschinenbau," *ZwF*, 2004, 99(4): 152–156.
6. Marshall, R., and G. Leaney, "Holonic Product Design: A Process for Modular Product Realization," *J. of Eng. Des.*, 2002, 13(4): 293–303.

Nomenclature

The contents of this book cover the modular design of machine tools to a various extent, ranging from the engineering and methodology for the modular design, through the joint stiffness, to the application of the ultrasonic waves. As a result, we must be aware of the importance of avoiding unnecessary confusion in the nomenclature. Thus, in this book, the leading nomenclature has been determined in each part in full consideration of that conventionally and widely used so far, and duplication has been eliminated as much as possible.

1. Nomenclature for Part 1

A, B, C: Rotational components in Cartesian coordinate
A_{ST}, B_{ST}: Structural patterns
B_S, C_S, D_S: GT codes
B_S', C_S', D_S': GT codes
D, E: Rotational components in auxiliary Cartesian coordinate
I_r: Rate of pattern similarity
L_f: Solution field
O_j: Connecting point of both structural bodies (components)
Q_d: Distribution rate
S_r: Rate of commonness
T_j: Transfer matrix
U, V, W: Rectangular components in auxiliary Cartesian coordinate
WP: Cutting point in functional description
X, Y, Z: Rectangular components in Cartesian coordinate
X_s, Y_s: Sets

a: Positioning movement
d_i: Deviation distance from diagonal line in reference matrix
dM: Maximum value in d_i
i, j, k: Local Cartesian coordinate
n_e: Number of entities consisting of part pattern

p_e: Number of identical entities
u: Feed movement
v: Cutting movement or movement in general

Λ_j: Deviation vector
δ_j: Vector
λ_{ji}: Linear deviation of point O_j
ψ_{ji}: Rotation of structural body component about axis passing through point O_j

2. Nomenclature for Part 2

A: Cross-sectional area, bearing area
A_a: Apparent contact area
A_r: Real contact area
C: Constant to represent joint stiffness, coefficient of contact compliance
C_k: Compliance
C_r: Load distribution factor
D: Damping ratio
E: Young's modulus, modulus
E_{loss}: Energy loss per cycle
F: Cutting force
F_r: Frictional force
F_{drv}: Driving force
H: Height of column, representative length or height
I: Cross-sectional moment of inertia (second moment of area)
K (K_R, K_{SB}, K_{ST}, K_{s0}, K_a, K_b, K_{drv}, K_{dyn}, K_e, K_{eq}, K_{sep}, K_{sh}, K_{sn}, K_r): Stiffness
K_0: Stiffness of joint surroundings
K_j: Joint stiffness
L (L_G, L_x, L_z): Representative dimension of guideway, protruded length
M: Moment
N: Rotational speed
P (P_{dyn}, P_h, P_r, P_{st}, P_r), ΔP: Load
Q: Normal load, normal preload, tightening force
Q_D: Magnification factor
R (R_{CLA}, R_T, R_a, R_{max}, R_{rms}): Surface roughness
R_A, R_B, R_C: Reaction forces
S: Constant
T: Torque
ΔT: Temperature difference at contact surface
V_T: Period of vibration
W_V: Volumetric weight of water
X, Y, Z: Cartesian coordinate

a: Vibration amplitude
b: Width of beam
b_δ: Power law distribution
b^*: Coefficient
c: Damping coefficient
d: Diameter
f: Frequency
f^*: Frictional supporting force per unit area
h: Height of beam, depth of dovetail, flange thickness
k (k_g, k_{m1}, k_{m2}), k^*: Static stiffness
l (l_a, l_b, l_c, l_p, l_w, l_{yc}): Length
Δl: Equivalent contact length
m: Exponent to represent joint stiffness
n: Jointing number
n_i: Number of rows
n_z: Number of rollers per row
p: Interface pressure
q: Applied stress
q^*: Heat flux
r: Radius
r_c: Thermal contact resistance
s: Sliding velocity
s_D: Specific damping capacity
t: Time
u, Δu: Microdisplacement
u_s, Δu_s: Microslip
v: Sliding velocity, traveling speed
x, y, z: Coordinate in relation to dimensions
x, x^*, Δx, y, z: Deflection, displacement
Δz: Thermal elongation

α: Vertical angle of pressure cone
α_T: Taper angle
α_λ: Thermal conductance, thermal conductivity
Δ_r: Fitting tolerance
δ: Deflection (joint deflection)
δ_D: Logarithmic damping decrement
ε: Machining error
ε_x, ε_y, ε_z: Strains
ζ: Coefficient to determine microslip
η: Loss factor
θ: Directional orientation angle
λ: Joint deflection
μ: Macroscopic coefficient of friction

μ_T: Tangential force ratio
ξ_g, ξ_k: Compensation factor related to slideway configuration
τ: Shear stress
$\varphi, \Delta\varphi$: Inclination angle
φ_T: Torsional angle
ψ: Damping factor, energy loss factor
ψ_T: Torsional angle
ω: Angular frequency

3. Nomenclature for Appendices

A: Cross-sectional area
B_0, B_2: Constants
C: Sound velocity in medium
E: Young's modulus
E_R: Echo height ratio ($= h_e/h_{e0}$)
E_R^*: $1 - E_R$
I: Cross-sectional moment of inertia (second moment of area)
K: Stiffness
L: Representative length
M: Moment
P_S: Sound pressure
R: Surface roughness
R_p: Reflection rate of sound pressure
S, ΔS: Area
W: Representative load
WDE: Energy of reflected echo

b: Width of beam
f: Frequency
g: Acceleration of gravity
h: Height (thickness) of beam
h_e: Height of echo on CRT
h_{e0}: Height of initial pulse on CRT
h_w: Transformed echo height
t: Time
x: Reflected echo
y: Deflection of beam, vibration amplitude of beam

γ: Specific weight
κ_s: Load scale factor
λ_s: Length scale factor
ρ: Density of medium
τ_s: Time scale factor

Conversion Table

The research and engineering development into the modular design and machine tool joint were first launched in Germany and the U.S.S.R., and then prevailed across the whole industrial nations through the activities of the United Kingdom. In addition, their history can be traced to the 1930s.

To respect the originality and priority of the materials to be referred to, the numerical unit within each material is not converted to the SI unit within this book, and thus the reader is requested to refer to the following conversion table, if necessary.

Length

$1\ \mu\text{in} = 0.0254\ \mu\text{m}$

Force

$1\ \text{kgf} = 9.80665\ \text{N}$
$1\ \text{tonf} \fallingdotseq 9.8\ \text{kN}$
$1\ \text{lb} = 4.448\ \text{N}$
$1\ \text{(short) ton} = 2000\ \text{lb} = 907.18\ \text{kgf}$
$1\ \text{(long) ton} = 2240\ \text{lb} = 1016.05\ \text{kgf}$

Pressure

$1\ \text{kgf/cm}^2 = 9.80665 \times 10^4\ \text{Pa} \fallingdotseq 0.098\ \text{MPa}$
$1\ \text{(long)ton/in}^2 = 157.\,48\ \text{kgf/cm}^2 \fallingdotseq 15.4\ \text{MPa}$
$1\ \text{lb/in}^2 = 0.07\ \text{kgf/cm}^2 \fallingdotseq 7\ \text{kPa}$

Torque and Moment

$1\ \text{kgf} \cdot \text{m} = 9.80665\ \text{N} \cdot \text{m}$
$1\ \text{lb} \cdot \text{in} = 0.113\ \text{N} \cdot \text{m}$

Stiffness

$1\ \text{kgf/mm} = 9.80665\ \text{N/mm}$
$1\ \text{kgf/}\mu\text{m} = 9.80665\ \text{N/}\mu\text{m}$
$1\ \text{lb/in} = 175.1\ \text{N/m}$

Engineering Guides of Modular Design and Description Methodology of Machine Tools

1

Basic Knowledge: What Is the Modular Design?

These days the modular principle is a very popular method in the design of the automobile, diesel engine, home appliances, information devices, industrial equipment, and so on. This trend can be considered as one of the great contributions of modular design of machine tools to those working in other industries. In retrospect, the predecessor of the current modular design appeared explicitly at the beginning of 1930s, and since then the related technologies have been duly advanced, revealing the remarkable impact of the modular principle not only on the machine tool design itself, but also on other products. The machine tool engineer is proud of the modular design. However, there are, in contrast, some difficulties in understanding exactly what modular design is and its historical background, which dates to the beginning of the 1930s. For example, the modular design can be classified into a considerable number of variants, depending on the idea, aims, and scope of the application, application area, expected advantages, and so on. In addition, the terminology of modular design itself has changed together with the hierarchical ramifications of its meaning, as already described in Terminologies and Abbreviations. It is thus very difficult to represent modular design in a simple sentence; however, we need a quick statement to understand the essential features of modular design. At present, employment of modular design in the manufacturing sphere ranges from the tool, jig, and fixture, through the machine tool, to the manufacturing system. In the following, several illustrations and some typical examples will be shown.

Figure 1-1 shows a representative modular tooling system proposed by Sandvik Co. in the middle of the 1980s. The tooling system was marketed under the commercial name *Block Tool System*, and it was duly

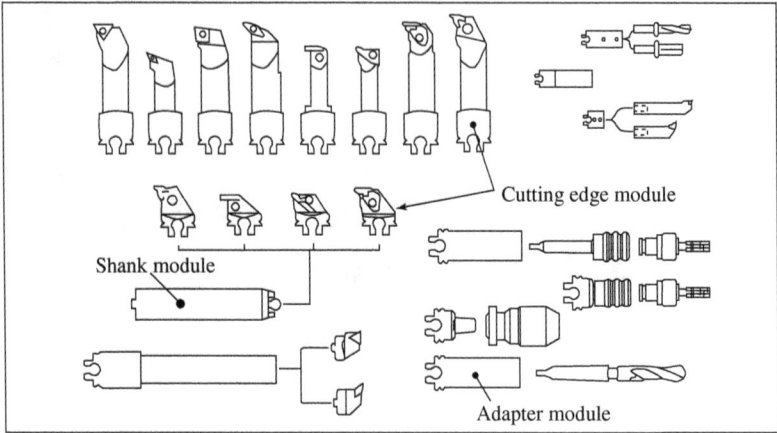

Assembly diagram

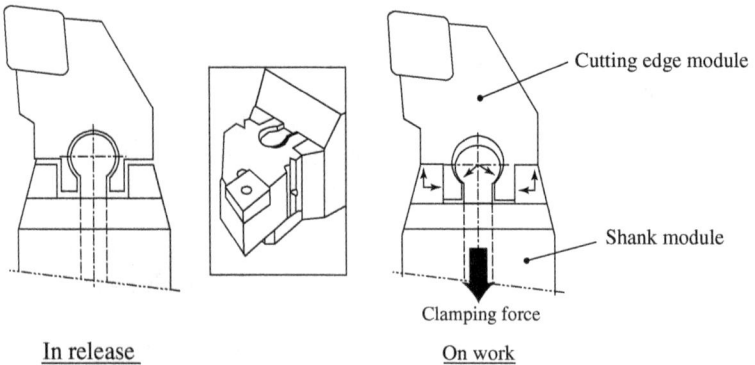

In release   On work

Clamping mechanism of cutting edge to shank modules

Figure 1-1   Modular tooling system in 1980s—basic ideas (courtesy of Sandvik Co.).

characterized by its wider tooling flexibility, which can be realized by exchanging the cutting edge module in accordance with the machining requirements. Actually, a tooling system consists of the shank (fixing), adapter (extension), and cutting edge modules, and the adapter module may assist to reinforce the further flexibility [1]. This tooling system was employed on a CNC (*computerized numerical control*) lathe of the George Fischer make (type NDM-16), in which the cutting edge modular was stored in the tool magazine of drum type and transferred to the machining space by the overhead traveling robot. Figure 1-2 displays another modular tooling system produced by Nikken Co., showing the effectiveness of the modular concept even in the year 2000. The modular principle was furthermore applied to the tool layout on the turret, e.g., base and tool holding blocks, as shown in Fig. 1-3 [2].

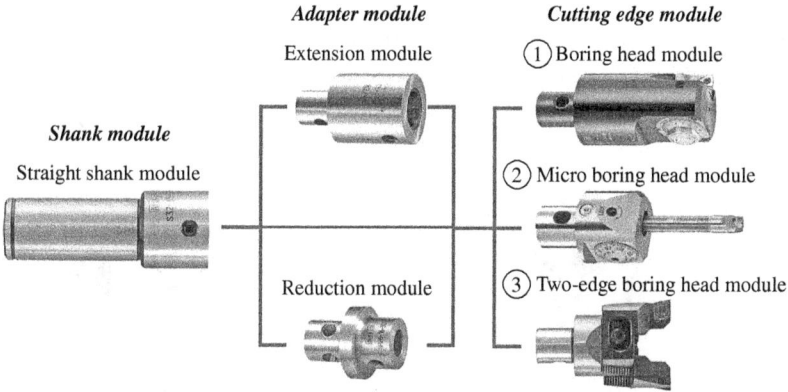

Figure 1-2 Modular boring system (courtesy of Nikken Kosakusho Works).

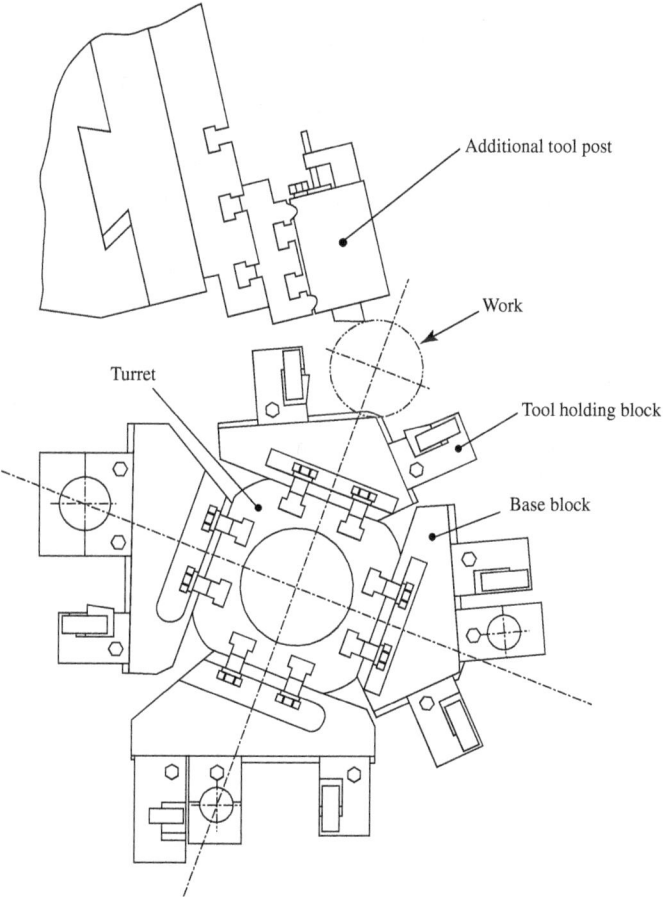

Figure 1-3 Modular principle applied to tool layout on turret (by Dietz, courtesy of Carl Hanser).

Figure 1-4 shows an advanced variant of MC (*machining center*) of Ikegai make (type MX3) in the beginning of the 1980s, which is a column traveling type. As can be seen, the modular design is preferably employed to produce 10 variants, ranging from the FMC (*flexible manufacturing cell*) of pallet pool type, through the station of FTL

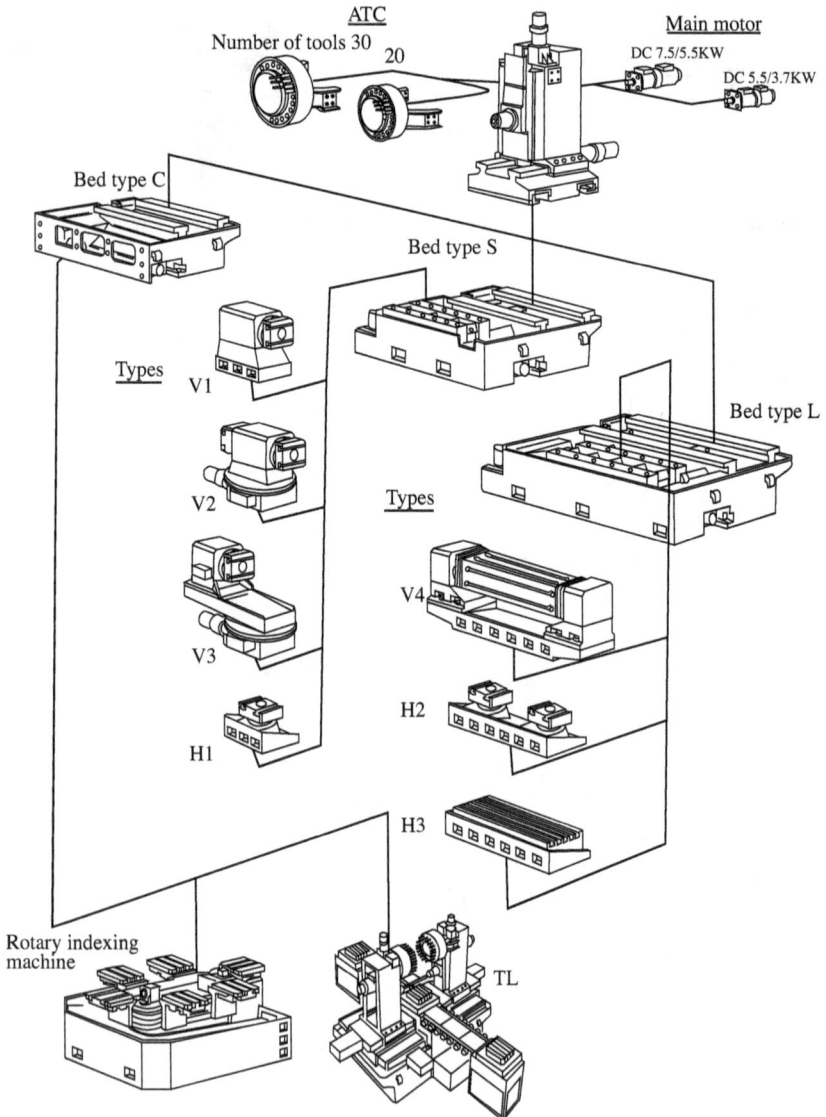

Type MX3 : Maximum spindle speed 4500 rpm. Allowable torque 3.6 kg·m. Spindle taper no. 40

Figure 1-1 Modular design in MC (courtesy of Ikegai Iron Works).

(*flexible transfer line*), to a five-face processing machine. In this case, the leading modules are the column, base, rotary table, tool magazine, main motor, and so on. This machine appears to be a typical predecessor of current five-face processing machines.

In the late 1990s, we can observe another eminent example of the application of modular design to the turning machine of the hanging spindle type (Index-Werke brand, commercial name: *Verticalline*). As shown in Fig. 1-5, the major modules of the machine are those for several structural body components, hanging spindle, turret head with either rotating tools or stationary tools, tool post fixed on the platter, and work pool stand. The platter can furthermore accommodate the motor-driven cutting tool and gang head on itself, resulting in greater flexibility in machining when varying the combination with the turret head. In addition, the machine can be characterized by some functions for laser welding, hardening, grinding, and assembly operation. The machine can be thus called the processing complex and appears to be a successor of the prototype named the "Complex Processing Cell of T-form." This prototype has been developed so far by the Japanese Big National Project entitled "Complex Production Systems Using High Efficiency Laser Processing" [3].

Within a system context, a typical modular design can be observed in the FTL of Diedesheim brand, as shown in Fig. 1-6 [4], and its core

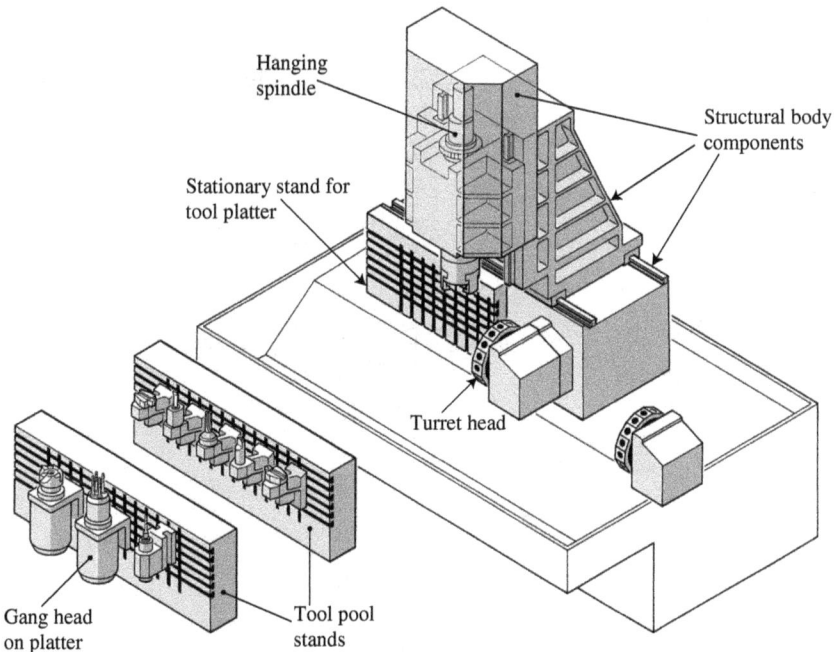

Figure 1-5   An application of modular design to the turning machine (type Vertical Line, 1999, courtesy of Index).

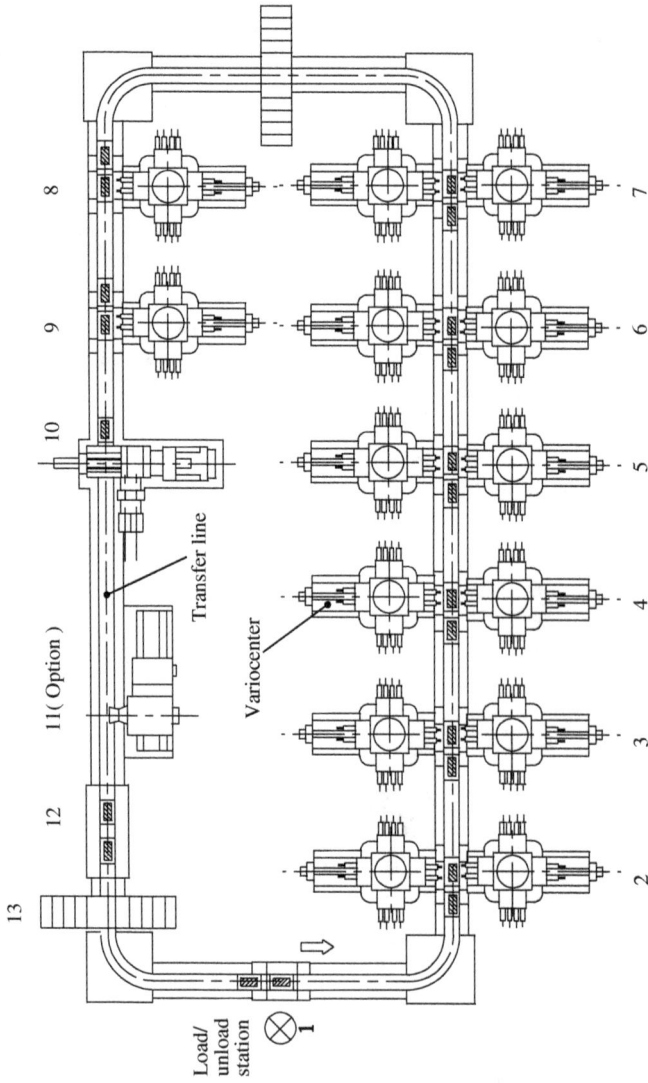

Figure 1-6 FTL for producing automobile parts (1980s, courtesy of Diedesheim): (a) Line flow type and (b) FMC-integrated type.

8

Main transfer line

Subtransfer line

Load/unload station

Buffer for waiting work

Variocenters

(b)

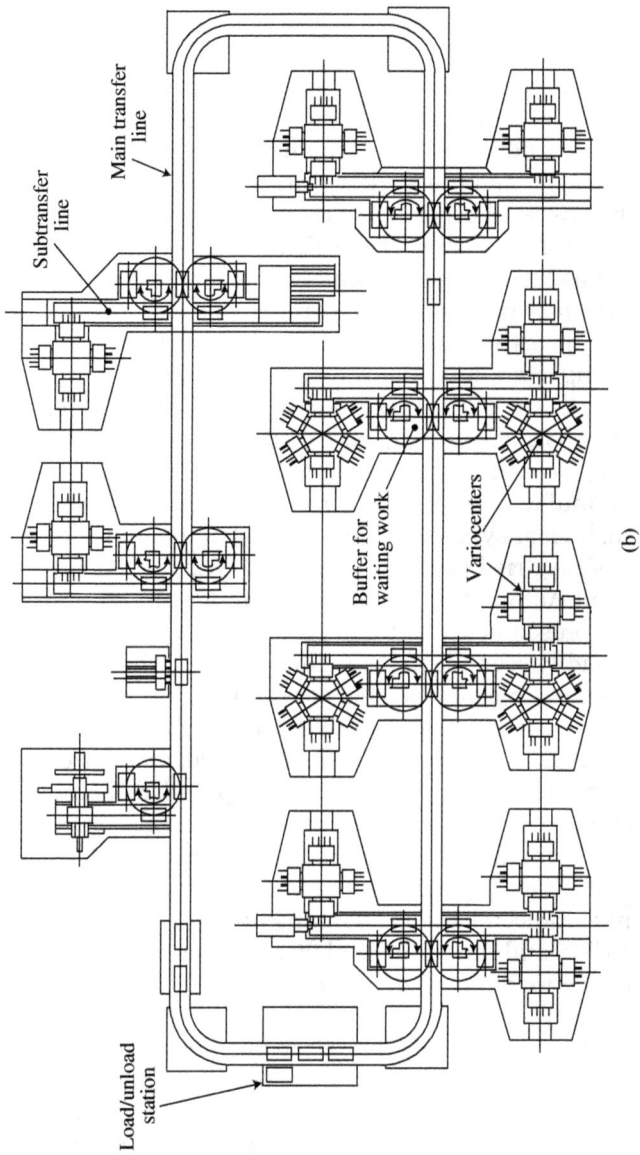

Figure 1-6 (*Continued*)

9

machining function consists of the head changers of modular type called *Variocenter*. Importantly, there are two types of FTL depending on the basic module and flexibility of the transfer line of asynchronous type. In the FTL of simple line flow type shown in Fig. 1-6(a), the basic module is that of *Variocenter* itself, resulting in less flexibility in the work transfer function, whereas in another FTL shown in Fig. 1-6(b) the basic module is of FMC to enhance the flexibility in both the machining and transfer functions. More specifically, the FMC can be formed from the *Variocenters* of various types in addition to having both the subtransfer line, i.e., transfer shuttle conveyor, and the work waiting station, which are capable of the leapfroglike work transportation, resulting in greater flexibility in the transfer function of the system. The FTL in Fig. 1-6(b) has been installed at Opel to handle the increasing number of engine varieties. In fact, the kernel of *Variocenter* is a hexagonal turret having a group of cutting tools to machine the objective work. The turret and work can be transported to the machining space by using the overhead crane and the carrier on the floor, respectively. Thus the system can facilitate drilling, deep hole drilling, counterboring, reaming, spot facing, tapping, precision milling, precision facing, and inspection. Then the system is available, for example, for the manufacture of cylinder heads and cylinder blocks made of gray cast iron, high alloy cast iron, and die cast aluminum alloy. According to the report of Siegfried at the International Symposium on Automotive Technology and Automation in 1984 held in Milan, 80 percent of the system can be reused in the event of product changes.

As can be readily seen, flexible manufacturing of the FMC-integrated type was an eminent contrivance from the modular design viewpoint, and even in the year 2000 it was the leading system design methodology. In due course, the FMS (*flexible manufacturing system*) of FMC-integrated type was to become a reality by the ZF (Zahnrad Fabrik Friedrichshafen GmbH), one of the representative mission gearbox manufacturers in Germany, for producing gears on that occasion. In the case of FMS of ZF make, one of the marked features was that it facilitated the inheritance of the craftsmanship by using the job rotation between the FMS and the traditional factory. This feature leads us later to an idea of the modular design of culture- and mindset-harmonized type (see Chap. 2).

These examples may help the reader to imagine what modular design is to some extent; however, we must discuss these in detail to best use the modular design. In this chapter, first we give a quick summary to deepen the reader's understanding of the essential features and to point out the advantages and disadvantages of modular design. Next we give a firsthand view of the long history of modular design by clarifying the epoch-making events and depicting some representative achievements made thus far.

1.1   Definition and Overall View
of Modular Design

The term *modular design* is very simple; however, its definition has been entangled and complicated. This is attributed to the ramifications of the engineering application of modular design during its long developing history, and to the noteworthy variants developed to various extents, even though the design principle remains the same. In short, it is desirable to accept the following definition in full consideration of nearly all the proposals so far suggested.

After having been determined a group of the modules, a machine tool with the required dimensional and performance specifications as well as required functionality can be designed and manufactured by choosing and combining the necessary modules from a predetermined group. In this case, a module must be standardized so as to have a functionality or performance including the interchangeability to other modules. In most cases, furthermore, a group of the standardized modules should be arranged in the dimensional specifications with the standardized numerical series together with maintaining qualitatively the same structural configuration.

Importantly, there have been a handful of proposals for the definition of the modular design, and all these definitions appear to be the same, as shown in the following. In other words, these proposals may verify the availability of the above-mentioned definition.

1. Up to the middle of the 1960s, modular design was applied to both the TL (*transfer line*) and the conventional machine tool under the common acronym of BBS (*building block system*). On that occasion, the BBS was defined as follows, and this definition is applicable even now by merely changing the terms *unit* and *module*.

   A machine tool with new function and structural configuration can be produced by choosing and integrating the units in full consideration of specified machining requirements, where a group of the standardized units are determined beforehand. In standardization of the unit, the following two aspects are, in principle, to be considered.

   (a) Each unit must have core or meaningful functionality and/or structural configuration.

   (b) Each unit must have the dimensional and configuration specifications to be joined to other units, i.e., guarantee of interchangeability.

2. Brankamp and Herrmann proposed, with wider scope, a definition of BBS that is reproduced in German here, to maintain the original sound and not introduce unnecessary confusion [5].

Das Baukastensystem ist ein Ordnungssystem, das den Aufbau ver-schiedener, zusammengesetzter Gebilde durch Kombination einer gewis-sen Anzahl vorhandener Bausteine aufgrund eines Bauprogramms oder Baumusterplans und eines bestimmten Anwendungsbereiches darstellt. (The BBS is an ordered system, by which various integrated structures can be formed on the basis of the "Structuring program or Structuring master plan" and also by combining certain number of predetermined modules in accordance with the objective application areas.)

Brankamp and Herrmann also suggested that the term *BBS* was originally used in relation to the bookshelf around 1900. On the extension of the bookshelf application, the predecessor in the machine tool sphere was the headstock of the engine lathe in the late 1920s, where 63 different gear trains were able to be produced on the basis of a group of 63 different gears and gear blocks. They furthermore envisaged that the modular design was based on the "eigen module" having the following functionalities.

(a) Well-defined interface to ensure the stiffness.

(b) Interchangeability.

3. In the late 1960s, Koenigsberger stated in his review paper [6] that modular design is a variant of standardization for the entity more than the functional complex, expecting the improvement of economic aspect in manufacturing from both the manufacturer's and user's viewpoints. In addition, he suggested that modular design facilitates the manu-facture of the machine tool in various sizes, ranges, and working capacities, as well as the scope and type of machining processes.

4. In offering his own suggestion, Koenigsberger introduced the proposal of Tlusty in relation to the following basic conditions for being a module.

(a) The alternative designs and combinations must cover the full range of requirements.

(b) The performance must meet the specifications.

(c) The connecting elements, e.g., slideway, shaft centers, clutch or coupling arrangements, and so on, must be so designed as to ensure interchangeability [6].

On the basis of these definitions, the whole concept of modular design can be depicted as shown in Fig. 1-7. Importantly, Brankamp and Herrmann, and Koenigsberger as well, proposed this concept around 1969 [5] and 1974 [7], respectively. More specifically, Brankamp and Herrmann proposed a modular design of different-kind generating type, simultaneously classifying it into the five variants; however, because they obviously neglected to mention the hierarchical aspect, there remains something to be seen in the

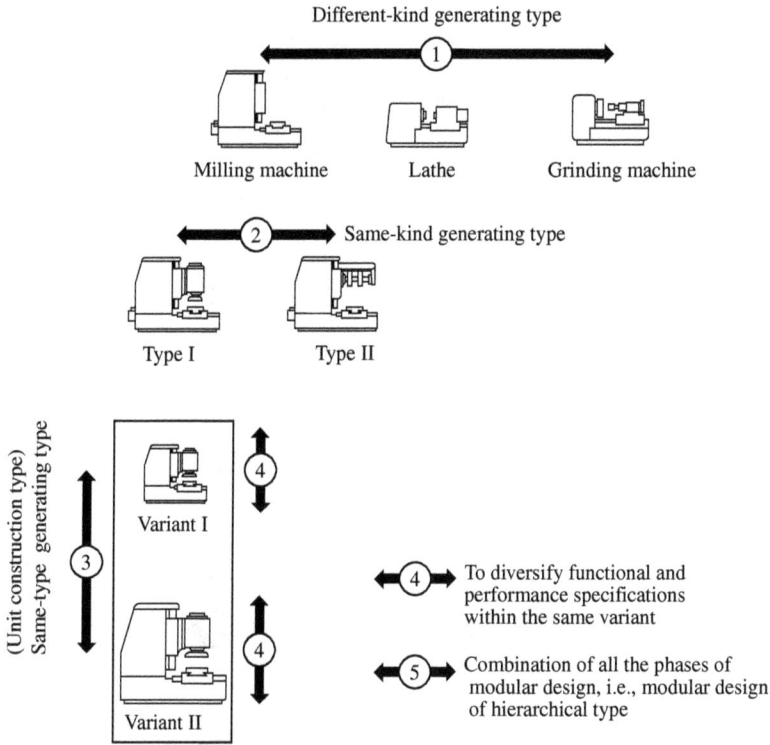

Figure 1-7 A whole concept of modular design proposed by Brankamp and Herrmann.

difference between variants 1 and 5. In fact, these variants are of different-kind generating modular design of monolayer and hierarchical types, respectively, although the practical use of hierarchical type is up to today far beyond from the fruition, apart from those of Ikegai Iron Works and VEB, which will be shown later. This might be attributed to the lack of a methodology for assisting the design, and Ito and his coworkers conducted the related research in 1979 (refer to Chap. 3). Although the difference between variants 1 and 5 remains uncertain, Fig. 1-7 is very helpful to understanding the overall view of the modular design being employed. In consequence, the modular design ranges from that available across the whole kinds, through all the types within the same kind and all the sizes within the same type, to one size within the same type.

Given such an uncertainty as that of Brankamp and Herrmann, another dire necessity is to understand the hierarchical features of the product. Figure 1-8 shows such a hierarchical feature in the MC [8]. For example, the machine consists of the unit complex, the unit consists of

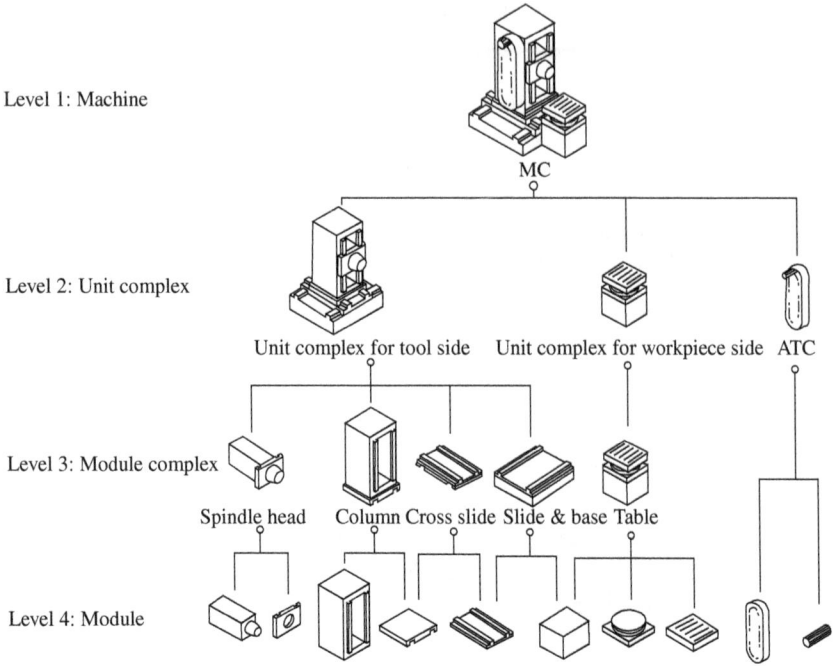

Level 1: Machine

MC

Level 2: Unit complex

Unit complex for tool side   Unit complex for workpiece side   ATC

Level 3: Module complex

Spindle head   Column Cross slide Slide & base Table

Level 4: Module

Figure 1-8   Hierarchical structure of MC with modular design.

the functional complex, and the functional complex is a combination of several parts. The question is thus how to determine the module in full consideration of the hierarchical attributes of the product. In other words, it is crucial to discuss which entity level within the hierarchical structure is suitable for the module, and this hierarchical feature in the product is one of the causes for the variants in the modular design. Intuitively, the overall view of the modular design may be delineated by the synergy of variants shown in Figs. 1-7 and 1-8. On this extension, we must be aware of the following.

1. The FMS is, in general, designed using the modular principle, where the basic module is an FMC or a machine as a whole, as shown together in Fig. 1-9 [9]. In due course, the FMC is also of modular type. As can be readily seen, the description in this book is, in principle, fully applicable to the design of flexible manufacturing.

2. The successor of compact FMC is the system machine, i.e., a machine tool compatible with flexible manufacturing. In fact, the system machine compactly integrates various machining methods or system functions within itself.

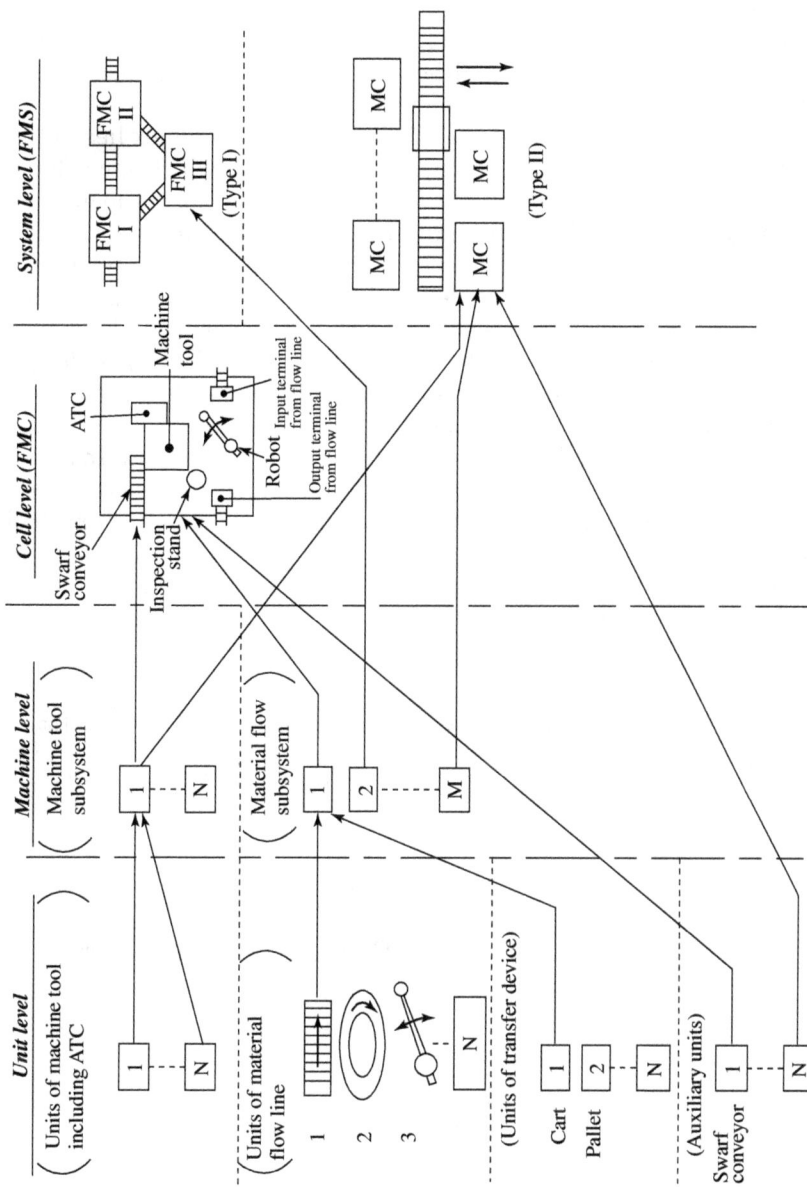

Figure 1-9 Concept of modular design of hierarchical type in FMS design.

Within a system machine context, the processing complex has been on the market since the late 1990s, as already shown in Fig. 1-5, and we must be aware that these cells and machines are of modular type. It is especially emphasized that the processing complex is expected to take over the role of the conventional MC and TC (*turning center*) of present day to a larger extent, although forcing perhaps some marked changes in the modular design (refer to Chap. 2).

Intuitively, an extreme problem in the processing complex lies in the design of the structural body component (module), which must have sufficient stiffness against all the resultant cutting forces loaded by various machining methods. With respect to the planer and planomiller, for example, the cross sections of their columns are, as widely known, of narrower width and larger depth, and rectangular, respectively, because the directions of the resultant cutting force are completely different from each other. As can be readily seen, the difficulty in design increases duly in the case of the column of the planer with milling head. The same scenario is a burning issue in designing the structural body component, e.g., bed and headstock, of the processing complex. In accordance with the experience so far, the processing complex shows very complicated thermal behavior beyond the prediction of the machine tool designer.

To deepen the design knowledge, Fig. 1-10 reproduces the proposal of Koenigsberger, in which machine tools of various kinds can, in principle,

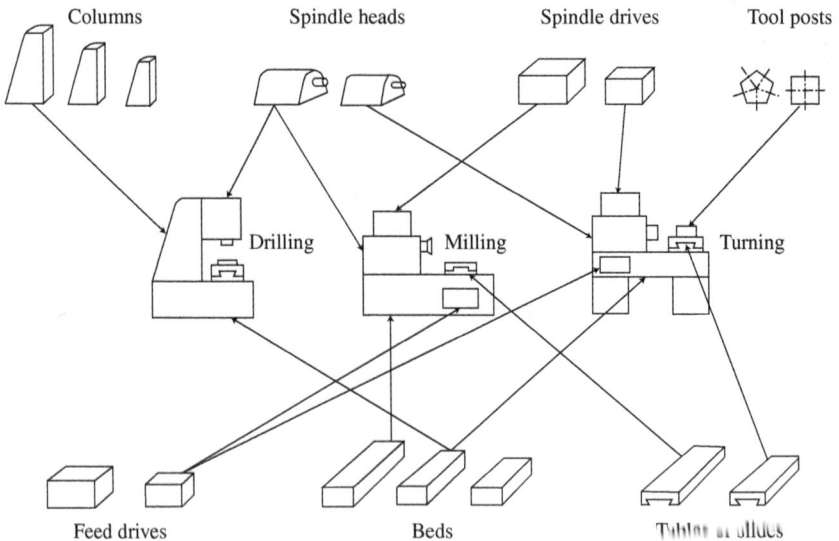

Figure 1-10 Concept of different-kind generating modular design (courtesy of Koenigsberger).

be manufactured from a group of the modules, where the module is in the form of a unit. For example, a group of the units can facilitate the manufacture of the drilling, milling, and turning machines [7]. Koenigsberger conceived this idea to apply the same design concept for the TL to the conventional machine tool available for the cell production [10]. In this proposal, the following design principles were furthermore stated.

1. Modules must be interchangeable without the use of measuring equipment.
2. Module must either be self-contained with its own power drive, feedback, and lubrication systems or form accessories for simply expanding such systems.
3. Each unit must have its own servo pack with electronic interfacing to digital input.
4. Modules must be usable in any orientation.
5. Modules must be interchangeable within about a half-hour.
6. Machining operations to be allowed in the first instance are turning, drilling, boring, and milling.

To summarize, the concept shown in Fig. 1-7 is considered the kernel of modular design; however, only a part of it will become reality, because of the hindrance from the technological and economic aspects. In due course, it is emphasized that the modular design appears to be very simple. However, there are still problems to be solved, as will be seen in the following chapters, even though the history of modular design is very long.

1.2　Advantageous and Disadvantageous Aspects of Modular Design

In discussing the advantageous and disadvantageous aspects of modular design, we must remember its successful application to both the TL and the conventional machine tool in the 1960s. Although the automotive manufacturer performs, generally speaking, mass production, frequent model changes are common to reinforce the marketability of the product, to respond quickly the users' demands, to introduce innovative technology, and so on. Importantly, the automotive industry began to use small batch production in the late 1990s to respond neatly to the individual customer's requirements. As a result, the machining facilities, i.e., machine tools and related production facilities, must renew their functionalities and performances at the factory floor of the car manufacturer. In retrospect, at burning issue in the 1950s was the lack of compatibility with such capabilities of the machine tool. Actually, the car manufacturer needed to pay larger installation and disposal costs, longer times and higher labor costs for the renewal of the production facilities. To minimize the renewal cost of the machining facility, the TL was in reality using the BBS principle, and as planned, the car manufacturer gained considerable economic advantage even while conducting the mass production and making frequent changes of the car model. Obviously, the machining facility was reconstructed at the factory of the automotive manufacturer, and the renewal at the machine tool manufacturer was no doubt far afield. Geminately, we observed evidence that the modular design for conventional machines provided both the manufacturer and the user with economic benefits together with a wide range of choices in accordance with the manufacturing requirements.

In short, the modular design was characterized by its capabilities for marked economical benefit and for wider flexibility in both the functionality and the structural configuration of a machine tool, notwithstanding the application area, and this characterization is applicable even in 2000s. Further, it is thus emphasized that the designer must obviously determine the aims, purpose, and scope of the modular design at the moment of its application. Within this context, some valuable suggestions are given in the following.

1. In 1960s, Koenigsberger suggested that the standardization of the module enables its full economic effect to be a reality, provided that the performance consistency and interchangeability in shape and size of the module are ensured. From the viewpoint of the reduction of manufacturing cost, he strongly recommended that the modular design be applied to the machine tool of the same kind on the basis of the principle to "increase batch size using the same module to produce various types." Obviously, the machine tool is relatively costly because of the small batch size production.

2. In the late 1960s, Brankamp and Herrmann suggested the advantages of the modular design for both the manufacturer and the user as follows [5].

For manufacturer

(a) Reduction of both the manufacturing cost and the required manufacturing time.

(b) Increase of the production volume and guarantee of preferable inventory derived from the repeated use of the module, simultaneously resulting in certain benefits beyond our expectations.

For user

(a) Shortening of delivery time and cost reduction.

(b) Reduction of inventory cost for spare modules by commissioning the preparatory storage of the module to the manufacturer, resulting in the savings of running cost.

Brankamp and Herrmann further exemplified the economic benefit, showing that the manufacturing cost and assembly time in the manufacturer saved about 50 and 40 percent, respectively, in the case of the drilling machine of multiple-head and automatic lathe of *Fronter* type.

3. In consideration of the successful application since the beginning of 1960s, the Maho Werkzeugmaschinen applied the modular design to the universal tool milling and boring machine in the mid-1970s. In this case, the three groups of basic modules (die Bausteine) were determined in closer relation to the basic component of machine, table configuration, and functional attachments, e.g., scale, quill, angle head, vertical head of quick traveling type, and slotting head [11]. In addition, the machine employed the same gear train for the main driving system and feed mechanism in accordance with the suggestion of Koenigsberger [6]. In this application, a noteworthy variant was a vertical milling machine of knee type, which was, e.g., capable of face milling and groove milling of the outer surface of the cylindrical work using the index head and tailstock located parallel to the spindle axis. Consequently, the emphasis lay in both the user's and the manufacturer's benefits as follows.

User's benefit

(a) Attachments were available for all the types.
(b) The function of the machine was expandable when user wanted.
(c) Ease of exchange and repair of the module.
(d) The considerable stock volume of modules enabled the ease of supply of the module with higher accuracy.

Manufacturer's benefit

(a) Notwithstanding the order volume, the standardized module was produced with economically reasonable cost.
(b) Ease of production planning and control together with cost-effective production.
(c) Improvement of assembly capability, because of manufacturing various types from a group of modules.
(d) Reconfiguration was simplified with less additional cost.

1.3 A Firsthand View of Developing History and Representative Applications

The first meaningful trial of the modular design is credited to the late Prof. Königsberger in the beginning of the 1930s on the basis of his first application of the unit construction to the milling machine of Wanderer brand [6].

Figure 1-11 is a firsthand view of the chronological history of the modular design, showing in part the change in terminology. As can be seen, the modular design has been developed continuously in responding to the changing requirements of the machine tool, and it was also deployed to various extents, encompassing the production system and machine tool accessories. As quickly described earlier and will be detailed later, the production system has often been converted to a compact machine, for instance, those from FMCs to system machines. In this context, the modular design has functioned very well to form the suitable structural configuration. In the following, the quick notes on the history of modular design will be stated using, if necessary, the historical nomenclature while emphasizing the epoch-making events and showing the representative applications of world class. Actually, the three milestones can be observed as follows, when we scrutinize the developing history up to now.

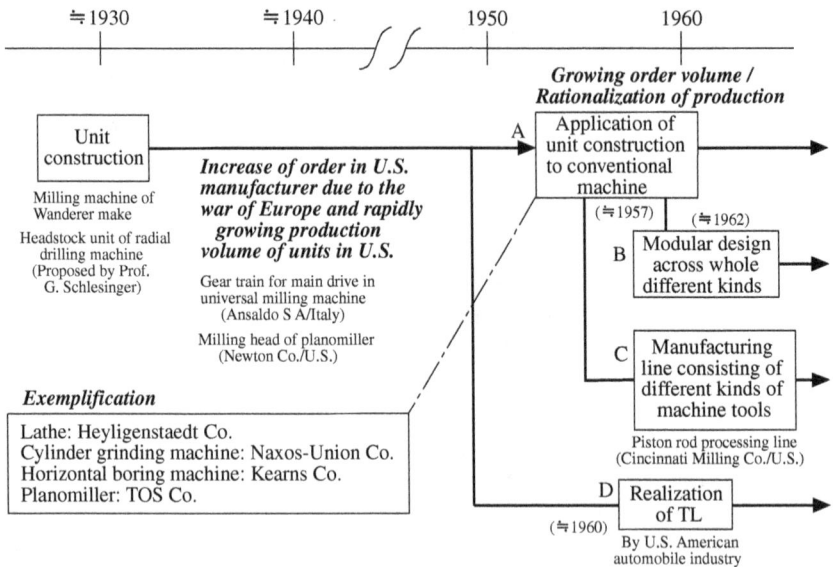

(a)

Figure 1-11 Developing history of modular design: (a) Between 1930s and 1965 and (b) between 1965 and 2000.

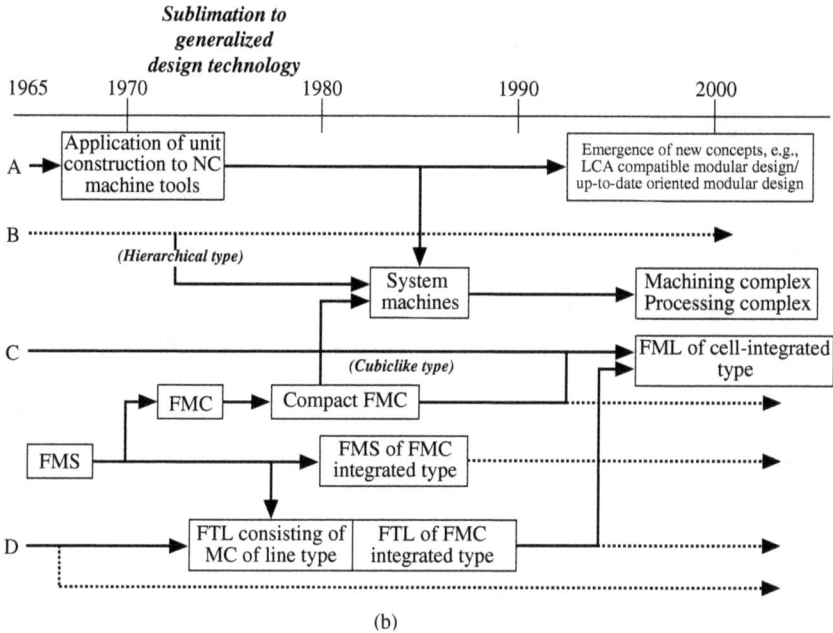

Sublimation to generalized design technology

Figure 1-11   (*Continued*)

1. In the beginning of the 1960s, the two-pronged application to both conventional and special-purpose machine tools. In the conventional machine tool, modular design can facilitate wider flexibility in the functional and performance specifications. However, for the special-purpose machine tool to be integrated in the TL, it is, e.g., expected to reduce the replacement cost when carrying out the renewal of the system.

2. In the beginning of the 1970s, the application to the conventional NC (*numerical control*) machine tool to reinforce the flexibility in machining capability from the hardware aspect. In this case, a broadcast trend was the application of the modular principle to the NC software, e.g., modular EXAPT (*extended subset of automatically programmed tools*) and CAPP (*computer-aided process planning*) of modular type. It is, however, noteworthy that even in the year 2000, the NC and CIM (*computer-integrated manufacturing*) software of modular type are not completely established as yet.

3. In the 1990s, the leading trend was to apply modular design to the five-face processing machine, system machine, and processing complex. Furthermore, there has been a symptom of some new concepts, i.e., deployment to those of LCA (*life cycle assessment*) oriented and preventive quality assurance types.

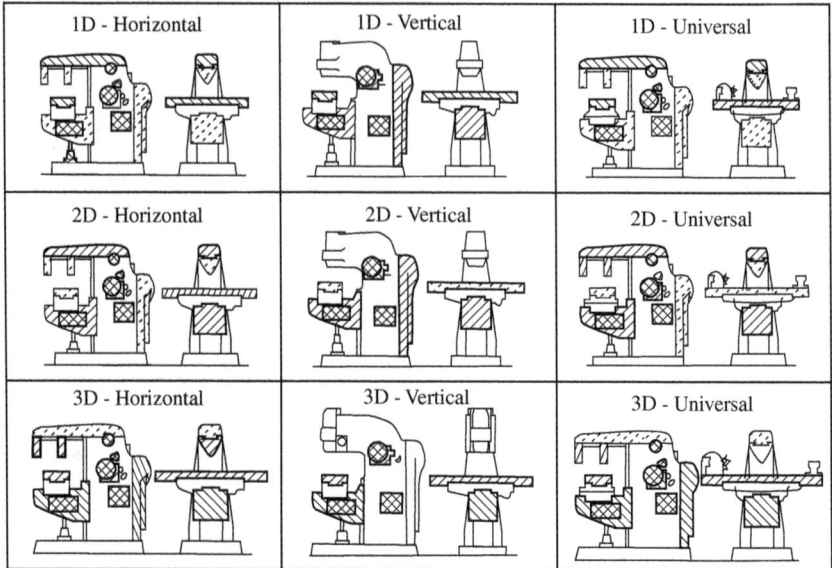

1D - Horizontal	1D - Vertical	1D - Universal
2D - Horizontal	2D - Vertical	2D - Universal
3D - Horizontal	3D - Vertical	3D - Universal

Figure 1-12   Unit construction of milling machines (Wanderer Co., about 1930).

The primary concern is that the fundamental feature of the modular design was established in the 1960s, i.e., its first developing phase, to a larger extent. To understand quickly what is the essential feature of the modular design, it is better to touch on the most representative applications in the 1960s. Figures 1-12 and 1-13 show two typical examples of the unit construction at the earlier stage. The former was that of Wanderer Co., where the horizontal and vertical milling machines of knee type were manufactured. The marked feature was that the body structural components varied according to the power capacity, although the feed driving unit and its control device were practically identical across the whole types and sizes as represented, for instance, by the hatched line in Fig. 1-12 [6]. Actually, the following preconditions were employed in this application.

1. The ratio between the feed power and cutting power in most machine tools is relatively small. As a result, the wastage due to having more feed drive power than necessary is not serious in smaller machines.

2. There is not such a great necessity to change the control unit, gear change mechanism, and so on.

In contrast, the latter was that of Newton Co., where the planomiller and milling machine of various configurations and types were manufactured gathering various units across the entire United States. This

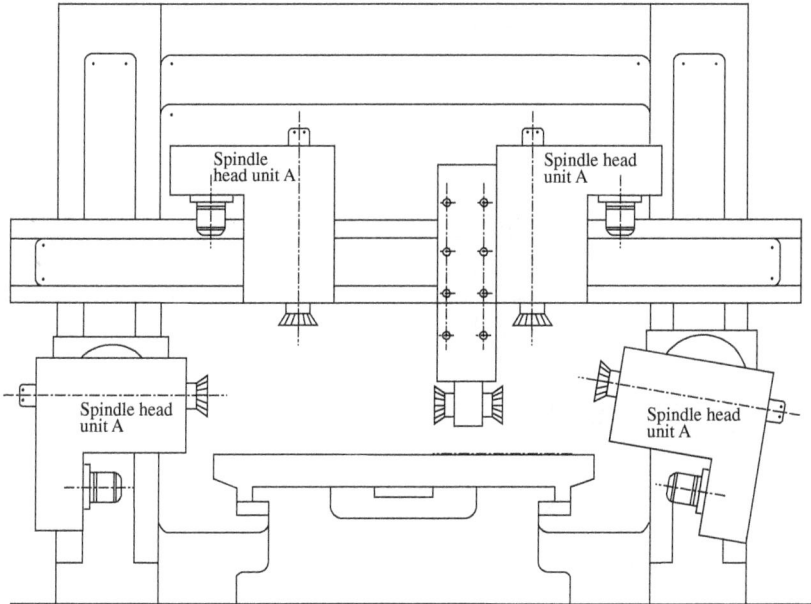

Case of four-head planomiller

Figure 1-13   Unit construction employed by Newton Co. in beginning of 1940.

unit construction was reported by the late Mr. Hayasaka of Ikegai Iron Works in the beginning of 1940 based on his on-the-spot investigation into the U.S. machine tool industry. In fact, the machine was announced to be capable of creating more than 10 variants to the milling machine and planomiller. On that occasion, there were strong influences derived from the war in Europe, and the U.S. machine tool manufacturer received a huge number of the orders from the United Kingdom and France in addition to assisting the war supplies production of the United States itself. As a result, the manufacturer concentrated its production activity on the specified unit, resulting in the increased production volume of the conventional machine tool of unit construction type. In fact, Hayasaka suggested at first that the division of works was very popular, and a fast-growing trend was to use the unit construction system. For instance, the manufacturer purchased easily the spindle head unit from 15 companies, the hydraulic feed unit from 8 companies, and the hydraulic pump from 4 companies. In addition, General Electric, Westinghouse, and Wesche Electric provided the motors of geared, flange-mounted, and torque types, respectively. Hayasaka also suggested that either a conventional or a special-purpose machine was able to produce by combining the commercialized units with an object-oriented basis. It is, furthermore, worth suggesting that the technological

Figure 1-14 Unit construction in milling machine of bed type (type LFD160/3090 of Bohle make).

resources accumulated on that occasion played a very important role in the continued existence of modular design in the late 1950s under the acronym *BBS*.

Although the unit construction within the same kind is obviously simple, the obtainable benefit is considerable. Thus, Reinhard Bohle KG applied it to the milling machine of bed type, as shown in Fig. 1-14 and Table 1-1, even in the 1960s [12]. So it is not surprising that the Index-Werke GmbH applied the unit construction to the NC turning machine in the 1990s [13]. In these cases, the module was determined on the basis of the fundamental design concept of the conventional machine tool, i.e., unit-based design, together with maintaining the applicability to both the conventional and special-purpose machine tools as Georg stated in the 1950s [14]. He suggested, e.g., a unique idea, in which a headstock was finish-machined on its four faces so that the designer could choose four different center heights. In addition, the headstock was available for turning, drilling and boring, milling, and grinding. This appears to be a predecessor showing the essential feature of the modular design, i.e., different-kind generating modular design.

In retrospect, modular design approached its development goal of the first stage in the mid-1960s, duly showing the establishment of the fundamental feature of modular design. It may be said that there have been no remarkable research and development activities thereafter, apart from the machine tool description and proposals for some novel

TABLE 1-1 Unit Construction in Milling Machine of Bed Type of Reinhard Bohler KG make (1960s)

Types	Milling heads		Table size
LFD LFE	T/H S/H	Spindle dia.: 85 mm Main motor: 5.5 kW	Width: 300 mm Length: 800, 1000, & 1200 mm
LFA LFD LFE LFP	S/H+S/V T/H S/H T/H	Spindle dia.: 100 mm Main motor: 7.5 kW	Width: 425 mm Length: 1600 & 2000 mm
LFA LFD	S/H+S/V T/H	Spindle dia.: 120 mm Main motor: 15 kW	Width: 560 mm Length: 1600, 2000, 2500, & 3000 mm
LFE LFP	S/H T/H	Spindle dia.: 160 mm Main motor: 22 kW	Width: 710 mm Length: 2000, 2500, 3000, & 4000 mm
LFD LFE LFP	T/H S/H T/H	Spindle dia.: 200 mm Main motor: 38 kW	Width: 900 mm Length: 2500, 3000, & 4000 mm

Note: H: horizontal spindle; V: vertical spindle; S: single head; T: two heads.

concepts (refer to Chap. 3). In other words, modular design was a fast-growing technology in the 1960s, especially in its application to the unit construction, and Koenigsberger publicized some excellent reports [6, 7].

In summary, it is very convenient to divide the developing history into the following three phases in consideration of the three milestones mentioned above.

Phase 1: around 1930 to 1965. In phase 1, the development of the BBS reached its climax at the beginning of the 1960s, when the BBS applied to both the TL and the conventional machine tool to a larger extent, although the application philosophies were different from each other. Table 1-2 shows the comparison of characteristic features of both applications. Importantly, the BBS was, on that occasion, classified into the three types shown in Fig. 1-15, emphasizing their advantageous aspects. In short, the particular emphasis is again that modular design was a fast-growing technology in the 1960s, especially in its application to the unit construction.

Phase 2: 1965 to 1985. The advent of NC technology can be regarded as more epoch-making than that of Wilkinson's cylinder boring machine in the long-standing history of the machine tool. Phase 2 began with merchandising the NC machine tool, and the modular design has been capable of providing the machine with greater flexibility from the hardware

TABLE 1-2   Comparison of Characteristic Features between Two Representative Applications in the 1960s

	Transfer line	Conventional machine
Application purpose	Enhancement of versatility or flexibility of production facilities	Quick response to user's order Enhancement of flexibility Rationaization of production processes
Place of use	Factory of machine tool user	Design and production processes in manufacturer
Effects	Reduction of renewal investment by decreasing the number of disposal units Increase of operating efficiency by shortening the renewal period of production facilities	Realization of machine tool with keen price by reduction of design time, cost reduction using identical part, and reduction of throughput time. Quick response to user's requirement Enhancement of competitiveness by shortening manufacturing time Reinforcement of R&D ability for new product Reinforcement of manufacturing capability

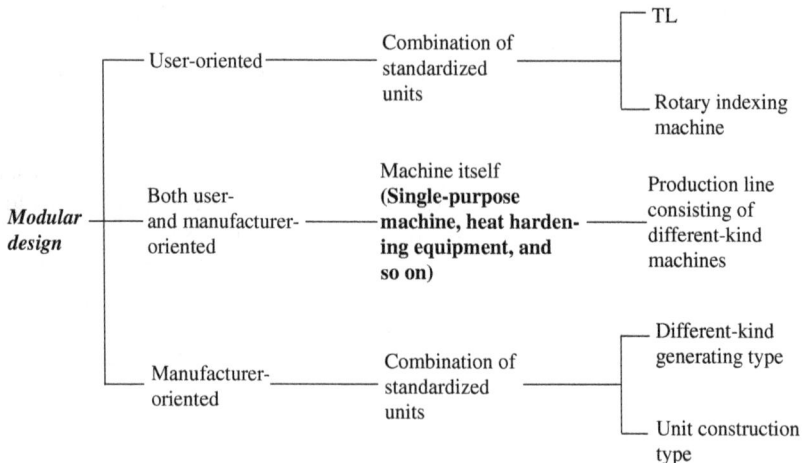

Modular design

- User-oriented — Combination of standardized units —
  - TL
  - Rotary indexing machine
- Both user- and manufacturer- oriented — Machine itself **(Single-purpose machine, heat hardening equipment, and so on)** — Production line consisting of different-kind machines
- Manufacturer- oriented — Combination of standardized units —
  - Different-kind generating type
  - Unit construction type

Note: The unit construction enables the machine tool within the same kind to be ramified in dimensional and performance specifications as well as structural configuration by varying, for example, the body component, spindle system, driving gear train, table width, and so on.

Figure 1-11   Classification of modular design in the 1960s.

aspect. More specifically, the NC machine tool itself has, in principle, considerable flexibility, which can be realized by only changing the NC information in accordance with machining requirements; however, in certain cases, such flexibility has certain limitations. To give certain remedies to the limited capability of NC software, an enabling technology is to apply the modular principle to the structural design, which can be considered as the utmost protruded evidence in phase 2.

Phase 3: 1985 to 2000 and beyond. In phase 3 of the developing history, a pronounced development of modular design has been realized by applying it to the FTL, FMS, and FMC. In fact, there are various applications as follows.

1. In the FMS, the basic module is that of the FMC, compact FMC, or system machine.
2. In the FMC and system machine, the basic modules are those of unit and unit complex.

Importantly, not only is the system itself of cell-based modular type, but also the cell has become popular as exemplified in the growing installation numbers of FMC within the SME (small-and medium-size enterprise). In fact, the FMC of flat allocation type[1] approached the compact FMC of cubic type, i.e., system machine [15], with the advance of flexible manufacturing. The system machine itself has been more developed to an advanced kind, i.e., machining complex, as shown in Fig. 1-16. The machining complex can be characterized by compactly integrating the various processing functions and enhanced performances within a machine as a whole. In due course, the system machine and machining complex are capable of working in stand-alone mode. Without exception, the SMEs are now very keen to install these new machines instead of the FMC of robot or pallet pool type shown in Fig. 1-17, which has so far been the most popular FMC (regarding the detail of phase 3, refer to Chap. 2).

Given such a firsthand view of history, some representations will be reproduced in the following.

1.3.1   Application to TL and FTL

After World War II, the automobile industry was the leading edge within the industrial nations, especially within the United States in the 1950s. The United States was requested to lead the world economy, because it

---

[1]The FMC consists of the five basic functions, i.e., those for machining (processing), transfer, storage, maintenance, and cell controller, and in the FMC of flat allocation type, all the system hardware is allocated within the two-dimensional space (see Fig. 1-17).

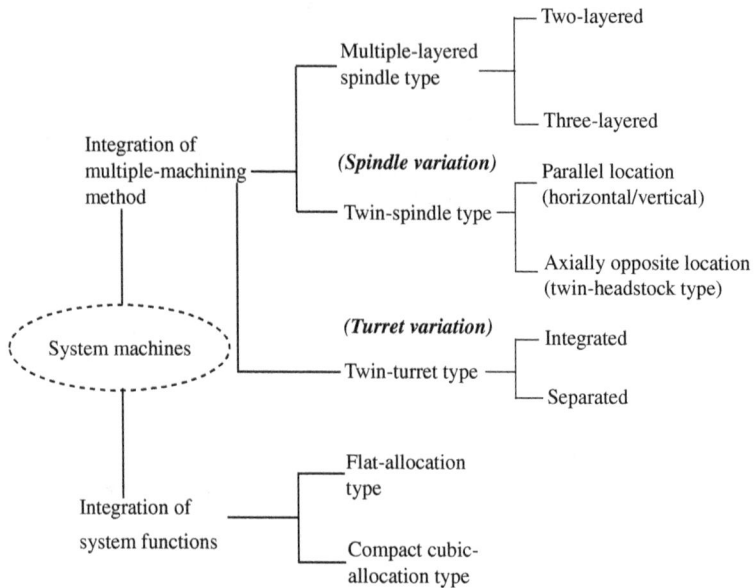

Figure 1-16 Classification of system machines.

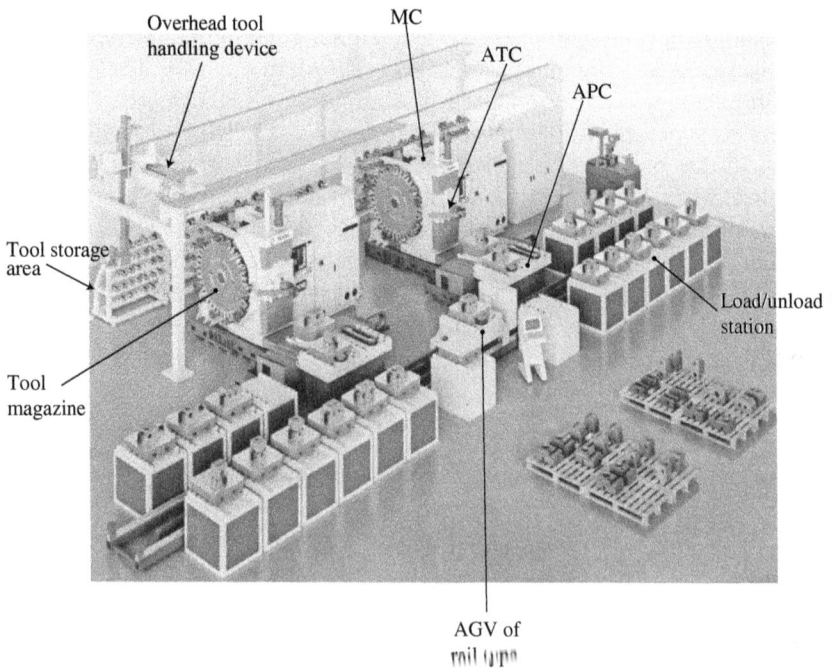

Figure 1-17 FMU of pallet pool type (type DFZ630, courtesy of Hammer of Fritz Werner).

had not received the devastation due to the war. More specifically, the production facilities had to be reinforced so that the production volume of cars increased with the keen economic benefits. In this task, a facing problem was how to rationalize the machining capability of the car parts, e.g., cylinder head, engine block, axle housing, and mission gearbox. Intuitively, a dominant hindrance derived from the mass production, for which the highly automatized machining system was suitable. For example, the work was transferred automatically among the machining stations and processed again automatically its corresponding portions at each station, maintaining always its preparatory setting condition. As can be readily seen, such a facility has less flexibility in machining capability instead of higher automatization, resulting in the requirement of the large amount of renewal cost, when the model change of car is made to reinforce the marketability.

In due course, U.S. car manufacturers proposed a concept of *machine tool construction with standardized units*, actually that of BBS, to enable the interchange of the unit with on-site manner, i.e., user-oriented modification at user's factory, and to minimize the renewal cost of the machining facilities. The car manufacturer then carried out the research and development necessary to make the concept a reality in cooperation with the machine tool manufacturer. As a result, the TL was realized, which was, in principle, classified as the line type, i.e., that of TL itself, and rotary indexing type including those of star-wing and trunnion types, depending on the machining requirements and size of parts to be processed. As planned, the car manufacturer gained considerable economic advantage even in carrying out the mass production and frequent change of the car model.

In the 1960s, Cross and Kingsbury Machine Tool were the leading manufacturers of the TL and rotary indexing machine across the world, and Figs. 1-18 and 1-19 show the typical TL and rotary indexing machine. From these, we can observe the basic entities (units) to consist of the TL and rotary indexing machine, e.g., spindle head, gang drilling head, power unit, feed unit, base, and adapter. In both the TL and the rotary indexing machine, the basic unit complex is called the station, which is actually the special-purpose machine tool. Importantly, the TL is for mass production of the specified work and can be defined as a production facility consisting of a group of the stations and related equipment, such as work turnover device, washing station, and measurement stand. In addition, the work is in line flow–like transferred from station to station with the constant tact time (*a kind of cycle time: work-in-station time in intermittent transfer line*). In contrast, the rotary indexing machine can be interpreted as a variant of the TL, where the work is in circular flow–like transferred from station to station. In due course, each station can process the work in accordance with the

(a)

Station no. 3

Station no. 2

Transfer line

Station no. 1
(consisting of
several modules)

(b)

Figure 1-18   Concept and configuration of TL: (a) For machining cylinder block (by Ex-Cell-O, courtesy of Carl Hanser) and (b) for machining electric motor frame (courtesy of Toshiba Machine).

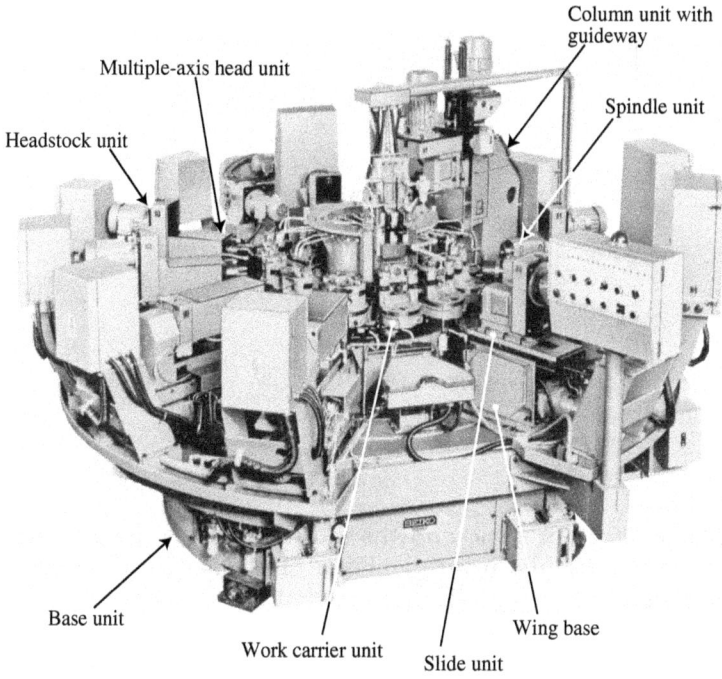

Column unit with guideway

Multiple-axis head unit

Spindle unit

Headstock unit

Base unit

Work carrier unit

Slide unit

Wing base

(a)

(b)

Figure 1-19   Rotary indexing machine and its variant in the 1960s: (a) Rotary index-ing machine for large-volume production of precision parts (by Seiko) and (b) three-way drilling machine—a variant of rotary indexing machine (by Hitachi).

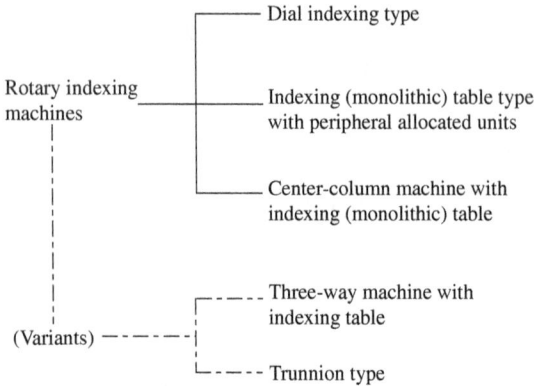

Figure 1-20   Classification of rotary indexing machines.

allocated machining process, and the station itself is, in principle, designed with the modular principle. Figure 1-20 shows the classification of the rotary indexing machine.

Following the success in U.S. car manufacturers beyond our expectations, the German industry launched, under the strong stimulus of the U.S. technology, the keen attempt to advance the BBS for the TL. Actually, a machine tool committee within VDI (*Vereinigte Deutscher Ingenier*) carried out these activities. As a result, VDI was able to standardize the base, sliding unit, spindle head, gang head, and so on, as shown in Fig. 1-21(a) and (b). In these units, the "*preferred numbers,*" i.e., series of R05, R10, R20, and R40, have been recommended to determine the dimensional specifications of the unit [16].

On the basis of these achievements, the design technology of TL was, to an extent, established in the beginning of the 1960s. It has been used up to today by modifying itself with the advance of the TL in accordance with the changing manufacturing requirements, e.g., FTL consisting of the NC station or the MC of line type, FTL of FMC-integrated type and FML (*flexible machining line*). More specifically, the automobile industry was faced with responding to the individual requirements of the customer while maintaining the mass production mode to a certain extent in the 1980s, and the FTL has been contrived. The developing history is thus shown in Fig. 1-22 for the sake of easily understanding the changing requirements for the station. For instance, the MC of line type is a kernel in realization of the most popular FTL at present [17]. Figure 1-23(a), (b), and (c) shows the most popular machine tools to systematize the FTL, i.e., MC of line type and head changer, together with a typical configuration of FTL. The same idea is also available for the FTL for grinding, as shown in Fig. 1-24, increasing the system flexibility by incorporating the rotating table for the workpiece, so that the simultaneous profile grinding for outer and inner surfaces can be done [18].

In short, the FTL can be characterized by both the considerably limited flexibility and the higher productivity, resulting in the growing importance of the system design with modular concept. In due course, at issue is how to determine the necessary number of stations in closer

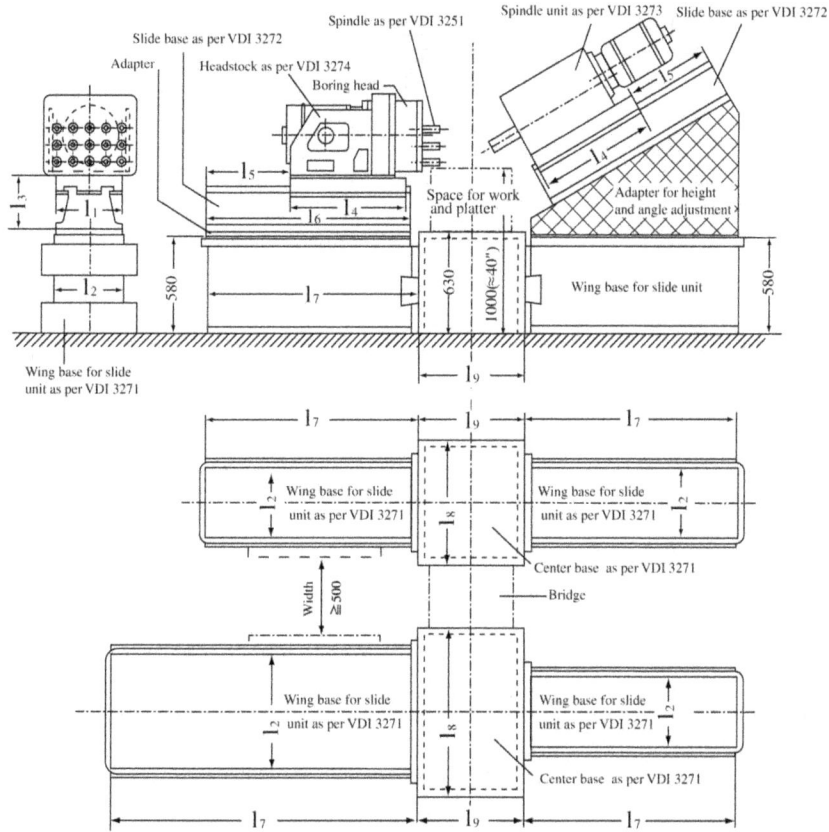

Leading dimensions

Sizes	l_1	l_2	l_3	l_4	l_5	l_6	l_7	l_8	a	l_9 b	c
B 1	320	340	260	560	400	1000	1050	630	450	630	800
B 2	400	420	320	710	500	1250	1300	710	450	630	800
B 3	500	520	320	900	500	1450	1500	850	450	630	800
B 4	630	650	350	1120	630	1800	1850	1000	450	630	800

(a)

Figure 1-21 VDI standard for TL.

$l_1 = 320$

200

$l_{10} = 280$

Spindle unit as per VDI 3273

Wing base for slide unit as per VDI 3273

Boring head coupler as per VDI 3274

320

$l_{10} = 280$

Wing base for slide unit as per VDI 3272

(b)

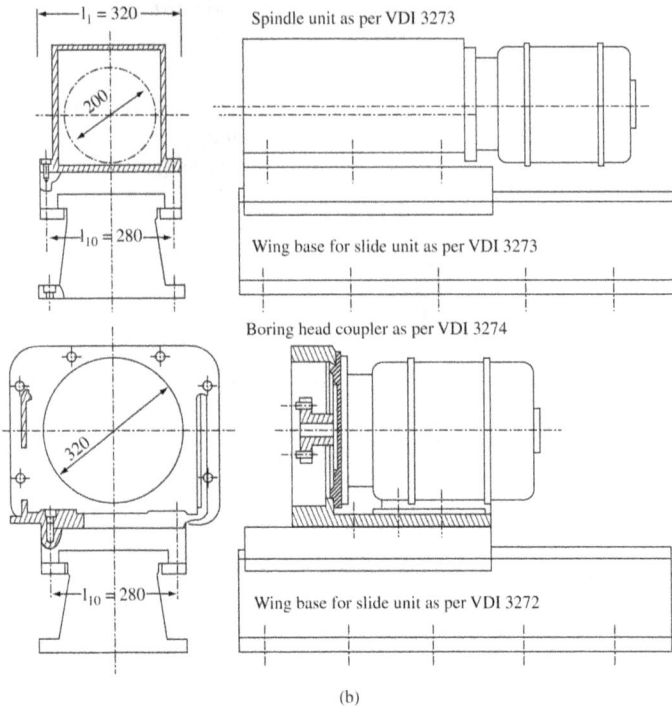

Figure 1-21 *(Continued)*

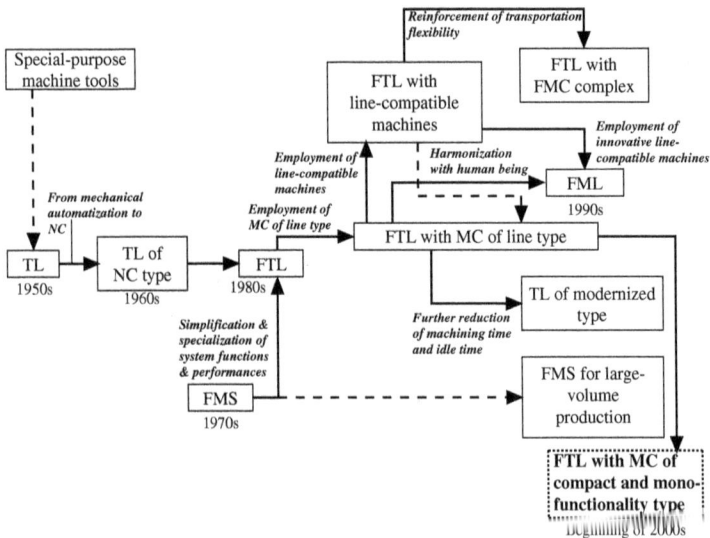

Reinforcement of transportation flexibility

Special-purpose machine tools

FTL with line-compatible machines

FTL with FMC complex

Employment of innovative line-compatible machines

Employment of line-compatible machines

Harmonization with human being

FML
1990s

From mechanical automatization to NC

Employment of MC of line type

FTL with MC of line type

TL
1950s

TL of NC type
1960s

FTL
1980s

Simplification & specialization of system functions & performances

Further reduction of machining time and idle time

TL of modernized type

FMS
1970s

FMS for large-volume production

FTL with MC of compact and mono-functionality type
beginning of 2000s

Figure 1-22 Development history of TL and FTL.

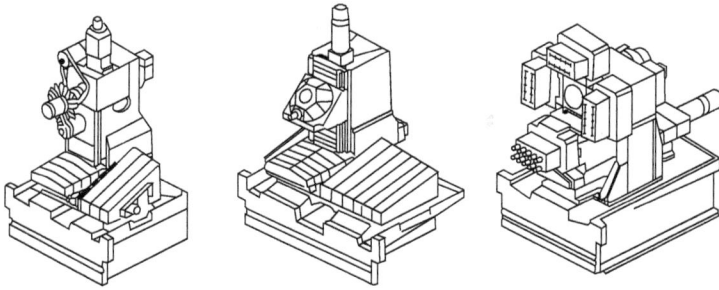

NC unit with ATC   NC unit with turret head   NC unit with ATC

(a)

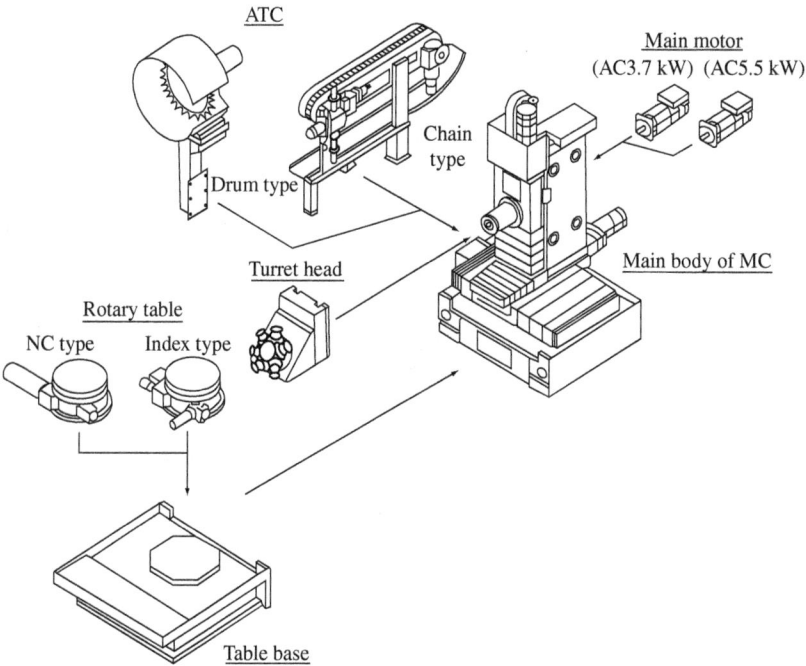

(b)

Figure 1-23 Core machines for and concept of FTL: (a) MC of line type and head changer (1989, courtesy of Toyoda Iron Works); (b) modules in MC of line type (type NF1-H, 1990, courtesy of Fujikoshi); and (c) FTL for machining cylinder head, cylinder block, mission case, and so on (1990s, courtesy of Toyama).

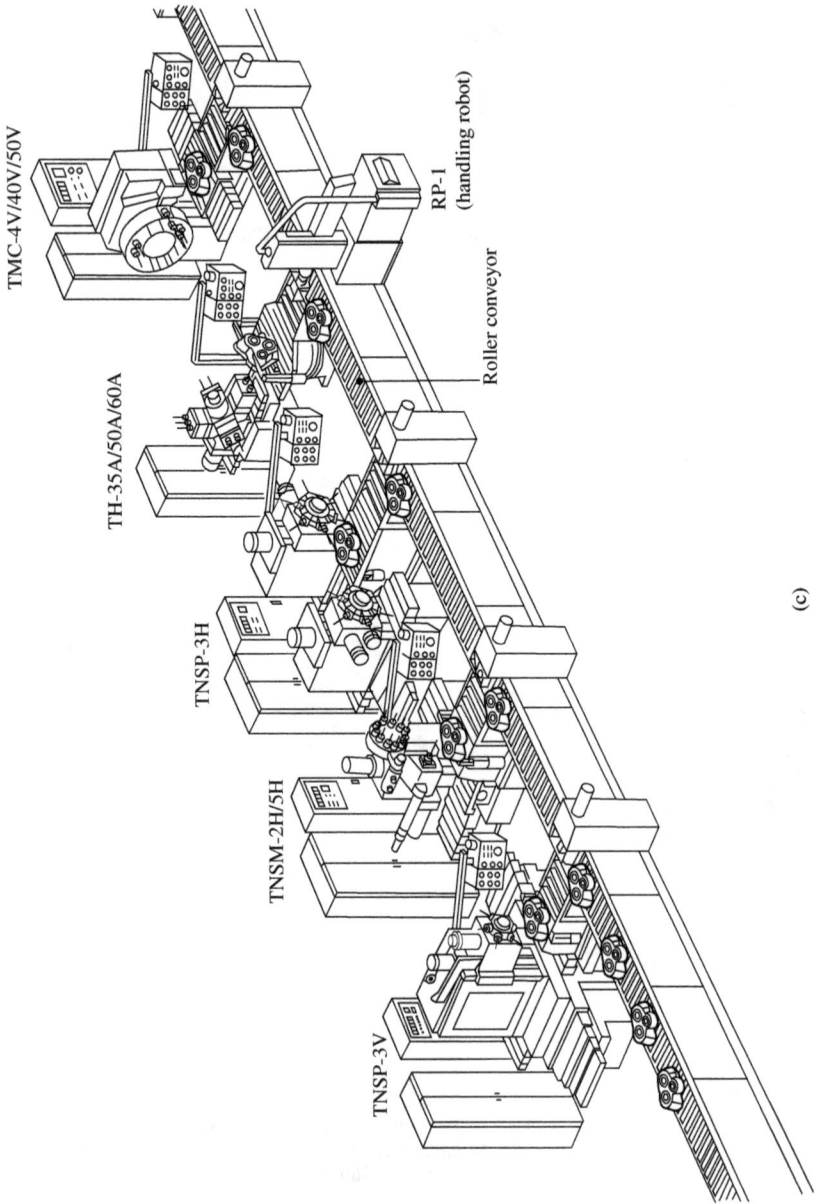

TMC-4V/40V/50V

TH-35A/50A/60A

TNSP-3H

TNSM-2H/5H

TNSP-3V

RP-1
(handling robot)

Roller conveyor

(c)

Figure 1-23 (Continued)

36

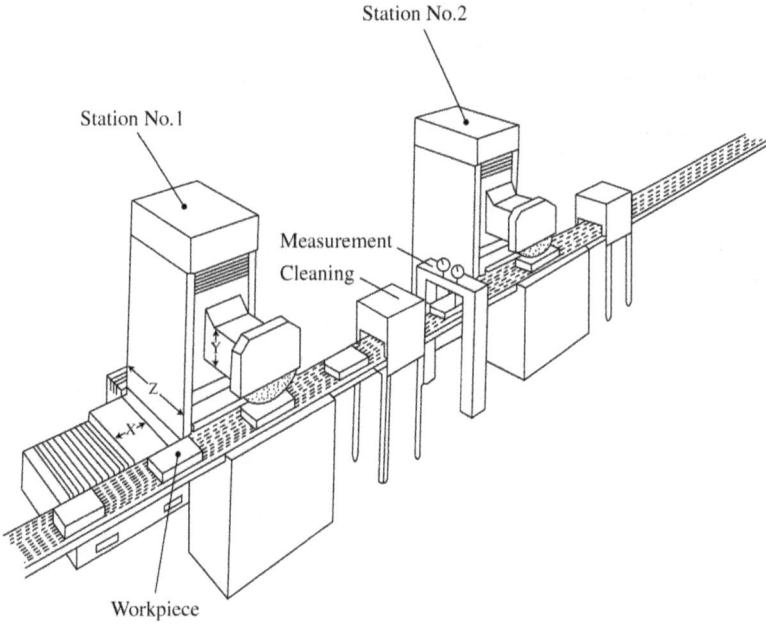

(a)

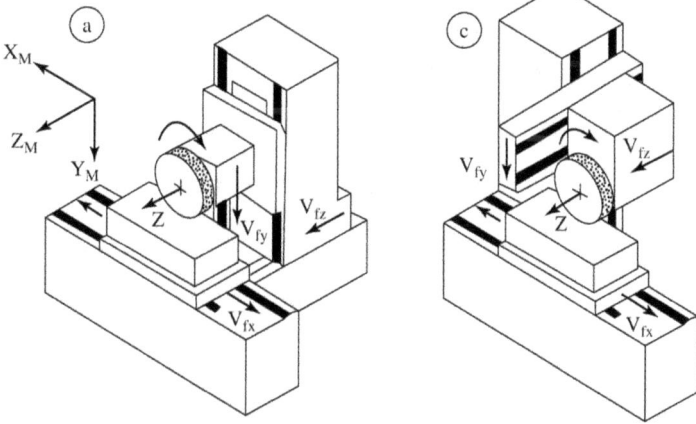

(b)

Figure 1-24   FTL for grinding (courtesy of Elb Co.): (a) Line configuration and (b) variants for line formation.

relation to the integration and disintegration of required machining processes for the objective works. In addition, the conventional FTL consisting of the MC of line type must have the following features.

1. Structural configuration of column traveling type with modular design, as shown in Fig. 1-23

2. Higher efficiency and reliability

3. Functionality for ease of maintenance and repair

4. Functionality for ease of swarf removal

5. Reduction of floor space in terms of machine width

In consideration of the characteristic features, furthermore, the units have been standardized across the whole nation and, in the preferable case, across the whole world to increase the obtainable benefit. In fact, the United States and Germany had enacted the national standard in the 1960s and on the extension of these national standards, the ISO (*International Standards Organization*) has legalized the due standards in the 1970s [19]. In short, the ISO enacted some international standards for both the TL and rotary indexing machine in the beginning of the 1970s. Figure 1-25 shows a firsthand view of these standards, and as can be seen, the designer must refer to a handful of standards. In addition,

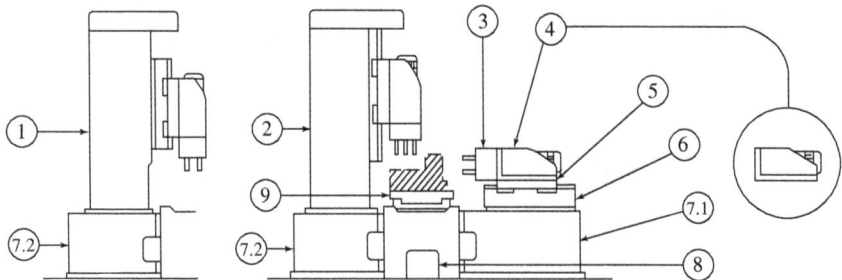

No.	Units
1	Column
2	Column—Integral way type
3	Multispindle head
4	Headstock
5	Saddle
6	Slide base
5 + 6	Slide unit
7.1	Wing base for slide unit
7.2	Wing base for column
8	Center base for platter
9	Platter

Figure 1-25 International Standard for TL (by ISO).

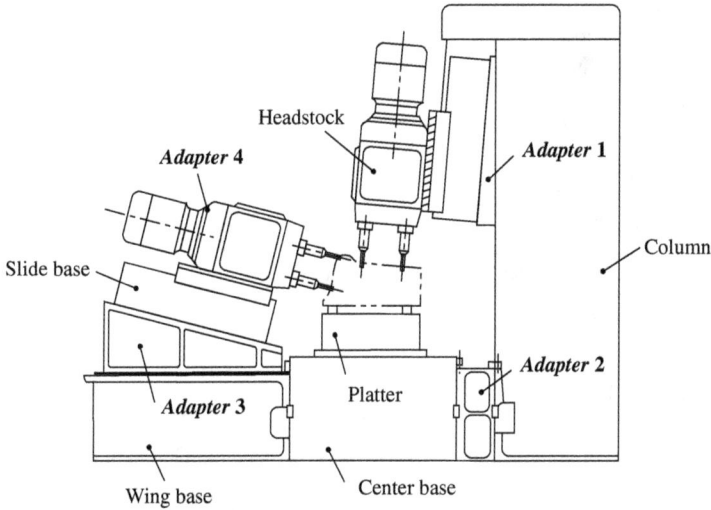

Figure 1-26   Allocation example of adapters.

the designer must pay special attention to adjusting the accumulation of the assembly error of units and to plugging the gap produced when realizing the desirable structural configuration with the standardized units. In fact, we need the supplementary entity, i.e., adapter, as shown, for example, in Fig. 1-26. Although it is imperative, the adapter is actually very inexpensive compared with the total price of TL, and thus it may be thrown away at the renewal of the TL.

To this end, Fig. 1-27 shows the advanced FTL for processing the car part, in which the station consists of the conventional MC of compact type instead of the MC of line type. In this case, the station length along the transfer line is as small as possible to reduce amazingly the idle time, where the transfer linkage is of roller conveyor, loader with circulating pallet, or loader with turntable type.

Meanwhile the TL can receive the raw material and output the finished work after processing the work at each station according to the predetermined machining information, and thus its configuration is in closer relation to process planning of the work. A burning issue even in the 2000s is, as already stated, the integration and disintegration of the processes to leverage between the tact time at each station and the number of the stations together with guaranteeing the machining efficiency. This is so because the larger the number of stations, the longer the throughput time is, but the simple machining process is allowed at each station [20]. It is envisaged that such process planning be carried out by the very mature engineer. In this context, a necessity is thus to contrive a new special-purpose machine tool as the station, which has

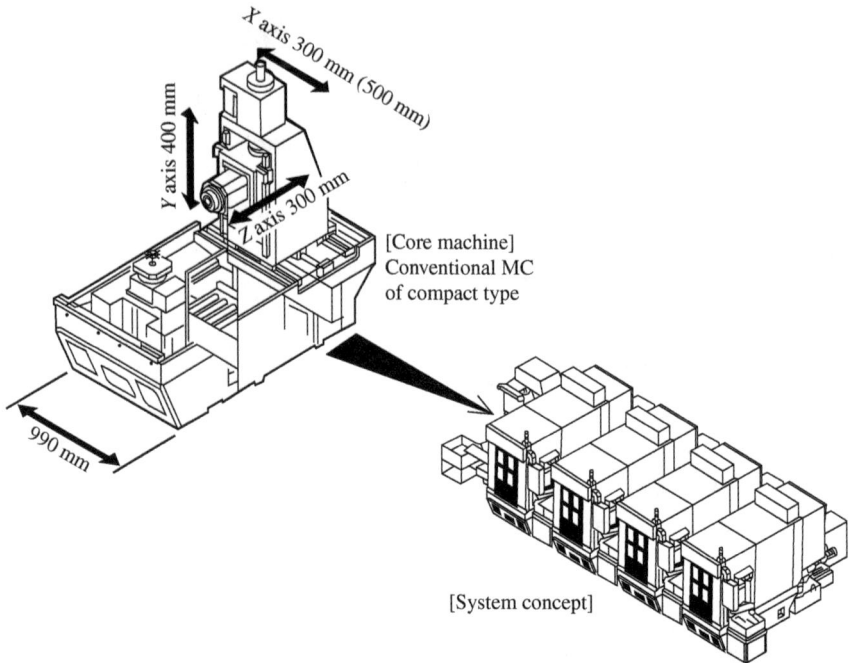

Figure 1-27   Advanced FTL for processing car parts (late 1990s, courtesy of Mori Seiki).

the simplified functional and performance specifications, duly resulting in simplification of the process planning. As a result, we can expect to realize an innovative FTL with high operability and ease of system design.

1.3.2   Application to conventional machine tools

Along with the successful application to the TL, the BBS was thrived in the sphere of the conventional machine tool, where the rationalization in manufacturing was aimed to respond to the growing demand for the machine tool. In short, the BBS for the conventional machine tool was capable of manufacturing the variants, which were commissioned to provide the wide flexibility in (1) the dimensional specifications, (2) the machining capabilities, and (3) the machining methods to the machine, once a group of the units was given. Accordingly, the encountered engineering problems differed from one another, depending on which was the major objective to be realized, and this is doubtlessly applicable for the machine tool of the present. In the following, some typical application cases ranging from the small-sized to the large-size machine tools will be demonstrated.

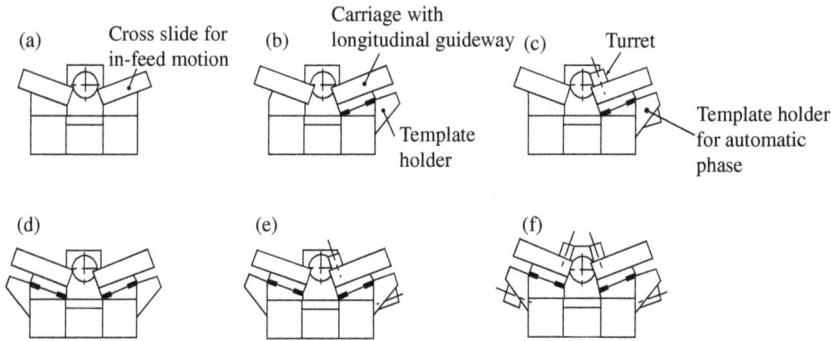

Figure 1-28 Modular design in turning machine of *Fronter* type (courtesy of Saljé).

Small- and medium-sized machine tools. The turning machine for the disklike work, i.e., face turning machine of shorter bed type, has been installed within the automobile industry to machine the brake disk and gear blanks. The machine is, in general, called the *Fronter* type, and it facilitates mass production together with providing the limited flexibility. In due course, the machine was designed with the modular principle from the old days to credit at least the necessary flexibility. Figure 1-28 shows a variation of *Fronter* type in the beginning of the 1960s, where the cross slide was hydraulic-driven and inclined 30° from the horizontal plane, and the guideway of carriage was of hybrid type, i.e., a combination of roller and sliding types [21].

In contrast, Heyligenstaedt applied the modular design to the machine for small batch production, i.e., medium-sized copying lathe (type: Heycomat 1) in order to respond quickly to the user's order in 1962 [22]. Figure 1-29 is one of the basic configurations in relation to the main spindle and its driving system, where the spindle speed can be changed with the manual shift of the gear and electromagnetic coupling. In this case, there are three basic spindle systems, which can produce the three variants; and if necessary, it is furthermore possible to vary the speed range by using the following methods.

1. Modification of reduction ratio in belt transmission from the main motor to headstock

2. Exchange of rotational speed of main motor of induction type, i.e., either 1500 or 3000 rpm

3. Use of pole-change motor

Moreover, the main spindle diameter at the front bearing is either 90 or 120 mm, and the machine with larger bearing is for the chuck work.

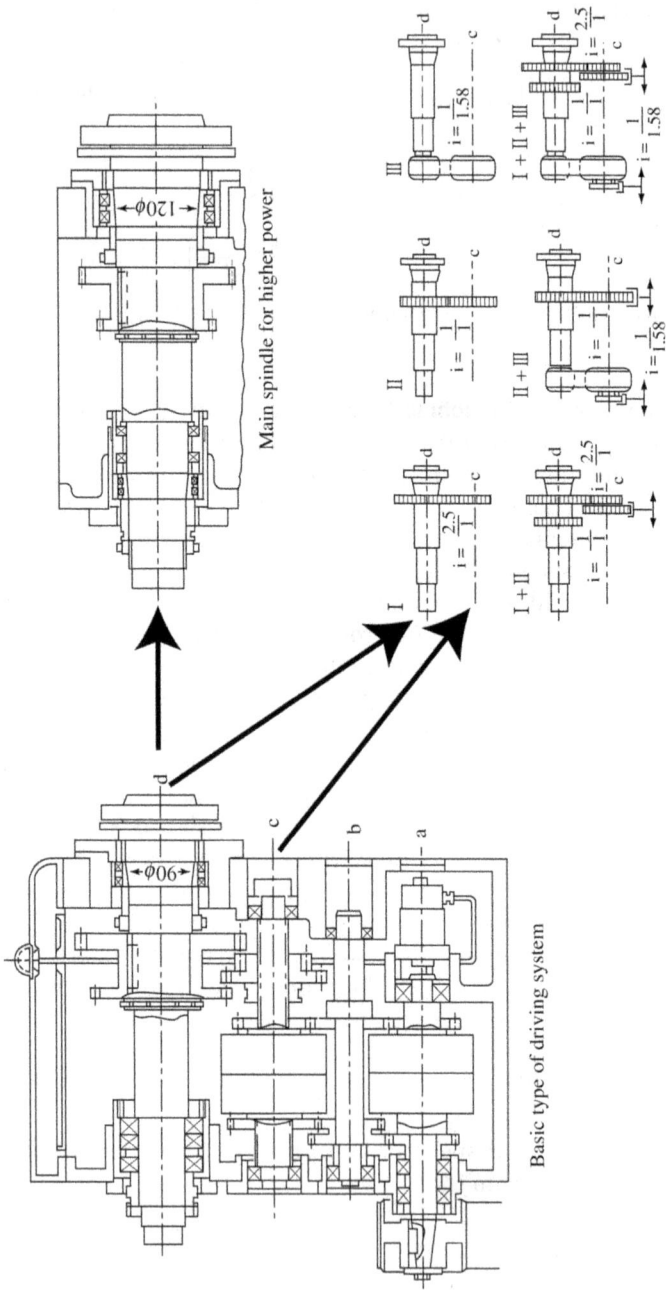

φ120

Main spindle for higher power

Developed types of driving system

III
$i = \frac{1}{1.58}$

II
$i = \frac{1}{1}$

I
$i = \frac{2.5}{1}$

I + II + III
$i = \frac{1}{1}$
$i = \frac{2.5}{1}$
$i = \frac{1}{1.58}$

II + III
$i = \frac{1}{1}$
$i = \frac{1}{1.58}$

I + II
$i = \frac{1}{1}$
$i = \frac{2.5}{1}$

φ90

Basic type of driving system

Figure 1-29 Modular design in main spindle and its driving system (type Heycomat of Heyligenstadt make, 1962).

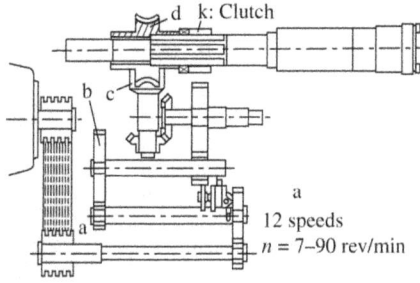

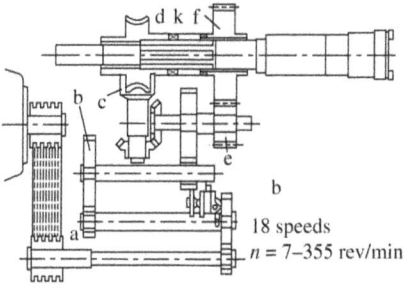

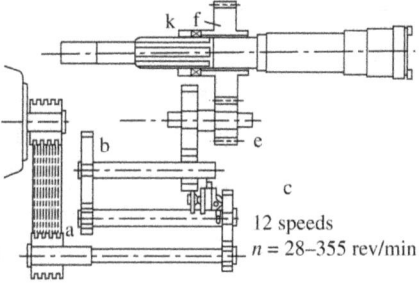

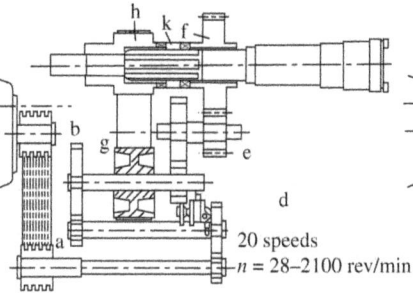

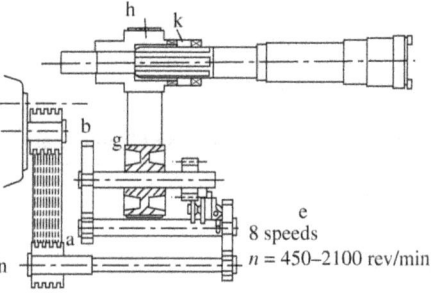

Figure 1-30 Modular design in main spindle driving system (Wanderer Co., 1960s).

The same idea was also applied to the milling machine of Wanderer make about 1960, as shown in Fig. 1-30. Furthermore regarding the size variation, Kearns merchandised the heavy-duty boring machine on the basis of the market survey. Actually, the wide ranges of the bed length, table width, and column height allowed the great variety of work size in that of Kearns. On this extension, an innovative idea was proposed by William Asquith. A monolithic basic pattern was predetermined as shown in Fig. 1-31, and a part selected from it was used as a column in accordance with the design requirements [6].

In short, the unit construction for the conventional machine tool of manual operation type was in salience in the 1960s as shown, furthermore, in other examples in Table 1-3.

(a)

(b)

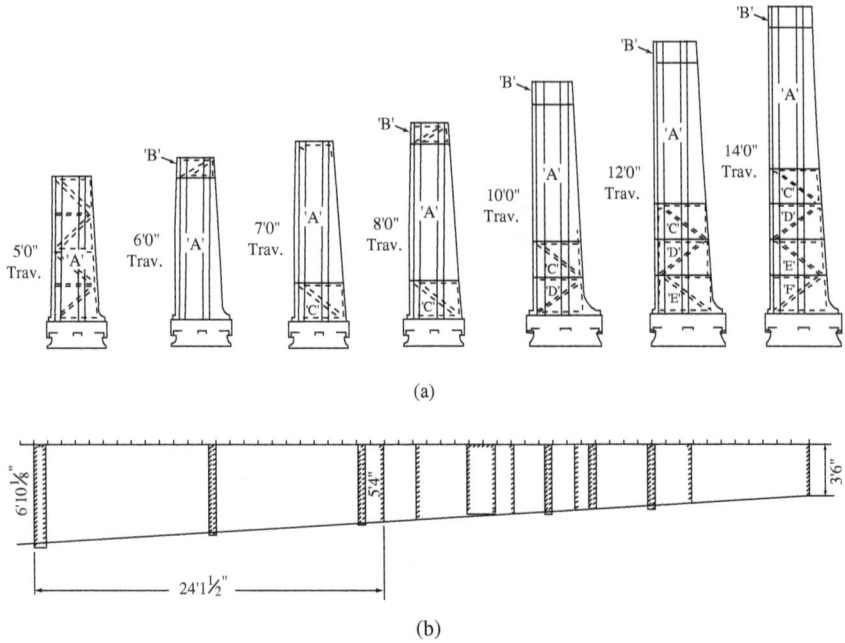

Figure 1-31   Modular design for column in horizontal boring and milling machine using monolithic pattern: (a) Modules of columns and (b) monolithic basic pattern (1960s, by William Asquith Co.).

TABLE 1-3   Modular Design of Traditional Machine Tools in the Beginning of the 1960s

Manufacturers	Kinds/Types	Objective units of modular design	Leading attributes
Clevel and Hobbing	Lathes of various types	Bed, cross slide, copying slide, feed gearbox, tailstock	
Dubied	Hydralic copying lathe (type: 517/500)	Spindle speed changing mechanism, copying attachment	
Kearney & Trecker	S-series milling machine	Base, column, knee, table	Plain type, ram head type, universal type, universal ram head type, vertical type
Makino Milling Machine	No. 1 vertical milling machine with turret head (type: KB) No. 1 jig boring & milling machine (type: KJ)	Knee, column, saddle	Dimensional unification
MAX Müller	Automatic lathe (types: AM & AME ELTROMATIC)	Cross slide, copying slide, turret slide, facing, & boring unit	
Shoun Kosakusho	Milling machine of production type (type: FP) Engine lathe (type: HB-725)	Spindle unit of quill type	

Large-size machine tools. In most cases, the large-size machine tool is manufactured with one-off or a kind of production mode when receiving the customer's order. The machine can thus be regarded as a suitable objective to apply the modular design for the same kind as well as for different kinds, with the expectation of considerable reduction of design work and time. In fact, there are two application methods of modular design to the large-size machine tool.

Extreme increase of flexibility with unit construction. This application aims at the reduction of the facility expense within the user's factory by reinforcing the flexibility of both the machining capacity and the method. Consequently, the primary concern is an extremely large-size machine, and TOS of Czechoslovakia manufactured the planomiller of various types by predetermining the 11 basic structural units as shown in Fig. 1-32(a) together with varying the table width across six dimensions. In fact, the BBS of TOS emphasized the variation of the structural configuration and allocation of the milling head together with varying the three different output powers. This BBS obviously shows the capability of producing 91 variants, provided that the commonness of the units across the whole design was about 80 percent [23]. More specifically, the noteworthy concept of the design lay in the increase of both the structural stiffness and the versatility. On this extension, it is worth suggesting that the milling heads of ram and quill types depicted, in the case of Butler, in Fig. 1-32(b) were allowed as the basic units to strengthen the machining variety.

Another representation was the horizontal boring and milling machine of Scharmann make in the mid-1960s. In this case, various structural configurations were produced as shown in Fig. 1-33 by combining the structural body units. Following that Toshiba Machine Manufacturing was marketing the vertical boring and turning machine with a table of more than 5 m in diameter (type TDP-105NC). This double column machine was characterized by its column of block built-up structure, double-table structure, i.e., stationary inner and rotatory outer tables, and cross rail of fixed type, resulting in the eight variants shown together with the overall view in Fig. 1-34 [24]. Importantly, the Cincinnati Milacron demonstrated the advantageous features of the modular design even in the 1990s. In fact, the CNC profiler (type L-series) with 10-axis control in maximum was designed by the modular principle, where 80 variants can be produced from the three basic types, i.e., those of bridge, bed, and rail types.

Manufacture of different kinds. In this context, Ikegai Iron Works is credited as an initiator in 1962 by manufacturing the planer, planomiller, vertical boring machine, vertical boring and turning machine, and bedway grinder from a group of the modules. In due course, a considerable number of the variants were able to be manufactured by varying the table width, spindle diameter, number of spindle heads, and so on. Following that of Ikegai, the same idea later became reality by the VEB of Karl-Marx-Stadt. The details of both trials will be stated in Sec. 1.3.4.

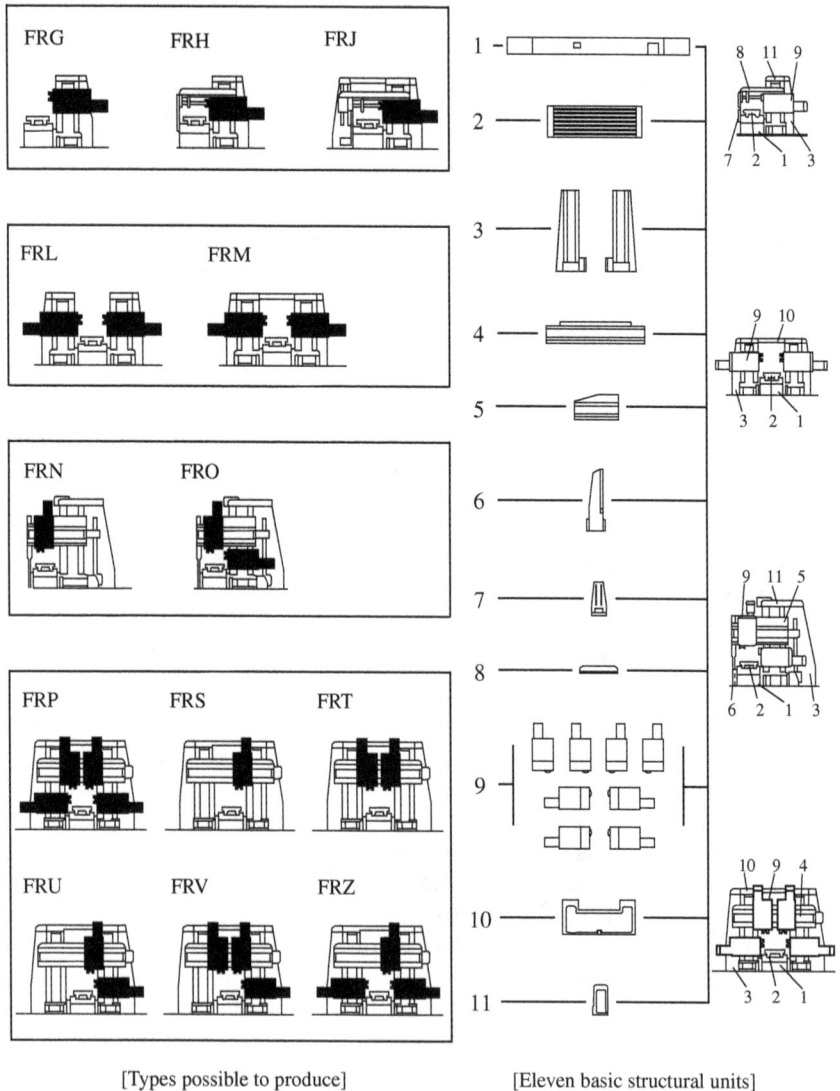

[Types possible to produce]   [Eleven basic structural units]

(a)

Figure 1-32 Modular design in planomiller: (a) In case of type FR of TOS make and (b) milling heads of quill and ram types (Butler Machine Tool Co.).

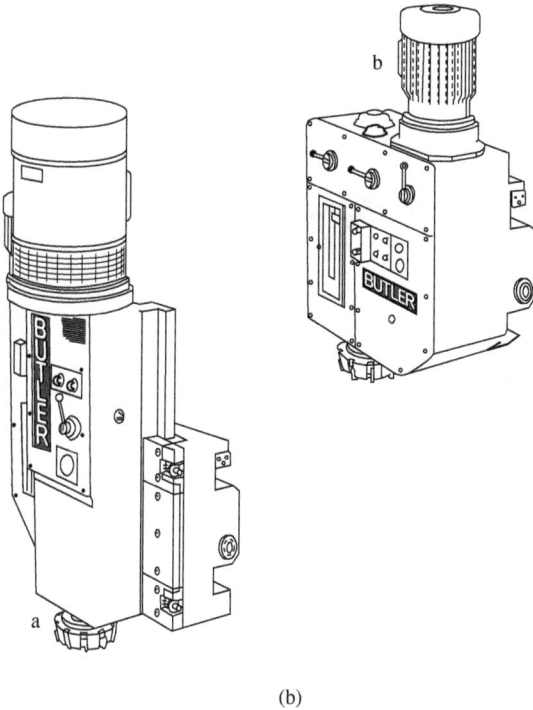

(b)

Figure 1-32 (*Continued*)

1.3.3 Application to NC machine tools

The NC is, as already suggested, considered the most epoch-making contrivance since Wilkinson's cylinder boring machine of the industrial revolution era. The NC was developed with amazing speed across the whole world and affected the enhancement of the machine tool technology to various extents. In retrospect, the tape-controlled NC boring and milling machine, and the drilling machine of planer type with ATC (*automatic tool changer*) were, e.g., on the market in the beginning of the 1960s by Gilbert and by Warner and Swasey, respectively. Consequently, the NC has continuously been developed as represented with several key terms, e.g., hardwired NC, modular NC, CNC, and open CNC. With the growing importance of the NC machine tool, the modular design was forced to change its concept and application method to a larger extent. Importantly, the modular design has been applied to the NC machine tool, particularly expecting to reinforce the flexibility of the machine from the hardware aspect. In fact, a concept having been once believed is the wider flexibility of the NC machine tool in machining patterns and methods, which can, in principle, be commissioned to

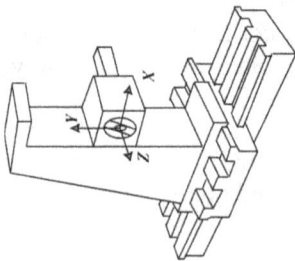

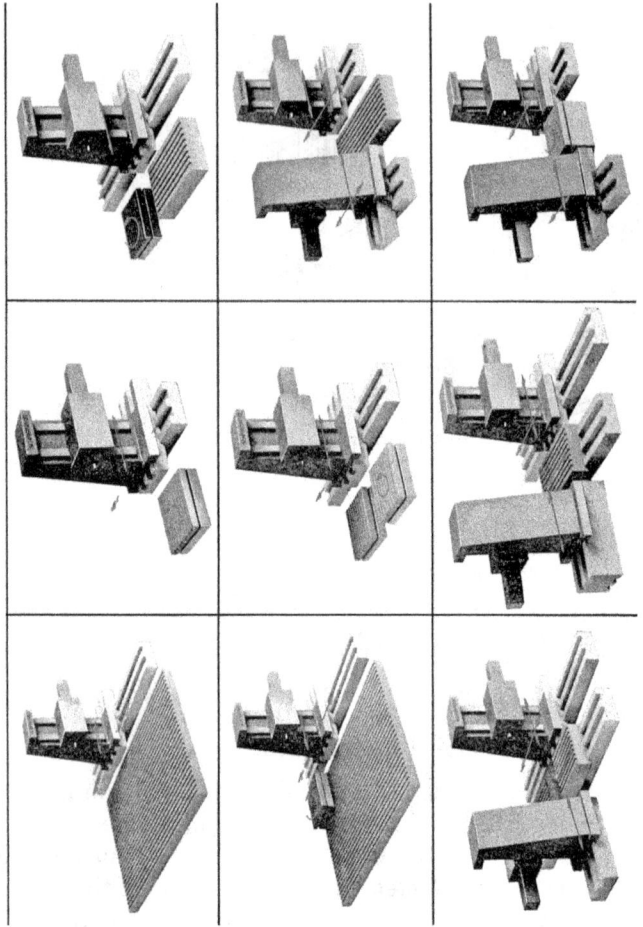

Figure 1-33   Horizontal boring and milling machine with modular design (by Scharmann Co.).

Built-up column
Cross rail
Turning head
Rotatory outer table
Stationary inner table

Milling head

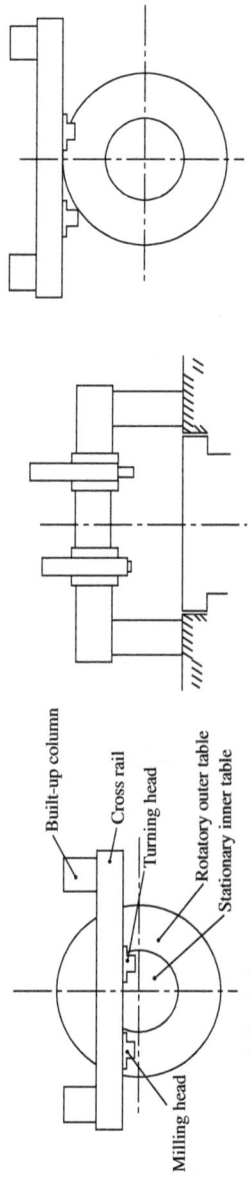

(1) Standard type

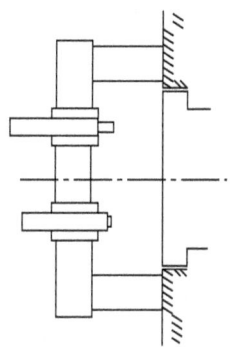

(2) With column higher than standard

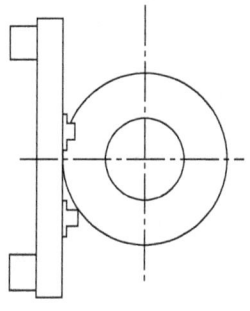

(3) For larger diameter workpiece

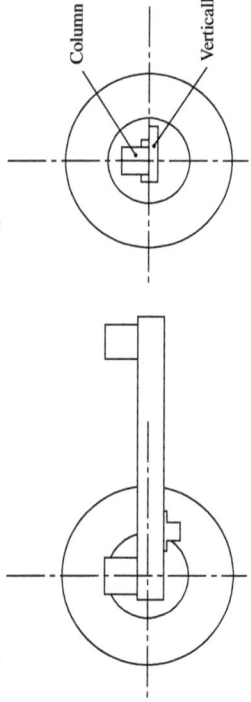

Column with guideway
Vertically traveling tool post

(4) For ring-shaped workpiece

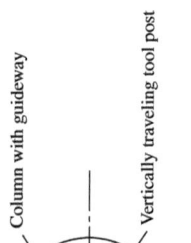

(5) For boring & facing of ring form workpiece

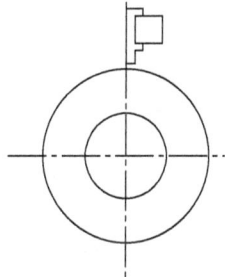

(6) For turning & facing

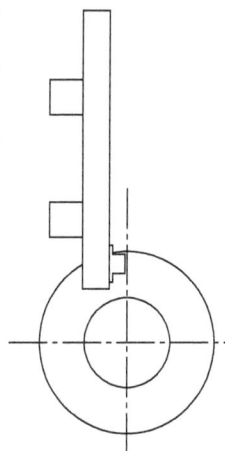

(7) For extremely large workpiece

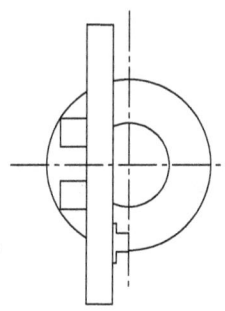

(8) For boring of extremely large workpiece

Figure 1-34 Eight variants possible in vertical boring and turning machine of Toshiba Machine make.

both the NC software and the number of control axes, although the software has certain limitations. When the required flexibility of the machine tool is far beyond that given by the software available, the machine must be restructured to reinforce the machine flexibility using the modular design.

In accordance with such a concept, Boehringer already produced, e.g., an NC lathe (product series PN 420) in the beginning of the 1970s [25]. This NC lathe has four variants configured by the combination of the slide, turrets of disk and universal types, and tailstock, in consideration of the compatibility with both the chuck and/or the bar work. To understand what was the NC machine tool on that day, Fig. 1-35 shows a protruded application of the BBS to the lathe of *Fronter* type, i.e., type DP 250 of VDF make with the commercial name *Machining System*. In fact, the basic machine called *Single-working Configuration* consisted of a group of units, i.e., those for main motor, gearbox, tool slide, and tool post, resulting totally in another two configurations. As can be seen

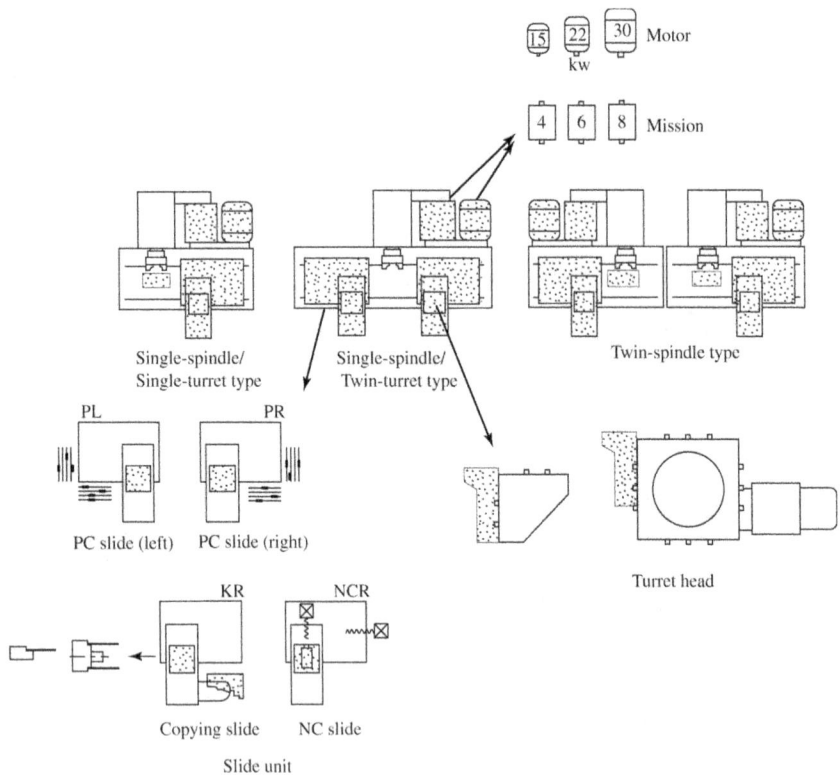

Figure 1-35   Unit construction of NC turning machine in the 1970s (type DP250 of VDF Boehringen Harbeck make).

from the variation of the tool slide, this case can be interpreted as being midway between the traditional (manual operation) and the NC machines, and from such a point of view the machine is worth recording in the developing history.

At this immature era, the NC machine tool was manufactured by simply attaching the NC controller to the machine tool and modifying the structure in part to equip, e.g., the ball screw and servomotor. In due course, the machine tool maintains the traditional appearance. In other words, the structural appearances were in mixed condition, i.e., that with built-in type NC or stand-alone NC, which simply attached the NC controller to the traditional machine. With the advance of the related technologies, the smart NC machine tool came to fruition, where the synergy of the NC technology and the structural design was skillfully used, and on this extension, the MC and TC have, at last, became a reality in the mid-1970s. To understand such a trend, Fig. 1-36 shows a developing history of TC originating with the engine lathe and its variants.

In short, the MC itself can manage various machining methods so far executed by the drilling, milling, and boring machines, whereas the TC includes, in principle, an advanced function of milling, although even the NC turning machine can manage milling in part by the special attachment fixed to its turret.

It is worth pointing out that even in the beginning of the 1970s, simultaneous quinary-axis control was already established, and obviously we conclude that the NC machine tool has enough flexibility beyond our expectation across the whole machining methods usually employed together with rendering the modular design useless. In fact, the machining function of TC is, in general, a synergy of turning, drilling, boring, and milling. Thus the machine is, in certain cases, equipped with a turret column having plural turret heads or tool magazines (see Fig. 2-17), resulting in the very flexible tooling layout, although turning is dominant. Consequently, between the middle and late 1970s, modular design was not often employed, because the TC and MC prevailed, and there was no necessity to provide the machine tool with the flexibility based on the modular design. In fact, the modular principle was applied to the design of the attachment in the case of TC, e.g., those for offset of rotating tool and twin-drilling, but not to the design of the structural configuration. Importantly, the flexibility of the NC machine tool was further reinforced by the skillful fusion of the NC software and tooling layout in the beginning of the 1980s, and the designer became progressively less interested in the modular design thereafter. With further due development of the TC and MC, the structural body configuration was scarcely designed by the modular principle, and such a trend was accelerated. Table 1-4 shows some examples of the NC turning machine

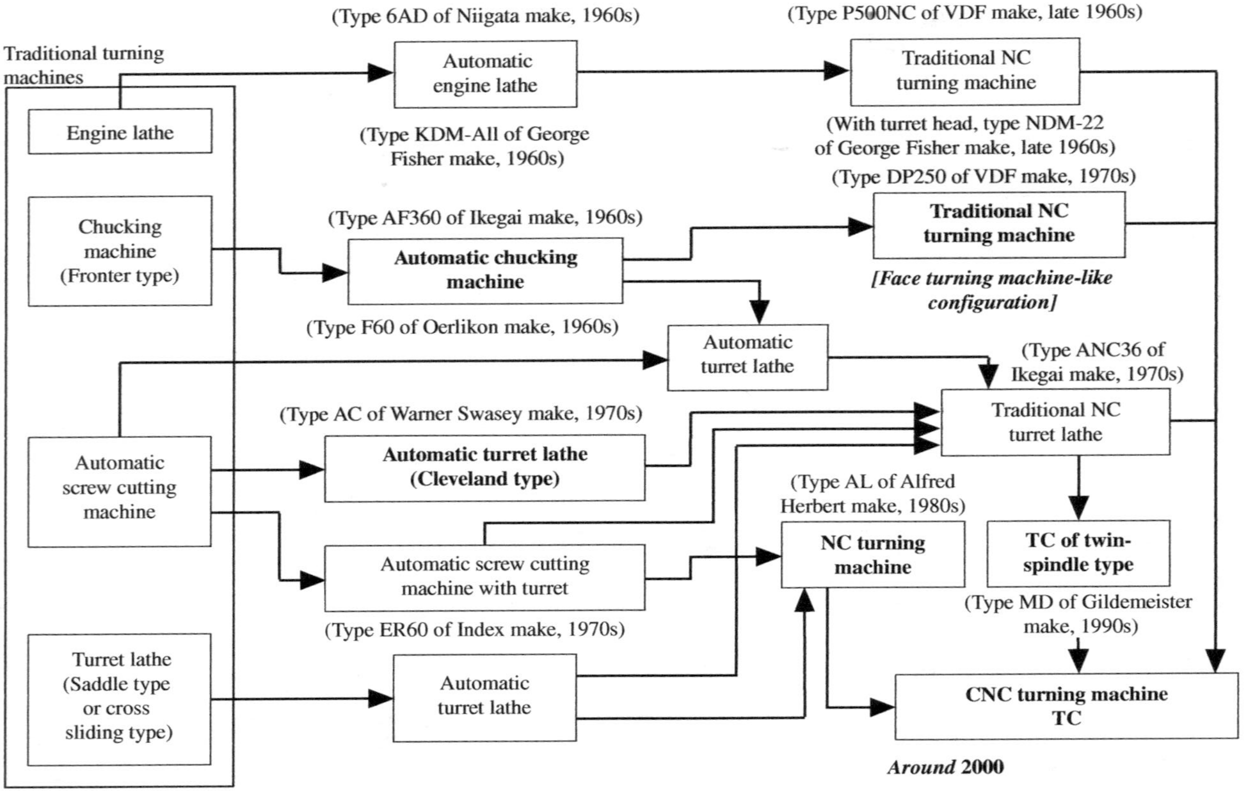

Figure 1-36 Milestones in development of TC from traditional turning machines.

Traditional turning machines

Engine lathe

Chucking machine (Fronter type)

Automatic screw cutting machine

Turret lathe (Saddle type or cross sliding type)

Automatic engine lathe (Type 6AD of Niigata make, 1960s)

(Type KDM-All of George Fisher make, 1960s)

Automatic chucking machine (Type AF360 of Ikegai make, 1960s)

(Type F60 of Oerlikon make, 1960s)

Automatic turret lathe (Cleveland type) (Type AC of Warner Swasey make, 1970s)

Automatic screw cutting machine with turret

Automatic turret lathe (Type ER60 of Index make, 1970s)

Traditional NC turning machine (Type P500NC of VDF make, late 1960s)

(With turret head, type NDM-22 of George Fisher make, late 1960s)

Traditional NC turning machine (Type DP250 of VDF make, 1970s)

[Face turning machine-like configuration]

Automatic turret lathe

Traditional NC turret lathe (Type ANC36 of Ikegai make, 1970s)

NC turning machine (Type AL of Alfred Herbert make, 1980s)

TC of twin-spindle type

CNC turning machine TC (Type MD of Gildemeister make, 1990s)

Around 2000

52

TABLE 1-4 Some Examples of Modular Design in NC Turning Machines (1970s and 1980s)

Manufacturers	Types	Leading specifications	Modular design aspects
Osaka Kikou (1976)	T55-N	Swing: 300 mm Max. spindle speed: 2000 rpm.	Turret: four types (square, Hexagon, drum & twin-turret)
Hitachi Seiki (1976)	NH-500	Swing over carriage: 515 mm Max. spindle speed: 2000 rpm.	Turret head: two types Tailstock (available)
Gildemeister (1977)	Fronter type	Max. work diameter: 250 mm	Main spindle: two types Cross slide: three types Tooling system: turret type or block tooling system
Traub (1987)	TNS-30/42D		Main spindle: two types Turret allocation: three types Pickup spindle for rear machining (available) Rotating tool (available) In case of one turret, tailstock and cutoff tool slide (available)

with modular design to have the firsthand view of their characteristic features on that day.

In addition, the underestimate of the modular design was accelerated with the advent of the GC (*grinding center*), although the flexibility of the manufacturing facilities was again a primary concern in the beginning of the 1980s. This trend was caused by the shortening of the product life and decreasing the batch size derived from the rapidly increasing speed of both the product and the production process innovations. In due course, the MC with larger rigidity can execute grinding with satisfactory quality, whereas the GC has greater capability of grinding the ceramics of Al_2O_3, Si_3N_4, and ZrO_2 types. The Mori Seiki has been on the market, an MC of simultaneous quinary-axis control type (type: M-400C1) for the users, e.g., turbine blade and propeller manufacturers. This machine has furthermore the function of GC by the increase of spindle speed and positive use of larger rigidity of the machine in the mid-1990s. The Rolls-Royce has also employed the MC of Makino make to produce the Inconel aircraft engine parts, e.g., compressor blade, turbine blade, and engine casing, instead of a creep-feed grinder in the late 1990s, showing major savings in capital investment, production cost, and lead times. The characteristic feature is that the small grinding wheels, each with the profile of a specific feature on the component, are held in the tool magazine [26].

Within a grinding machine context, the NC machine was belatedly on the market, and thus the GC itself and the transfer grinding line were contrived in the late 1980s. This backwardness is caused by the requirement for higher machining accuracy in grinding than in cutting. For example, Yamazaki Mazak merchandised the GC in the beginning of the 1990s on the basis of the TC. The GC in the 1990s can be classified into (1) TC-based and (2) MC-based types, and furthermore (3) turret column type, in which the grinding wheel can be exchanged by indexing the column. In due course, both the GCs based on TC and MC are dominant, and in these GCs, the basic necessity is to provide the adaptive control function, continuous truing and dressing device of between-process type, tool changer. and so on. In the tool changer, the ATC for cutting tools is apparently replaced to that for grinding wheels, although the tapered shank to mount the grinding wheel to the spindle is in contact at both the side surface of flange and the tapered surface.

On the extension of the TC, MC, and primitive GC, the system machine and machining complex have been contrived as their advanced kinds in the mid-1980s and later, respectively. The system machine can be interpreted as a cubiclike compact FMC to reduce the floor space, simultaneously maintaining the function and performance of the FMC. As already clarified elsewhere, the capability of FMC can be compared to that of FMS, and thus the SME has been very keen to install the FMC, because the SME faces, in general, the acute shortage of factory space. In this context, the system machine is one of the further solutions for facing such a problem of SME together with responding to the various machining requirements. In contrast to the conventional and advanced NC machine tools, the advent of system machine and machining complex has thus evoked the growing importance of the modular design once again. It is very interesting that the system machine is typical evidence for "Developing trajectory of a technology being 'Divergence and Convergence' processes" in engineering field (see Chap. 2).

1.3.4 Different-kind generating modular design

This concept is considered to be modular design by birth, and machine tools of various kinds can, in principle, be manufactured from a group of modules, where the module is in the form of a unit. For example, a group of units facilitates the manufacture of the drilling, milling, and turning machines as already shown in Fig. 1-10, which was displayed by Koenigsberger in 1974 [7, 10].

On that occasion, Koenigsberger stated that there was no practical evidence of this modular design concept and that the due research was being conducted by the UMIST (*University of Manchester Institute of*

Science and Technology). Importantly, Ikegai Iron Works already applied it to the large-size machine tool in the beginning of the 1960s, although there were multiple obstacles derived from the optimization of the structural configuration. In fact, the optimization is in larger dependence upon the differing magnitude and direction of external loads, difficulties to provide rationally the subform generating function to each module, and also the functionality to be provided to each module. Taking into account the difficulties in the unit construction within the same kind, the obstacles mentioned above are easily imaginable and acceptable.

In fact, Ikegai Iron Works tried to manufacture the planomiller, bedway grinder, planer, vertical turning machine, and vertical boring machine, from a group of the units in 1962, as shown in Fig. 1-37. Regarding this trial, it was reported that the manufacturer was able to reduce the design time and throughput time to a great extent. Consequently, several vertical turning machines and planomillers were actually installed in the factories of Toyota Motor Car and Nissan Motor Car. Even in the late 1990s, a planomiller is used on work at the die manufacturing factory within Nissan Motor Car, after making the retrofit compatible with the modern NC.

The same idea was later employed by the VEB of Karl-Marx-Stadt. In this case, the planer, planomiller, boring machine, and guideway grinder were produced from a group of modules, which were classified into those for structural configuration, form-generating movement, and additives. The machines can be characterized by the driving mechanism consisting of the crossed helical pinion-rack mechanism of built-in unit type, which was available for all the kinds possible. In contrast, the driving source was varied depending on the kind; in addition, the electric equipment was of modular type. As can readily be seen, the user and manufacturer could expect high machine effectiveness and high productivity, respectively. In fact, the manufacturer saved the developing cost up to 10 percent in standard type and up to 20 percent in special type by the employment of the modular design. Furthermore, it is worth stating that the planomiller was used as a machining entity within a system *PRISMA II*, which is the well-known FMS produced by East Germany [27].

Following that of VEB, a machine with modular design of different-kind generating type was conceptualized, as shown in Fig. 1-38, by the University of Strathclyde in accordance with the ASP Plan [28], although the responsible committee concluded that the machine was not to be a reality. The machine aimed to be a kernel of the manufacturing system for small batch size. In retrospect, it is very interesting that Mauser-schaerer tried to produce an FMS called *Produktionssystem 2000* using the MC of column traveling type and vertical turning machine. These machines were produced with the modular design of different-kind generating

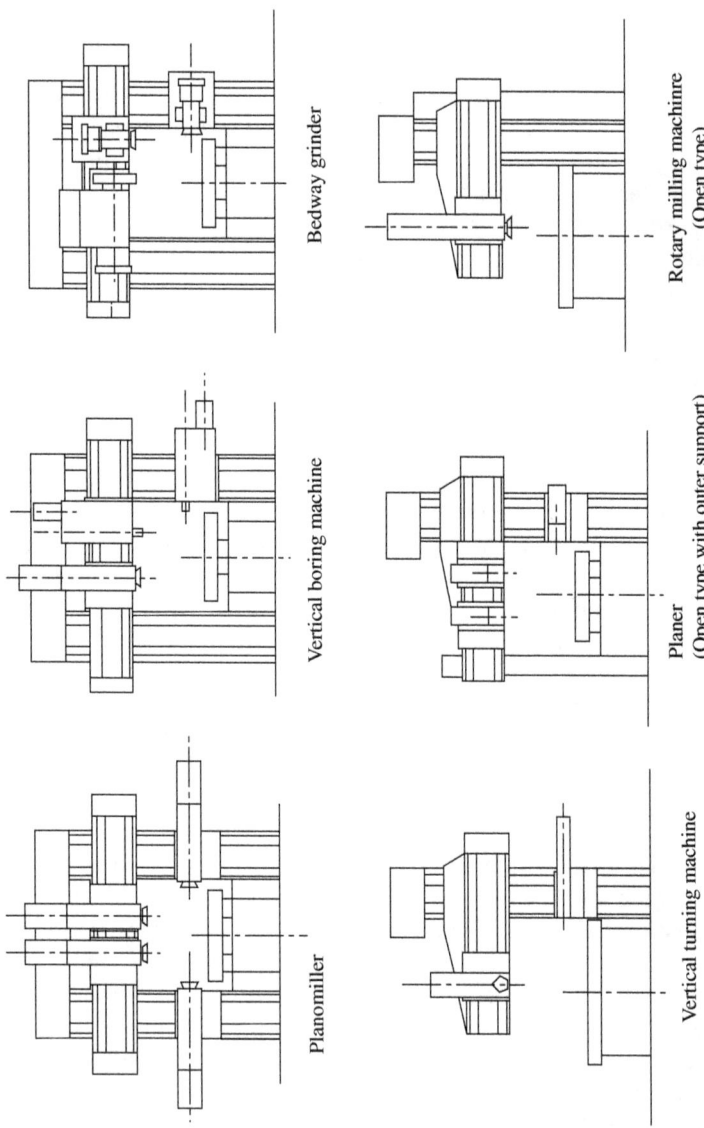

Figure 1-37 Some examples of structural configuration with different-kind generating modular design in 1962–standard combination of common modules (courtesy of Ikegai Iron Works).

Bedway grinder

Rotary milling machine
(Open type)

Vertical boring machine

Planer
(Open type with outer support)

Planomiller

Vertical turning machine

Tool magazine
of chain type

Tool changing arm

Tool pallet

Indexing table

Additional tool magazine

Headstock with
twin-spindle

Turret

Replaceable disk
turret of star type

Turret indexing unit

Tailstock

Bed

Machine concept

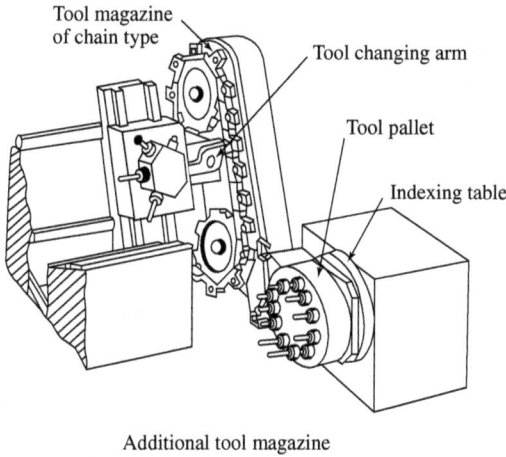

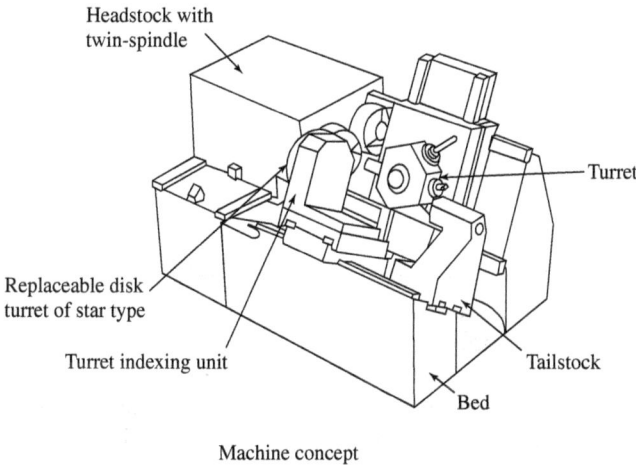

Figure 1-38 Machine tools with modular design in ASP plan (by
Astrop, courtesy of *Machinery & Prod. Eng.*)

type [29] as shown in Fig. 1-39, although the on-site exchange of the
module is impossible.

Given the evidence mentioned above, it is furthermore worth suggesting that there was an aftermath to applying the modular design to
the manufacturing line in the beginning of the 1960s. Actually, the manufacturing line consisting of different kinds of the machine was used on
work to aim at the enhancement of economic benefit. For example, the
Cincinnati Milling Machine merchandised a manufacturing line for processing the piston rod with 0.0001 in for roundness, 0.0002 in for
straightness, ±0.0002 in for size, and so on. The line consisted of five

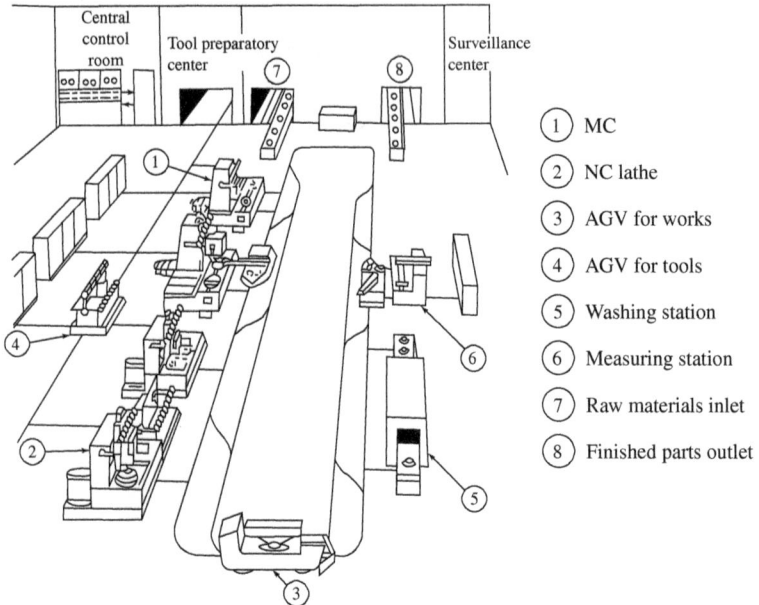

Figure 1-39   FMS consisting of machine tools with modular design of different-kind generating type (Produktionssytem 2000, courtesy of Mauser-Schaerer).

centerless grinders, each grinder having its own simplified function, and one induction hardening station; and between the first and second grinders the rod rotated through the induction coil and quenching ring. On this extension, Cincinnati Milling advertised its· ability to supply other grinding lines together with various processing functions, e.g., milling, heat treating, broaching, turning, electrical machining, and/or drilling. A similar idea, as shown in Fig.1-40, was applied to an FMS for the manufacture of spiral bevel gearing by Oerlikon in 1982.

In 1991, Rock Drill Factory of the Tamrock in Finland installed an FMS of Yasuda Kogyo make to produce the rock drill, which consists of the lathe for premachining, three MCs, heat treatment equipment, and a grinding machine. This FMS is thus regarded as a successor of that of Cincinnati in the 1960s.

To this end, it is again emphasized that Brankamp and Herrmann proposed, as already shown in Fig. 1-7, the valuable concept for different-kind generating modular design in 1969, providing us with a total view of modular design, although there remains something uncertain. Importantly, the modular design appears to be very simple; however, there are various crucial problems to be solved in the near future, as will be seen from the following chapters, even though the history of the modular design is very long.

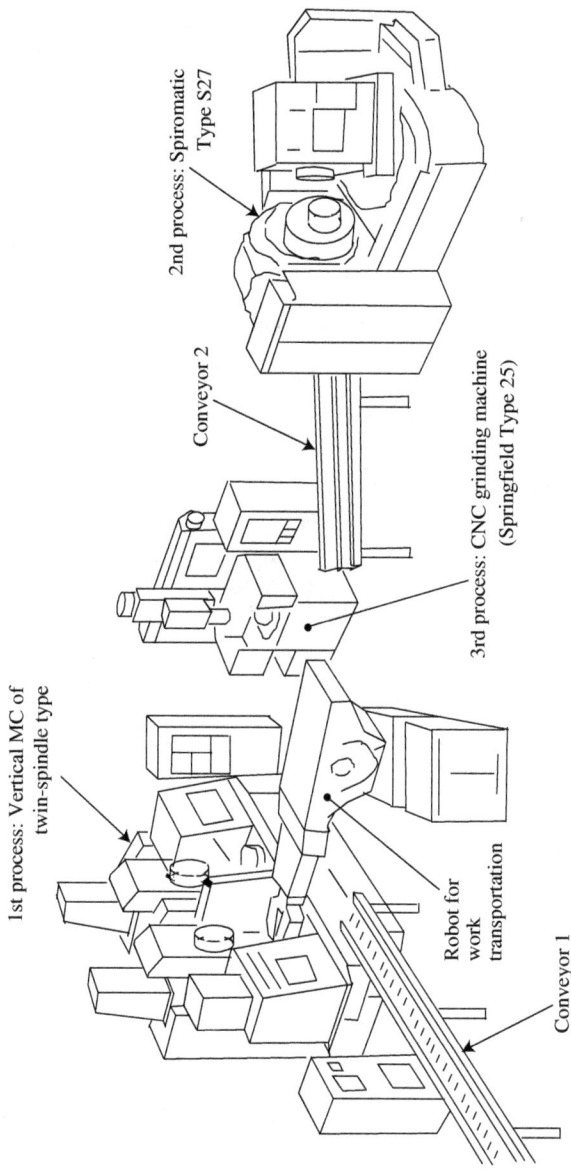

1st process: Vertical MC of twin-spindle type

2nd process: Spiromatic Type S27

Conveyor 2

3rd process: CNC grinding machine (Springfield Type 25)

Robot for work transportation

Conveyor 1

Figure 1-40 FMS for producing spiral bevel gear in 1982 (courtesy of Oerlikon Contraves AG).

References

1. "Sandvik Launches New Interchangeable 'Cutting Unit' System." *The Production Engineer*, December 1980, pp. 20–21.
2. Dietz, P., "Baukastensystematik und methodisches Konstruieren im Werkzeugmaschinenbau." *Werkstatt und Betrieb*, 1983, 116(4): 185–189.
3. Ito, Y., "Flexible Manufacturing System Complex Provided with Laser—Part 3 Application System Design," *Proc. of 5th ICPE*, 1984, pp. 28–36, JSPE Tokyo.
4. Zeh, K-P., and H. E. Frank, "Simulationsgestützte Planung einer flexiblen Fertigungsanlage," *tz für Metallbearbeitung*, 1984, 78(5): 11–17.
5. Brankamp, K., and J. Herrmann, "Baukastensystematik—Grundlagen und Anwendung in Technik und Organisation," *Industrie-Anzeiger*, 1969; 91(31): 693–697 und 91(50): 133–138.
6. Koenigsberger, F., "Modular Design of Machine Tools," *Proc. of Inter. Conf. on Manufacturing Technology*, University of Michigan Ann Arbor, 1967, ASTME, pp. 35–54.
7. Koenigsberger, F., "Trends in the Design of Metal Cutting Machine Tools," *Proc. of 1st ICPE*, 1974, JSPE Tokyo.
8. Lee, H. S., H. Shinno, and Y. Ito, "Structural Configuration Design of Machining Center—On the Variant Method Using Conjunction Pattern," *J. of JSPE*, 1986, 52(8): 1393–1398.
9. Ito, Y., "System Configuration and Design of FMS in Next Generation," *Advanced Robotics*, 1987, 2(2):103–120.
10. Koenigsberger, F., "Private Draft Proposal for Research Project—Modular Design of Machine Tools," *University of Manchester Institute of Science and Technology*, July 29, 1975.
11. Schwarz, W., "Universal-Werkzeugfräs-und-bohrmaschinen nach den Grundprinzipien des Baukastensystems," *wt-Z. ind. Fertig.*, 1975, 65(1): 9–12.
12. Product Catalogue of Reinhard Bohle KG, 1960s.
13. "Baukastensystem für Drehmaschinen," *fertigung*, Dezember 1993, pp. 28–30.
14. Georg, O., "Ein allgemein anwendbares Baukastensystem für Werkzeugmaschinen," *Werkstattstechnik und Maschinenbau*, 1950, 40(3): 65–70.
15. Ito, Y., "The Production Environment of an SME in the Year 2000," in K. McGuigan (ed.), *Flexible Manufacturing for Small to Medium Enterprises—A European Conf.*, 1988, pp. 207–234, EOLAS, Dublin.
16. Ropohl, G., and F. Schreiber, F. "Grenzen der Flexibilität von Aufbaumaschinen aus genormten Baueinheiten," *Werkstattstechnik*, 1968, 58(7): 301–306.
17. Ito, Y., "Desirable System Configurations of FTL for Automotive Industry," *Proc. of Inter. Conf. on Auto Technology*, 1990, pp. 254–262, Chulalongkorn, Univsity of Bangkok.
18. Elb-Schliff, "Modular aufgebaute, CD-gerechte flexible Schleifsysteme," in *Schleifen, Läppen, Honen Jahrbuch*, 1984, pp. 335–356, Vulkan Verlag.
19. ISO 2562, 2727, 2769, 2891, 2912, and 2934 (Modular Units for Machine Tool Construction), 1973.
20. Ito, Y., H. Shinno, and A. Sacaguti, "Grouping and Decomposition Methodology for Machining Process to Assist the Design of Flexible Transfer Line," *Proc. of 26th Inter. Symposium on Automotive Technology and Automation—Dedicated Conf. on Lean Manufacturing in the Automotive Industries*, ISATA, Aachen, 1993, pp. 375–382, Automotive Automation Ltd., Croydon, U.K.
21. Saljé, E., "Eine Baukasten-Drehmaschine für scheibenförmige Werkstücke," *Werkstatt und Betrieb*, 1961, 94(10): 757–764.
22. Hölscher, W., "Die Anwendung des Baukastenprinzips bei der Konstruktion und Herstellung von Werkzeugmaschinen," *Werkstatt und Betrieb*, 1962, 95(9): 586–589.
23. Kvetoslav, E., "Building-Block Design of Milling Machines with Regard to Design of Different Machining Variants," *ASTME Technical Paper* No. MS68-203, 1968.
24. Ito, Y., "New Technology and Its Problem in Japanese Machine Tools," *Digest of Japanese Industry*, 1973, 66:10–14.
25. Schuler, H., "Numerisch gesteuerte Drehmaschinen nach dem Baukastenprinzip," *wt-Z. ind. Fertig.*, 1970, 60(11). 120–124.

26. "Grinding on a Machining Centre," *Manufacturing Engineer,* 2000, 793: 93.
27. Helmeck, W., "Projektierende Arbeitsweise nach einem Baukastensystem bei Großteilbearbeitungsmaschinen," *Maschinenbautechnik* 1976, 25(5): 212–217.
28. Astrop, A., "Time to Take Action on the ASP Report," *Machinery and Prod. Eng.,* Aug. 2, 1978, p. 17.
29. Mönkemöller, H., "Werkzeugmaschinen für ein automatisches Fertigungssystem," *Werkstatt und Betrieb,* 1972, 105(3): 213.

2

Engineering Guides and Future Perspectives of Modular Design

Apart from the research into machine tool description and joint stiffness, modular design has not been vigorously studied so far in the academic sphere, but has been developed on the basis of longstanding practical experience together with use of the trial-and-error method. Within a modular design context, we must always remember the utmost valuable proposal of Doi [1], in which he laid out the four principles of modular design based on his extensive experience. In fact, these principles—the principles of separation, unification (standardization), connection, and adaptation—are very valuable in rationally applying modular design, to quickly grasp the facing problems, to predict further perspectives, and so on; however, they are not detailed in the form of guides or a design handbook as yet. In addition, each design principle must take into account the common requirements along with the specific ones according to the kind of machine tool.

Importantly, it is worth suggesting that the machine tool description can facilitate the choice of preferable structural configurations from a group of modules, which is one of the design methodologies and subject to the principle of adaptation. In contrast, the joint between both the modules must be designed to have enough stiffness, i.e., high joint stiffness on the basis of the principle of connection. At burning issue is to establish the detailed design guide for each principle by amalgamating the practical experience with academic knowledge. Regarding joint stiffness, we have already established a sphere called engineering problems in machine tool joints.

2.1   Four Principles and Further Related Subjects

In this section, the four principles are discussed in detail to quickly understand what basic engineering knowledge is used in the application of modular design to the machine tool. In fact, the four principles are basically worth following even now. It is very important, however, to be aware of the changing trends of modular design resulting from the enlargement and enrichment of the application areas. For example, the modular design concept must be modernized so as to be compatible with the production environments at present and in the near future. Thus, in this section we furthermore touch on the new variants of the modular design concept in the year 2000 and beyond.

Principle of separation.   The principle of separation is in closer relation to how to determine the module, and of the four principles, this principle has the greatest difficulty being sublimated to a preferable technology. In fact, the basic modules have been determined so far by using the trial-and-error method. The *principle of separation* can thus be defined as follows.

> A module is allowed to have only a specified function and/or structural configuration in full consideration of the following.

1. The user does not mind whether the machine is designed using the modular principle or not, apart from machines of customer-oriented type, and thus the module must have the least function and configuration acceptable, but not be overspecified.
2. The module must have satisfactory stiffness as well as high joint stiffness.
3. The machining accuracy of the module should be within allowable tolerances to achieve the required assembly accuracy under any joint conditions.

In this context, the crucial issues are how to disintegrate a machine tool as a whole into the proper number of modules and how to determine a group of standardized modules according to the design purpose. For instance, the machining space should be investigated in the unit construction of an NC turning machine such as that of Feldmann [2]. Figure 2-1(a) and (b) shows the frequency distribution of the traveling ranges of the carriage and cross slide, as well as the structural configuration and rotating axis of the turret head of the NC turning machine. It is obvious that the cross slide travels mainly between 300 and 600 mm notwithstanding the type of machine, and the turret head of disk type is protruded. Importantly, we must be aware that the machining space is in good agreement with the work spectrum, which must be determined

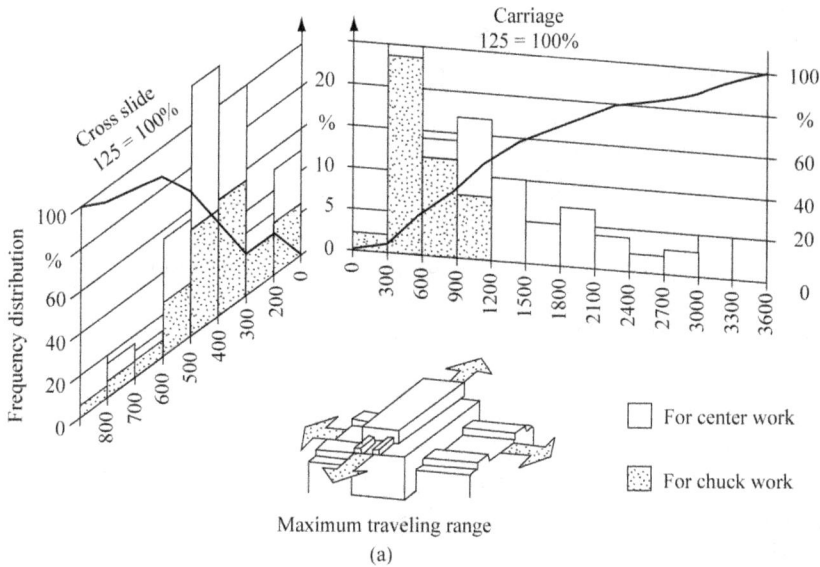

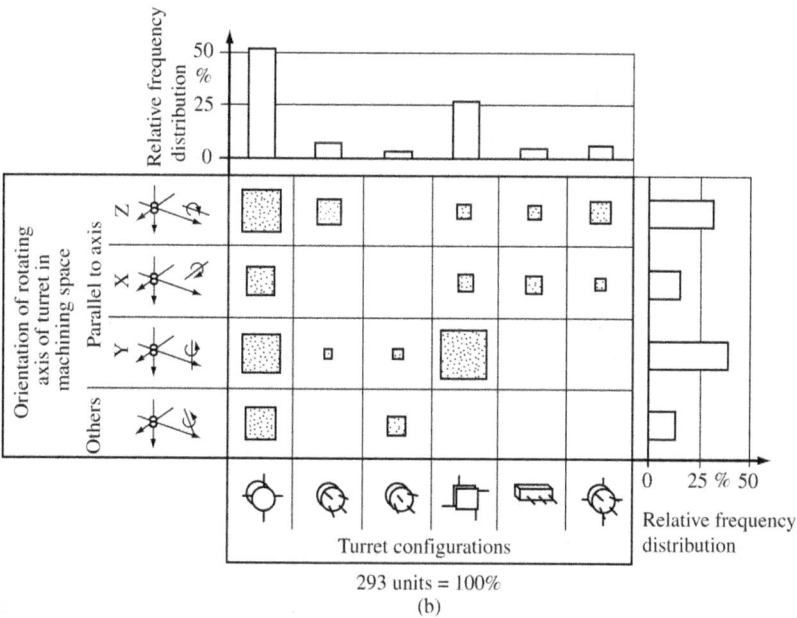

Figure 2-1 Frequency distributions for: (a) Traveling ranges and (b) structural configuration and rotating axis of turret in NC turning machines (by Feldmann).

on the basis of market needs. In fact, the work spectrum is another dominant factor for the principles of both separation and standardization. Ito and Yoshida [3] later carried out a further similar activity to establish a modular design guide for large-size machine tools. As can be readily seen, the module may be rationally determined from these data.

Regarding this principle, a new subject that is concerned with the LCA and remanufacturing of the product and that has actually been proposed elsewhere is the concept of a *platform* [4]. A group of platforms is capable of manufacturing the individual product configuration with higher reusability, where reusability means the applicability of each platform up to several cycles within its life to other products after necessary modifications have been made. In accordance with the proposal, the platform appears to be a variant of a module; and if this is so, what is the difference between the module and the platform? Importantly, the final goal of the platform is that of modular design so far, where the platform is a combination of several common modules defined in the manufacture of the different kinds of machine tool from a group of modules. In other words, the platform is an entity of higher level than the module, which consists of a certain number of modules commonly used in the manufacture of different kinds of machines. As a result, the platform concept is a clue to sublimate the principle of separation to the design guides. As can be seen from Fig. 2-2, the platform is of hierarchical type for manufacturing the different kinds in full consideration of its availability for the variant of the machining complex.

This interpretation may be supported in referring to the case study of Metternich and Würsching [4]. Gleason Pfauter Hurth has manufactured the hobbing machine, gear grinder, and gear shaping machine for work up to 2400 mm in diameter using the common base and same column since the beginning of the 1990s. In this case, the base is the platform; in addition, the joints of the base to the column and the base to the table are standardized. Actually, the candidate for the platform is a group of such modules that are, in most cases, in the same combination across the whole different kinds.

Principle of unification.   The design flexibility increases with the increasing number of modules predetermined mostly on the basis of the principle of separation; however, to reduce, for instance, the asset tax, the total number of modules should be minimized. This is a typical trade-off or ill-defined problem, resulting in the most crucial issue when detailing the principle of unification. Thus up to today the principle of unification has not been defined with wide acceptance among machine tool engineers. In retrospect, we have considerable experience in this context through the design of the TL in the 1960s. At that time, the principle of unification was more concerned with how to formulate a group

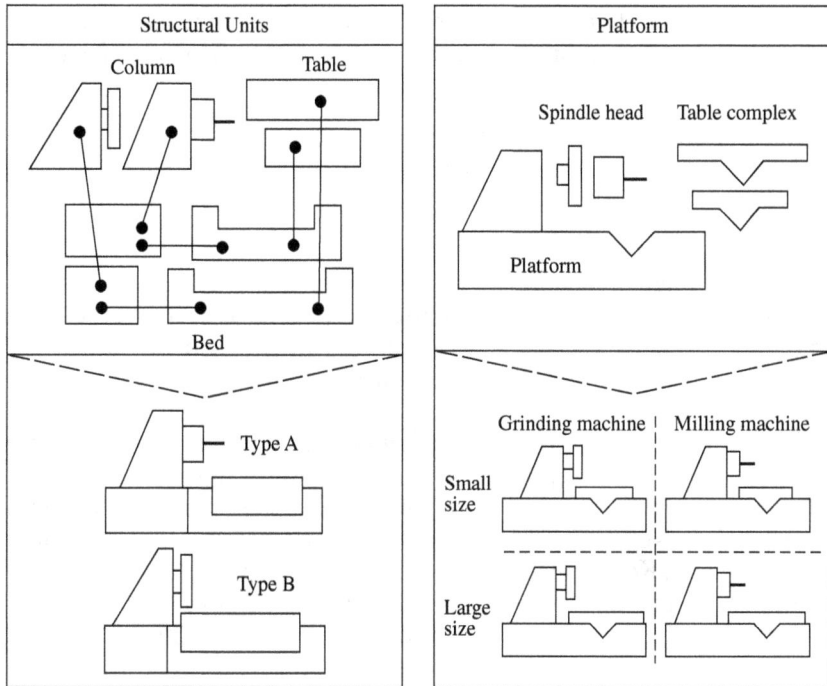

Figure 2-2 Concept of platform to enhance reusability of structural entities—an advanced concept of modular design by Metternich and Würsching (courtesy of Carl Hanser).

of modules with special reference to a size series of the units. Within a TL context, the principle of unification can thus be delineated as follows.

A group of units should be standardized with special respect to their dimensions, preferably using the preferred numbers such as R10 and R20 as already standardized.

In fact, the TL can be designed with satisfactory rationality by using the modular principle as exemplified by the ISO. One question, however, is whether the principle of unification so far established is, e.g., applicable for even the FTL. The answer is not given as yet and it is necessary to establish a new definition of the principle of unification, in which the module should be standardized in full consideration of not only its dimensional specifications, but also its functionality, capability, and structural configuration. In this regard, we must remember that the standardization technology for the TL was converted to that of the conventional machine tool to some extent, but not in satisfactory states. At burning issue is thus to develop a modified or new definition and concerns together with the related design technology, which are available across the whole kinds of the machine tool.

Principle of connection.   In modular design, the machine tool as a whole has a considerable number of joints, and thus the principle of connection should be interpreted as follows.

> When modular design is employed, the jointing method and joint surface should be unified at least or standardized in the most preferable case while maintaining allowable assembly accuracy and acceptable joint stiffness under the repeated use of the module.

In this principle, a basic necessity is to consider the mutual effect of the jointing accuracy and joint stiffness. In addition, it is preferable that the reference surface be separated from the bearing surface for loading, to realize the allowable magnitudes in both assembly accuracy and joint stiffness.

Within this context, the engineering problems in machine tool joints are, as already stated, a primary concern, and the designer is often asked to contrive a new jointing method. For example, the connecting bolt with locating pin or key is dominant in the bolted joint, i.e., a kind of stationary joint; however, Koenigsberg [5] suggested a functional entity, i.e., a combination of connecting bolt and wedge, in 1975. This functional entity has both the locating and tightening functions, as shown in Fig. 2-3.

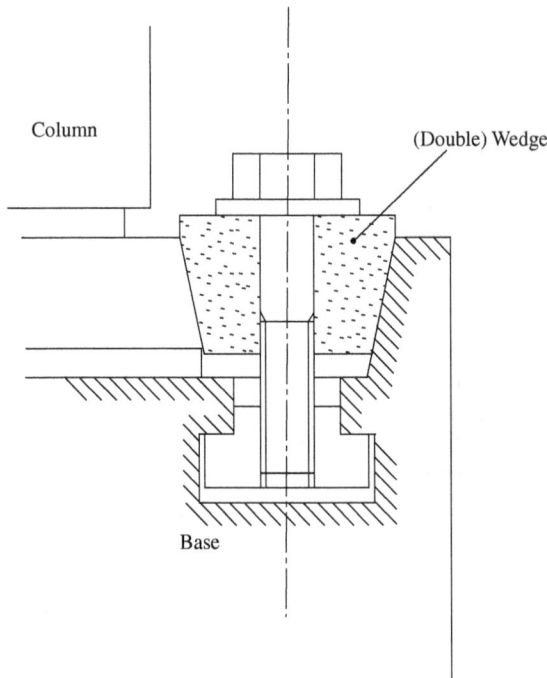

Figure 2-3   A functional entity (courtesy of Koenigsberger).

Principle of adaptation.   The principle of adaptation can be defined as follows.

Various structural configurations with the multifarious functionalities, performances, and dimensional specifications should be arbitrarily produced from a group of modules. The crucial problems are to establish the preferable combination method and interfacing method among modules and to evaluate the compatibility of the generated configuration with the design requirements.

For example, the adapter is very popular as an interfacial module in the special-purpose machine tool; however, the conventional machine renders this remedy useless, because of the uncountable combinations of modules. In addition, so far we have no reliable and effective methods and methodologies to evaluate the dimensional and performance specifications as well as the functionality of the machine tool at the design stage. In this regard, at further burning issue is to establish a conversion method of uncertain design attributes, e.g., ease of operation, compatibility with individual differences and penchant for configuration, and customer satisfaction and delight, into the quantified design specifications.

In the late 1990s, Tönshoff and his colleagues were actively involved in research into the remaining problems related to the four principles of modular design, especially those of separation and adaptation. More specifically, Tönshoff et al. developed a guideline for modular machine tools in a subproject "MAREA (*Study and Definition of Machining Workstation Reference Architecture*)" within the BRITE EuRam II[1] commenced in 1993, as well as a configuration method based on the functional modularity in a subproject "MOSYN (*Modular SYNthesis of advanced machine tools*)" within BRITE EuRam III [6]. In the MAREA, the reference architecture consists of (1) the formalization of component specifications, (2) estimation of possible interfaces among components, and (3) enhancement of configurability. As can easily be imagined, this architecture appears to be very similar to that of Doi of Toyoda Iron Works in the 1960s. In the MOSYN, they have dealt with the generation of the structural configuration based on the functional module. This appears again to be similar to that of Shinno of the Tokyo Institute of Technology apart from the use of the QFD (*quality function deployment*) [7].

Importantly, Tönshoff and Böger have proposed an idea for the principle of separation, in which the entity is the functional module [8]. The functional module can be determined on the basis of the analysis of the

---

[1]BRITE EuRam is one of the EU Projects, and its firsthand view can be obtained by searching the Web with a key term "BRITE EuRam"

machining process in consideration of form-generating movement. Although the proposal is far from being realized and remains a serious problem to be solved afterward, i.e., guarantee of the one-to-one conversion of functionality to structural configuration,[2] the separation into the reasonable module becomes easy together with the guaranty of the greater adaptability of each module at smaller expenditure. In fact, the proposal of Tönshoff et al. can be characterized by the use of both more function-oriented modules and the conversion method based on the QFD of hierarchical type. In other words, the user's requirement can be converted first to the function to be provided to the machine and then to the structural configuration module. A key issue is thus how to actually carry out the QFD. In due course, Höft and Ito [9, 10] proposed a similar idea in the design of the culture- and mindset-harmonized product (localized community-oriented product) including the machine tool with modular design, where the basic necessity is also to convert the uncertain attributes related to the culture and mindset to the quantified engineering specifications in consideration of the superiority order of each attribute, or by weighing the relative importance of each attribute. In the proposal of Höft and Ito, this weighing procedure can be displayed by using the radar chart.

In addition, the proposal of Tönshoff et al. emphasizes both the considerable benefits to the manufacturer, which can respond to users' requirements in wider scope than ever before, and the higher exchangeability of the module at the user's factory. In fact, the functional module should be determined in consideration of the manufacturing requirements of the user as follows.

1. The future product spectrum, which could be dealt with for the machine tool being conceptualized

2. Predictive production capacity

3. Organizational and investment limitations

It is worth pointing out that the modular selection procedure is innovative in Tönshoff et al., because of arranging the modules first so as to

---

[2] The utmost difficulty lies in the differing properties in the related information between the functionality and structural configuration, e.g., those of uncertain and qualitative design attributes versus quantified engineering specifications. This conversion is as same as that in the CAPP where the part (geometric) information on drawings must be converted to completely different (machining method) information. The same problem can be observed in the work of Abele and Wörn, in which a modular design has been proposed ranging from the functional entity, through the structural body component, to the work grasping and cutting tool.

Abele, E., and A. Wörn, "Chamäleon im Werkzeugmaschinenbau   Rekonfigurierbare Mehrtechnologiemaschinen," *ZwF*, 2004, 99(4): 152–156.

easily form the basic structural configuration by using the FOF (*flow of force*) (der Kraftfluß) and then adding the auxiliary modules. This can be regarded as a valuable proposal for the methodology related to the principle of adaptation. Importantly, the noteworthy feature is the application of the proposed modular design across the whole European small machine tool manufacturers. This intention is similar to that of TL as per ISO in order to design the customized machine tool with dedicated specifications; however, the method has been claimed to be in practical use only within the two big European manufacturers.

To this end, it is emphasized that the four principles proposed by Doi are even now very valuable and applicable; however, some variants are also needed. Actually, variants by Koren, Tönshoff, and Koenigsberger exist. For example, Koenigsberger emphasized, as already stated in Chap. 1, the growing importance of the machine tool with modularity, which was for cellular manufacturing, resulting in the user-oriented type, and he detailed these principles in his research plan in 1975 [5] (see Chap. 1).

In addition, Table 2-1 suggests some leading subjects to be investigated together with clarifying their relationships with the four principles of modular design.

TABLE 2-1   Some Leading R&D Subjects Regarding the Four Principles

Principles	Methodology and engineering tools at present	Leading-edge research subjects and engineering problems
Separation	(Far from completion)	Reasonable separation methodology including economic viewpoint
	Methodology based on frequency distribution of work spectrum and/or machining space	Applicability of platform concept proposed by Metternich and Würsching
Unification	(Far from completion apart from that for TL)	Reconsideration of effectiveness of modular design of hierarchical type Evaluation method for preferred module determination
	Module standardization for TL	Establishment of unification methodology available across the whole kinds
Connection	(Nearly to established)	Simple connecting method with multiple functionality, e.g., (1) joint with higher static stiffness with higher damping capacity, (2) simple joint compatible with complex and multidirectional loading, and (3) jointing method with higher stiffness and better locating accuracy
	Design guides have been nearly established such as "Engineering Problems in Machine Tool Joints"	
Adaptation	(Design guides being established)	Evaluation method for availability of modular design Methodology for choosing preferable configuration from generated results including performance simulation on drawing Determination methodology for a group of modules
	Design guides are being established such as "Machine Tool Description and Its Application"	

2.2   Effective Tools and Methodology
for Modular Design

Admitting that the principles of both separation and unification are facilitated by the trial-and-error method employed by learned and mature designers, an extremely crucial problem in the modular design is how to execute the principle of adaptation. As can be readily seen from Sec. 2.1 and will be stated in Chaps. 3 and 4, some computer-aided methodologies are now available with the assistance of the machine tool description. Although the computer-aided methodology is very powerful, the manually based method is, in certain cases, very effective and aids in the understanding of the essential features of modular design. Thus in the following discussion, some manually based methods of old are reproduced.

In the late 1970s, Ikegai Iron Works proposed a variant of modular design applicable to the TC and MC, which can promptly respond to the machining requirements of users. In this case, the adaptation of the modules was simulated by using wooden blocks of different colors, as shown in Fig. 2-4, after classifying users' requirements into the shaft-like part, flange and gear blank, boxlike and flat-like parts. As can be imagined, this simulation can now be performed by the three-dimensional CAD (*computer-aided design*).

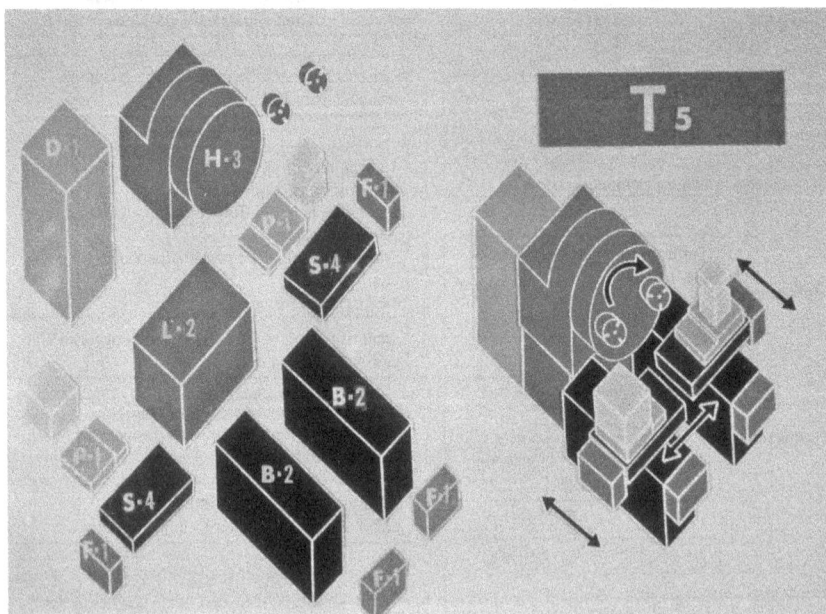

Figure 2-4   Simulation for structural configurations possible by using colored wooden blocks (courtesy of Ikegai Iron Works).

In modular design, a two-dimensional decision table is another effective tool, in which two dominant design factors are allocated to the vertical and horizontal axes, e.g., versatility of structural configuration and machining capability, e.g., power, stiffness and metal removal, as shown for the case of a planomiller of TOS make (Czechoslovakia) in Fig. 2-5. In this case, the machining capacity within the same structure can be varied by the output power of the main motor; also the type variety can be extended by adding the milling head of ram type to that of quill type. As a simple method, an one-dimensional table is helpful in rationally managing the design, as shown in Fig. 2-6, where the structure of the milling machine is reinforced by the overarm and stay, and furthermore the auxiliary column.

Having in mind these predecessors, Brankamp and Herrmann proposed furthermore the idea of a *function chain* (das Funktionskette), shown in Fig. 2-7, for its ease in finding suitable combinations of the modules. Importantly, the functions that must be provided and the modules, e.g., units, functional complexes, and parts, that can be possibly realized are allocated in the two-dimensional table. The structural configuration can be generated by choosing the suitable module from a group of the modules, which are allocated in every line in the table, showing a zigzag choice trajectory, i.e., chain. On this topic, Dietz proposed two similar ideas called *structure* (der Baukasten)-*structural entity* (die Baureihe) and *connecting diagram* (das Verknupfungs-Diagramm) [11]. These facilitate the estimation of possible variants that can be produced from a group of modules. In the work of Dietz, the row shows the unit and unit complex in the order of assembly, as seen in Fig. 2-8. It can thus be interpreted as a predecessor of the design methodology proposed by Lee et al. [12], in which the connecting diagram is represented by the direct graph for ease of computer processing. In addition, that of Dietz can be

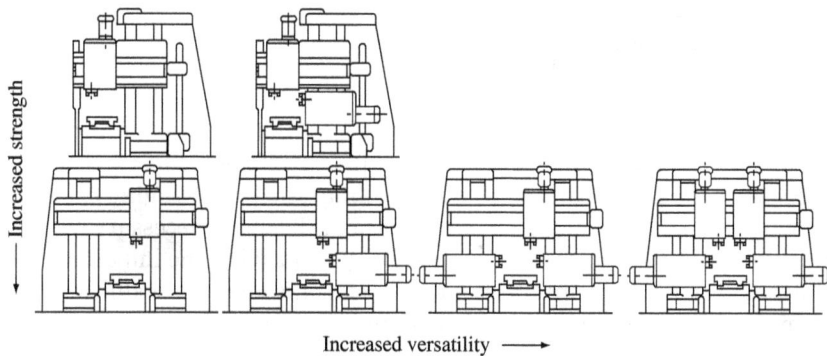

Increased versatility ⟶

In case of planomiller of TOS make (maximum clamping width of table 1600 mm; spindle drive power approximately 10, 20, and 40 HP).

Figure 2-5 Two-dimensional decision table for modular design.

Clamping width of table is maximum: 800 mm;
spindle drive motor: 10, 20, and 40 HP.

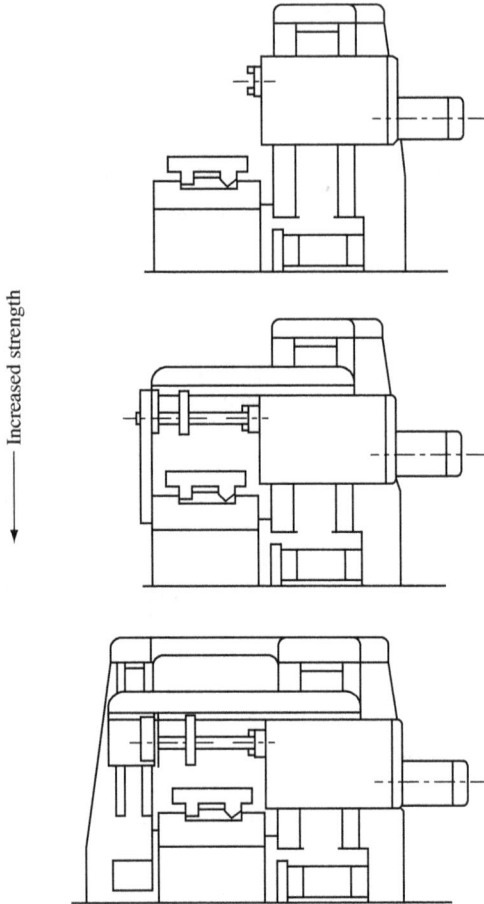

Increased strength →

Figure 2-6 Application of one-dimensional decision
table (planomiller of TOS make, Czechoslovakia)

characterized by the hierarchical allocation of the module. More specifically, the function of the machine was first classified into the turning
process, tooling, measurement, work loading, and tool changing from the
viewpoint of the automatized flexible turning. Then the necessary module
was determined in the hierarchical way by identifying the importance of
the modules, i.e., principal, associate, special, and adapter modules, when
manufacturing the machine. The same hierarchical system was also
applied to the tooling system of quick changing type, which is regarded
as a predecessor of that of Sandvick, already shown in Fig. 1-1.

The methodology proposed by Dietz was applied to the design of the turning machine of *Frontier* type, so that each user could have a cost-effective

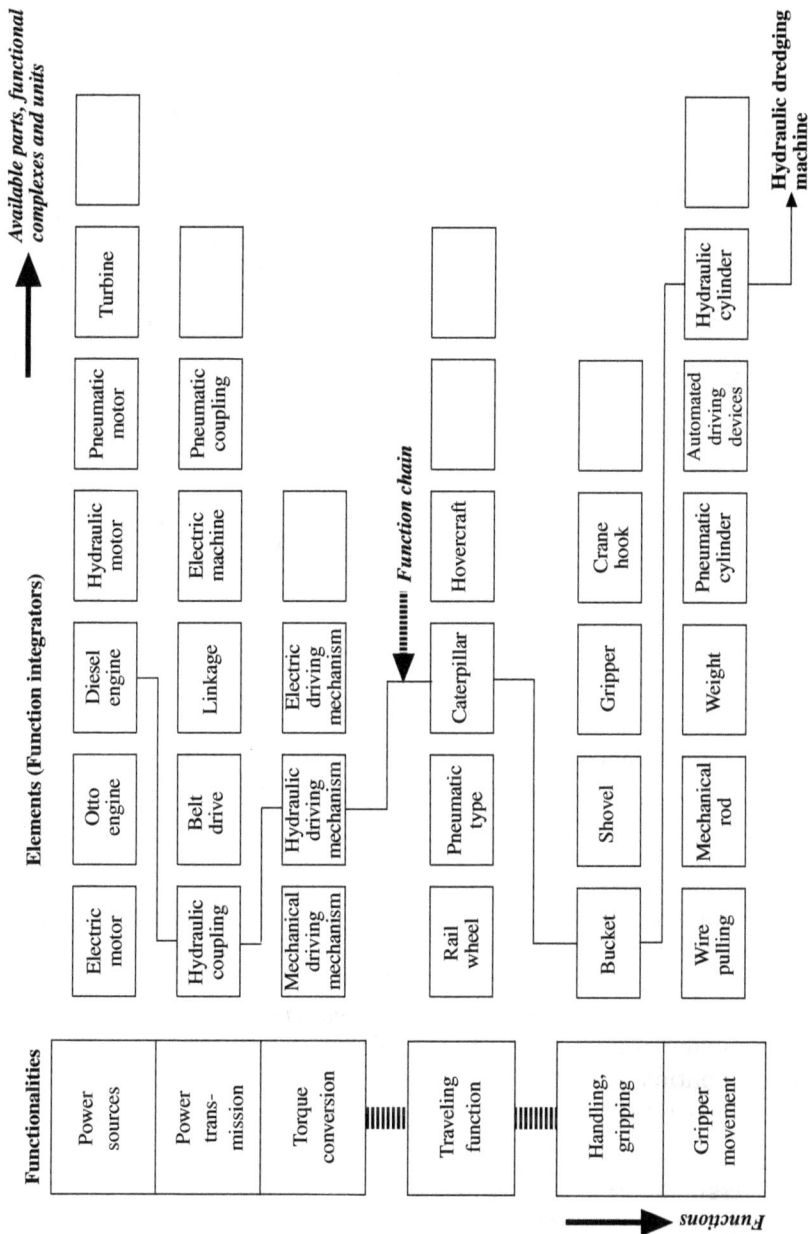

Figure 2-7 Concept of function chain in case of dredging machine (by Brankamp and Herrmann).

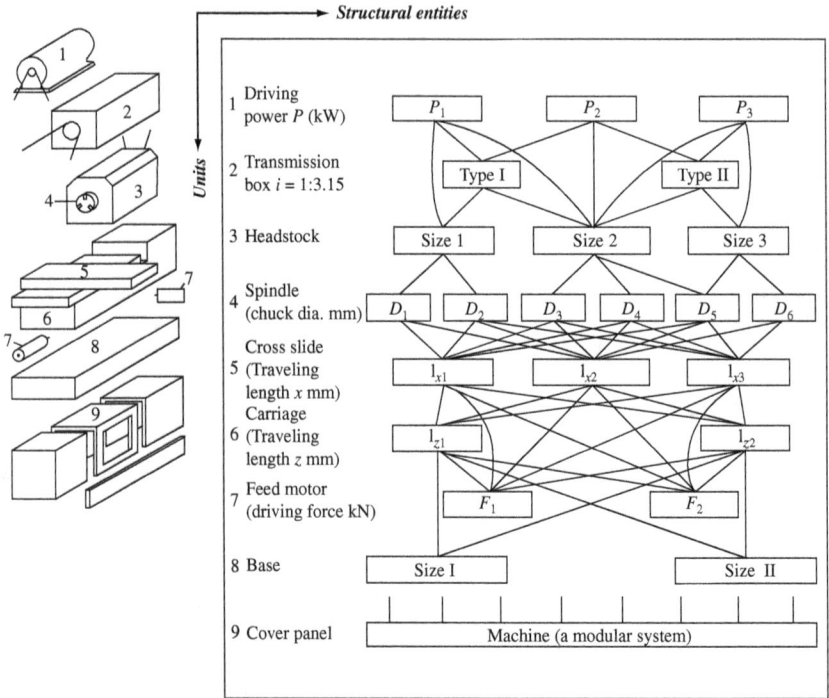

Figure 2-8 Connecting diagram proposed by Dietz in case of chucking machine (courtesy of Carl Hanser).

machine in accordance with her or his requirements. The *Fronter* type can be characterized by machining the work with either simultaneous processing or "by turns" processing. For instance, by-turns processing is suitable for the medium batch size ranging from 1500 to 5000 parts. Thus in the design first the connecting diagram was employed to seek the variants possible to manufacture, and then a design guide was arranged in the form of the structural master plan shown in Fig. 2-9. The master plan can show the typology of the variant using both the attributes, i.e., number of spindles and slides, and can indicate the variants for the practical use. In determination of the available variant, the designer must pay special attention to the importance of tooling layout and interface, on which the operating efficiency is largely dependent [13].

2.3 Classification of Modular Design Including Future Perspectives

Modular design has been developed to various extents, as shown in Chap. 1, and thus there is a need to rationally classify modular designs

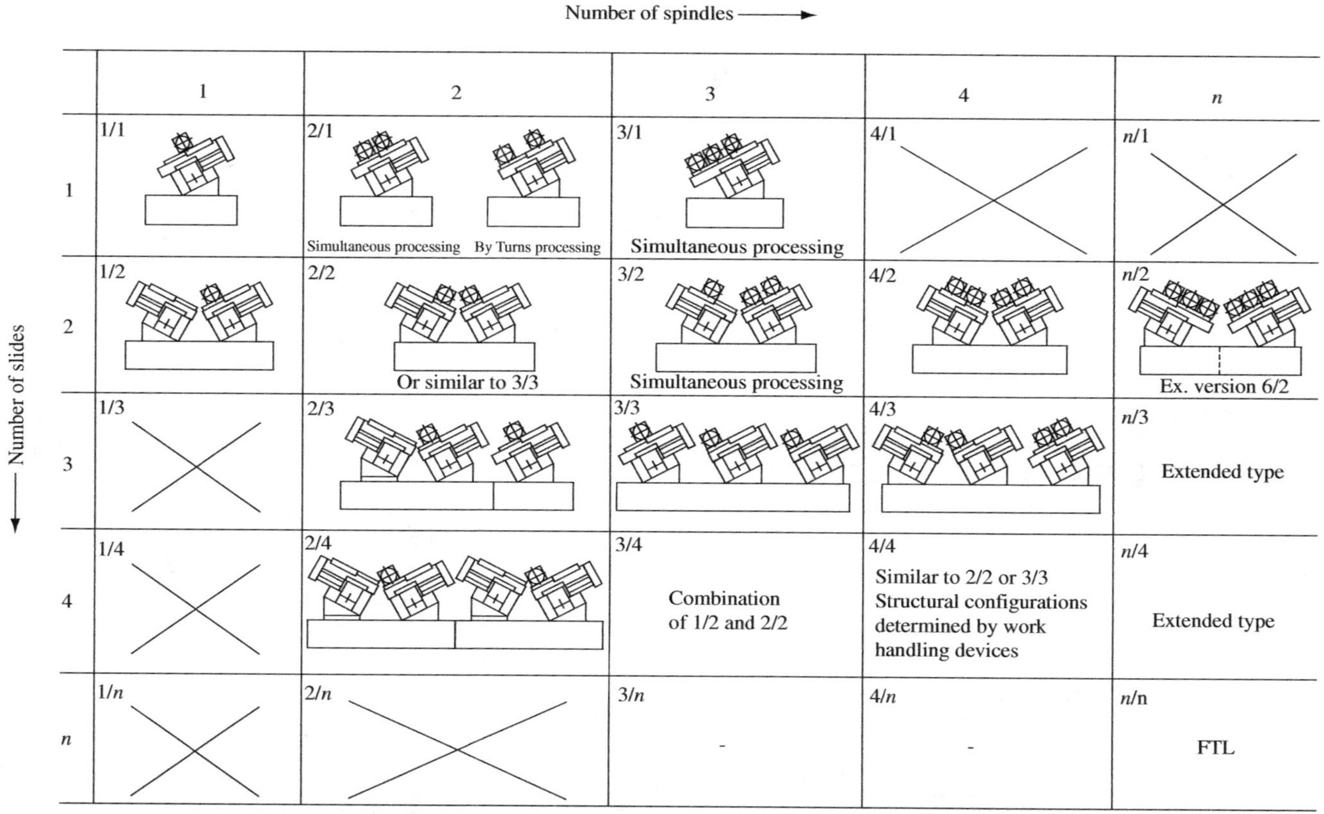

Number of spindles →

Number of slides →

	1	2	3	4	n
1	1/1	2/1	3/1	4/1	n/1
2	1/2	2/2 — Simultaneous processing / By Turns processing / Or similar to 3/3	3/2 — Simultaneous processing	4/2	n/2 — Ex. version 6/2
3	1/3	2/3	3/3 — Simultaneous processing	4/3	n/3 — Extended type
4	1/4	2/4	3/4 — Combination of 1/2 and 2/2	4/4 — Similar to 2/2 or 3/3 Structural configurations determined by work handling devices	n/4 — Extended type
n	1/n	2/n	3/n	4/n	n/n — FTL

Simultaneous processing (2/1)

Figure 2-9 Structural master plan to produce chucking machine of Fronter type (by Dietz, courtesy of Industrie-Anzeiger).

77

created up to now so that the designer can apply the modular principle properly in accordance with the design requirements. There are, as shown below, various classification systems depending on the core factors to be considered.

1. *From the viewpoint of the utilization method:* This classification system was established in the early 1960s and up to today has been widely used. In this case, modular design can be classified into that already shown in Fig. 1-15, depending upon either the user-oriented or the manufacturer-oriented type.

2. *From the viewpoint of design methodology (those for free and variant designs):* In discussing the classification method of modular design, there is an idea compatible with both the free and variant designs. In other words, this idea is based on the principle of adaptation. In the free design, the machine tool as a whole can be reconfigured according to the various aspects of its functionalities, performances, and configurations with the modular design, whereas the machine tool can be only modified within the machining space in the variant design even when the modular design is used (refer to Chap. 3).

Because of the simplicity of the principles in modular design, the basic classification system of modular design is not especially complex, although the application of modular design has apparently been spread over a wider area, resulting in a handful of variants. In the following, the classification system available around the year 2000 is stated in full consideration of the present and future perspectives of modular design.

2.3.1 Modular design being widely employed

Importantly, modular design has a handful of representative variants, and they must be applied to the design work so as to effectively function their characteristic features. With special respect to the modular design being employed, there are three representations, and thus their firsthand views will be stated in the following.

1. *Variant 1: accommodation of versatile performance and dimensional specifications.* This modular design is the unit construction type of old, and it is capable of ramifying the performance and dimensional specifications within the same kind, e.g., main motor power and table size, although the variation of machining methods is relatively constrained. Obviously, this is very popular even now and has been applied to the general-purpose machine tool such as the conventional MC and TC. In short, the primary concern is the economic advantage together with the ease of use including the higher compatibility with

Turret head available
`for rotating tool

Headstock with
index function

Turret head for
stationary tool

Bye-headstock

Tailstock

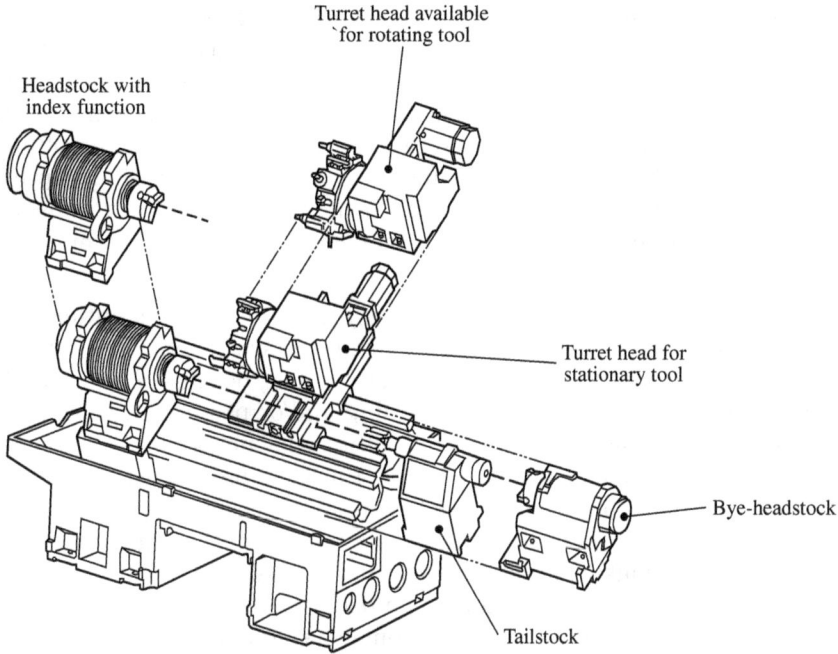

Figure 2-10   A unit construction for TC (type SL, courtesy of Mori Seiki).

the production system, and thus this type is very effective even now. For example, Fig. 2-10 shows a unit construction employed in the TC of Mori Seiki make around 1995, and in the late 1990s, a marked observation in EMO Shows in Hannover has been a user-oriented modular design applied to the CNC turning machine by Boehringer, Index, and Gildemeister.

2. *Variant 2: accommodation of multifarious machining methods.* This modular design aims at the enhancement of the various machining methods within the same kind, maintaining the same dimensional specifications. Actually, the machine can facilitate various machining methods by only changing, e.g., the head attachment in the five-face processing machine. As another kind of this category, at burning issue is the machining complex, in which the multiple machining functions are, in principle, integrated within a machine such as already shown in Fig. 1-5 so as to machine the work to be required of multiple-stage processing. A root cause of difficulties lies in how to design the structural body unit under complex loading and temperature distributions, e.g., those derived from turning and milling using the tool on the turret. Eventually, more difficulties can be recognized when the four principles of modular design are applied.

Importantly, the machining complex is, in part, an extension of compact FMC, which is capable of turning, milling, grinding, gear cutting, and so on and has substantially greater flexibility in the machining method apparently showing no need to use modular design. In general, this interpretation is acceptable; however, often the manufacturer needs to use modular design to respond promptly and appropriately to the manufacturing requirements of the user, although there are some crucial problems in designing the module under the complex loading.

3. *Variant 3: different-kind generating type.* In the past, the different-kind generating type was applied to the large-size machine tool; however, nowadays this type is no longer popular, because a new trend is to integrate multifarious machining and/or processing functions within a machine itself. As a result, the renowned kind of this variant is that of the machining complex or processing complex. In short, a factory may consist of a machining or processing complex only instead of a group of machines having different machining and/or processing functions or even the conventional MC and TC, simultaneously rendering the modular design useless. At present, the concept and methodology of this type are somewhat applicable to the design of the processing complex.

2.3.2   Modular design in the very near future— a symptom of upheaval of new concepts

It appears that modular design is now approaching its fourth developing phase, in which the driving force is, in wider scope, the growing importance of localized globalization in manufacturing. Obviously, the production system can be run by the operator of multiple nationalities, and thus the "ease of use" facet in the modular design under sustainable growth should be emphasized. In fact, there have been a handful of innovative proposals to enhance the compatibility of modular design with the production environment under localized globalization. It is thus vital to describe in detail the concept so far proposed and recently seen in the design guides.

Remanufacturing-oriented type.   With the growing importance of worldwide environmental problems, the disposability of the production facility becomes a crucial subject more than ever before, and modular design is obviously required to enhance reasonable disposability. As can be readily seen, disposability is one of the core branches within remanufacturing. At burning issue is thus how to enhance the remanufacturability of the machine tool, where remanufacturability can be defined as a synergy of production, reuse, and recycling. Within a remanufacturability context, two variants exhibit high future potential: one is the LCA-oriented type and the other is the "maintaining up-to-date specifications" type. In the former, the module obtained from the machine tool in the end of life might be used again as a new module after readjustment or modification, if necessary. In other words, a group of modules is, in preferable cases, guaranteed reusability to a larger extent, because the reuse is more desirable than recycling. Importantly, this LCA-oriented type has not yet been realized, but it was in the conceptual stage at the beginning of 2000, although the same concept is being employed in the case of the automobile, as already proposed by Neumann [14].

In the latter (maintaining up-to-date specifications) type, some modules can be replaced on-site by the user depending upon their deterioration rates, so that a machine tool can maintain its functionality and performance in constant standard. In an advanced case, users may properly determine by themselves the machine life desired in replacing the objective modules given the leverage of the lives of all the modules [15]. A predecessor of this type is a machine designed according to the principle of *preventive quality assurance*, which often has been used in designing the main spindle of quill type. In preventive quality assurance, the modules being incorporated, e.g., unit and functional complex, should be replaced with new ones after operating a certain time, even when the module has no trouble or damage.

More specifically, in a cost-effective machine tool for civil supplies production, at burning issue is to develop a modular design coping with the production in which the module life is determinable by the user. This modular design differentiates itself from others in the following ways.

1. The user can carry out the on-site replacement of the module at any time, to maintain the functionality, performance, and quality of the machine in preferable standard.

2. The user can determine the machine life by himself or herself by looking for the label *modular design of selective quality assurance type*.

In contrast, the customer could apply this idea also to the durables, home appliances, daily supplies, and so on. This trend will create another requirement for developing a new machine tool, i.e., machine tool capable of providing the customer with the spare part and unit, which should be replaced in accordance with the due life determined by the customer. In fact, this implies the necessity and inevitability of developing a "dexterous machine tool" that is compatible with one-off production as well as enabling quick response to customer demands with the reasonable production cost.

In the remanufacturing-oriented type, furthermore, we must be aware of the following.

1. With the advances of technology or appearance of innovative technology, the module often seems old-fashioned, even though it is within its product life from a technological viewpoint and still has remaining design life. This situation is a variant of remanufacturing, and in the future the proper design technology should be developed. In addition, the same situation can apply for a certain kind of product, e.g., a health care product, when a new law is enacted. In due course, the product currently on the market must be withdrawn, even though it has no defects and is far from the end of its life.

2. Within a remanufacturing context, it is necessary to have a methodology for life cycle management, and one is *configuration management* [16]. There are four roles of configuration management: (1) configuration identification, (2) change management, (3) configuration status accounting, and (4) configuration audits and review. For example, the configuration integrity can be maintained throughout the product life cycle, i.e., to provide the ability to manage the as-defined, as-planned, as-built, as-delivered, and as-maintained configurations. In fact, the Advanced Configuration Management System Project was a 5-year-long study in the case of the European aerospace industry, and thus the availability of configuration management to the machine tool must be investigated next.

Use of module obtainable from supply chain of world class.   Within this context, the target is a machine consisting of a certain number of modules, which are, without any hindrance, capable of purchasing through the supply chain of world class. The machine aims at cost reduction together with realization of higher functionality and performance even when using the low-quality modules indigenously produced by the industrializing nations. The crucial problem is thus how to maintain the satisfactory quality of a machine as a whole. A variant of predecessor can be observed in the gear cutting machine of P. R. China brand, which has been used in Indonesia from the mid-1990s. In this case, the gear cutting machine can produce the gear itself one year after installation and in turn a sprocket wheel in the succeeding year with the deterioration of machine quality. In Taiwan, all the parts and unit production is commissioned to other companies, and the machine tool manufacturer performs only the final assembly; i.e., there is a horizontal division of work, and thus such a modular design would be a must. For example, a remedy is to measure the necessary engineering properties, e.g., spring constant of the rolling bearing with lower quality, and then the product quality can be commissioned to the NC information, which can correct the unfavorable deterioration in performance.

In contrast, one might expect that the highly value-added machine tool could be manufactured by assembling the modules with higher quality and obtainable through the world-class supply chain. In this case, the machine design should be carried out so that the modules function very well as exemplified in the case of the grinding complex of Taiyo Kouki brand in the 1990s.

Multiple-function integrated entity-based type.   As exemplified in the fierce competition among the Chinese, Japanese, and Korean molding die manufacturers, the focus now lies in how to differentiate the function, performance, and quality of the product. In this context, the Japanese enterprise has been very keen to conduct product innovation, where the product consists of multiple-function integrated entities. In general, the product has hierarchical structure in the order of the part, functional complex, unit, unit complex, and product itself. The part itself thus has no functions, but single function even in the preferable case. The multiple-function integrated element can be interpreted as the entity of translayer type and can be expected to be a core when the more multiple-function integrated product becomes a reality. In fact, a typical example of the multiple-function integrated element is an innovative rolling bearing called CARB of SKF make. This bearing has been contrived by integrating advantageous features of the cylindrical, spherical, and needle roller bearings. Importantly, highly sophisticated and skillful technology enabled the multiple-function integrated element to be realized, and thus only industrial nations such as Japan can produce it, resulting in the desirable remedy to the differentiation between Japan and other Asian nations. The machine tool design with multiple-function integrated module is based on such an idea, and thus it appears to be called the platform, which was recently proposed by Metternich and Würsching [4].

Culture- and mindset-harmonized type.   The culture- and mindset-harmonized machine tool can be characterized by its design specifications. In this type, the culture- and mindset-harmonized attributes are, in fact, newly considered as having the same weight as those related to the engineering and shipping destination, and to the commodity in part so far used in the design, as shown in Fig. 2-11. In this case, knowledge about the "culture of manufacturing" [17] dominates the machine design, where the culture of manufacturing is a synergy of the manufacturing technology and cultural issues, e.g., economics, social and labor sciences, geopolitics, and folklore. In short, a handful of representative machine tools, shown in Table 2-2 [18, 19], were designed by placing particular emphasis upon the culture- and mindset-harmonized attributes.

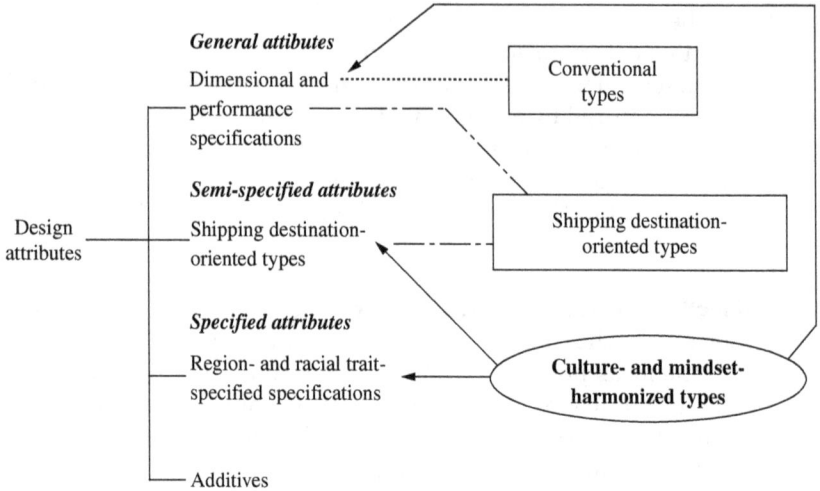

Figure 2-11 Tentative definition for design specifications of culture- and mindset-harmonized machine tools.

Importantly, an underlying hypothesis in the culture of manufacturing is that a technological entity, e.g., system and machine, should consist of two subentities, i.e., those commonly available across the whole world and at the specified region. As can be readily seen, the culture- and mindset-harmonized type may be regarded as the most suitable variant to which to apply the modular principle, and a crucial problem lies in how to determine the modules in relation to each regional environment. In this context, Höft [10] has already proposed an idea in her dissertation.

Furthermore, a customized product for the local community is expected in accordance with the predictive research into the future production environment [20–24]. This customized product is, in principle, a culture- and mindset-harmonized product, and indeed it is available for the machine tool. In short, it is vital to create the culture- and mindset-harmonized machine tool with the modular principle.

To this end, a new challenge is recently required of modular design from the area of process innovation instead of product innovation. This can be observed in the conventional MC and NC turning machine including the TC. In fact, the conventional MC and NC turning machine show amazing market share across the whole industrial and industrializing nations to produce, e.g., home appliances, information devices, industrial equipment, plastic injection molds, automobiles, and so on. Importantly, the MC and NC turning machines of Japanese make have been very powerful so far; however, the MC of Korean make has become very competitive recently based on the continual improvement of the related technology and has compelled the MC of Japanese make to be enhanced

TABLE 2-2   Machine Tools of Culture- and Mindset-Harmonized Type so far Merchandised

Design attributes / Products	Compatibility with regional infrastructure			Standard & qualification of human & technological resources					Sensitivity response		
	Use of indigenous supplies	Ease of transportation/ Preferable plant location	Harmony with environment	Level of production technology	Qualification of worker	Skills of user	Simplification of structure	Adhesion to brand & its image	Comfortable operability	Ease of operation	Increase of purchasing motivation
Turning center — Leadwell/Taiwan		○									
Hitachi Seiki/Japan			○						○	○	
Tatung Okuma/Taiwan		○								○	
Okuma America		○					○		○	○	○
Daewoo/Korea				○	○		○				
MC — Makino Asia/Singapore	○	○		○	○	○	○	○			
Milling machine — Long Chang/Taiwan		○							○		○

Note: In case of Tatung Okuma, ROI is within 3 years of not conducting maintenance somewhat. As a result, Mori Seiki, one of the leading Japanese manufacturers, seeks a new strategic technology, i.e., modular design available for larger volume and smaller batch size production to reinforce the competitiveness. In certain cases, furthermore, the conventional MC and NC turning machines are designed with unit construction (BBS), and thus another remedy is simplification and ease of assembly by improving the machining accuracy of each unit. As is well known, the bedway grinding machine of German make can finish the guideway with acceptable accuracy, which is a reference surface for linear form-generating movement. In contrast, we do not have a machine for line and face grinding of the main bearing seats in the headstock so far. This means that we have no sophisticated grinding machines to finish a reference surface for rotational form-generating movement. Thus, Taiyo Koki has, with the best reputation, contrived a grinding complex in the late 1990s to respond to such a requirement. As can be readily seen, a combination of both the bedway grinding machine and the grinding complex renders the adjustment work nearly needless and can facilitate the process innovation in the assembly of the conventional MC and NC turning machines with unit construction.

2.4   Characteristic Features of Modular Design Being Used in Machine Tools of the Most Advanced Type

The application of modular design ranges from the conventional NC machine tool and five-face processing machine, through the system machine including MC of line type, to the machining and processing complexes at the beginning of 2000. Of these, both the five-face processing machine and the MC of line type have, in principle, been designed using the modular design of well-known type from the old days. In contrast, the system machine and its successors, i.e., machining complex including the processing complex, have been designed either not using modular design or using modular design of the advanced type.

When we consider the future potential, the emphasis in this section must be on the delineation of the system machine and its successors. In other words, claims about the machine tool have become multifarious, simultaneously emphasizing the realization of the highly integrated functionality with the advance of human society. Currently it appears that machine tool technology is about to return to the era of the engine lathe, which reveals the growing importance of the machining and processing complexes. In due course, the modular design being requested may be assumed to be the modernized version.

Given such present and future perspectives, Figs. 2-12 and 2-13 show the modular construction in the five-face processing machine and MC

(a)

Ball end milling by
45° angle head   Drilling

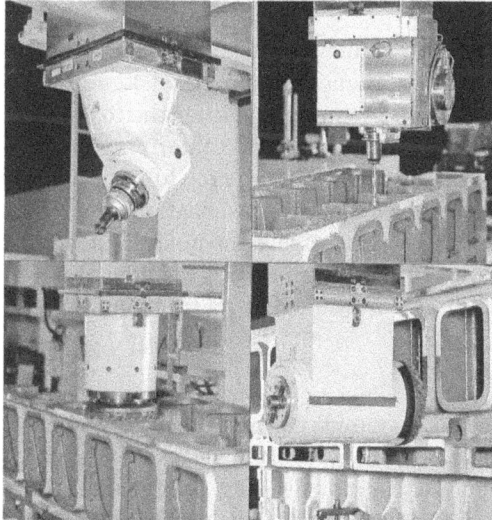

Face milling by snaut   Side face milling by angle head

(b)

Figure 2-12 Modular design applied to five-face processing machine: (a) Overall view and (b) various attachments equipped at quill (type MPC, 1980s, courtesy of
Toshiba Machine).

Main spindle speed: 10,000 rpm maximum. Main motor: AC 3.7/5.5 kW

Figure 2-13   MC of line type (type HMC-40LS, 1985, courtesy of Enshu Co.)

of line type. In the former, the various attachments are acceptable at the quill of spindlestock to realize versatile machining methods, and in the latter such versatilities can be allowed by changing either the tool magazine or turret head, maintaining the configuration of column traveling. To deepen the understanding of what is underway in the modular design for the system machine and its successors, some representations and their characteristic features are stated below. Intuitively, the modular design of the present apparently becomes very complicated with the advent of the system machine and machining complex, although it confronts new facets.

2.4.1 System machines

The term *system machine* sounds very new; however, the concept had already been suggested in the beginning of the 1980s, when the system machine was defined as "a machine compatible with flexible manufacturing," i.e., that of either machining method-integrated type or system function-integrated type. On that occasion, there were three basic machines to develop the system machine, i.e., traditional machine tools, MC and TC, and furthermore the newly conceptualized machine tool In fact, Yamazaki Mazak merchandised, with wider scope, an MC of modular construction in 1976, variants of which were as follows.

1. Extended MC capable of turning and broaching
2. Core machine for NC TL (*numerically controlled transfer line*) and FMS by attaching the cluster head
3. Five-axis controlled NC machine with the NC rotary tilting table

In the beginning of the 1980s, the system machine was promptly launched to its practical application in accordance with the second-stage development of flexible manufacturing, where the system design of top-down type became dominant. As can be readily seen, the system machine grew in importance with the thriving trend of the system design of top-down type. In retrospect, there were, as shown in Fig. 2-14, multifarious trials to develop the system machine from the late 1970s, paying special attention to the dominant design attributes to be provided. It is surprising that very few trials were, as shown together by framing in Fig. 2-14, as successful as those for practical use [25].

Figure 2-15 delineates a longstanding developing map of system machines by both clarifying the weighing functions to be compatible with the system and identifying the basic machines mentioned above, from which the system machines were contrived. From Fig. 2-15, it can be observed that the system machines in the early 1980s were designed so as to reinforce the accessibility to work and tooling flexibility. The former is represented by the column or outer column traveling type, and the latter aims at the leverage between the productivity and the tooling flexibility. In short, with the increasing number of tools, the necessity is to provide, e.g., the auxiliary tool conveyor, additional magazine, and tool magazine of carrying type. From this, the head changer was realized.

Obviously, a crucial issue has been whether the modular design is a must for the system machine, when we considered the compatibility enhancement of the machine tool with the system. For example, we can suggest the following two machines as predecessors of the system machine, i.e., " that of machining method-integrated type."

1. Type FT 600 of Ikegai make in 1981 (maximum allowable work diameter 590 mm), which is an MC of extended type by facilitating the turning function, i.e., turret column with two octangular turrets for stationary tools and rotating tools
2. Type LM70-AT of Okuma make in 1978 (swing over bed: 700 mm, main motor 15 kW), which is a TC of extended type with the turret column to facilitate milling and drilling functions, as shown in Fig. 2-16

In addition, we can recognize that there were, as shown in Fig. 2-15, some representative system machines in the mid-1980s, i.e., those for FTL and FMC. The former and latter can be regarded as the machining method-integrated and system function-integrated machines, respectively.

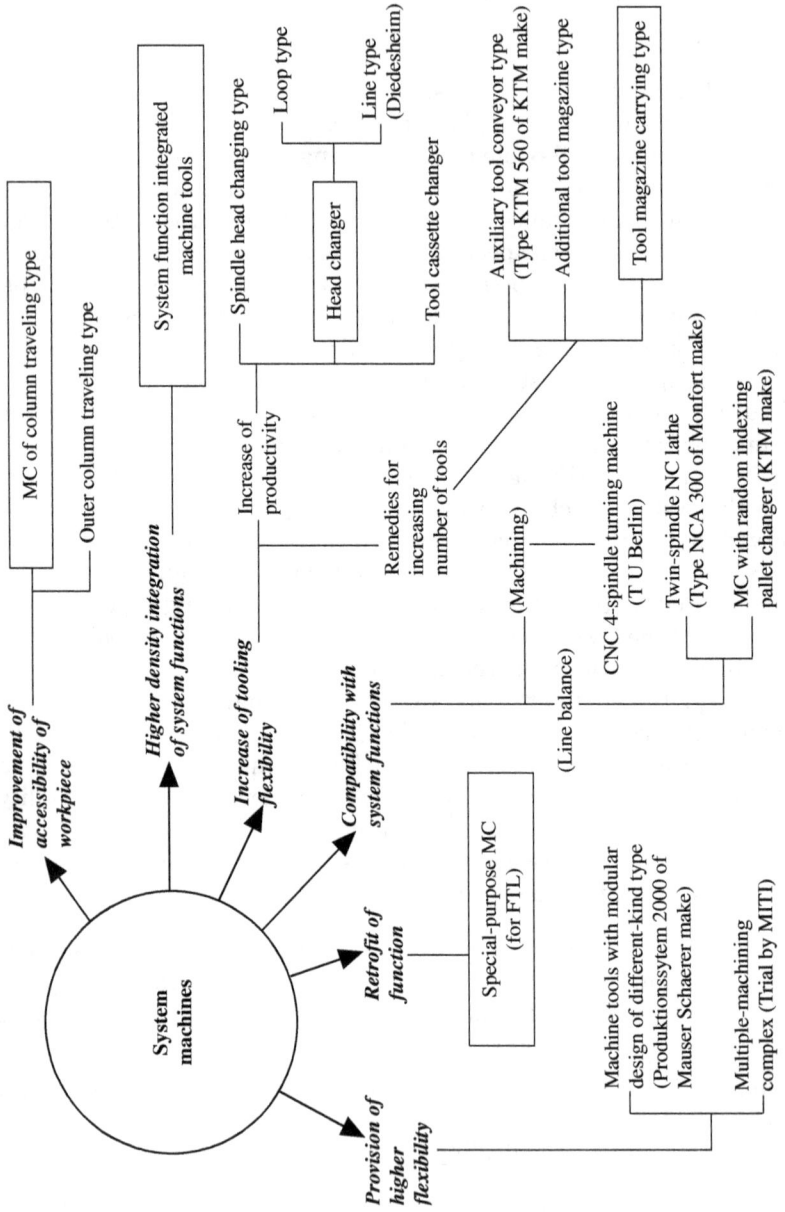

Figure 2-14 Trials for developing system machines.

For TL & FTL

Special-purpose machine tools for TL → NC special-purpose machine tools → MC of line type

MC → Head changing type / Head indexing type

Simplified function type

Special type: Enhancement of flexibility for machining methods (e.g., innovative turret configuration and/or spindle allocation)

TC

System machines → System machines for flexible manufacturing

Culture- and mindset-harmonized type

Multiple-function integrated type

FMC of flat type ⋯ Compact FMC of cubic type

Traditional horizontal boring & milling machine ⋯⋯(Modernization)⋯⋯

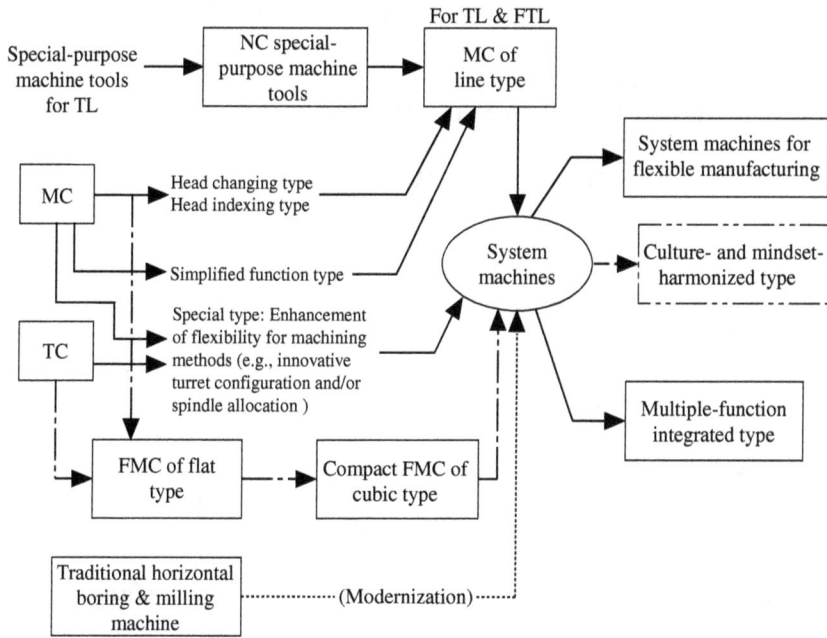

Figure 2-15 History of system machines.

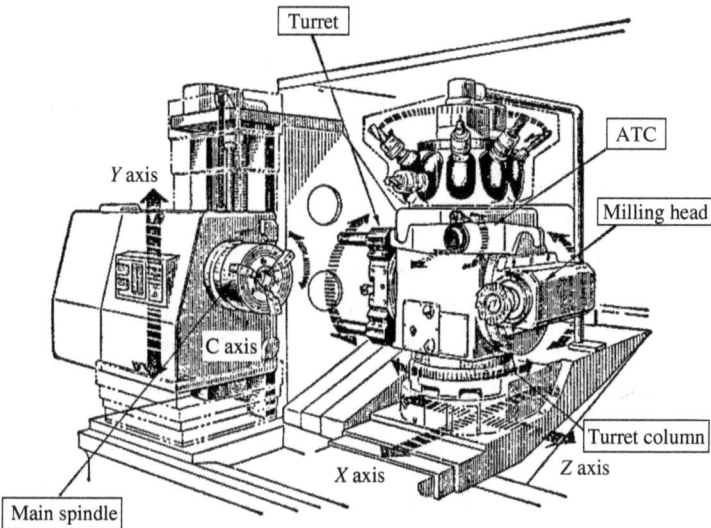

Figure 2-16 A TC of extended type in the late 1970s (courtesy of Okuma Co.).

Importantly, the system function-integrated machine was contrived with the prevailing trend of using the modular design of the production system. In fact, the FMC was, in general, to play the role of the module within both the FTL and FMS, provided that the FMC was of compact type and modified to the machinelike configuration.

Nowadays, the system machine can be classified into (1) the advanced system machine, (2) the machining complex (machining function-integrated type for conventional use), and (3) the processing complex. Table 2-3 displays the system machine sublimated to that for practical use between the 1980s and 1990s. The machining complex can be characterized by its greater flexibility, or rather agility, which is achieved in reality by combining the turning function and the milling function, so as to endow the machine with time- and space-dependent flexibility. As a result, there is no need to use modular design in the case of the system machine, even when the machining requirements are very versatile and ramified. As will be clear from the above, there is a two-pronged design concept for the system machine and its successor, depending upon whether the modular design principle is employed or not. Intuitively, the present modular design becomes very complicated with the advent of the system machine and machining complex, simultaneously approaching a new

TABLE 2-3   Representative System Machines in Late 1980s and 1990s

Manufacturers	Kinds/Types	Functionalities	Remarks
Bernhard Steinel	System function-integrated (type: EFZ 200)		Spindle speed: 30–12,000 rpm. Main motor: 20 kW
Kearney & Trecker Marwin	MC of head changing type (type: Fleximatic CNC Multi-headchanger)		For FTL to machine service parts of motor car
Kopp Werkzeug-maschinen	Kurvenbearbeitungs-zentrum (type: FSK25G)	Cylindrical grinding, non-circular grinding, surface grinding, 5-axis machining & high-speed machining	Main spindle speed: 5000 rpm in maximum Main motor: 11 kW Displayed at EMO Show '97
Nippon Kokan	Multiple-purpose MC (type: MMC-30)	Synergy of MC and NC turning machine	User-oriented modular design
SIG Swiss Industrial	Contour milling & grinding machine (type: CF)	Drilling, countersinking, reaming, threading & grinding	NC planetary grinding unit attachment facilitating direct mount to milling spindle
Yamazaki Mazak	TC of twin-headstock type (type: Multiplex)	Combination of turning & milling	Oppositely allocated twin-headstock Twin-turret of drum type

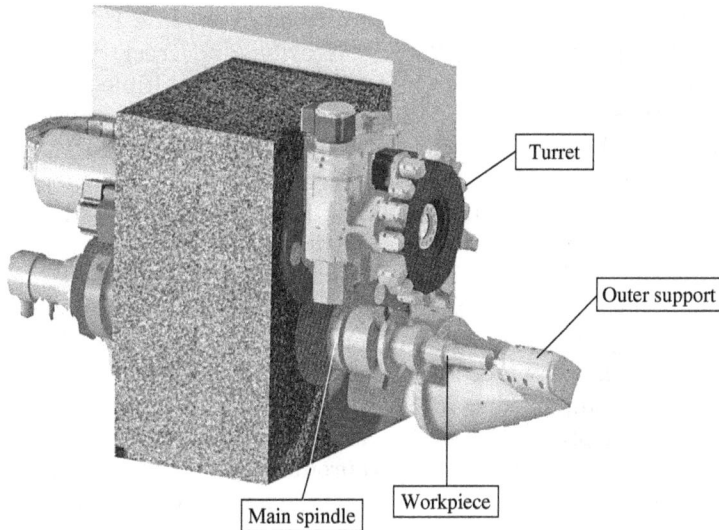

Figure 2-17   A new modular concept with greater accessibility—two-pillar central unit (patented by Accim).

front. In this regard, Ito [26] asserted in the mid-1980s the need to develop a system machine of on-site changeable and hierarchical modular design type to enhance the system design, based on his observation of the related perspectives. Alternately, it is worth paying special attention to the machining complex of Accim make in 1999. This machining complex is a representative showcase displaying the importance of modular design even in the system machine. In fact, the machining complex of Accim make can be characterized by its interesting modular design, in which the core entity patented, as shown in Fig. 2-17, has the following features [27].

1. Wider structural configuration flexibility to generate those ranging from NC lathe to turning milling center

2. Expandability to turning center of twin-spindle type
   a. The form-generating movement appears to be of *Cleveland type.*[3]
   b. The spindle is mounted within the spindlestock made of resin concrete and driven by the vector flux control motor of built-in type (8.2 kW) so as to greatly reduce the heat generation of the motor rotor and vibration.

In the following, some characteristic system machines will be stated.

---

[3]The *Cleveland type* is one of the representations of the automatic turret lathe of the 1970s, where the turret bar (turret over spindle with pentagonal cross section to mount the tool blocks), which can travel axially, is allocated right upward of the main spindle in the vertical front wall of the main frame. A representative is type 3AC of Warner Swasey make.

Head changer. The head changer was contrived in the late 1970s and in fact was already patented in 1978 by Kearney and Trecker Co., as shown in Fig. 2-18(a), under the commercial name of Multiple Spindle Machine Tool (U.S. Patent No. 4,125,932, Nov. 21, 1978). As already shown in Fig. 1-6, the head changer can machine the work by changing the gang head and single spindle head in accordance with the machining sequence, where the head is stored in the head magazine located in both flanks, rear side or rear top of the machining space. A crucial problem is that the machining flexibility is dependent upon the preparation plan and changing program of the head. In addition, the head itself is very expensive, e.g., at a cost of several million Japanese yen per head. In contrast, the head changer shows very high machining efficiency, and thus it can be employed as an entity of the FMS for large volume and large batch size production as well as medium volume and medium batch size production. Figure 2-18(b) reproduces the head changer of Heller make in the mid-1970s. This head changer displays the typical structural configuration of that time, and it was widely applied to the production of automobile parts.

Figure 2-19 is an interesting head changer (commercial name: CNC Station) developed by Ford Co. The machine is applicable to even small batch size production and can be characterized by the function for the single tool changing system. The square turret facilitates the layout of the head with the assistance of a robotlike changer, and some tools within a head can be changed by another ATC between the square turret and the tool magazine. In short, the ATC enables the single tool within a gang tool cassette to be changed independently to reinforce the tooling flexibility. In addition, the machine is equipped with both the tool life control function achieved by detecting the feed component of cutting force and the tool damage detection with TV camera or ultrared beam [28].

On that occasion, KTM (Kearney Trecker and Marwin) employed a strategy by which the machining function in the factory was able to expand from the stand-alone operation to FMC of pallet pool type, and even to the FMS. Based on such an idea similar to the modular principle, KTM produced a head changer called the KTM Multiple-Head changer, which was capable of stand-alone machining and of playing the role of increasing the productivity in the FMS, as shown in Fig. 2-20. In addition, this head changer can be characterized by an additional 40-tool magazine, which is on the top of the column, although not shown in Fig. 2-20. As a variant of these head changers, an interesting concept is that of tool cassette changing type, which is, in principle, the smaller size of head changer, i.e., the changing head being compact [29].

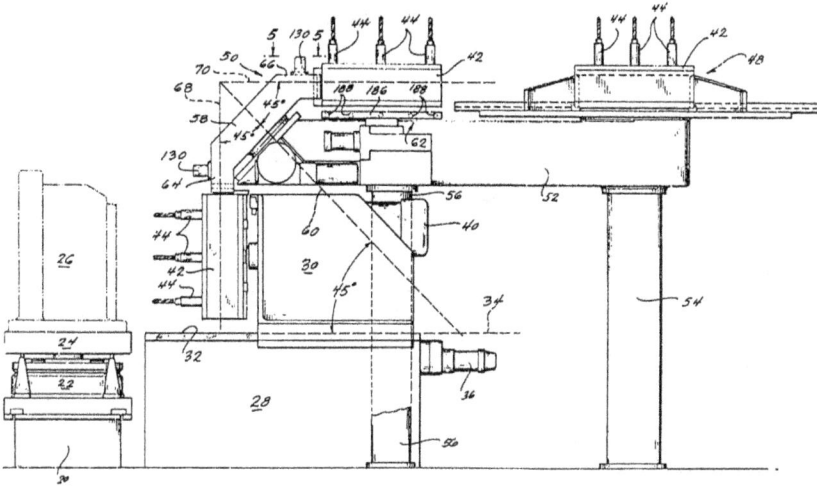

(a)

(b)

Figure 2-18   Head changer at first phase: (a) Head changer patented by Kerney & Trecker Co. and (b) head changer in mid-1970s (courtesy of Heller Maschinenfabrik).

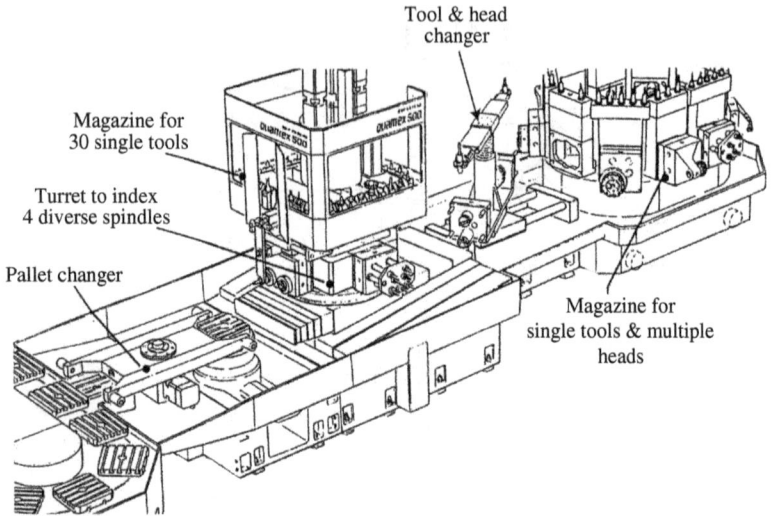

Figure 2-19   CNC station in FTL developed by Ford Co. in mid-1980s (courtesy of Brose Fahrzeugteile GmbH & Co.).

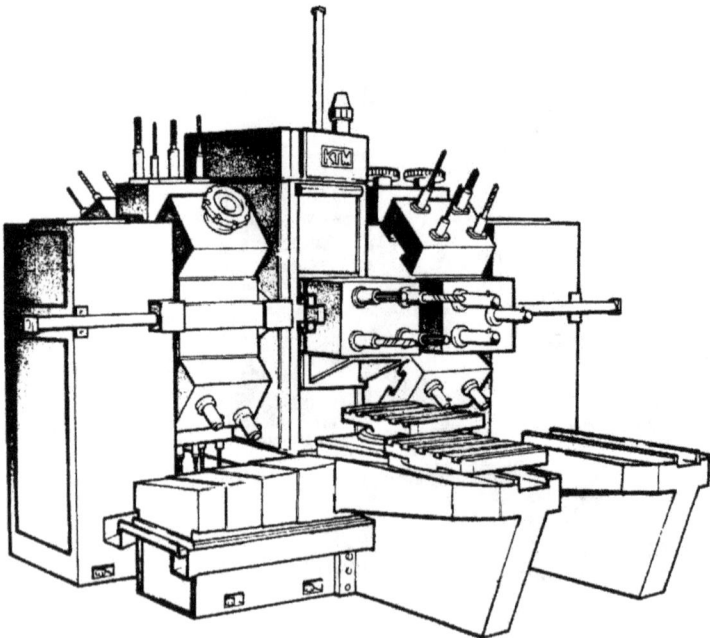

Figure 2-20   CNC multiple-head changer around 1980 (courtesy of KTM Co.).

For the ease of further understanding, some characteristic features of a *Variocenter*, shown already in Fig. 1-6, are depicted here.

1. To be applicable for the small and medium batch size production, the modular design has been realized by the combination of the following functionality and performance.
   a. Stationary or movement function of machining unit along X axis
   b. Traveling function of machining unit along Z axis
   c. Stationary or movement function of table along Z axis
   d. Machining unit with quick changing tool cassette of triangle, square, hexagonal, or octagonal type
   e. Steel welded table unit with swiveling base and tool driving DC motor of 20, 30, or 50 kW

2. Each *Variocenter* is, in general, available for facing, milling, angle milling, single and gang head drilling, threading and fine turning, and furthermore for five-face processing with angle cassette, broaching, assembly, and inspection with special cartridge.

3. It is worth pointing out that the table unit of *Variocenter* was already a twin-ball screw driving type in 1984 so as to carry smoothly the work up to 4000 kg in weight. Such a driving method became popular in the late 1990s.

4. The tool monitoring and work changing can be carried out at the outside of the machining space while machining.

5. In due course, the *Variocenter* aims at utilization within the FTL and also the stand-alone FMC.

System function-integrated machine.   The system function-integrated machine may be defined as a compact "cubiclike FMC" having the configuration similar to that of a machine as a whole. Consequently, the system function-integrated MC and TC can be considered as the utmost representatives of the system machine. In other words, these machine tools have multiple and various functions with higher-density integration, showing a preferable configuration as a machine for SME and as a basic module for the production system. Obviously, such machine tools have been developed on the basis of the flatlike FMC, in which all the system components, e.g., machine tool, robot, automatic warehouse and cleaning station, are allocated in the two-dimensional space (see Fig. 1-17).

Marwin Production Machines developed, as shown in Fig. 2-21, a system machine through joint work under the subject of the British Aerospace patent. In fact, the machine was designed to operate both in a stand-alone mode and for use in the FMS by dealing a random mix of

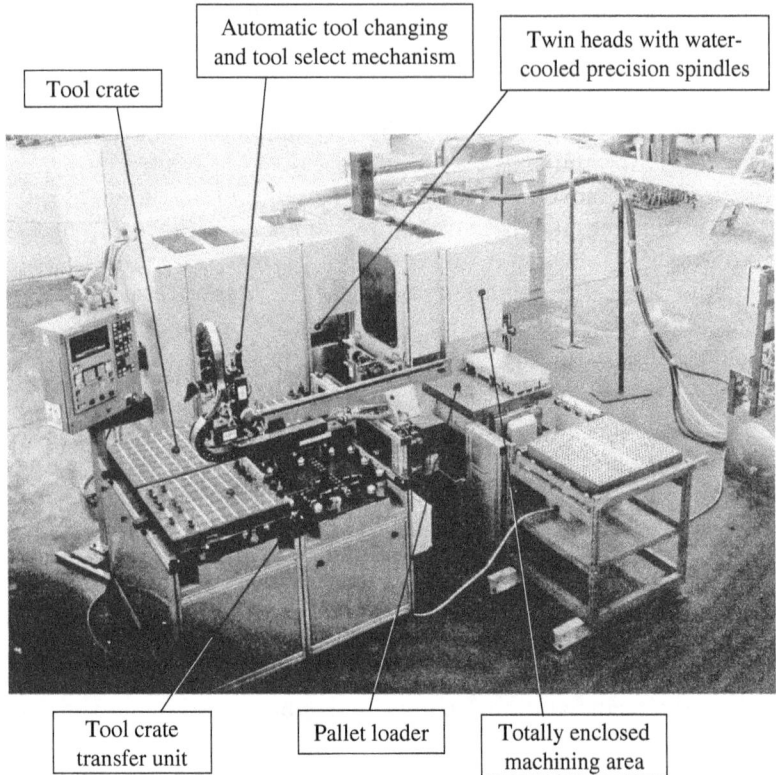

Tool crate

Automatic tool changing and tool select mechanism

Twin heads with water-cooled precision spindles

Tool crate transfer unit

Pallet loader

Totally enclosed machining area

Figure 2-21   A system machine (type AUTOMAX I, courtesy of MPM Co.).

components of unit batch size, and in due course the machine has the following characteristic features [30].

1. Tool storage system, in which the tool is transferred through the crate type pallet (tool crate) carried by the AGV (*automatic guided vehicle*).

2. Vertically allocated twin spindle with water cooling, each of which is mounted on the individual Z axis slide unit positioned one above the other.

3. For light alloy machining, the speed of main spindle with tapered hole of No. 30 ISO is 12,000 rpm maximum, for which the main motor has 15 kW in output.

4. The double-sided pallet loader supporting the pallet with the work rotates through 90° anticlockwise to position the pallet vertically and then transfers the pallet between the machining area and the pallet loader.

It is furthermore worth stating that the system machine of Marwin Production Machines make was a core entity in the cell for demonstrating the achievement of ESPRIT Project No. 955 CNMA (Communication Network for Manufacturing Application) of EC.

A similar configuration can be observed in the cubiclike FMC of Steinel make, although it is not sophisticated yet. In fact, Steinel Co. developed the FMCs of modular design based on the MC of BZ type, which can be characterized by its compact configuration and pallet pool integrated with the work transfer function, as shown in Fig. 2-22 [31]. It is worth suggesting that Steinel's FMC originated with modular-configured MC on the basis of the ASP Plan of the United Kingdom [32]. Based on similar idea, Cincinnati Milacron produced the GC called "Total Production Team" in 1984, which consisted of the universal grinding machine (type: Cinternal 3) and gantry crane.

Afterward, the compactness of the machine prevailed, and from the beginning of the 1980s, Tsugami has been producing the system machine (MA type) shown together in Fig. 2-23 which is appraised by a very good reputation, because it has the following characteristic features.

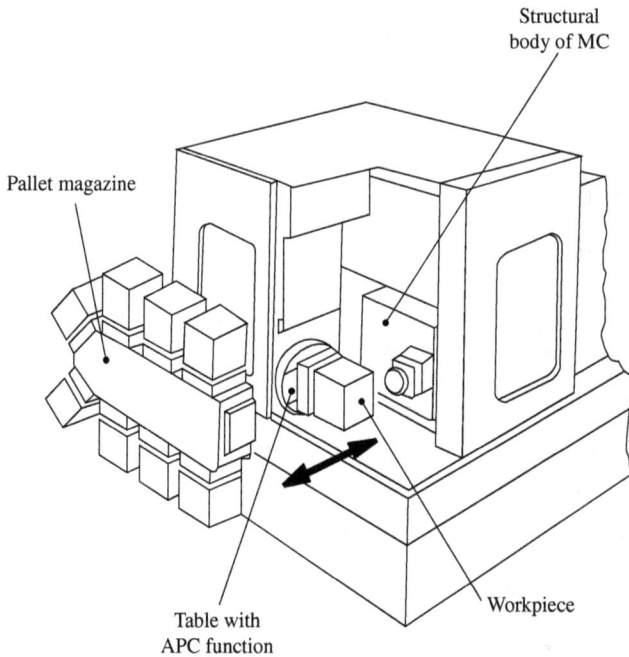

Figure 2-22   Cubiclike FMC of Steinel make (type BZ24FFZ).

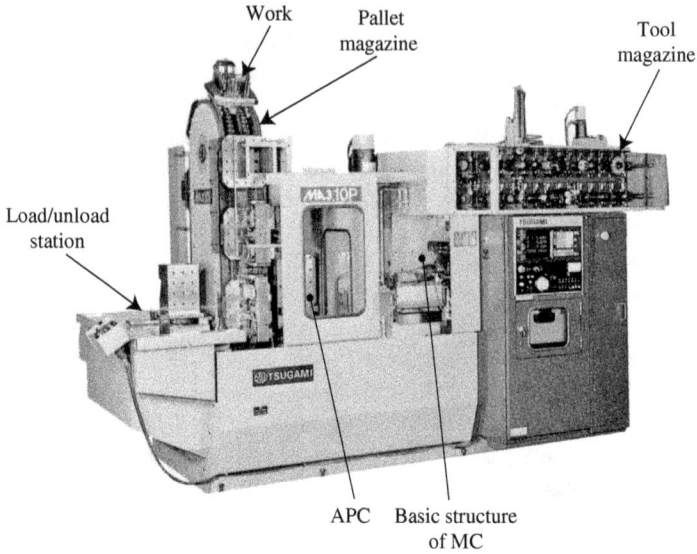

(a)

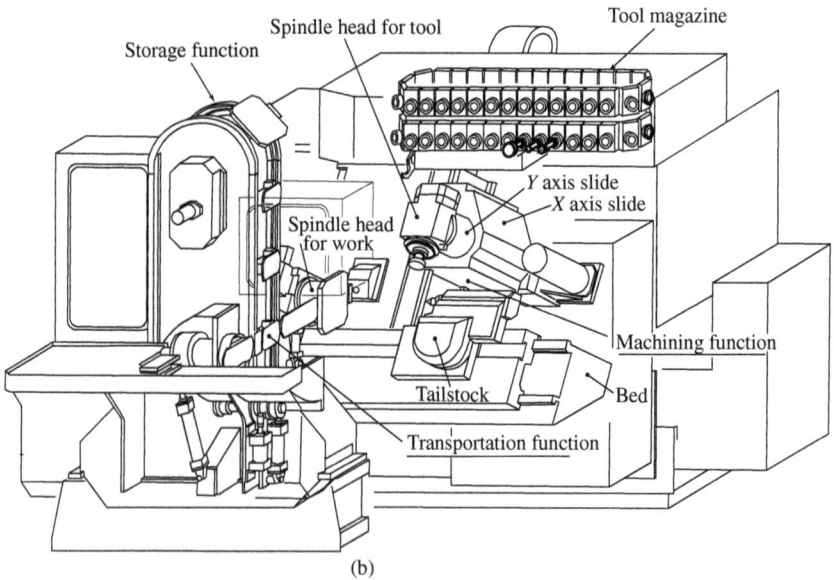

(b)

Figure 2-23 System machine of function-integrated type: (a) MA type (on market from 1982) and (b) TA3 type (not on market) (courtesy of Tsugami Co.).

1. The repeatability of pallet positioning accuracy is better than 2 μm.
2. The thermal elongation of the main spindle is 10 μm after running 7 h at 1500 rpm rotational speed.
3. The Y axis quill for pallet fixing and traveling is preadjusted and assembled in consideration of the applied loads in operation.

On the basis of the good reputation of the MA type, Tsugami once developed a prototype of TC-based system machine, i.e., type TA3 shown in Fig. 2-23, which can be characterized by its chucking mechanism and automatic chuck changing; however, it was not on the market.

Modernization of traditional machine tools.   With the growing importance of the system machine, the NC horizontal boring and milling machine of floor type has been converted to a system machine. All the traverses are located within the column, and greater machining flexibility due to the double- or triple-layered main spindle provides the machine with higher compatibility with flexible manufacturing. Figure 2-24 is such a machine of Kearns-Richards make in 1990 (boring spindle diameter: 110 or 130 mm), which can also grind the work using the planetary guiding spindle attached to the facing plate. According to the same idea, Vigel (Italy) and Steinel (Germany) merchandised a three-layered spindle and two-layered spindle machine in 1991 and 1994, respectively, in which the inner spindle has eccentricity to the outer spindle, so that machining flexibility is enhanced.

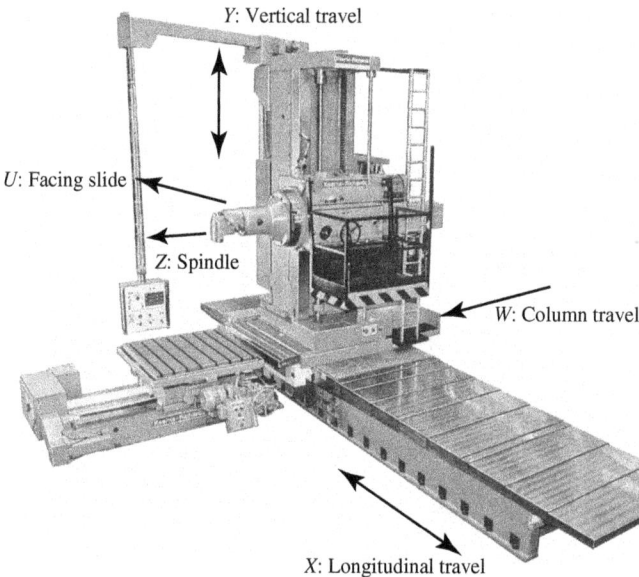

Figure 2-24 System machine originated from traditional horizontal boring and milling machine (type F, around 1990, courtesy of Kearns-Richards).

2.4.2   Machining complex and processing complex

In retrospect, an ancestor of the machining complex may be the engine lathe of Ramo make in the mid-1960s. The lathe was, in addition to performing various turning operations, capable of even key-slotting and broaching using the devices attached to the tailstock. Consequently, this idea has been keeping its value in machine tool design, and duly functioning on the occasion of the Japanese Big National Project conducted between 1977 and 1984 [33, 34]. In this project, a "Complex Machining Centre" was developed as shown in Fig. 2-25, which is not so compact, but of very modular type, and can be characterized by the following functions.

1. The machine can deal with turning, drilling, milling, gear cutting, grinding as well as laser welding and surface hardening, and thus it appears to be a predecessor of the processing complex.

2. The machine is equipped with the measurement function.

3. The machine has the automatic spindle unit changer, automatic work changer, work hands-off function, subassembly function, ATC, tool magazine changer, automatic chuck changer, and measuring probe changing system.

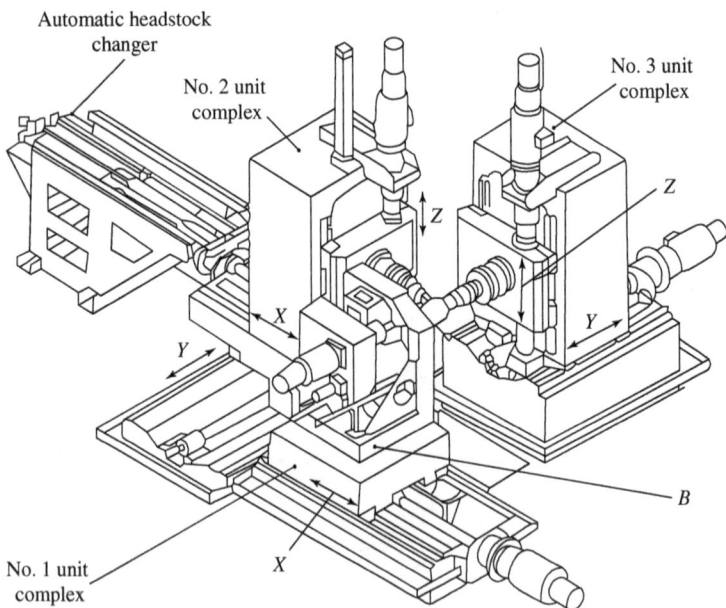

Figure 2-25   Concept of complex machining center developed by Japanese Big National Project in the 1970s.

This idea has later merchandised by Index Co. in 1999, and the machine named "*Vertical Line*" was exhibited on the occasion of EMO Shows in Paris (see Fig. 1-5). As can be readily seen, the successors of the engine lathe of Ramo make and complex machining center are the machining complex and processing complex, respectively. For the ease of understanding, some representative machines will be quickly shown in the following.

The Boley displayed the machining complex in 9th EMO Show (1991), which can be characterized by its twin-spindle opposite and offset located type, as well as by a couple of spindle heads and turret heads of traveling type. Figure 2-26 is another machining complex of Nihon Kokan make, which has simultaneously both the functionalities of TC and MC, and which is of modular type, so that the user's requirements can be fulfilled to various extents. Actually, the designer is capable of choosing the various combinations among the turning, milling, and boring functions, number of tools in turret, number of pallets, and number of tools in ATC. It appears that the machine has overspecifications; however, the machine is, in the extreme case, allowed to change from the TC to MC, i.e., that with different kind-oriented modular design. In accordance with this evidence, it is envisaged that the different kind-oriented type is a "must" when applying the modular design to the machining complex. The modular design is furthermore applied to the machining complex including the CNC turning machine by Ikegai Iron Works, as shown in Fig. 2-27 in the late 1990s, and Table 2-4 is a firsthand view of machining complexes that have been on the market so far with better reputation.

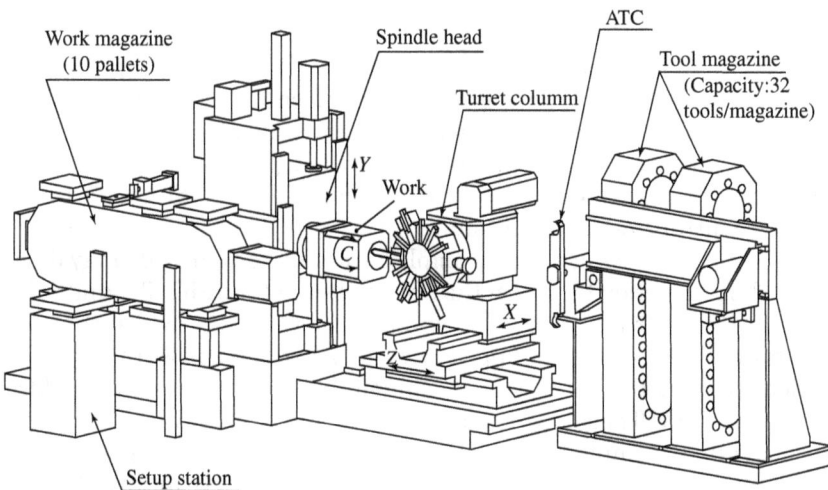

Figure 2-26 Concept of TC—MC Complex (type MMC-30, courtesy of Nihon Kokan).

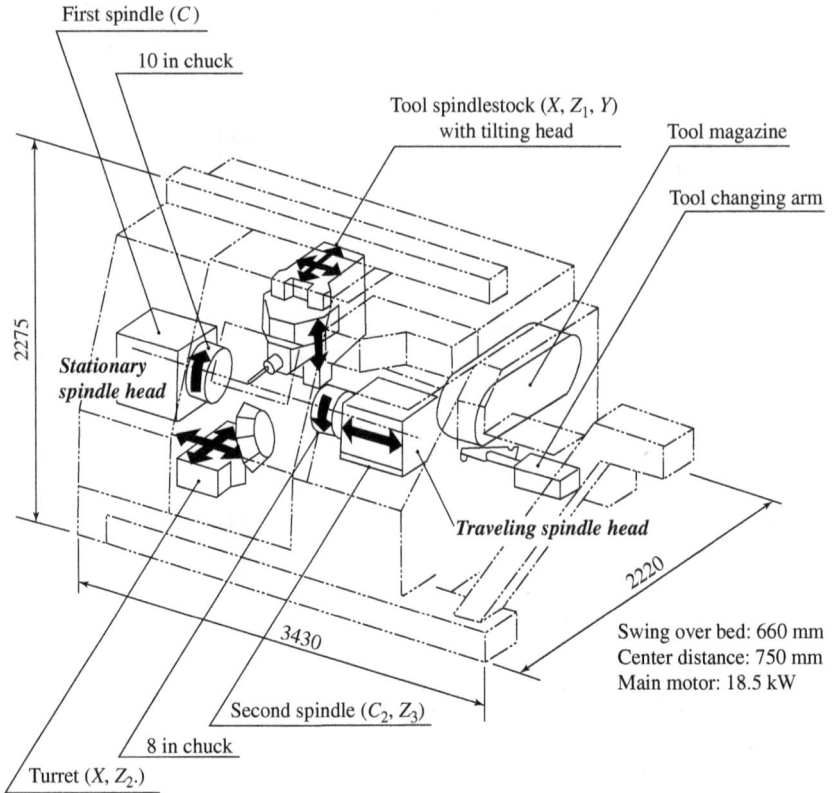

Figure 2-27   Machining complex with modular design—eight-axis control (type TM25YS, courtesy of Ikegai Iron Works).

In consideration of the developing history, the machining complex can be regarded as originating the system machine of multiple machining method-integrated type in the embryonic stage. In due course, such a system machine has become the general-purpose type with compact and cubic configuration and with a one-workpiece set, although multiple NC control axes are necessitated.

Importantly, the machining complex around the year 2000 may dominate the production facility of the SME instead of the MC, TC, and cubiclike FMC so far installed. This will become reality, provided that the machining complex is equipped with at least the software package, which specifies the parts to be machined and functions as an interface to the network. These are considered to be the preconditions to install the machining complex to the SME (*small and medium-size enterprise*). In addition, the machining complex is expected to serve as an entity for the island automation, e.g., FML, for large enterprises, in which the

TABLE 2-4   Classification of Machining Complexes between 1995 and 2000

Work branch	Tool branch	Products on market & remarks
Opposite-located twin-spindle (second spindle: traveling type)	Rotary tool spindle head with swiveling type	Type MT-250S of Mori Seiki (8-axis controlled type): **Tool spindle with curvic coupling for fixing turning tool, roller linear guide of German-make, 7.5 kW built-in-motor (8000 rpm.) and maximum allowable diameter of work 570 mm**
	Upwards: Tool spindle head of swiveling type Underneath: Turret head	Type Integrex 100 II of Yamazaki Mazak:**Main motor 7.5 kW (6000 rpm.), tool spindle motor 5.5 kW (10,000 rpm.) in max., and minimum swiveling angle 0.001 deg.**
		Type TM25YS of Ikegai (8-axis controlled type): **Swing over bed 660 mm, primary spindle with 10 in chuck (438 N-m in max. allowable torque), secondary spindle with 8 in chuck (105 N-m in max. allowable torque), tool spindle (45 N-m in max. allowable torque) and bore diameter of inner ring of main bearing 120 mm**
	Twin-turret of opposite-located type (Upwards and underneath)	Type B56 of Biglia: Turret head allows to attach milling cutter
Single spindle with tailstock	Twin-turret	Type NST-40M of Shin Nippon Koki: 520 mm in max. allowable diameter of work, and turret allows to attach milling cutter
Single spindle	Single turret column (with octagonal turret head and milling head)	Type MT-25A/500 of Mori Seiki
Twin-head (each head has twin-spindle)	Four-turret head	Type PPC of Pittler: To minimize idle time, simultaneous machining for the same work

basic necessity is to intermediate between the preferable role of human beings with multiple nationalities and the fully automatized production facilities. In short, the machining complex can, at present, be defined as the system machine of multifarious machining function-integrated type having the configuration of combining the TC and MC with twin-spindle allocated oppositely.

In due course, the processing complex can be regarded as an advanced variant of the machining complex, because it has more processing functions than machining. In the mid-1980s, the Oerlikon tried to produce a portal machine of multiple-processing-integrated type called *Multitechnologie-Zentrum* (spindle tapered hole: ISO 50), which can deal with drilling, milling, jig boring, jig grinding, and coordinate measurement, although the machine was designed not using the modular principle. In addition, the TC of trommel type was on the market, where the trommel with twin-spindle has a function of work holding and transportation.

Another example is, as shown in Fig. 2-28, a CNC vertical turning machine with twin-spindle of Hüller Hille make (type DVT) in 2002,

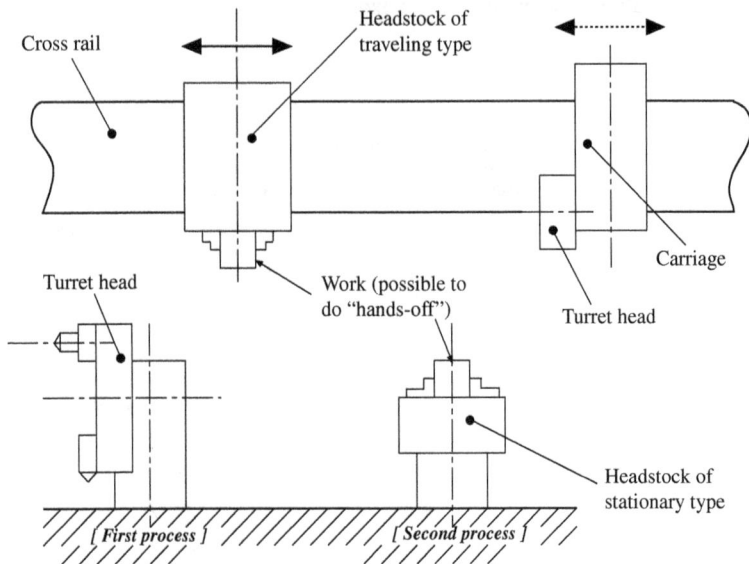

Note: For machining of flangelike parts, e.g., gear blank and brake disk.

Figure 2-28   NC vertical machine of Hüller Hille make.

which is furthermore capable of grinding and laser processing. Although the machine itself is not of modular type, it is designed as a basic module for the production cell, expecting to realize the compact manufacturing line. It is very interesting that the machine originated with the "turning-boring-centre (Dreh-Bohr-Zentrum)", i.e., type DV 62 of Hessap make, which was on the market as a machining complex in 1984.

As can be seen from these complexes, at burning issue is whether the modular design is mandatory. Importantly, the complex has duly greater flexibility from the viewpoints of both the structural configuration and the NC software; and thus in its design, it appears that the modular principle renders it useless. In contrast, some manufacturers have been very keen to design the complex using the modular design, although modular design has become apparently very complicated with the advent of both the system machine and the complex. It is thus emphasized that we need to scrutinize in the very near future the substantial necessity of the modular principle in designing the machining and processing complexes.

In discussion of which kinds are suitable for modular design, at least, we can assert that the processing complex is a protruded objective. For example, de Vicq of AMTRI (Advanced Manufacturing Technology Research Institute) in United Kingdom has viewed a test bed with 13 controlled axes together with twin-headstock TC with three- or five-axis milling modules. In addition, the test bed facilitates 3 kW laser heat treatment and grinding [35]. In this processing complex, must we employ modular design?

For the sake of further consideration of whether the modular design is mandatory in designing the complex, a quick note is given below. Consequently, in viewing that a protruded development of the modular design has been realized in the third phase of its developing history by applying it to flexible manufacturing, the necessity is, in part, touched on the modular design in flexible manufacturing. In fact, (1) in the FMS, the basic module is that of the FMC, compact FMC, or system machine, which is duly of modular construction in certain cases, and (2) in the FMC and system machine, the basic modules are those of unit and unit complex.

More specifically, not only is the system itself of cell-based modular type, but also the cell has begun to prevail as the growing installation number of FMC in the SME. In addition, we must be aware that the system machine and machining complex are capable of working in stand-alone mode [36]. In due course, the SMEs are, as already stated and with no exception, very keen to install these new machines at the beginning of the 21st century instead of the FMC of robot or pallet pool type (see Fig. 1-17). These FMCs are of typical conventional type and have so far been as the most popular FMC.

Within a system sphere, recently the Cincinnati Milacron has merchandised a kit for in-house building cell called *Profit Shop—Concept,* This idea has been well known for a long time and employed widely by the enterprise; however, such a commercial-based kit has not been on the market. For example, in the EMO Show of 1997 in Hannover, the company displayed a system consisting of vertical MC and CNC turning machine. Actually, people can purchase separately the machine tool, robot, automated warehouse, briquetting press for swarf disposal, and so on from the related manufacturers, and integrated them into a system. This is of course one of the variants of modular design.

Regarding the future perspective of the modular principle in designing the machining and processing complexes, furthermore, the basic necessity is to scrutinize the foresight of the future machine tools [37]. In accordance with the prediction, it may be said that the modular principle is not decaying, but is maintaining its importance with certain necessities. In short, the modular design is, at least, requested in the manufacture of the system machine for the island automation, dexterous machine tool to be compatible with remanufacturing issues, and also the culture- and mindset-harmonized machine tools. Although there remains something to be seen, we could need an innovative variant of the modular design for a leading machine tool in the future, which can handle simultaneously the better accuracy and higher-speed machining together with heavy cutting by preferably leveraging the related dimensional and performance specifications. Obviously, these specifications cannot be fulfilled simultaneously at present in the design of the machine tool, which is one of the leading causes for the ill-defined problem in the machine tool design.

To this end, it is again envisaged that the complexity in the applied loads due to multifarious machining methods is one of the leading structural design factors in the machining complex; however, the utmost serious problem is how to equalize the temperature distribution and minimize the thermal deformation. Another difficulty lies in the preparation of the NC software, because of the complexity of the part shape and one-chucking machining including a hands-off procedure. In other words, at issue is the leverage between the production volume and the production cost of the NC software.

References

1. Doi, Y., "On Application of BBS," *Toyoda Technical Report,* 1963, 4(3): 22–32.
2. Feldmann, K., "Analyse der Gestaltung von automatischen Drehmaschinen," *Industrie-Anzeiger,* 1975, 97(67):1467–1468.
3. Ito, Y., and Y. Yoshida, "Design Conception of Hierarchical Modular Construction—Manufacturing Different Kinds of Machine Tools by Using Common Modules." In S. A. Tobias and F. Koenigsberger (eds.), *Proc. of 19th Int. Machine Tool Design and Res. Conf.,* Macmillan, 1979, pp. 147–153.
4. Metternich, J., and B. Würsching, "Plattformkonzepte im Werkzeugmaschinenbau," *Werkstatt und Betrieb,* 2000, 133(6): 22–29.
5. Koenigsberger, F., "Modular Design of Machine Tools," private draft proposal, July 29, 1975, The University of Manchester Institute of Science and Technology, United Kingdom.
6. Tönshoff, H. K., M. Mey, and A. Schnülle, "An Approach for the Concurrent Development and Manufacturing of Modular Machine Tools," *Production Engineering,* 1998, 5(1): 63–66.
7. Shinno, H., and Y. Ito, "Computer Aided Concept Design for Structural Configuration of Machine Tools—Variant Design Using Directed Graph," *Trans. ASME J. Mechanisms, Transmissions and Automation in Design,* 1987, 109: 372–376.
8. Tönshoff, H. K., and F. Böger, "Kundenspezifishe Konfigurierung modularer Werkzeugmaschinen," *ZwF,* 1996, 91(9): 433–436.
9. Höft, K., and Y. Ito, "A Method for Culture- and Mindset-Harmonised Design," in Poster Session of ICED '99 (International Conference on Engineering Design), München, August 24–26, 1999.
10. Höft, K., "Culture- and Mindset-Harmonised Manufacturing in Sustainable Global Environments," Dissertation, Tokyo Institute of Technology, March 1999.
11. Dietz, P., "Baukastensystematik und methodisches Konstruieren in Werkzeugmaschinenbau," *Werkstatt und Betrieb,* 1983, 116(4): 185–189.
12. Lee, H. S., H. Shinno, and Y. Ito, "Structural Configuration Design of Machining Center—On the Variant Method Using Conjunction Pattern," *J. JSPE,* 1986, 52(8): 1393–1398.
13. Dietz, P., "Pendelbearbeitung und Baukasten-Maschinensysteme steigern die Produktivität," *Industrie-Anzeiger,* 1983, 105(17): 42–47.
14. Neumann, P., "Entwicklung von Qualitätskriterien für die Weiter-oder Wiederverwendung angepasster Produkte und Komponenten," Matr-Nr. 146123, Technische Universität Berlin, Dec. 21, 2001.
15. Jones, D. T., "The Route to the Future," *Manufacturing Engineer (IEE),* 2001, 80(1): 33–37.

16. Kidd, C., "The Case for Configuration Management," *IEE Rev.*, September 2001, pp. 37–41.
17. Ito, Y., and K. Höft, "A System Concept for Culture- and Mindset-Harmonized Manufacturing Systems and Its Core Machining Function," in F.-L. Krause and E. Uhlmann (eds.), *Innovative Produktionstechnik*, 1998, pp. 175–186 (For the Celebration of Professor Spur's 70th birthday), Carl Hanser Verlag, München, Wien.
18. Ito, Y., and K. Höft, "A Proposal of Region- and Racial Traits-Harmonised Products for Future Society: Culture and Mindset-Related Design Attributes for Highly Value-Added Products," *Int. J. Adv. Manufacturing Technol.*, 1997, 13: 502–512.
19. Ito, Y., "A Symptom of Growing Importance and Impacts of Manufacturing Culture for Strategic Production Environments in the Future," in M. Fischer, G. Heidegger, W. Petersen, and G. Spöttl (Hrsg.) Gestalten statt Anpassen in Arbeit, Technik und Beruf—Festschrift zum 60. Geburtstag von Felix Rauner, 2001, S. 376–391, W. Bertelsmann Verlag, Bielefeld.
20. Steering Committee on Mechanical Engineering Research within the Japan Science Council, "Research Guides for Production Science and Engineering in the Beginning of 21st Century—Contribution of Production Science and Engineering to Future Society and Facing Academic Problems to Be Solved," June 1994, Tokyo, Japan.
21. Handouts distributed on the occasion of Workshop Produktion 2000 on Sep. 11 and 12 1997, Forschungszentrum Karlsruhe.
22. Grant, D., "2015 Vision," *Manufacturing Engineer (IEE)*, 1998, 77(5): 237–241.
23. Committee of Visionary Manufacturing Challenges et al., *Visionary Manufacturing Challenges for 2020*, National Academy Press, Washington, 1998.
24. Ito, Y., "Technology Prediction 11—Production Systems and Machining Technologies," in N. Makino and L. Ezaki (eds.), *Technology Innovation in 21st Century,* Nov. 2000, pp.152–170, Kougyo Chousakai, Tokyo.
25. Ito, Y., "Developments of System Machines Compatible with Flexible Manufacturing System,"*J. JSPE,* 1982, 48(6):794–800.
26. Ito, Y., "System Configuration and Design of FMS in Next Generation," *Adv. Robotics,* 1987, 2(2): 103–120.
27. "The Machine Tool of the Future: A Concept Patented by Accim," *Atouts,* May 1999, 27: 43.
28. "FMS in the Automotive Industry," *The FMS Magazine,* 1985, 3(1): 54–55.
29. Maier, D., "Umstellbare Mehrspindlelbohrköpfe für numerisch gesteuerte Bearbeitungszentrum," *wt-Z. ind. Fertig.*, 1976,66(4): 197–200.
30. Baxter, R., "Manufacturing System Is 'Revolutionary,'" *The Production Engineer*, November 1984, pp. 45–46.
31. Schütz, W., and R. Steinhilper, "Kostengünstiger Palettenspeicher für Bearbeitungszentrum," *wt-Z. ind. Fertig.*, 1982, 72(3):151–155.
32. Astrop, A., "Time to Take Action on the ASP Report," *Machinery and Prod. Eng.*, Aug. 2, 1978, p. 17.
33. Ito, Y., "Flexible Manufacturing System Complex Provided with Laser—Part 3 Application System Design," in *Proc. of 5th ICPE*, JSPE, Tokyo, 1984, pp. 28–36.
34. Kimura, M., et al., "Flexible Manufacturing System Complex Provided with Laser (FMC)—A National R & D Program of Japan," in B. J. Davies (ed.), *Proc. of 23d Int. MTDR Conf.*, Macmillan, 1983, pp. 475–481.
35. de Vicq, A., "21st Century Machinery," *Manufacturing Engineer (IEE)*, 2001, 80(3): 104–109.
36. Ito, Y., "The Production Environment of an SME in the Year 2000," in K. McGuigan (ed.), *Flexible Manufacturing for Small to Medium Enterprises—A European Conf.,* 1988, pp. 207–234, EOLAS, Dublin.
37. Ito, Y., "Predictive Research into Desirable Features of Machine Tools in the Year 2020 and Beyond—Private Viewpoints and Assertion," in *Proc. of Int. Machine Tool Technical Seminar*, 2000, pp. 3–18, Korean Machine Tool Manufacturers' Association and Korean Society of Precision Engineering, Seoul.

3

Description of Machine Tools

Within a modular design context of machine tools, we need both the design technology and the design methodology; however, the design methodology is far from the completion compared to the design technology, e.g., computational method of static and dynamic stiffness. In fact, the design methodology can assist the systematization of the design data related to three of the four principles of modular design, i.e., the principles of separation, standardization, and adaptation (see Chap. 2).

Regarding these principles, at issue is the establishment of a methodology for the principle of adaptation, i.e., rational combination of the modules from a group of predetermined modules in accordance with the design specifications. Such a combination problem could be solved easily using the computer, once the machine tool could be represented with a certain description method, i.e., machine tool description, which is understandable by the computer in the same way as the parts description within the computer-aided drafting. This means that a core technology for the principle of adaptation is the machine tool description. In addition, we must be aware that the machine tool description is one of the preconditions to promote the efficient use of the computerized design of machine tools.

In contrast, another intake objective is the system design of flexible manufacturing including agile manufacturing. After the establishment of its first stage, flexible manufacturing can be classified into the (1) FMS, (2) FMC, (3) FTL, and (4) FML. In general, the FMS, FMC, FTL, and FML are designed by the modular principle, and with the advent of flexible manufacturing, the NC machine tool launched another noteworthy development, i.e., that for the system machine. In due course, the system consists often of the system machine, which is of modular design type to some extent. In short, we need to develop the system description including the description for the system machine.

In retrospect, a crucial problem was to find some compromise solutions with respect to, e.g., the shapes and sizes of modules from the viewpoint of the casting pattern as well as the allocation of slideways and location faces in consideration of machining in the mid-1970s. At that time, the designer must be qualified for seeking the compromise solution with her or his high abilities and skills. As will be discussed later, the machine tool description may furthermore add something definite to this problem, and that of Redeker and Saljé is associated with it.

To summarize, the methodology for the modular design must, as mentioned above, deal with several issues related to three of the four principles of modular design, especially emphasizing the principle of adaptation. This emphasizes directly the fast-growing importance of the machine tool description.

3.1 Basic Knowledge about Functional and Structural Description Methods

It is very interesting that Stau is credited with being the first engineer to propose the functional description in 1963, although he had no intention of applying it to modular design. It appears that he tried to provide a clear idea for a form-generating function in turning [1]. The description method proposed by Stau is very simple, but its basic idea has been employed within various functional description methods developed since then. In his book entitled *Die Drehmaschinen*, he tried, as shown in Table 3-1, to classify the form-generating function of the machine tool by using the symbolic representation and decision table. Actually, the machining method is classified using the combination of traveling and rotational movements in both the work and tool branches.

Intuitively, it is desirable in the machine tool description that the machine tool be represented using only one method; however, the machine tool can be represented in various ways, depending upon what feature is emphasized in the description. At present, there are the two methods: one is the functional description (movement description) and the other is the structural description. In a machine as a whole, a one-to-one relationship between the function and the structural configuration is obviously not guaranteed; and a function can, in general, be realized by myriad structural configurations, although a structural configuration can provide us with a single form-generating function. Conceptually, the functional description is in a higher level than the structural description, resulting in the obvious difference in the description method, difficulties in the description, and application among as

TABLE 3-1   Preliminary Proposal of Functional Description (by Stau)

			Tool branch							
			On bed or base				Spindle including ram and quill			
		No. / Movement patterns	a	b	c	d	e	f	g	h
			•	→	↻	→↻	•	→	↻	→↻
Work branch — Spindle including ram and quill	1	•				Trepan boring				
	2	→								
	3	↻			Turning, Drilling	Cutting-off by turret lathe of drum type, Roto mill	Screw cutting by Mach methed, Trepan boring			
	4	→↻	Turning by automatic screw cutting machine of Swiss type							
Work branch — On bed or base	5	•						Key-way machining, Broaching	Sawing	Boring, Drilling
	6	→						Planing, Shaping, Slotting	A	
	7	↻							Gear shaving	Hobbing
	8	→↻							Thread milling	Gear shaping

Notes: • : Stationary   → : Linear movement

↻ : Rotational movement   →↻ : Linear and rotational movements

A : Cylindrical milling, face milling, line boring by horizontal boring, and milling machine of table type

shown in Table 3-2. More specifically, the machine tool description, as will be stated in Chap. 4, can facilitate effectively some leading design work as follows.

1. Evaluation of structural similarity

2. Prediction of variants possible to create from the basic configuration

3. Procurement of the principle of adaptation, i.e., estimation of machine tools possible to generate from a group of predetermined modules

4. Determination of the most suitable structural configuration for a group of the workpieces to be machined

5. Functional and structural configuration analyses of machine tools

6. Compatibility analysis of structural configuration with human amenity

7. Analysis of the market competitiveness

TABLE 3-2 Comparison of Characteristic Features between Functional and Structural Descriptions

Description method	Description procedure	Simplicity	Application areas
Functional	Implicit representation of flow of force Representation using linear and rotational movements in direction of X, Y and Z axes, and around them	The very ease of description: Only elementary knowledge about machine tools & manufacturing procedures is required	Functional analysis of machine tools Decision of qualitative configuration similarity Prediction of variants from basic structure Computer-aided drafting for concept drawing Automatized process planning Structure analysis from ergonomics aspect
Structural	Explicit representation of flow of force Representation using GT codes and flow of force (structural pattern)	Certain difficulties in description: Deep knowledge about machine tool structures is required	Classification of machine tools Structural analysis of machine tools Evaluation of structural similarity Generation of structural configuaration (variant and free types)

As a result, the designer must choose either the functional or the structural description depending on the purpose of the application, as already shown in Table 3-2.

The functional description can, in principle, be handled more easily than the structural description, because it consists of the combination of the leading traveling and rotational movements of the machine tool, i.e., linear (X, Y, Z) and rotational (A, B, C) motions in Cartesian coordinates. Thus when describing the machine tool by the functional description, we are required to have only very simple knowledge about machining, whereas we are required to have some detailed knowledge about the machine tool structure, when representing the machine tool with the structural description. In the structural description, a root cause of difficulties lies in the correct recognition of dimensional, functional, and performance specifications of each structural body component, to represent it with a proper GT (*group technology*) code.

In the machine tool description, furthermore, the concept of FOF (*flow of force, Der Kraftfluß*) is very important, although the concept itself is very simple. In fact, as will be shown, the FOF is employed implicitly and explicitly in the functional and structural descriptions, respectively. In this context, Jäger suggested that the definitions of the FOF so far proposed, e.g., by Schöpke, Saljé, Königsberger, and others differ from one another, and Jäger proposed to use the term *der Wirkkreis (effective circle)* [2]. His suggestion did not catch on, and we have used the term FOF up to now. It appears that the FOF can be absolutely defined by the well-known proposal made by Schlesinger [3] in the 1930s.

It is, in principle, desirable that the machine tool be represented with only one absolute description, or that the functional description be in one-to-one relation to the structural description, and vice versa. Such a requirement can be fulfilled in the case of the part and functional complex, but is far from being fulfilled in the case of the machine as a whole, i.e., entity belonging to higher layer within the hierarchical structure of a product. In fact, the structural entity of higher layer has complicated properties in both the functional and structural aspects. In the design procedure, it is thus imperative for the time being that a functional description may accordingly correspond with various structural descriptions, although a structural description is in one-to-one relation with a functional description.

3.2 Details of Functional Description

The functional description can be defined as a machine tool representation method with the leading traveling and rotational movements or form-generating movement, where the latter is a chosen combination of the former to especially represent the form-generating functions possible. As already mentioned, Stau proposed the preliminary idea of functional description, and later Vragov of the U.S.S.R. publicized a noteworthy description method in 1972 [4]. He suggested that the machine tool structure can logically be represented with the AND and OR connection between both structural entities, resulting in structural formulas based on the logic algebra and multiple-factor theory. Figure 3-1 shows some examples of the description, and in the following, the description is stated in steps.

1. Coordinates are determined by distinguishing the linear and rotational movements and identifying the leading and auxiliary movements, provided that the X axis is for horizontal longitudinal movement and the Z axis is parallel to the main spindle axis.

   Leading movements: Linear X, Y, and Z
                                         Rotational A, B, and C
   Auxiliary movements: Linear U, V, and W
                                         Rotational D and E

2. Affix the subscripts h and v to Y and Z to distinguish the horizontal and vertical movements.

3. Represent stationary units (modules) with "O" and moving units with coordinates determined in step 1. In principle, the machine tool can be represented with the symbolized combination of the rectilinear

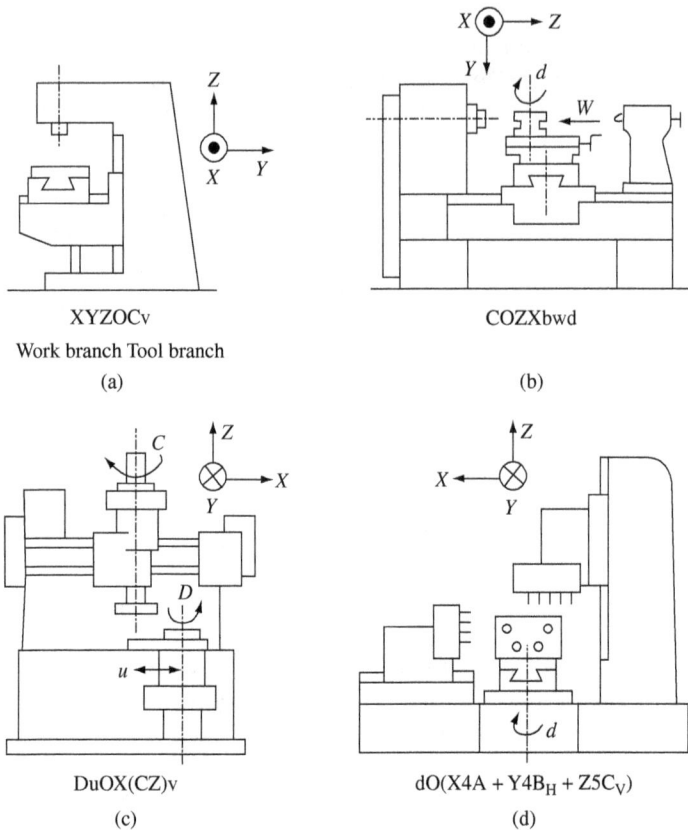

XYZOCv

Work branch Tool branch

(a)

COZXbwd

(b)

DuOX(CZ)v

(c)

dO(X4A + Y4B$_H$ + Z5C$_V$)

(d)

Figure 3-1 Functional description: (a) Vertical milling machine; (b) engine lathe; (c) gear shaper; and (d) rotary indexing machine (by Vragov).

movements in the direction of coordinates (X, Y, Z) and rotating movements (A, B, C) around them. If necessary, the auxiliary movements can be used additionally.

4. Coordinates represented by capital and small letters indicate the movements in closer relation to the machining process and auxiliary movement, respectively.

5. Representation of unit connection is as follows, and in this process the FOF is used to determine the order of the symbolized movements.

Series connection ∧ or •

Parallel connection ∨ or +

6. The parallel connection, i.e., simultaneous movements for form-generating, is in the round bracket together with indicating the number of identical units with the corresponding number.

7. The units that loaded the workpiece and fixed the tool are allocated to the far left and far right in the description, respectively.

Vragov tried to apply the proposed method to the classification of the machine tools, to choose an optimum structural configuration of MC, and to estimate the variants capable of generating from a basic structure by rearranging the alphabetical symbols, as shown in Fig. 3-2.

Following that of Vragov, Saljé and Redeker investigated the functional description, intending to apply it to the basic layout design of the lathe, especially regarding which allocation of the guideway is suitable for the operator and swarf disposal. They classified systematically the movements of the cutting tool and work as shown in Fig. 3-3, where v, u, and a are the cutting, feed, and positioning movements, respectively, in cylindrical and face turning. In analyzing the form-generating movement, furthermore, these alphabetic symbols

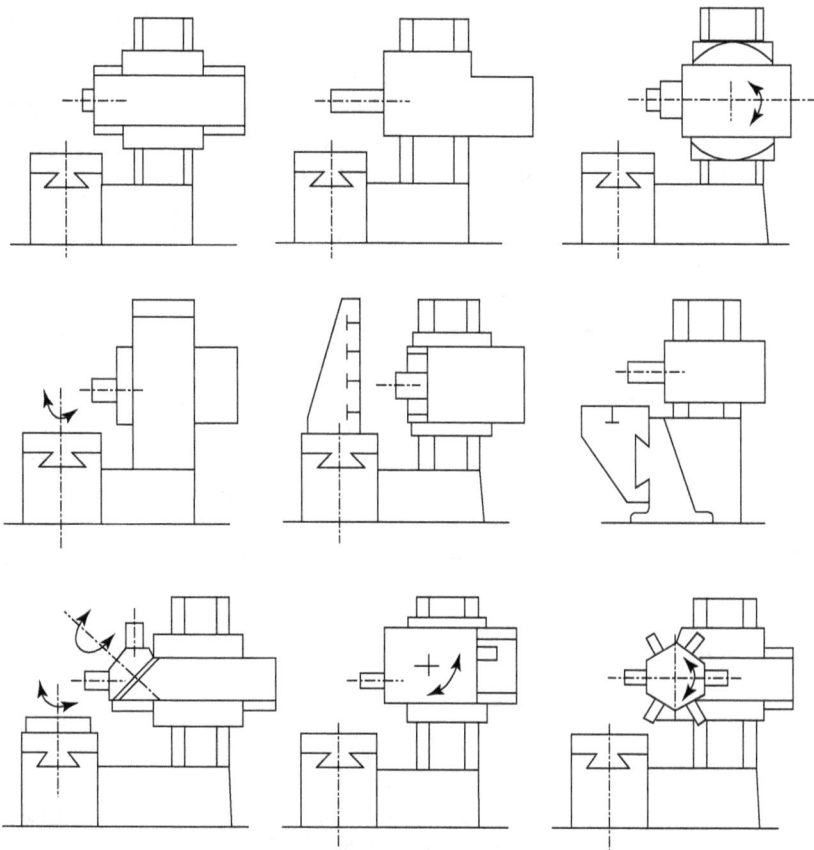

Figure 3-2   Examples of design arrangement of $XOYZ$ (by Vragov).

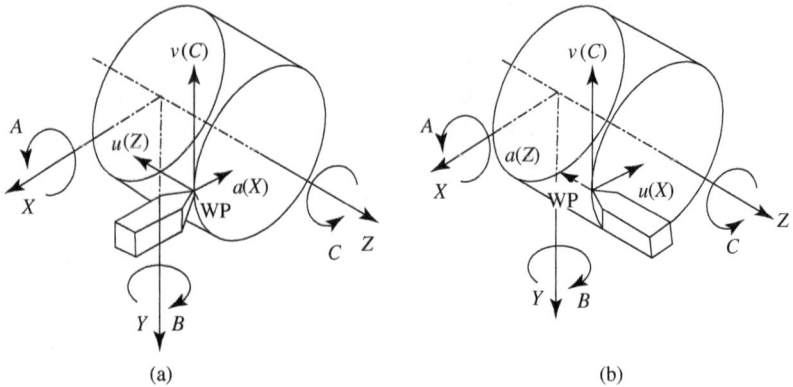

(a)   (b)

Figure 3-3   Relative movements of cutting tool and work: (a) Cylindrical turning and (b) face turning (courtesy of Saljé).

are integrated with the coordinates to clarify their spatial allocation, resulting, e.g., in $a(Z)$, $u(x)$, and $v(C)$ in face turning. As a result, Saljé and Redeker suggested that the functional description enables potential configurations to be brought in relief, some of which are often difficult to envision by the designer. On the basis of such knowledge, furthermore, they asserted that the structural configuration may be varied by the relative allocation of the guideway to the floor and by both the size and the weight of the work to be machined [5]. Figure 3-4 shows some layout drawings and related functional descriptions after all possible variants have been chosen [6], and as can be seen, the description method is very similar to that of Vragov, except that it uses WP instead of O. In addition, the coordinates were determined in accordance with the VDI-Richtlinie 3255, and a solidus (slash) is employed to distinguish the movement of the workpiece branch. More specifically, the analysis of the form-generating movement was carried out together with consideration of a suitable combination of the form entity of work and the cutting tool.

These predecessors have, however, a serious problem: The order of the symbolized movement cannot be fixed absolutely, and thus Ito and colleagues improved the method of Vragov by including obviously the idea of FOF in the functional description, although it is not explicitly displayed. In fact, a machine tool can be represented by ordering the movements of the structural units along the FOF, which starts from the cutting point and flows in the structure through the tool and work. In addition, the slash is allocated at the position from which the force flows out to the floor and distinguishes the kinds of movements, i.e., those belonging to either the tool or the work branch [7]. Figure 3-5 is a reproduction of a description for TC by Saito et al., and afterward they applied

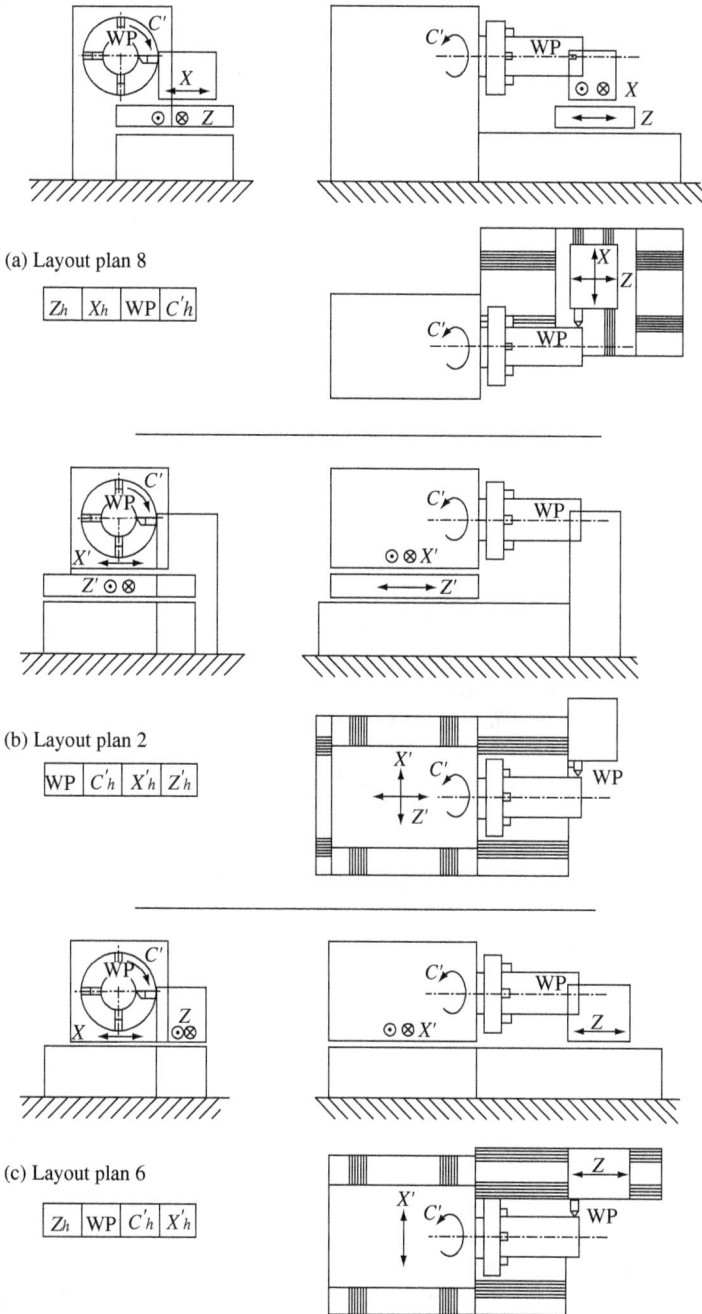

(a) Layout plan 8

Z_h | X_h | WP | C'_h

(b) Layout plan 2

WP | C'_h | X'_h | Z'_h

(c) Layout plan 6

Z_h | WP | C'_h | X'_h

Figure 3-4 Some layout drawings and related functional descriptions for a turning machine (courtesy of Saljé).

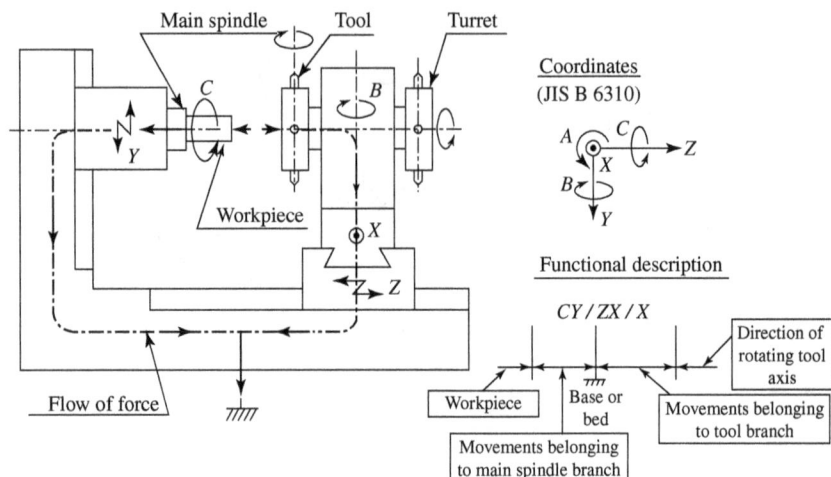

Figure 3-5 Functional description of turning center.

the proposed method to the production of the concept drawing [8] and marketability analyses [9], as will be described in Chap. 4. In Saito and Ito [8], the tool post structure and tool rotational axis can be comprehensively described because of the versatility of the tooling layout in the TC, which is often very dominant for the expected application of the functional description.

As can be readily seen, the functional description appears not to be in direct relation to the modular design; however, it is actually a supplementary design tool with the ease of use derived from its simplicity. To further deepen one's understanding, some marked issues will be discussed in the following within the context of the historical point of view.

In this context, we must touch on those researches carried out in the U.S.S.R. so far. Actually, there have been a considerable number of researches into the functional description following that of Vragov, e.g., those of Aver'yanov, Portman, and Davydov. In Khomyakov and Davydov [10], a new comprehensive description consisting of both the machine and layout codes has been proposed on the basis of those of Vragov and Ito and Shinno [11]. In short, the machine code may include the information block (e.g., model and production year), technological block (e.g., relating to drive ratings, range of spindle speeds, and range of NC system indicators), common parameter block (e.g., cost, dimensions, and weight), and the layout and unit blocks. Consequently, the layout code can eminently represent the basic functional parts of the layout, i.e., the load-carrying components of the machine, which is often identical with those related to the form-generating function, automatic tool changer, and automatic work changer. As can be readily seen, Khomyakov and Davydov tried to amalgamate the functional description with the structural description to enhance the design methodology for the machine tool. Importantly, the earlier achievement performed within the U.S.S.R. has been finally compiled into a book written by Reshetov and Portman [12].

To this end, it emphasizes that the research into the functional description was completed in the mid-1980s, apart from that of Iwata and Sugimura. Figure 3-6 shows the basic coordinate systems used to describe the machine tool and a description example proposed by them [13, 14]. Because of their focus on the application of the functional description to automatic process planning, their method is more logical than those mentioned above. One noteworthy aspect is, furthermore, the incorporation of the shape of the cutting edge into the form-generating movement, where the geometry of the cutting edge can be defined in a perpendicular plane to the direction of the cutting motion. Importantly, the form-generating function is affected by not only the movement function of the machine tool, but also the tool geometry to a larger extent, as can be exemplified by the gang or monolithic formed cutting tool. They have thus conceptualized the form-generating movement by a sweep action of the cutting tool, where the principal cutting movement can produce the form entity, which becomes a geometric form by the travel of the cutting tool within a three-dimensional space. A characteristic feature is to classify the form entity into the three types shown in Fig. 3-7.

To deepen the related knowledge, an interesting trial performed by Yoshikawa [15] will quickly be reviewed. He tried to represent the machine and its units with the bond graph, so that the machine reliability could be analyzed. In the bond graph representation, e.g., the geometric connecting relation of parts within a gearbox can be depicted as in Fig. 3-8, where 8 parts and 11 contracts among them are the nodes and branches, respectively. This bond graph can be regarded as a functional description containing valuable information. For example, there are three kinds of branches, i.e., those for transmitting motion and/or signal, for constraining the freedom of motion, and for fixing; and the compound-connectivity graph and the functional path can be produced from the "part-connectivity graph" by choosing the adequate branch. In due course, we use the concept of functional paths to discuss the structural reliability. In contrast, the bond graph was produced without considering the FOF, and both the path and the subpath are not defined absolutely, but are determined by the designer's arbitrary choice. As a result, there remains something uncertain to determine the path.

Coordinate systems

(a) Shaper

(b) Horizontal type machining center

(c) Lathe

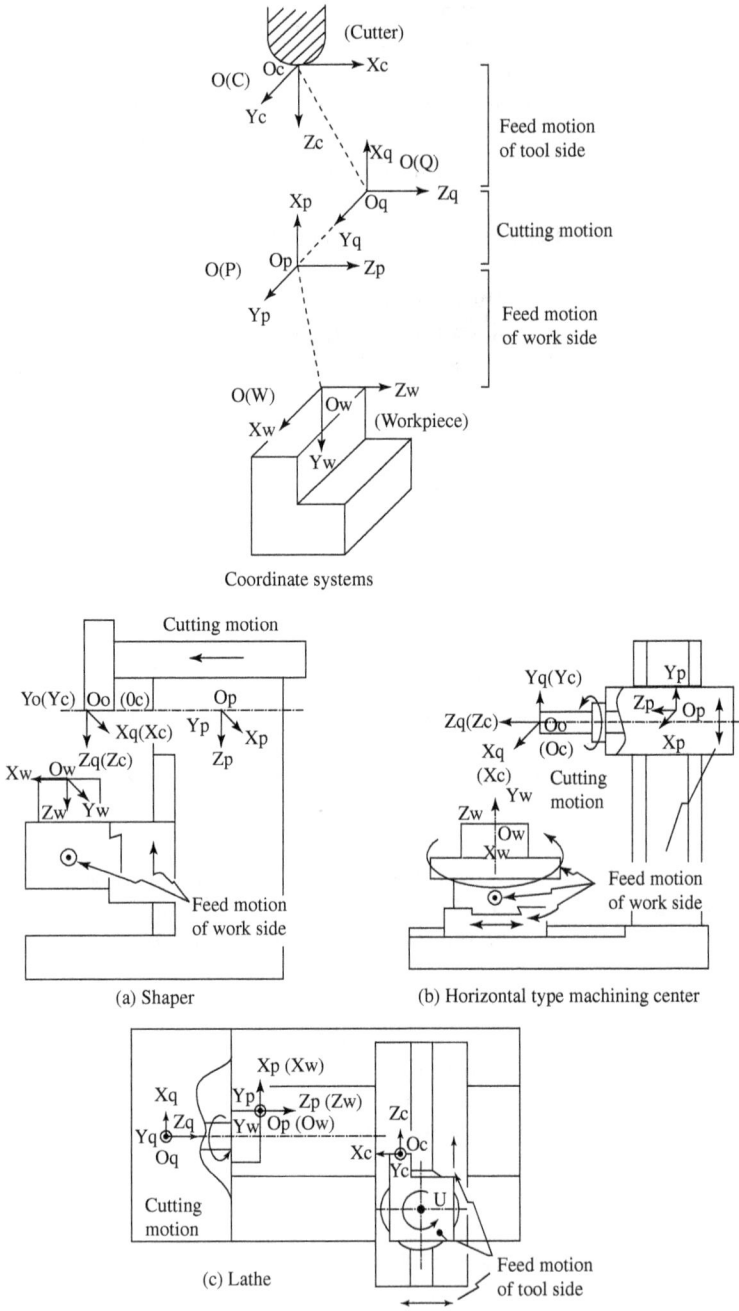

Description for some kinds of machine tools

Figure 3-6 A variant of functional description (courtesy of Iwata).

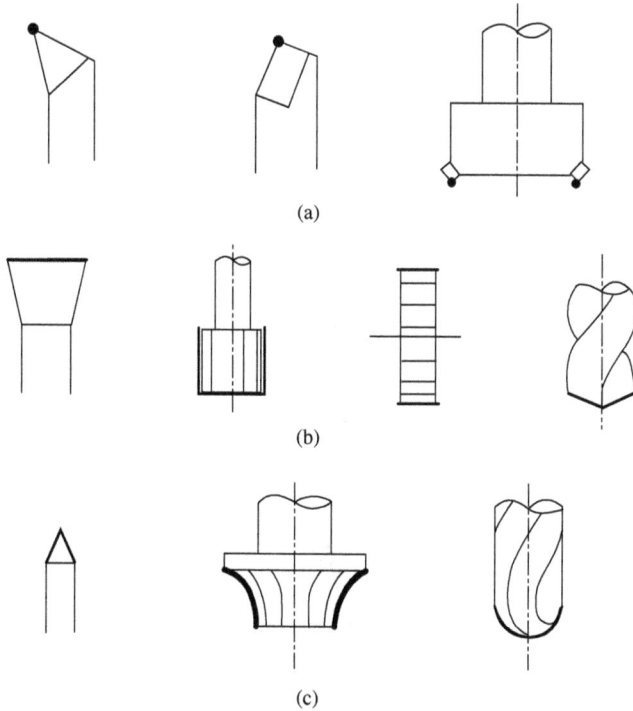

(a)

(b)

(c)

Figure 3-7 Classification of form entities: (a) Pointlike cutting edge; (b) linelike cutting edge; and (c) formed cutting edge (courtesy of Iwata).

3.3 Details of Structural Description

Compared to the functional description, the structural description may be considered as a complicated representation method, and thus the basic requirement for its use is to have deep knowledge about and long-standing experience with the machine tool structure. Because of such limitations, the structural description has not been investigated popularly and is not dominant yet. In fact, the most representative researches are credited to those of Ito and Shinno in the beginning of the 1980s [11, 16]. In their proposed method, the machine tool can be represented with a group of the GT codes affixed to structural units (structural body components) and their allocation within the FOF, resulting in the *structural pattern*. More specifically, a machine tool can be represented by first affixing the corresponding GT code to each unit and then ordering all the GT codes along the FOF. In the description, the GT code and FOF represent the characteristic features of the structural unit and structural configuration, including the bearing condition of the external load, respectively. Figure 3-9 shows a structural pattern of the planner of

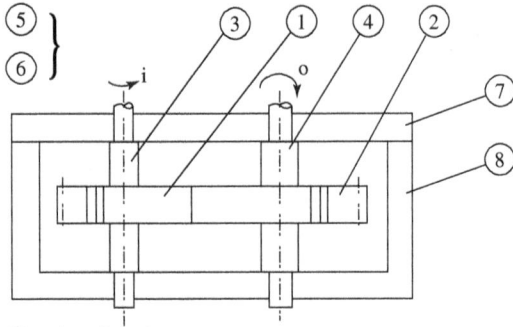

Drawing of gearbox

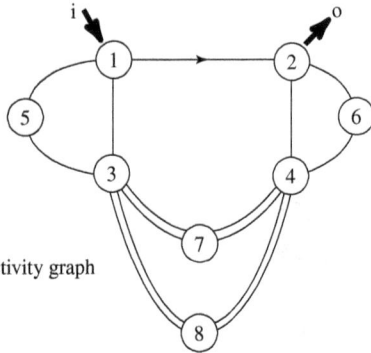

Part-connectivity graph

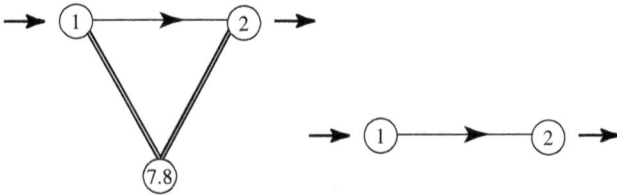

Part-connectivity graph and functional path

Figure 3-8 Bond graph representation of gearbox (courtesy of Yoshikawa).

portal (double column) type. As can be readily seen, the structural pattern appears to be very complicated, although it contains, as enumerated below, much more valuable information than that in the functional description.

1. Fundamental shapes and main functions of structural units

2. Adjacent relationships between both structural units, i.e., inflow and outflow surfaces of force in a structural unit

3. Number of structural units configuring a whole structure and their allocation in three-dimensional space

Flow of force
25082
GT code
(1/2/4)4080
22015
22015
06085
02015
2508(3/4/0)
2401(5/5)
20575
(2/3)408(1/0)

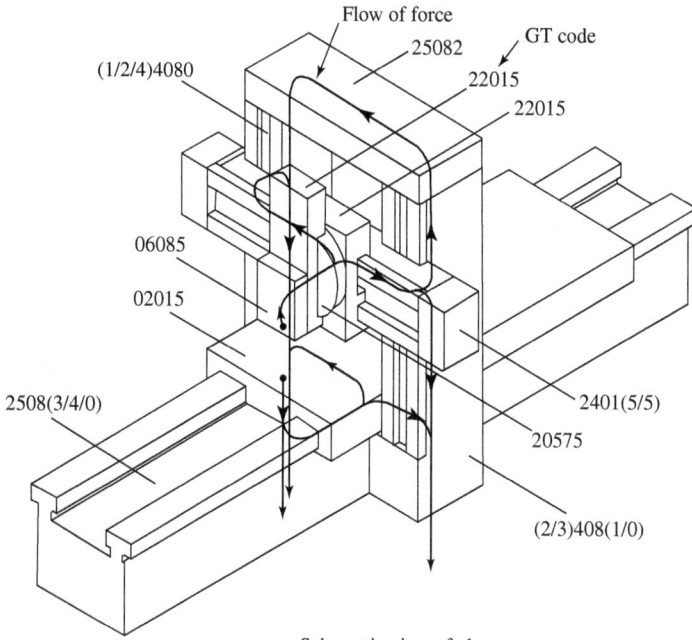

Schematic view of planer

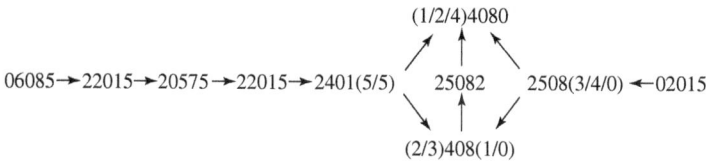

(1/2/4)4080

06085 → 22015 → 20575 → 22015 → 2401(5/5)   25082   2508(3/4/0) ← 02015

(2/3)408(1/0)

Structural pattern

Figure 3-9 Structural description for planer of double column type.

4. Machining point and supporting points of the machine tool, i.e., starting and terminal points in the FOF

It is worth suggesting that the structural pattern corresponds actually to the directed graph in the field of graph theory, showing obviously
the property of each point (vertex) in the graph by the GT code.
Consequently, the structural pattern can be used to evaluate the structural similarities among various kinds, to estimate structural configurations possible to manufacture from a predetermined group of
structural units, to generate a new structural configuration by modifying and deploying a graph, to produce the concept drawing by reconfig-
uring the graph, and so on [16]. These applications can claim as the

leading areas to be carried out by the methodology of modular design, i.e., the principle of adaptation.

In the following, the details of the description and its procedure will be stated.

1. *Description of FOF*. Under the classification of the FOF into the three types, i.e., main, bye, and virtual FOFs, the description rules have been determined as shown in Table 3-3, where each FOF is defined as follows.

   Main FOF: Flowing in the structural body from the machining point of tool branch and flowing out to the factory floor.

   Bye FOF: Flowing in the structural body from the machining point of the work branch and merging, in principle, into the main FOF within a machine.

   Virtual FOF: Diverging from main or bye FOF and stopping at the terminal structural unit. The virtual FOF is not required of the description procedure; however, it could be necessary when the basic layout drawing will be produced from the structural pattern.

2. *Coding of the structural units within the FOF*. Here the GT code of five figures is predetermined as shown in Fig. 3-10 and Table 3-4. In coding, the second, third, and fourth figures represent the shape, main cross-sectional shape, and dominant function of the structural unit, respectively. In addition, the first and fifth figures show the inflow and outflow surfaces in each structural unit of the FOF.

TABLE 3-3   Description Rules of Flow of Force

1. Machine tools to be described are operated in the usual ways.
2. The flow of force starts from the machining point and is terminated at the foundation or at the structural module located just before the foundation.
3. The force in a structure flows in only one direction: it flows from the machining point to the foundation, and it does not flow in a backward direction.
4. In principle, the force flows to the positive direction of the local coordinate (i, j, k) with a right-hand system. The priority of each coordinate axis is in the order of i, j, k; and considering this, the coordinate axes should be determined.
   a. In the case of platelike or beamlike structural module, the i axis is fixed in the perpendicular direction to the BH plane.
   b. In the case of a boxlike structural module, the i axis is fixed in the inflow direction of the force.
5. When the flows of force are overlapped such as is observed in a structure of portal type, the counterclockwise flow is given the priority, and the opposite flow is deleted.
6. When the number of the mainflows of forces in a structure is more than two, the mainflow nearest to the foundation is superior to others.

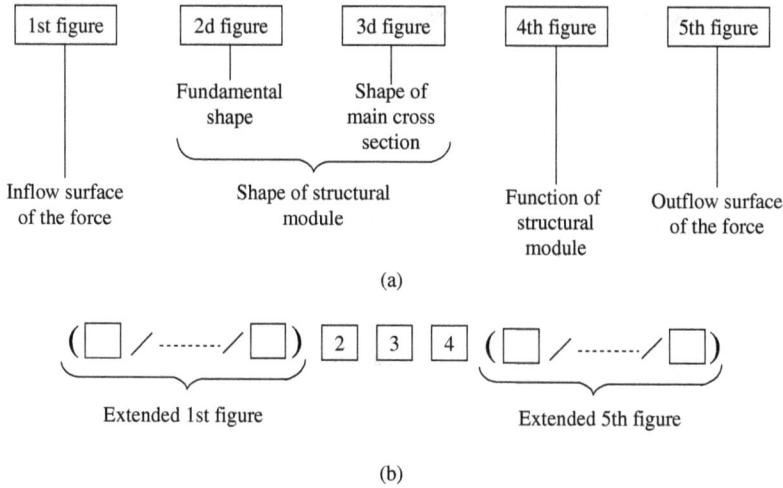

Figure 3-10   Description of structural modules by GT code: (a) Basic coding system for the description of structural module and (b) extended coding system for the description of structural module.

Accordingly, there are multiple inflow and outflow surfaces, and thus first and fifth figures can be extended by adding the slash.

3. *Generation of structural pattern.* The structural pattern can be generated by arranging all the structural units with GT codes in accordance with the FOF. In this case, other representation rules are determined as follows.

   a. The vertices, i.e., structural units, corresponding to the starting points of the main and bye FOFs must be allocated at the far left and right, respectively.

   b. The vertices must be duly connected with the arrow lines, i.e., edges.

In extension of the work of Shinno, Huang and Brandon [17] proposed a further application of graph theory to the modular design of hierarchical type and suggested its effectiveness in classifying the machine tool of present. Apart from the classification of Ito et al., in which the machine tool can be classified systematically by using the functional description [8, 9], in fact, there have been no rational classification systems of the machine tool since the 1960s. We must be aware that the machine tool at present is completely different from that of the 1960s as exemplified by the MC and TC (refer to Chap. 2).

In the work of Huang and Brandon, the method is called a GMRM (*Graphical Machine Representation Model*) and consists of two levels of the module, i.e., the recursive graph (description of individual machine) and attributed graph (set of machine configuration alternatives). In short, the recursive graph can be probabilistically produced from the attributed graph by using the random graph theory.

TABLE 3-4   GT Codes for Describing Inflow and Outflow Planes of Forces

	1st figure and 5th figure	
	Inflow or outflow surface of force	
0	None	
1	1	
2	2	
3	3	
4	4	
5	5	
6	6	
7	Inside surface of the hole its longitudinal axis being normal to the BH plane.	
8	Inside surface of the hole, its longitudinal axis being normal to the HL plane.	
9	Others	

Note: L > H > B

References

1. Stau, C. H., *Die Drehmaschinen*, p. 3, Springer-Verlag, Berlin, 1963.
2. Jäger, H., "Zum Begriff 'Kraftfluß' bei Werkzeugmaschinen," *wt-Z. ind. Fertig.*, 1977, 67: 605–606.
3. Schlesinger, G., *Die Werkzeugmaschinen*, Vol. 2. Springer-Verlag, Berlin, 1936.
4. Vragov, Yu D., "Structural Analysis of Machine Tool Layouts," *Machines and Tooling*, 1972, 43(8): 5–8.
5. Saljé, E., et al., "Eine Systematik für Relativbewegungen bei spanender Bearbeitung in Abhängigkeit von Werkstück und Werkzeug," *ZwF*, 1973, 68(8): 404–408.
6. Saljé, E., and W. Redeker, "Konzipieren von Drehmaschinen—Ein Beispiel für das methodische Konstruieren spanender Werkzeugmaschinen," *Konstruktion*, 1975, 27(6): 240–245.
7. Saito, Y., Y. Ito, and T. Ohtsuka, "Automatisierte Darstellung von Entwurfszeichungen für Werkzeugmaschinen Konstruktionen," *ZwF*, 1980, 75(10): 100 106.

8. Saito, Y., and Y. Ito, "Computer-Aided Draughting System 'ALODS' for Machine Tool Structures," in F. Koenigsberger and S. A. Tobias (eds.), *Proc. of 22d Int. Machine Tool Design and Research Conf.*, Macmillan, 1982, pp. 69–76.

9. Lee, H. S., E. F. Moritz, and Y. Ito, "Functional Description of Machine Tools and Its Application to Marketability Analyses," *J. Engineering Design,* 1996, 7(1): 83–94.

10. Khomyakov, V. S., and I. I. Davydov, "Coding of Machine Layouts for Computer-Aided Design Work," *Stanki i Instrument*, 1989, 60(9): 8–11.

11. Ito, Y., and H. Shinno, "Structural Description of Machine Tools—1st Report Description Method and Some Applications," *Bulletin of JSME*, 1981, 24(187): 251–258.

12. Reshetov, D. N., and V. T. Portman, *Accuracy of Machine Tools*, ASME Press, 1988. (Russian edition was published by Mashinostroenie Publishers, Moscow, in 1986.)

13. Sugimura, N., et al., "An Analytical Approach to Machine Tool Design (Modeling and Analysis of Shape Generation Processes of Machine Tools)," *Trans. JSME*, 1981, 47(418): 793–802.

14. Sugimura, N., K. Iwata, and F. Oba, "Formulation of Shape Generation Processes of Machine Tools," *Preprints of 8th Triennial World Congress*, Aug. 24–28, 1981, 14: 158–163.

15. Yoshikawa, H., "Fundamentals of Mechanical Reliability and Its Application to Computer Aided Machine Design," *Annals of CIRP,* 1975, 24(1): 297–302.

16. Ito, Y., and H. Shinno, "Structural Description and Similarity Evaluation of the Structural Configuration in Machine Tools," *Int. J. Mach. Tool Des.*, 1982, 22(2): 97–110.

17. Huang, G. Q., and J. A. Brandon, "Topological Representations for Machine Tool Structures," in B. J. Davies (ed.), *Proc. of 27th Int. MATADOR Conf.*, Macmillan, 1992, pp. 173–178.

Application of Machine Tool Description to Engineering Design

The structural design must be the synergy of the engineering calculation and computation and the configuration methodology. The former must deal with the analyses of the static, dynamic, and thermal behavior, whereas the latter must deal with the systematic and rational generation of the structural configuration, allocation of traveling function, and so on, resulting in the determination of preferable or optimum structural configuration. In due course, the machine tool description is the utmost kernel of the configuration methodology, and it has been considered to have higher potential for design innovation, as reported elsewhere.

Importantly, the research works conducted so far have validated the higher value and wider application areas of the machine tool description. In fact, the application areas range from the classification of machine tools, through the analysis of the machining function in flexible manufacturing and basic layout production, to the guide for the modular design from the similarity evaluation of structural configuration. In the following, thus, some representative research works will be quickly summarized.

4.1 Application of Functional Description

4.1.1 Classification of machining centers and its application to marketability analysis

Nowadays, the most popular machine tools are the conventional TC and MC, and of these the MC has considerable versatility. Such a versatility observed in the MC crucially appeals to the urgent necessity of establishing a new classification system. For example, the MC can machine the die mold with complicated cavity and monolithic parts of both the aircraft and the computer, resulting in the growing requirement of more versatile types ever than before. In retrospect, there have been no trials or proposals for the classification system of the machine tool since the mid-1960s. With the advent of the NC machine tool, the classification systems used so far became obsolete, reflecting their poor abilities to cope with the amazing development of machine tools. Obviously, we have inconvenience in, e.g., the compilation of national statistics for import and export volumes per kind or type, in which uncertainties are often included, because of the lack of reliable classification system available across the whole world. Within this context, Lee et al. proposed a classification system using the GT code, in which the kernel is to describe the movement function of MCs [1]. Figure 4-1 reproduces an investigated result of typology of the MCs on the market, which comprise more than 200 types from about 50 manufacturers. From Fig. 4-1 we can obtain clearly the important information regarding which type is salable. In fact, the vertical MC with movement of XY/ZC and horizontal MC with movement of BX/ZYC or BXZ/YC are very attractive. In addition, clearly the functional description can contribute to the establishment of a new classification system to a large extent.

Importantly, a simplified description has been used to classify the MC of various types by the VDI-Z-Datenbank, where the coordinate axes used to describe the machine are in accordance with DIN 66217 [2]. In addition, a German manufacturer has claimed the validity of the functional description to classify the grinding machine, as already shown in Fig. 1-24, together with justifying the applicability of each variant for a specified machining requirement [3].

In general, of some methodologies we require the systematic and rational determination of the new product concept from the strategic and tactical points of view including the marketability evaluation, and the functional description provides us with an effective modern tool for such activity. The MCs of German, Japanese, and Korean make were thus classified using the functional description, and then their characteristic features were analyzed, resulting in the obvious comparison of their competitiveness in the world market [4]. Reportedly, these comparisons can particularly provide us with some guidelines regarding how to find the market niche, i.e., given that competition is still sparse in some types of MC as shown in Fig. 4-2. In addition, it is possible to clarify the market strategy of each nation from the comparison. For example, we can characterize the German marketing strategy by the following two facets.

1. Together with the manufacture of MCs in fierce competition in the world market, e.g., X/ZYC type with horizontal single spindle and

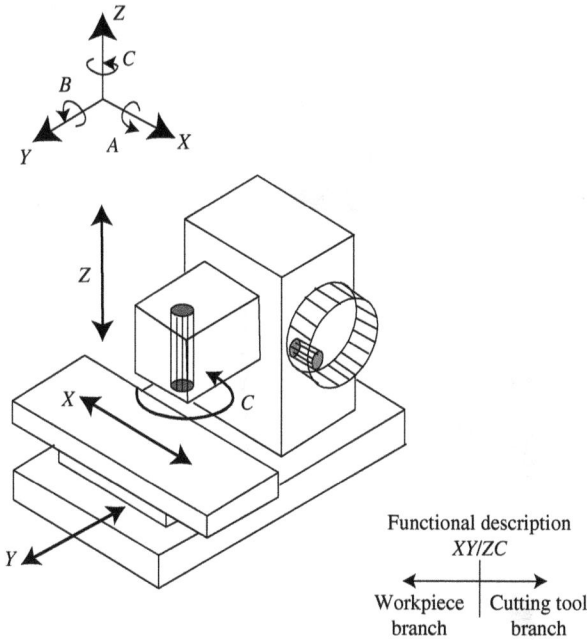

Figure 4-1 Typology of MCs on market around 1985.

Spindle type	Functional description (movement function)							
	XY/ZC	Y/XZC	X/YZC	B/XZYC	BX/ZYC	BXZ/YC	BZX/YC	Others
Single spindle of horizontal type				J 3	J 24 F 16	J 25 F 2	J 1 F 3	J 11 F 8
Single spindle of vertical type	J 44	J 8	J 5 F 2					J 2 F 4
Complex spindle (horizontal & vertical types)		J 4						
Plural spindles of vertical type	J 3							

Note: J: MC of Japanese make; F: MC of foreign make.

XY/ZC type with vertical single spindle, the manufacturers endeavor to develop new types of MC, e.g., vertical MC of /XYZC type. This type appears to satisfy the requirement of the German machine tool market, especially that for application to the FTL. It is very interesting that the Japanese manufacturers tend to respond to such a requirement using the Y/XZC type.

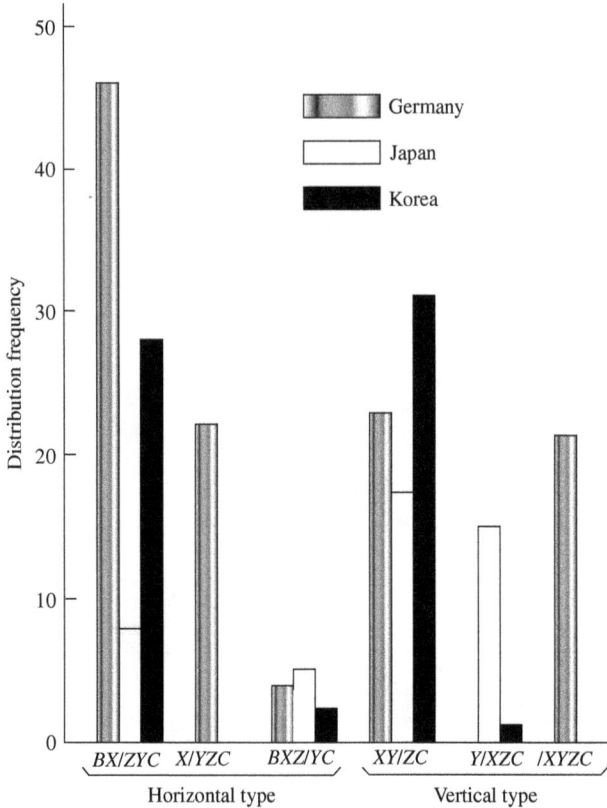

Figure 4-2 International comparison of kinds of MC (single-spindle type).

2. The manufacturers have considered the special-purpose MC as one of the strategic products, because this MC is not "on the shelf" for the Japanese manufacturer.

It is furthermore interesting that the time series–like analysis of MCs made in a nation may imply the changing trends of technology innovation, market strategy, technology inheritance, and so on. For example, Table 4-1 shows the trends of Japanese make MCs in 1993, and by comparing it with Fig. 4-1 we find the following:

1. In general, the same trend as that of 1983 can be observed, i.e., the continued production of both the vertical MC of XY/ZC type and the horizontal MC of BX/ZYC type. It can be deduced that the increase in the sales volume of the vertical MC is attributed to the continuous growth of the molding die sector of Japanese industry.

TABLE 4-1 Typology of Japanese-Make MCs in 1993

Spindle type	Functional description							
	XY/ZC	*Y/XZC*	*BX/ZYC*	*X/YZC*	*Y/ZXC*	*IXYZC* (/*YXZC*)	*BXZ/YC* (*BZX/YC*)	Others
Single-spindle of horizontal type	1		8				5	7
Single-spindle of vertical type	17	15		5	4	6		11

2. The newcomer is the vertical MC of *Y/XZC* type: this trend could be derived from the positive application of the vertical MC to FMC and FTL. In this context, Lee et al. have suggested a shortcoming: The functional description cannot distinguish the differences in certain structural configurations, as shown in Fig. 4-3 due to its simplicity. In retrospect, Vragov already pointed out this problem, taking a description *XOYZ* shown already in Fig. 3-2 as an example in the early 1970s [5]. At present, there is no research being done into this problem.

4.1.2 Analysis of machining function and its application to evaluate compatibility with production systems

Although the description becomes relatively complex, an extended functional description has been proposed to associate the analysis of the machining functions possible within a machine. In the extended description, the

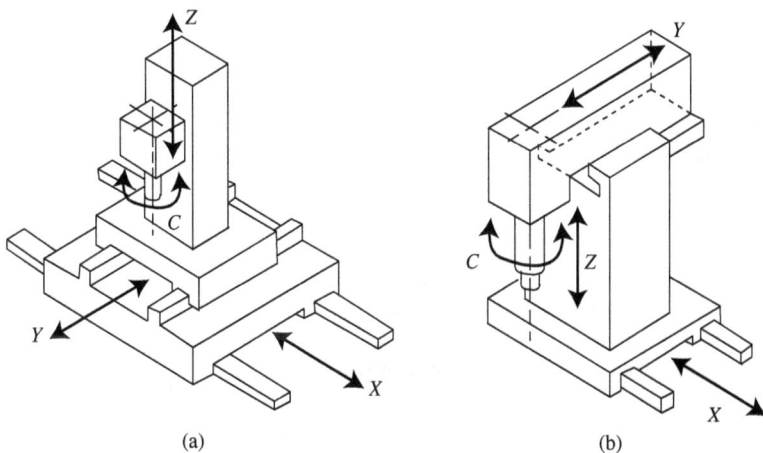

(a)   (b)

Figure 4-3 Structural configurations corresponding to functional description /*XYZC*—certain constraint due to description simplicity.

subscripts and symbols are added to the generalized description such as follows to enlarge and enrich the obtainable information.

1. Kinds of the movement, i.e., principal (e.g., rotational movement in TC), auxiliary (e.g., dividing motion), and complex (e.g., table rotating motion in NC milling machine) movements using the subscripts

2. Multiple form-generating functional units being connected with the same adjacency units using + and ( )

3. Combination of rotational and linear movements, e.g., spindle of quill type in milling head using the symbol ∩ (round hat)

4. Number of tools in the magazine and tool changing motion

In accordance with this description rule, the horizontal boring and milling machine can be represented as shown in Fig. 4-4. Obviously the

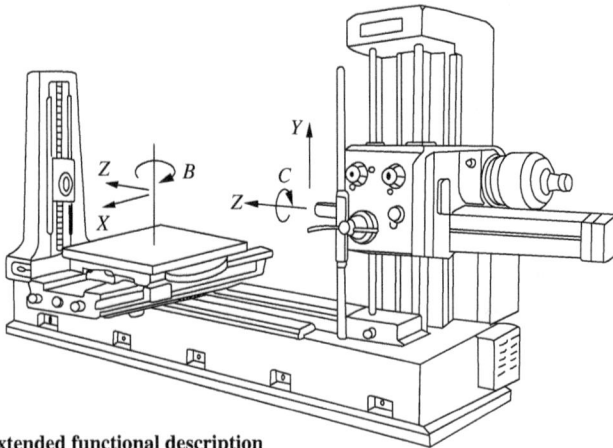

Extended functional description

$B_0 X_1 Z_1 / Y_1 \widehat{Z_1} C_1$

Various machining methods possible

Z_1 / C_1 : Boring
$/ \widehat{Z_1} C_1$: Drilling

$\left. \begin{matrix} X_1/C_1 \\ /Y_1 C_1 \end{matrix} \right\}$: Face milling, groove milling (one-dimensional milling)

$\left. \begin{matrix} Z_1/Y_1 C_1 \\ X_1/Y_1 C_1 \\ X_1 X_1 / C_1 \end{matrix} \right\}$: Contour milling (two-dimensional milling)

$X_1 Z_1 / Y_1 C_1$: Engrave milling (three-dimensional milling)

Suffixes 0 : Auxiliary motion
  1 : Principal motion

Figure 4-4 Extraction of various machining methods from an extended functional description.

extended functional description contains more detailed information for the possible form-generating functions [6]. As shown in Fig. 4-4, the machining procedures, which are carried out by the horizontal boring and milling machine, can be completely extracted from the extended functional description. In fact, the extended description can, in principle, represent more details of the form-generating function by considering the combination of movements and clarifying the direct relation to the well-known machining methods. In addition, a variant of extended description is available for the hypoid gear generator, as shown in Fig. 4-5, in which the dexterous combination of main and auxiliary movement axes is required.

In addition to the analysis of the machining function, the compatibility of the system machine with the FMS and FMC can be rationally verified by using the extended functional description. Importantly, the hardware aspect in the FMS and FMC can be represented with the directed graph, i.e., system pattern, as that of structural description [6]. The machine tool, storage, transfer device, and so on are replaced with the point (vertex) having the corresponding property, and these points are connected to one another according to the material flow. Thus the system pattern enables not only the validity of the machining function to be decided, but also the similarity of the system to be evaluated. Obviously, the system machine can be designed rationally using the system pattern, provided that the property of the point can be given by either the functional or the structural description. Figure 4-6 is a description for MC of column traveling type, which is one of the typical system machines for FTL.

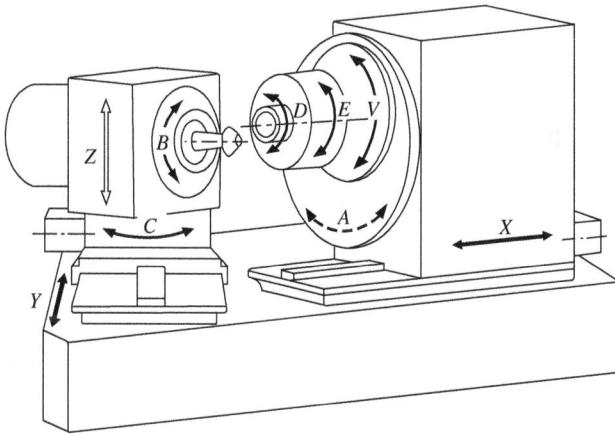

Figure 4-5 A variant of extended functional description—form-generating function in hypoid gear generator.

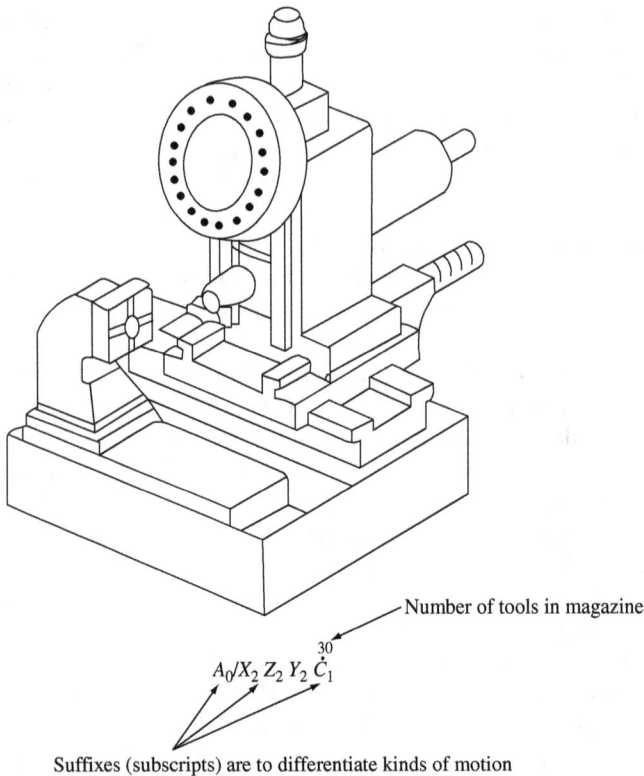

Number of tools in magazine

$$A_0/X_2\ Z_2\ Y_2\ \overset{30}{\overset{\cdot}{C}_1}$$

Suffixes (subscripts) are to differentiate kinds of motion

Figure 4-6 Description of MC of column traveling type (MX3, Ikegai Iron Works make, refer to Fig. 1-4).

4.1.3   Automated generation of concept drawing

The functional description can be applied to the design methodology to a certain extent, and a representative application is to produce the basic layout drawing when inputting a group of parts to be processed. The concept and layout designs are very tedious work in order to find a suitable solution compatible with the machining requirements of the client, because the design typically is one of the ill-defined problems. In machine tool design, the closed-loop and "chicken and egg" features are extremely prominent. The functional description enables such a problem to be solved with less difficulty, as proposed by Saito et al. [7]. In their proposal, the basic layout design can be carried out by inputting either the shape and size of the work or the machining method in accordance with the manufacturing requirements of the user.

Figure 4-7 reproduces a design system to produce the basic layout drawing of the TU, in which the input is either the machining process

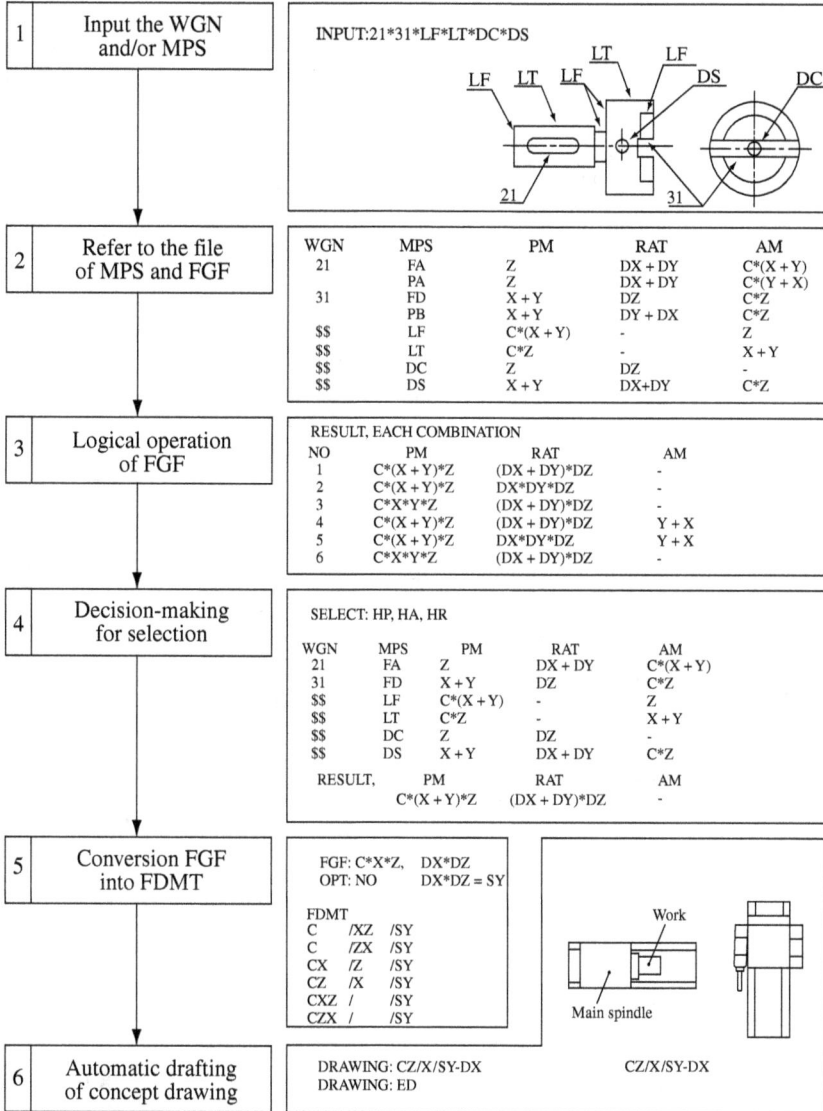

1	Input the WGN and/or MPS

INPUT:21*31*LF*LT*DC*DS

2	Refer to the file of MPS and FGF

WGN	MPS	PM	RAT	AM
21	FA	Z	DX + DY	C*(X + Y)
	PA	Z	DX + DY	C*(Y + X)
31	FD	X + Y	DZ	C*Z
	PB	X + Y	DY + DX	C*Z
$$	LF	C*(X + Y)	-	Z
$$	LT	C*Z	-	X + Y
$$	DC	Z	DZ	-
$$	DS	X + Y	DX+DY	C*Z

3	Logical operation of FGF

RESULT, EACH COMBINATION

NO	PM	RAT	AM
1	C*(X + Y)*Z	(DX + DY)*DZ	-
2	C*(X + Y)*Z	DX*DY*DZ	-
3	C*X*Y*Z	(DX + DY)*DZ	-
4	C*(X + Y)*Z	(DX + DY)*DZ	Y + X
5	C*(X + Y)*Z	DX*DY*DZ	Y + X
6	C*X*Y*Z	(DX + DY)*DZ	-

4	Decision-making for selection

SELECT: HP, HA, HR

WGN	MPS	PM	RAT	AM
21	FA	Z	DX + DY	C*(X + Y)
31	FD	X + Y	DZ	C*Z
$$	LF	C*(X + Y)	-	Z
$$	LT	C*Z	-	X + Y
$$	DC	Z	DZ	-
$$	DS	X + Y	DX + DY	C*Z

RESULT,	PM	RAT	AM
	C*(X + Y)*Z	(DX + DY)*DZ	-

5	Conversion FGF into FDMT

FGF: C*X*Z,   DX*DZ
OPT: NO   DX*DZ = SY

FDMT
C	/XZ	/SY
C	/ZX	/SY
CX	/Z	/SY
CZ	/X	/SY
CXZ	/	/SY
CZX	/	/SY

Main spindle

6	Automatic drafting of concept drawing

DRAWING: CZ/X/SY-DX   CZ/X/SY-DX
DRAWING: ED

(a)   (b)

WGN : Workpiece geometry number   FGF : Form-generating function
MPS : Machining process symbol   FDMT : Functional description of machine tool

Figure 4-7   Flowchart for producing concept drawing: (a) Operational steps and (b) example of execution.

MPS	LT	LF	LP	LC	LK
PM	C*Z	C*(X+Y)	C*Z*(X+Y)	C*(X+Y)+C*Z	C*(X+Y)
RAT	-	-	-	-	-
AM	X+Y	Z	-	Z*(X+Y)	Z*(Y+X)

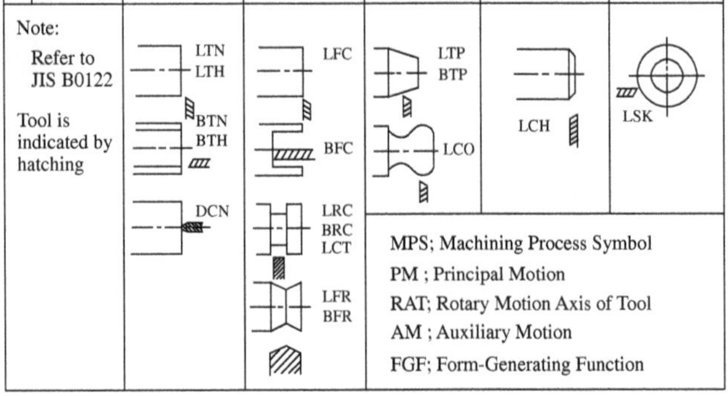

Note:

Refer to JIS B0122

Tool is indicated by hatching

LTN, LTH, BTN, BTH, DCN, LFC, BFC, LRC, BRC, LCT, LFR, BFR, LTP, BTP, LCO, LCH, LSK

MPS; Machining Process Symbol
PM ; Principal Motion
RAT; Rotary Motion Axis of Tool
AM ; Auxiliary Motion
FGF; Form-Generating Function

(a)

MPS	FA	FB	FC	FD	FE	FF	FG	FH	FI	FJ
PM	Z	Z	X+Y	X+Y	X+Y	C	C	C*(X+Y)	C*(X+Y)+X*Y	C*Z
RAT	DX+DY	DX+DY	DX+DY	DZ	DZ	DX+DY	DZ	DX+DY	DZ	DX+DZ
AM	C*(X+Y)	C*X*Y	C*Z*(Y+X)	C*Z	C*Z*(Y+X)	Z*(X+Y)	Z*(X+Y)	Z	Z	X+Y

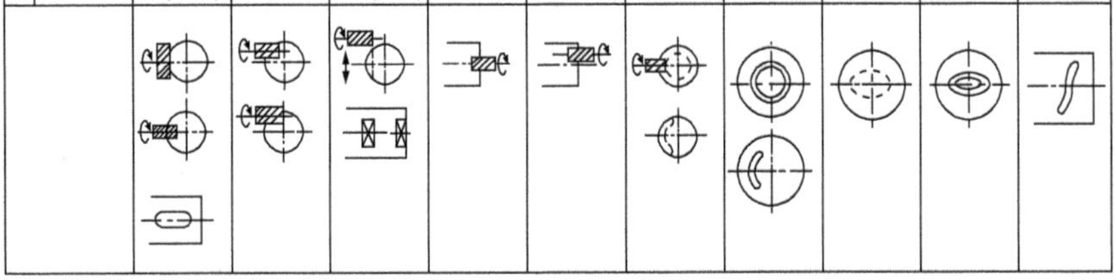

(b)

Figure 4-8 Input data by machining process symbol—logical description of FGF: (a) Turning [L] and (b) milling [F].

symbol (symbol in process planning), as shown in Fig. 4-8, or the workpiece geometry number (pattern number), as shown in Fig. 4-9. In these figures, the symbols • and + mean AND and OR in the logic operation. For example, *X* + *Y* means that the form-generating movement can be realized using either *X* or *Y* movement. As can be readily seen, the basic layout drawing can be produced by referring to the machining method symbol, from which the form-generating movement necessary can be extracted, and then carrying out the very simple logic operation so as to determine the minimum combination of necessary movements. More specifically, the procedure to produce the layout drawing is as follows.

1. Input of the necessary information, using the machining method symbol in principle and also the pattern number in special cases after referring to the part drawing.

2. Reference to the data file for comparing the MPS (*machining process symbol*) to the FGF (*form-generating function*), which was prepared beforehand.

WGN	21 (Workpiece geometry number)			
Workpiece geometry 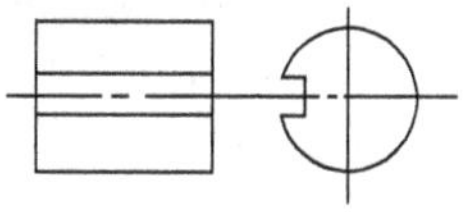				
MPS	FA	(FB)	(FJ)	PA
PM	Z	Z	C*Z	Z
RAT	DX + DY	DX + DY	DX + DY	DX + DY
AM	C*(X + Y)	C*X*Y	X + Y	C*(Y + X)

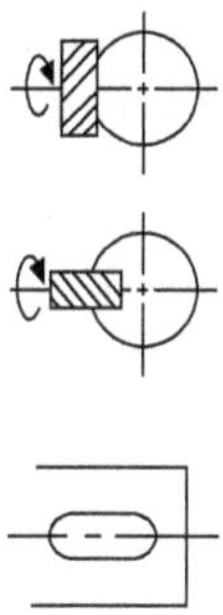

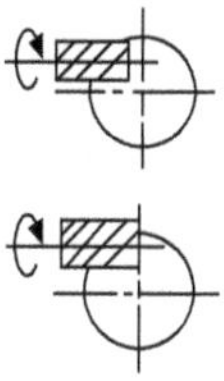

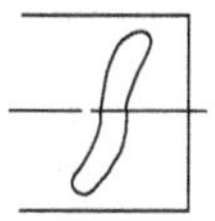

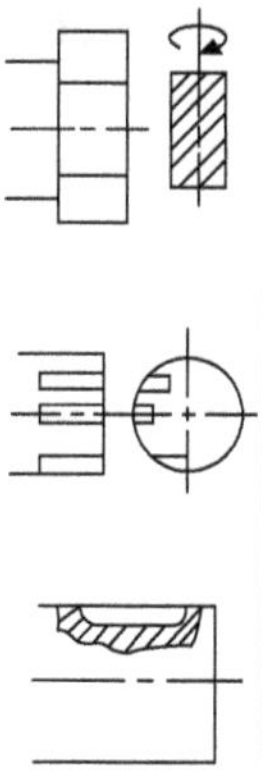

Figure 4-9 Input data by workpiece geometry number—relationships between workpiece geometry and machining process.

3. Logic operation of all the FGFs to be performed by a machine being planned.

4. Decision making with special respect to the choice of the preferable design output from the result of logic operation to minimize the movement axes in FGF.

5. Conversion of FGF to the functional description of the machine tool.

6. Automated drafting of concept drawing.

The core of the design system proposed is the evaluation of the obtained combinations according to certain constraints, e.g., allocation of minimum number of functions to main movement. In addition, the system can take into account the rotational axes of the cutting tool by using the symbols *DX, DY,* and *DZ*. After the final combination is obtained, the necessity is how to determine the work and tool branches.

In short, the concept and layout design can be performed by even immature designers using the functional description, data file to convert the machining method symbol to FGF, and logic operation, although the dimensional specifications cannot be given. In this context, it is worth considering the research of Saljé and Redeker [8], which can deal with the generation of a group of the machine tools available for the manufacturing system design. Figure 4-10 reproduces a design flow of a manufacturing system, and as can be seen, Saljé and Redeker proposed a forerunning idea similar to that of Saito and Ito [7]. In addition, the above-mentioned idea is very convenient for people in the sales department, because they can display the image of the machine promptly in front of the client.

Importantly, as a result of information exchange through private letters together with leapfroglike competition to research work conducted by Ito and Saito, Redeker developed a computer-aided methodology of structural configuration in the concept design stage [9]. In the developed methodology, the following design specifications can be considered.

1. Shapes, dimensions, and tolerances of both the blank and the finished part

2. Shapes and dimensions of the tool

3. Clamping methods and concerns

4. Form-generating movement of the machine

Work description to be machined	Analyses of machining sequences possible	Choice of desirable order of guideway
Disintegration of work into machining entities	Choice of desirable machining sequence	Allocation analyses of guideway to floor
Analyses of machining methods for entities possible	Determination of work holding methods necessary	Choice of desirable allocation of guideway
Choice of desirable machining method for entity	Analyses of form-generating movements possible as per holding method	Analyses of structural body configurations
	Choice of desirable movement	Choice of desirable structural configuration
Combination of machining methods possible	Order of guideways to movements	Concept generation of machine tools
Choice of desirable machining methods for work	Analyses of order combinations in guideways possible	*Concept generation of manufacturing systems*

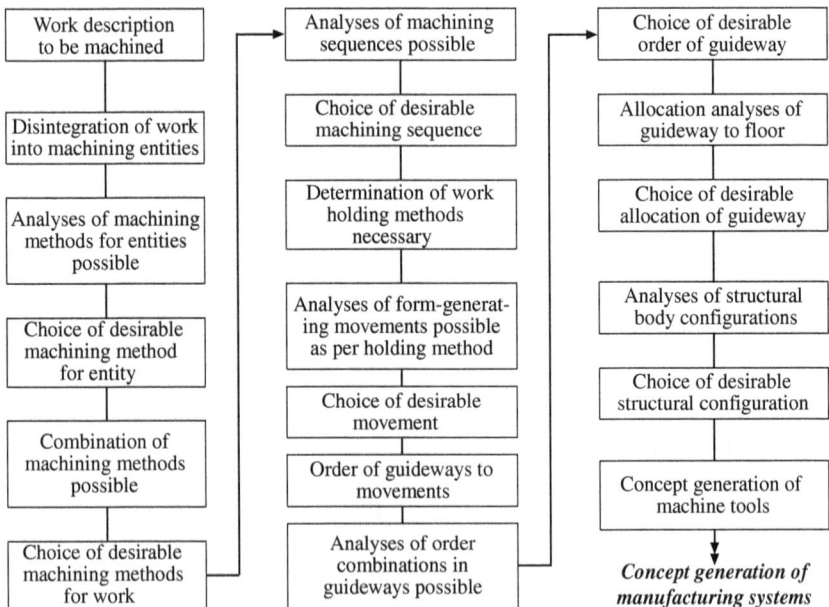

Figure 4-10 Concept design processes for manufacturing systems.

The allowable structural configuration can be ascertained after judging the reality of all the possible form-generating movements to various extents, such as shown for the case of three-axis controlled turning machine in Fig. 4-11. In fact, the corresponding functional descriptions for practical use are shown in the column "Solution no.," and some layout drawings are also shown in Fig. 4-12. In conducting such a judgment, as shown in Fig. 4-13, Redeker employed the idea of FOF as similar as that of Ito together with longstanding knowledge obtained from research into the functional analysis of form-generating movements. More specifically, Redeker suggested the necessity of more detailed representation to apply the functional description to the CAD. Thus he proposed, e.g., the symbols = and O to represent the primary form-generating movement and $\neq$ and ϕ to represent the adjustment movement. In addition, he proposed to use, e.g., the catalog of slideways shown in Fig. 4-13. The catalog includes the leading dimensions, if necessary, to assist the computerized drafting, expecting to extend it to the database of modules with dimensional specifications.

Scheme no.	Workpiece			(Cutting point)	Tool			Solution no.
1				WP	C	Z	X	1
2				WP	C	X	Z	2
3			X'	WP	C	Z		3
4			Z'	WP	C	X		4
5		X'	Z'	WP	C			5
6		Z'	X'	WP	C			6
7			~~C'~~	~~WP~~	~~Z~~	~~X~~		
8			~~C'~~	~~WP~~	~~X~~	~~Z~~		
9		~~X'~~	~~C'~~	~~WP~~	~~Z~~			
10		~~Z'~~	~~C'~~	~~WP~~	~~X~~			
11	~~X'~~	~~Z'~~	~~C'~~	~~WP~~				
12	~~Z'~~	~~X'~~	~~C'~~	~~WP~~				
13				WP	Z	C	X	7
14				WP	X	C	Z	
15			X'	WP	Z	C		8
16			~~Z'~~	~~WP~~	~~X~~		~~C~~	
17				~~WP~~	~~X~~	~~Z~~	~~C~~	
18				~~WP~~	~~Z~~	~~X~~	~~C~~	
19	~~X'~~	~~C'~~	~~Z'~~	~~WP~~				
20	Z'	C'	X'	WP				9
21		~~C'~~	~~Z'~~	~~WP~~	~~X~~			
22		C'	X'	WP	Z			10
23	C'	Z'	X'	WP				11
24	C'	X'	Z'	WP				12

Figure 4-11 Concept schemes for turning machines with three-degree of movement freedom (courtesy of Saljé).

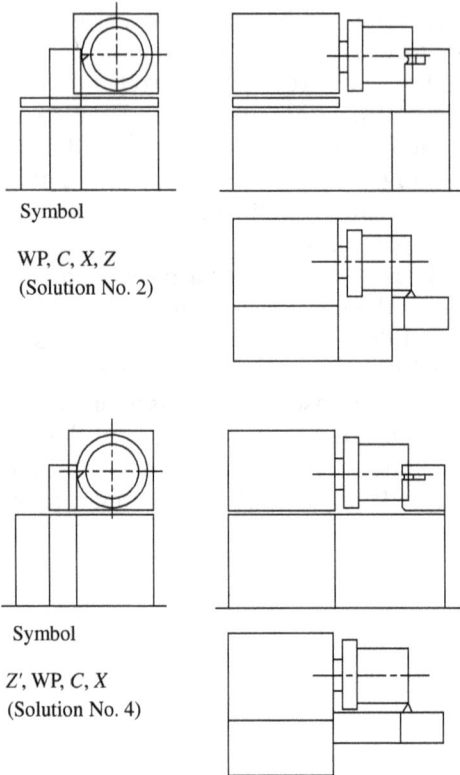

Symbol

WP, *C, X, Z*
(Solution No. 2)

Symbol

Z', WP, *C, X*
(Solution No. 4)

Workpiece length: 500 mm
Diameter: 550 mm

Figure 4-12 Examples of layout drawings (courtesy of Saljé).

In due course, Saljé compiled later their works into a private report [10], which consists of the following.

1. A firsthand view of the research works into the functional description with dimensional analysis

2. Methods to distinguish the kind of form-generating motion using the appropriate subscripts

3. A CAD system for the basic layout drawing of the guideway structure with the association of the catalog of modular type for guideways

He exemplified a total view of description for some quinaxial-controlled milling machines, as shown in Fig. 4-14. More specifically, each item can be detailed as follows.

1. The machining space can be determined on the basis of the functional description together with direct consideration of the dimensional

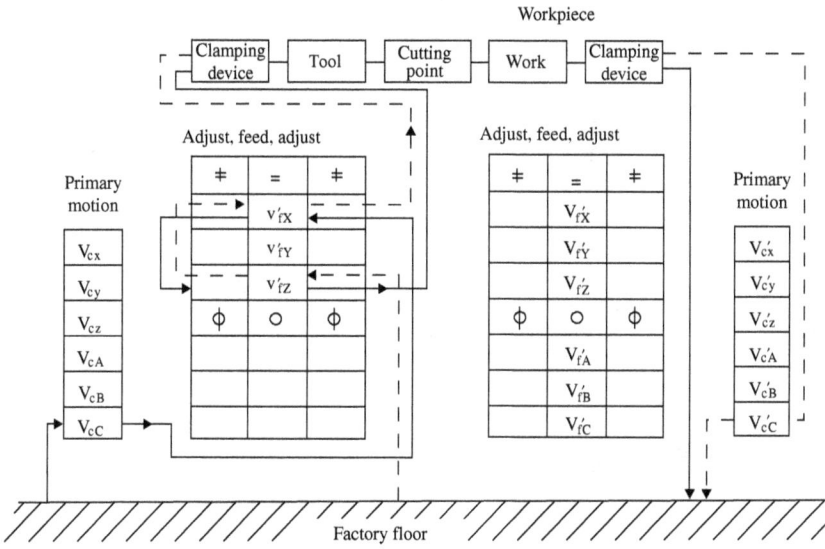

(a)

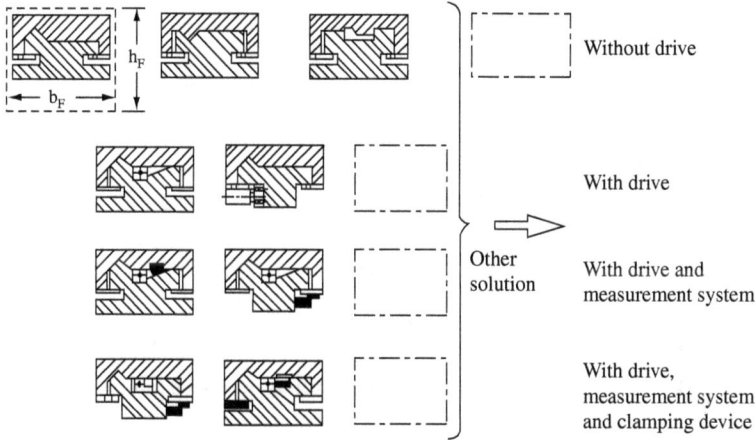

(b)

Figure 4-13 Detailed functional description aiming at CAD: (a) General concept representation and (b) catalog of guideways (courtesy of Saljé).

specifications of the work and tool, as shown in Fig. 4-15, and also indirectly some attention is paid to the production cost. Importantly, the functional description of Saljé and Redeker can distinguish the work and tool branches, as well as differentiate the primary and fixed motions, depth setting and tool approach motions, and furthermore the reset motions.

(a)

| Nr102 | Y | Z | A | CS | WP | C' | X' |

(b)

| Nr116 | Y | X | Z | A | CS | WP | C' |

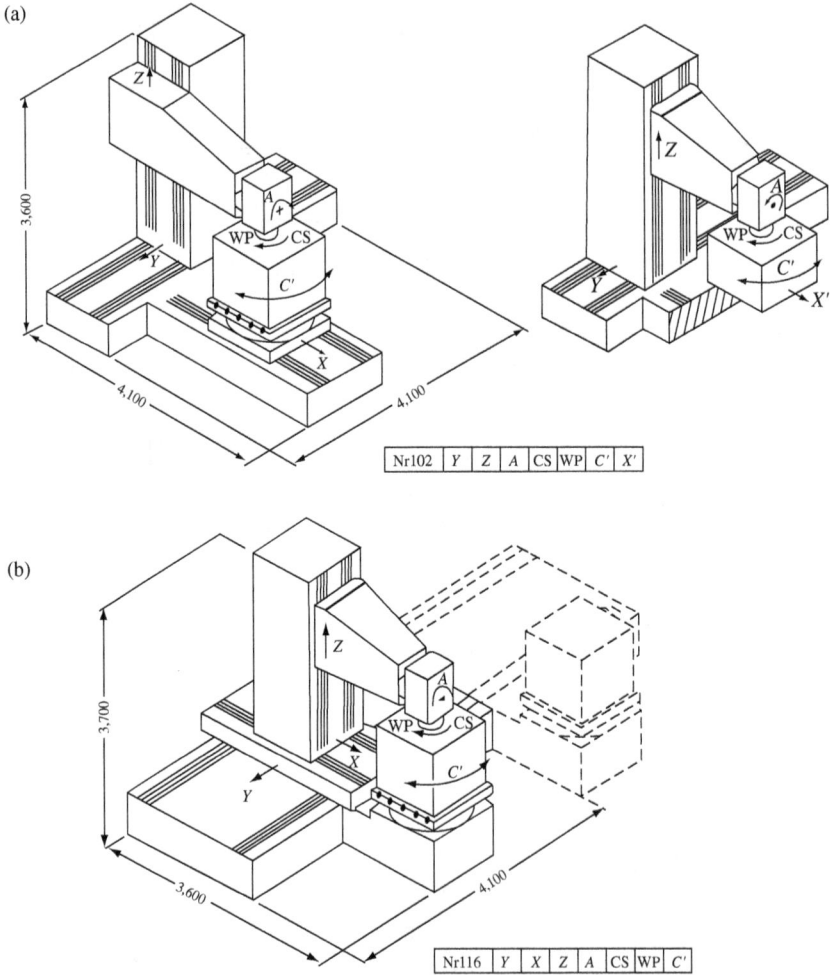

Figure 4-14 Design results of quinaxial controlled milling machines (courtesy of Saljé).

2. When the arbitrary exchangeable degree of freedom is three in the traveling function, there are 24 obtainable structural configurations. In due course, the possible variants to use, as already shown in Fig. 4-11, can be chosen to a various extent under certain constraints, e.g., arrangement and location of the guideway in relation to both the work traveling and the machining space.

Following these earlier works, Ratchev and Gindy [11] conducted a research into the rationalization of the choice process of the preferable modular-configured machine tool. The process commences from the specifications of machining requirements for a set of the components,

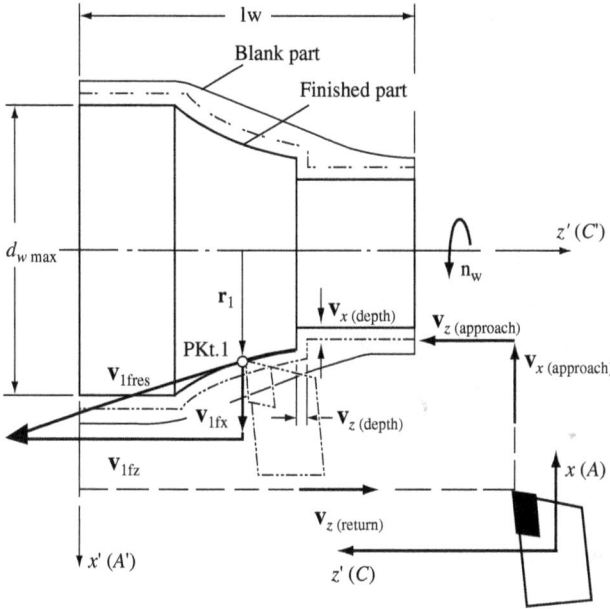

Figure 4-15 Advanced functional description with leading dimensions (courtesy of Saljé).

through the comparison between the form-generating scheme extracted from the machining requirement and the machine tool configurations permissible, to the matching procedure based on linguistic theory. In addition, the structural configuration can be generated from the combination of the tool and workpiece subpatterns, and as can be readily seen, the flow for the configuration and selection of the structural pattern is similar to that of Saito and Ito. A marked feature is considered as a use of the syntactic structure of language for describing the machine tool so as to simplify the matching process between a set of machining requirements and the feasible processing machine tools of various configurations.

In the basic layout design, at burning issue is a choice of the preferable or optimum structural configuration; however, there have been fewer research work in this area. In this context, Herrmann has proposed a decision methodology using the *morphological method* in the biology sphere. Consequently, an optimum structural configuration can be determined from the possible number of basic layout drawings, i.e., the solution field (das Lösungsfeld). In the proposed method, first a solution field $L_f = \Pi\, ei$ (ei: solution entities, $i = 1$ to m) is given. Then in full consideration of the object-oriented evaluation attributes with weighing factor to each solution drawing, a preferable or an optimum structure can be determined. In fact, Herrmann applied this method to

the NC lathe by choosing both the inclination angle of guideways ranging from horizontal, through slant, to vertical types and the relative location of the guideway to the spindlestock as the solution entities. In
addition, he considered the following objective-oriented attributes [12].

1. Swarf dropping ability

2. Approachability to work

3. Approachability to tool head

4. Loading and unloading capability by crane

5. Possibility to allocate many tool heads

6. Possibility to cover machining space with safety panel

7. Observation of machining space

8. Influence of self-weight of work to machining accuracy

4.1.4   Estimation of assembly accuracy in design stage

As an extensive application of the functional description, the assembly
accuracy of the machine tool can be evaluated, at an early design stage,
in full consideration of the position errors of the structural body components located within an FOF. In fact, the assembly accuracy results
in the relative variation within the coordinates of the machining point [13].
In this case, the layout pattern of structural body components, i.e.,
roughly the kinds and types of the machine, affects also the assembly
accuracy of the machine tool, because of changing susceptibility of the
error in each structural body component to the machining accuracy.

Assuming the jth structural body component within a machine to be
in connection with the other at a point O_j, which is taken as the origin
of coordinates, the errors of its position can be characterized by the
vector $\{\Lambda_j\} = \{\lambda_{jX}, \lambda_{jY}, \lambda_{jZ}, \Psi_{jX}, \Psi_{jY}, \Psi_{jZ}\}^T$, where λ_{ji} and Ψ_{ji} ($i = X, Y, Z$)
are linear deviations of the point O_j, and rotation of the structural body
component about the axes passing through this point.

In short, the vector $\{\delta_j\}$ of deviation components at the machining
point caused by $\{\Lambda_j\}$ can be written as

$$\{\delta_j\} = [T_j]\{\Lambda_j\} \qquad (4\text{-}1)$$

where $[T_j]$ is the transfer matrix.

In due course, Moriwaki and his colleagues applied their description
method to the kinematics error analysis and structure generation. In
short, the total relative kinematics error at the machining point can be
computed as the sum of transfer errors at each traveling and rotational
movement concerned with form generation by the matrix operation [14].

In this computation, a further problem is to identify the leading errors within the traveling and rotational movements depending on the kinds and types of machine tool so as to simplify the computation. In retrospect, this transfer matrix operation was used to estimate the cutter location in the universal head in the traditional milling machine in the 1960s. In addition, Yokoyama and his colleagues conducted a similar research into the generation of the machine structure [15].

4.2 Application of Structural Description

Although the functional description guarantees the ease of use, it is difficult to provide the detail of the structural configuration, apart from that of Saljé and Redeker, together with the dimensional specifications. To overcome this disadvantageous feature, the structural description is recommended rather than the functional description. Actually, we need to have deep (tacit) knowledge about the machine tool structure and to manage some tedious work when using the structural description. In contrast, the structural description contains more design information than the functional description: The structural description has higher potential to establish the methodology for the structural design of machine tools.

To understand such a potential, a simple application will be stated below in which the structural configuration is classified on the basis of qualitative structural pattern. In this case, the vertices allocated at the starts, branches, and terminal are stressed, and as a result, three representative structural patterns can be observed in Fig. 4-16. Pattern A, i.e., open type, has greater flexibility for producing the structural configuration,

Type	Prominent features of structural pattern		Example of machine tools belonging to each type of structural pattern
A	Open type (Line arrangement type)	 o—•—o	Gear shaper, slotter, surface grinder, internal grinder, face turning machine, table-type horizontal boring and milling machine (in boring), radial drilling machine (with square table)
B	Closed type	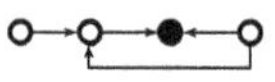 o—o—•—o	Upright drilling machine
C	Double closed type	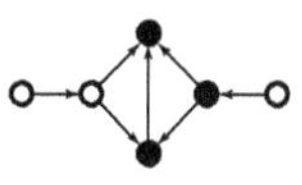 o—o—o	Large-size machine tools with double column type, such as planer, vertical turning machine, planomiller, bedway grinder, vertical boring and milling machine

Figure 4-16 Three representative structural patterns in machine tools.

resulting in a considerable number of variants as shown together in Fig. 4-16, although the structural stiffness is not so high. In contrast, pattern C, i.e., double closed type, shows completely opposite behavior.

In fact, there have been a considerable number of research works to exemplify the higher potential of the structural description. Such research work range from the similarity evaluation of the structural configuration for investigating the possibility of modular design to the variant and free designs of the machine tool.

4.2.1 Similarity evaluation of structural configuration—availability constraints of modular design

The structural pattern can provide us with valuable leading information as follows.

1. Fundamental shape of each structural module

2. Leading function of each structural module

3. Adjacency relationships between both modules

4. Starting and terminal vertices within FOF

5. Total number of structural modules

6. Pattern of FOF

On the basis of these information, the structural similarities of both machine tools can be calculated, using both the rates of commonness and pattern similarity, as typically proposed by Ito and Shinno [16].

More specifically, the *rate of commonness* can be defined as the relative value of the identical to whole numbers of the structural modules between both structural patterns. Then it can be calculated by using the information for the fundamental shape of each structural module, leading function of each structural module, and total number of structural modules in the structural pattern.

By assuming a set to be a machine tool as a whole, the rate of commonness can be represented with a graph such as shown in Fig. 4-17. Then, by defining $|X_s|$ and $|Y_s|$ as the kinds of structural modules in sets X_s and Y_s, respectively, and after eliminating the duplicate structural modules in both sets, structural modules in both sets are in one-to-one correspondence, as shown in Fig. 4-17. Given that one structural module, mathematically called the vertex, in the set $|X_s|$ cannot be connected with more than one structural module in the set $|Y_s|$, the rate of commonness S_r can be written as

$$S_r = (|R(X_s)| + |R(Y_s)|)/(|X_s| + |Y_s|) = 2|R(X_s)|/(|X_s| + |Y_s|)$$
$$= 2|R(Y_s)|/(|X_s| + |Y_s|)   (0 \leq S \leq 1)   (4\text{-}2)$$

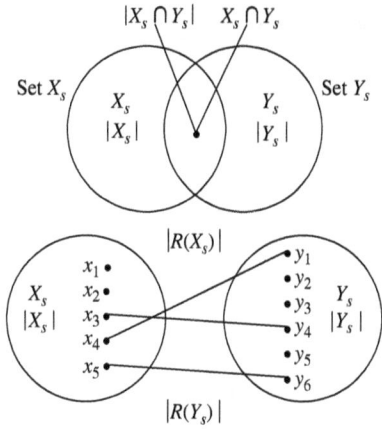

Figure 4-17 Definition of rate of commonness.

where $|R(X_s)|$ is the number of vertices in $|Y_s|$ connecting with those in $|X_s|$, $|R(Y_s)|$ is the number of vertices in $|X_s|$ connecting with those in $|Y_s|$, and in general $|R(X_s)| = |R(Y_s)|$.

In the calculation, the evaluation criteria are as follows, when the second, third, and fourth figures within the GT codes for both structural modules can be written as (B_s, C_s, D_s) and (B'_s, C'_s, D'_s).

1. When $B_s = B'_s$, $C_s = C'_s$, and $D_s = D'_s$, both modules are identical, and the evaluation value is unit.

2. When $B_s = B'_s$, $C_s \neq C'_s$, and $D_s = D'_s$, the evaluation value is 0.5

3. In other cases, the evaluation value is nil.

Actually, in the evaluation procedure, the reference matrix is produced by allocating the vertices of both sets to the line and row, and by determining the (i, j) component using the evaluated value obtained from the above-mentioned criteria, accordingly, the pattern similarity can be calculated by using the information for the adjacency relationships between both modules, starting and terminal vertices within FOF, and pattern of FOF in the structural pattern. Considering the structural pattern is that of the directed graph, the pattern similarity yields to the identical evaluation of a couple of adjacency matrices derived from both structural patterns. In fact, the calculation procedure is as follows.

1. Arrangement of the structural pattern on the disposition plane [X, Y], using the converse directed graph and rooted-directed tree, as shown in Fig. 4-18.

2. Determination of the arranging order of [X, Y] coordinates on the row and column of the adjacency matrix. As shown on the far left

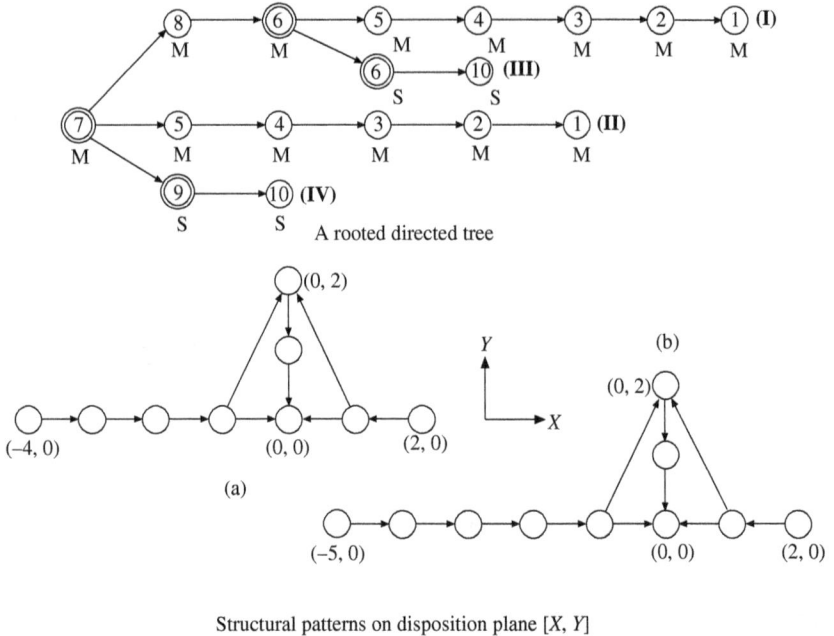

A rooted directed tree

Structural patterns on disposition plane [X, Y]

Figure 4-18   Rooted directed tree and disposition plane: (a) Planomiller of double column type and (b) planer of double column type.

of Fig. 4-19, the [X, Y] coordinates are allocated so that the increasing order of the X coordinate values corresponds with the increasing order of row and column.

3. Assuming that A_{ST} and B_{ST} are the structural patterns to be compared, their adjacency matrices can be expressed by $[A_{ST}]$ and $[B_{ST}]$, where the components δA_{STij} and δB_{STij} of the matrices are given by

$$\delta A_{STij}, \delta B_{STij} = 1 \text{ in the case in which the directed edge exists from } i\text{th (row) to } j\text{th (column) vertices}$$
$$= 0 \text{ other cases}$$

In consideration of the common space where either δA_{STij} or δB_{STij} is unity, the rate of pattern similarity I_r can be written as

$$I_r = u_{ST}/(u_{ST} + v_{ST} + w_{ST}) \qquad (0 \leq I \leq 1) \qquad (4\text{-}3)$$

where $u_{ST} = \Sigma \Sigma \, \delta A_{STij} \cdot \delta B_{STij}$
$v = \Sigma \Sigma \, \delta A_{STij} \cdot (1 - \delta B_{STij})$
$w = \Sigma \Sigma \, (1 - \delta A_{STij}) \cdot \delta B_{STij}$

Although the evaluation standard should be improved to enhance reliability, the structural similarity can be given by a combination of the rates

	X	Y		1	2	3	4	5	6	7	8	9	10		1	2	3	4	5	6	7	8	9	10
1	−5	0	1											1	1									
2	−4	0	2	1										2		1								
3	−3	0	3		1									3			1							
4	−2	0	4			1								4				1						
5	−1	0	5					1		1				5						1		1		
6	0	0	6											6										
7	0	1	7			1								7				1						
8	0	2	8					1						8						1				
9	1	0	9					1		1				9						1		1		
10	2	0	10								1			10										1

(a)   (b)

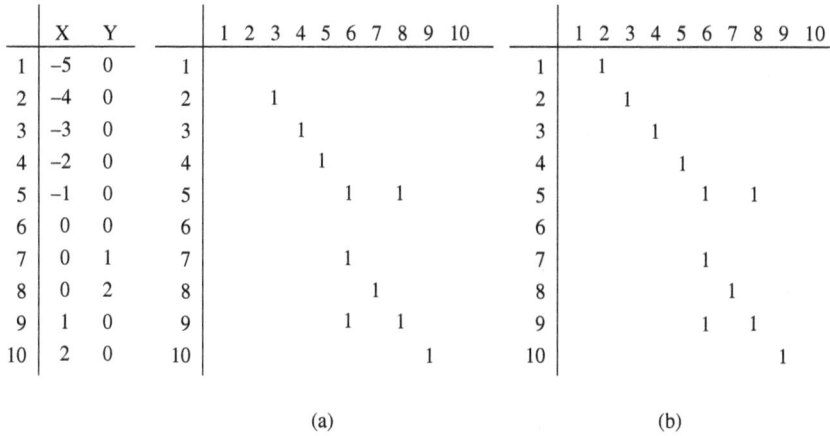

Figure 4-19 Adjacency matrices of structural patterns: (a) Planomiller of double column type and (b) planer of double column type.

of commonness and pattern similarity, where the rates of commonness and pattern similarity correspond to the evaluation attributes of having the same units and structural configuration, respectively. Figure 4-20 shows an evaluation result when the planomiller of portal (double column) type is the reference kind. As can be readily seen, the structural pattern is very effective for estimating the similarity of structure. For example, the vertical boring machine, planer, bedway grinder, and vertical lathe show higher pattern similarity along with relatively high commonness. In contrast, the planomiller of gantry type, horizontal boring and milling machine of floor type, and vertical boring machine of portal type show higher commonness, but lower pattern similarity.

In consequence, this structural similarity evaluation is available for a methodology to judge whether a group of kinds is suitable for the modular design, i.e., the feasibility of the modular design of different-kind generating type. Figure 4-21 is a decision diagram used to judge the available region of modular design, simultaneously differentiating the region for either the unit construction or the different-kind generating type.

In retrospect, Maeda et al. conducted an interesting study of the similarity evaluation of the part to increase the classification efficiency and to advance the generation of the GT code [17]. In their method, two attributes were considered and symbolized to represent the part, i.e., geometric entity to determine the shape (shape entity) and specific function added to the shape entity (function entity): the former and latter were symbolized by using, for example, C_e (cylindrical surface) and P_e (side surface), and also S_e (screw) and K_e (keyway), respectively. Figure 4-22 reproduces the part representation (part pattern), emphasizing the

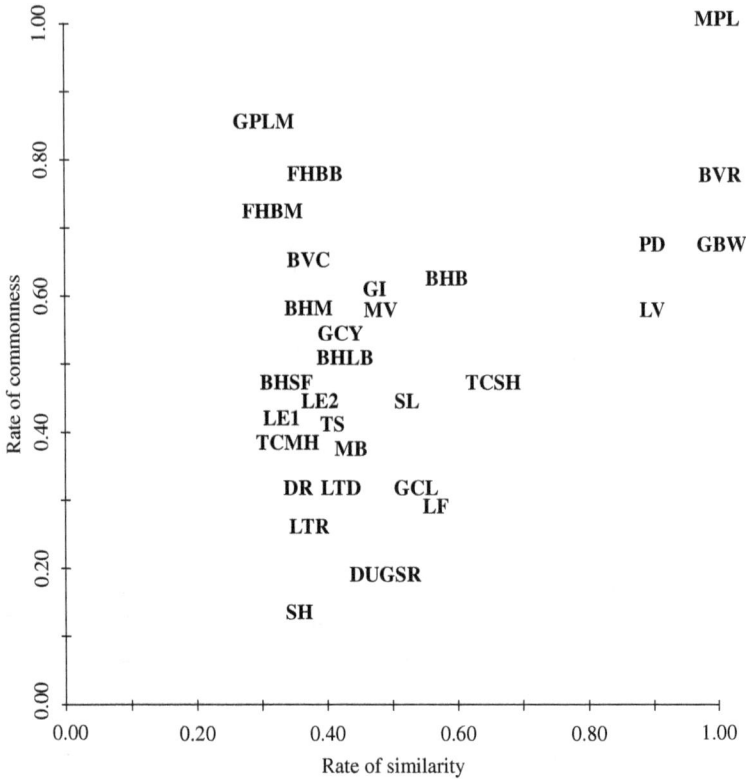

Figure 4-20 Similarity evaluation of machine tools using the structural pattern.

MPL	Double column type planomiller	MH	Horizontal milling machine
GBW	Bedway grinder	MB	Bed type milling machine
PD	Double column type planer	BHB	Table type horizontal boring and milling
BVC	Double column type vertical boring		machine (in milling process)
	machine (round column)	BHM	Table type horizontal boring and milling
BVR	Double column type vertical boring		machine (in milling process)
	machine (rectangular column)	BHLB	Table type horizontal boring and milling
LV	Double column type vertical lathe		machine (in line boring)
GPLM	Gantry miller	TCH	Production type hobbing machine
FHBM	Floor type horizontal boring and milling	GSR	Surface grinder
	machine (in milling process)	LE 1	Engine lathe (in center work)
FHBB	Floor type horizontal boring and milling	LE 2	Engine lathe (in chuck work)
	machine (in milling process)	LF	Facing lathe
DU	Upright drilling machine	LTR	Ram type turret lathe
SL	Slotting machine	LTS	Saddle type turret lathe
SH	Shaping machine	LTD	Drum type turret lathe
TCSH	Gear shaping machine	GCY	Cylindrical grinding mahine
DR	Radial drilling machine	GI	Interless grinding machine
MV	Vertical milling machine	GCL	Centerless grinding machine

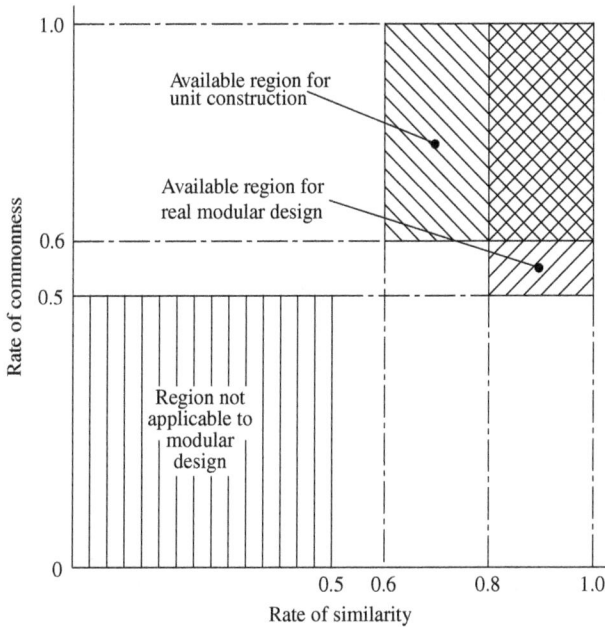

Figure 4-21 Schematic representation of available range of modular design systems.

functional aspect of the part, where upper and lower rows show the function and shape entities.

The similarity evaluation can be carried out using the two characteristics, i.e., commonness rate to evaluate the number of identical shape entities and the allocating order rate to evaluate the similarity of the allocating order of the common entities between both parts. More specifically, the commonness rate can be calculated by using the reference matrix shown together in Fig. 4-22, where the (i, j) component yields the following:

Identical shape entity with identical function = unit

Entity in part identical = 0.5

Entity completely not identical = nil

Consequently, from the reference matrix, the distribution rate Q_d and the commonness rate S_r yield

$$Q_d = \Sigma di/(p_e \cdot dM) \qquad (0 \leqq Q \leqq \text{unit}, i = 1 - p) \qquad (4\text{-}4)$$

where p_e = number of identical entities
 di = deviation distance from diagonal line in reference matrix
 dM = maximum value of di

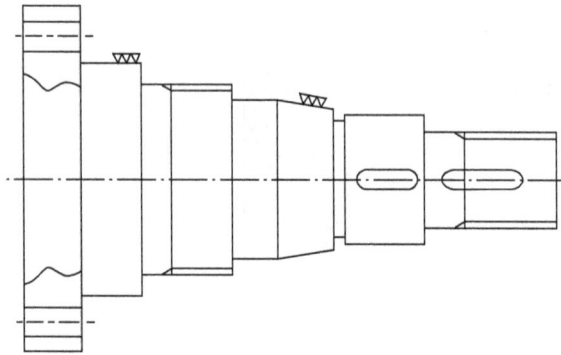

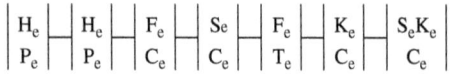

| H_e | | H_e | | F_e | | Se | | F_e | | K_e | | S_eK_e |
| P_e | | P_e | | C_e | | C_e | | T_e | | C_e | | C_e |

(a)

	1 H_eH_e P_e	2 F_e C_e	3 S_eF_e C_e	4 F_e T_e	5 S_eK_e C_e	6 S_eK_e C_e
1 H_eP_e	0.5	*	*	*	*	*
2 H_eP_e	0.5	*	*	*	*	*
3 F_eC_e	*	1.0	0.5	*	*	*
4 S_eC_e	*	*	0.5	*	0.5	0.5
5 F_eT_e	*	*	*	1.0	*	*
6 K_eC_e	*	*	*	*	0.5	0.5
7 $S_eK_eC_e$	*	*	0.5	*	1.0	1.0

(b)

Figure 4-22 Part pattern and reference matrix: (a) Sample part and its representation and (b) reference matrix (by Maeda, courtesy of JSPE).

$$S_r = 1 - n_e1 \cap n_e2 / n_e1 \cup n_e2 \qquad (0 \leqq S \leqq \text{unit}) \qquad (4\text{-}5)$$

where

n_e1 and n_e2 = number of entities consisting of each part pattern

$n_e1 \cap n_e2$ = number of common entities between both part patterns

$n_e1 \cup n_e2 = n_e1 + n_e2 - n_e1 \cap n_e2$

4.2.2 Variant design for structural configuration

Apart from the prototype design of a new innovative kind or a design based on a novel product concept, the machine tool has been manufactured using the variant design. In the variant design, the designer can use longstanding experience; results of the analytical, achievement, and acceptance tests; and knowledge obtained from remedies to cope with the daily improvement and so on to enhance the product quality. Consequently, the quality of the machine tool is guaranteed to a larger extent when it is manufactured by the variant design.

Having in mind the beneficial features of the variant design, Shinno and Ito proposed the generating methods of structural configurations, in which the possible variants can be produced from the basic structural pattern. Figure 4-23 shows a basic structural pattern for milling machine of knee type, and in the variant design, the machine should be divided into the two subpatterns, i.e., common and variant branches, as

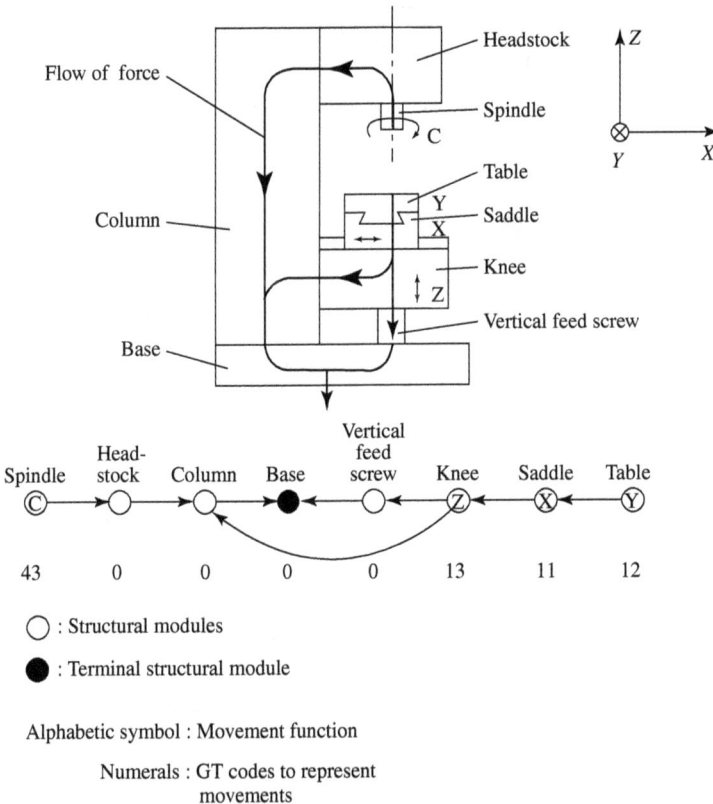

Figure 4-23 Basic structural pattern for milling machine of knee type.

shown in Fig. 4-24(a). In general, the machining space corresponds with the variant branch, and then the movement functions at the variant branch are reallocated considering possible combinations of the structural unit [18]. In an actual case, such combinations are stored in the computer in the form of a pair of units, which can be chosen through the evaluation of its relative movement and dynamic performance. Figure 4-24(b) shows some variants produced from the basic pattern.

In contrast to a simple variant design mentioned above, Shinno and Ito proposed another variant design for structural configuration using

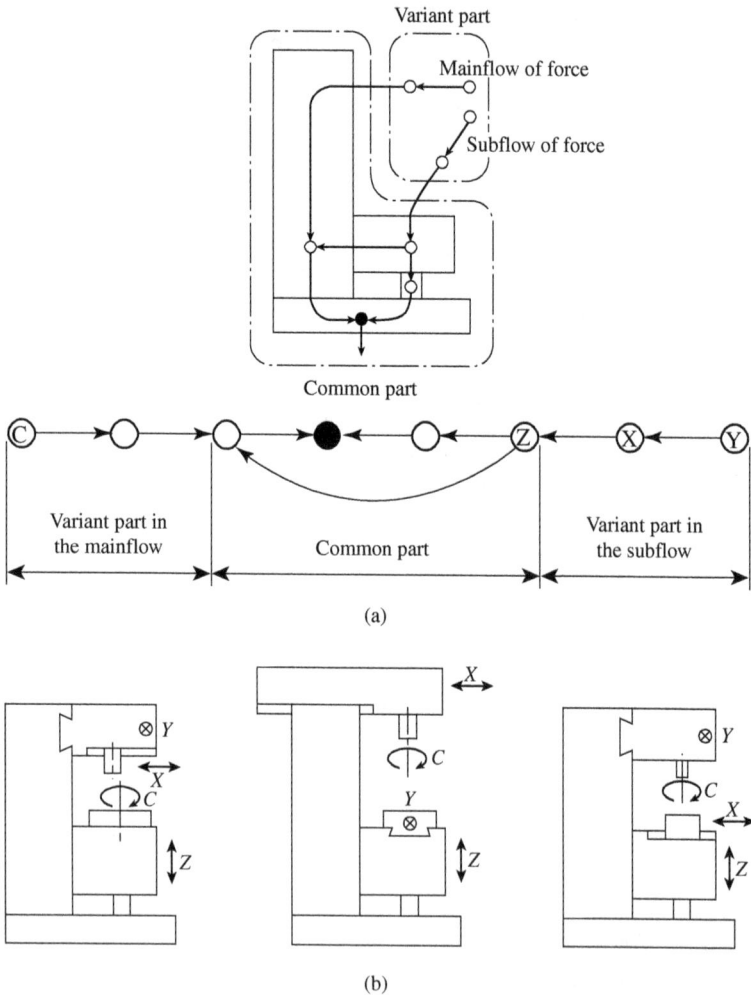

(a)

(b)

Figure 4-24 Variant design method and its application to milling machine of knee type. (a) Idea of variant design and (b) some examples of variants produced.

the directed graph, which is one of the converted forms of the structural description [19]. Importantly, various configurations can be produced from the directed graph representing the data possible to combine among the already known structural modules, i.e., a group of predetermined modules, through the necessary decision making. In addition, the proposed methodology is based on the modular principle of hierarchical type consisting of (1) whole structure (structural pattern), (2) substructural patterns corresponding to main and sub-FOFs, (3) structural module complex, and (4) structural module.

In short, the procedure for generating structural pattern is as follows.

1. Determination of a group of primitives (structural modules) as shown in Fig. 4-25.

2. Generation of single modular complex and its representation with the vertices and directed edge in graph theory. Within a group of structural

	Primitive	Name of primitive
1	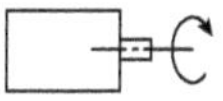	Spindle head
2	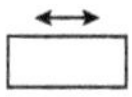	Slide unit
3	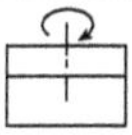	Swivel slide
4		Column
5	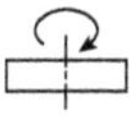	Rotary table
6		Cross slide unit
7	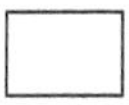	Base (stationary unit)
8	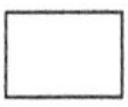	Column base
9	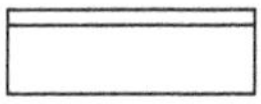	Bed (with longitudinal slide)

Figure 4-25   A group of primitives.

modules, there are some constraints on the module combination, and thus the combinations possible must be examined and represented by the basic graph, i.e., connecting pattern complex, as shown in Fig. 4-26. By integrating the single modular complex, an adjacency matrix can be obtained as shown in Table 4-2, which is in correspondence with Fig. 4-26.

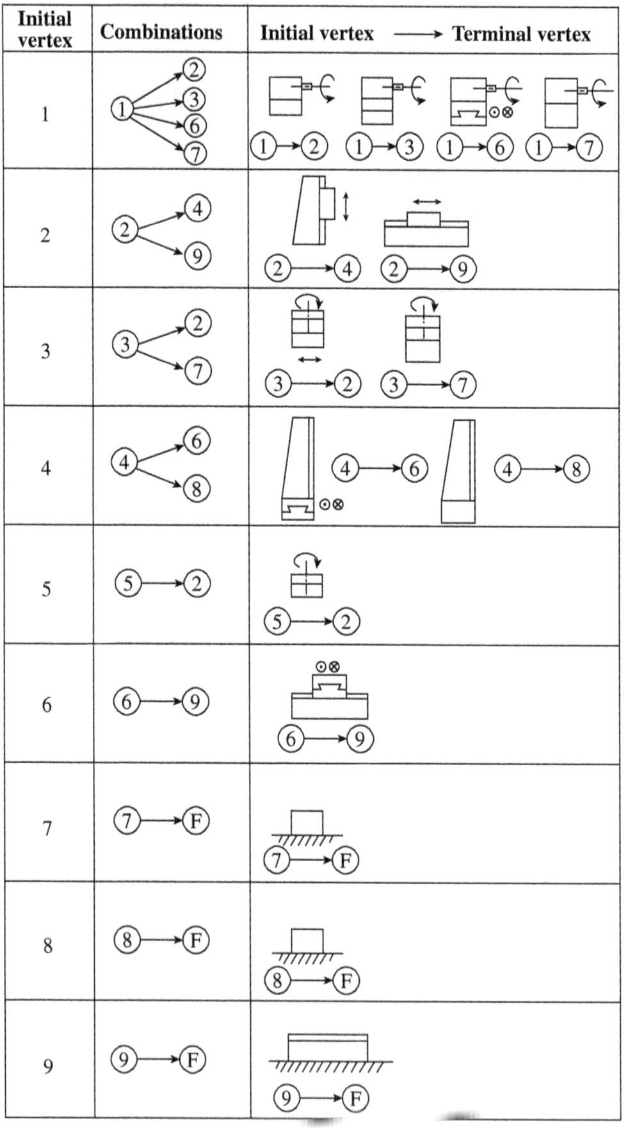

Initial vertex	Combinations	Initial vertex ⟶ Terminal vertex
1	①→②③⑥⑦	①→② ①→③ ①→⑥ ①→⑦
2	②→④⑨	②→④ ②→⑨
3	③→②⑦	③→② ③→⑦
4	④→⑥⑧	④→⑥ ④→⑧
5	⑤→②	⑤→②
6	⑥→⑨	⑥→⑨
7	⑦→Ⓕ	⑦→Ⓕ
8	⑧→Ⓕ	⑧→Ⓕ
9	⑨→Ⓕ	⑨→Ⓕ

Figure 4-26 Connecting pattern complex.

TABLE 4-2   Adjacency Matrix of Connecting Pattern Complex

		Terminal vertex								
		1	2	3	4	5	6	7	8	9
Initial vertex	1	0	1	1	0	0	1	1	0	0
	2	0	0	0	1	0	0	0	0	1
	3	0	1	0	0	0	0	1	0	0
	4	0	0	0	0	0	1	0	1	0
	5	0	1	0	0	0	0	0	0	0
	6	0	0	0	0	0	0	0	0	1
	7	0	0	0	0	0	0	0	0	0
	8	0	0	0	0	0	0	0	0	0
	9	0	0	0	0	0	0	0	0	0

3. From the basic connecting pattern, both connecting patterns for the mainflow and subflow of FOF can be produced by adding the vertices regarding the tool (T), workpiece (W), and floor (F), as shown, e.g., in Fig. 4-27.

4. By using the rooted-directed tree of both the main and sub-FOFs, all the possible structural configurations can be given in the form of a matrix, as shown in Fig. 4-28.

Lee and his colleagues applied this methodology later with special reference to the MC [20]. More specifically, first the modules popularly

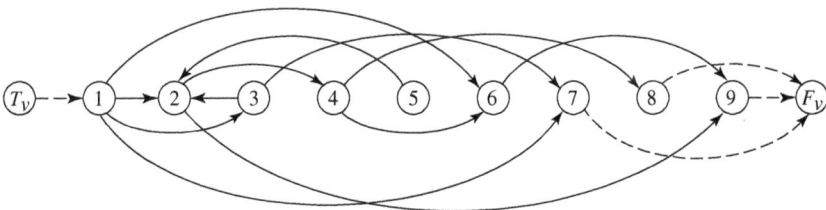

Figure 4-27 Connecting pattern complex for generating substructural pattern regarding mainflow of force.

Generated results of structural configuration

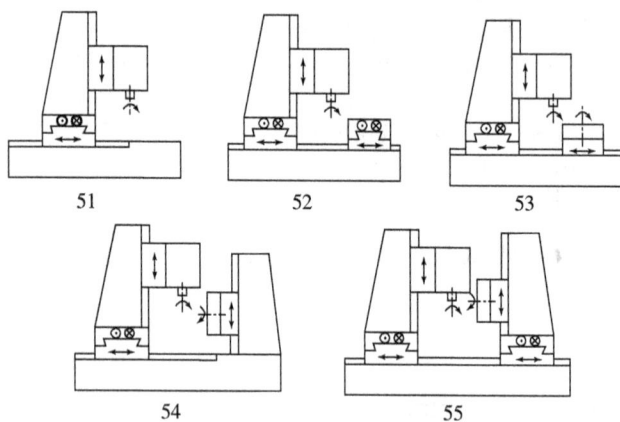

Some examples of structural pattern

Note: Numeral indicates the corresponding substructural
patterns in the order of the line to row in the matrix

Figure 4-28 All possible structural configurations generated.

observed are extracted from the MCs on the market, as shown in Fig. 4-29(a), and then the connecting possibilities among modules are analyzed, resulting in the conjunction pattern shown in Fig. 4-29(b). In other words, the conjunction pattern is that representing the FOF. In due course, a group of module complexes can be determined from the conjunction

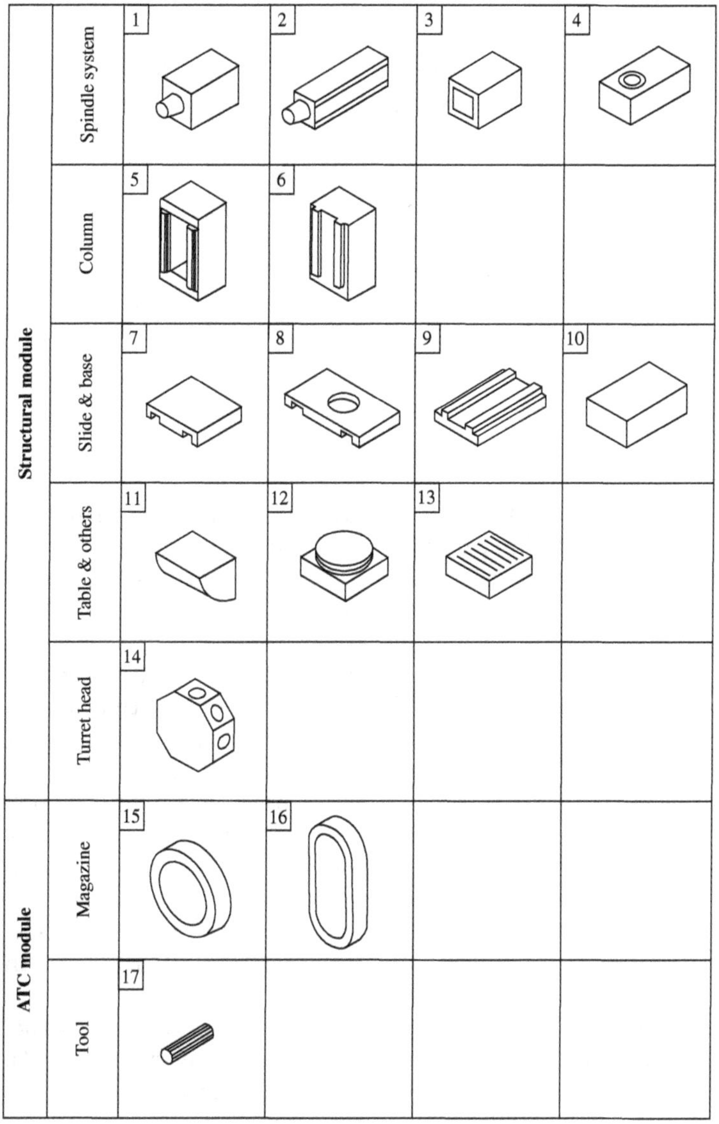

(a)

Figure 4-29   Modules and conjunction pattern of MCs: (a) Modules popularly observed in traditional MCs and (b) conjunction pattern of structural modules.

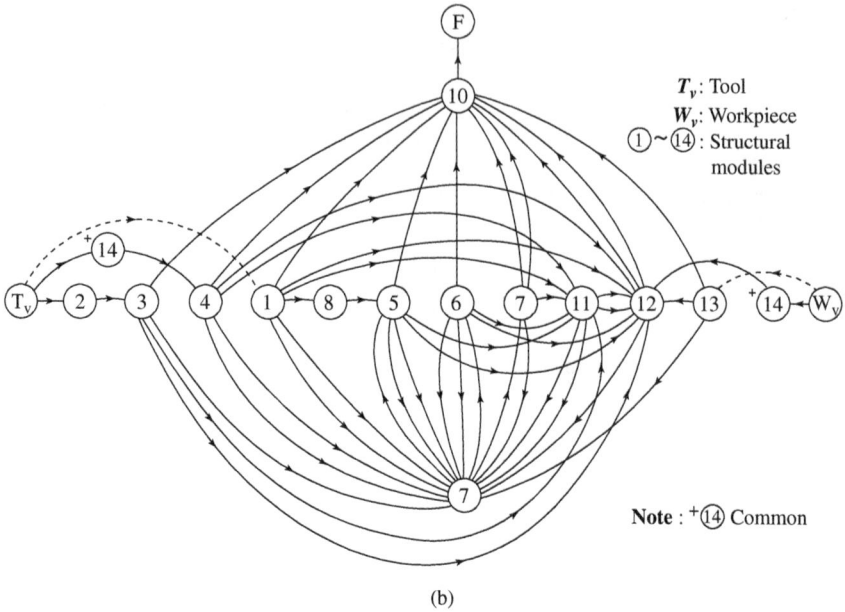

T_y: Tool
W_y: Workpiece
①~⑭ : Structural modules

Note : $^+$⑭ Common

(b)

Figure 4-29 (*Continued*)

pattern after the paths compatible with the design requirements are chosen. To simplify the design procedure, Lee proposed a design method whereby the module complexes are regarded as the entities (unit complexes) within the design data file such as files I and II in Fig. 4-30. In fact, the method of Lee can be considered as a forerunning trial of the platform proposed by Metternich and Würsching. Importantly, the conjunction pattern of the conventional MC yields in general that shown in Fig. 4-31(a) and duly deployed into the subpatterns for tool and work branches, as shown in Fig. 4-31(b) and (c). These subpatterns can be determined in consideration of the conjunction patterns possible, which are verified using the concept of FOF.

Figure 4-32 shows some of unit complexes generated from the design data file I. In short, the unit complexes for tool and work are 62 and 40, respectively, and by combining these unit complexes for both branches, the structural configuration as a whole can be obtained, as shown in Fig. 4-33, provided that the movement degrees of freedom are 5.

In addition, Iwata et al. proposed a method to produce the basic structural configuration on the basis of the form-generating function [21]. Their method is similar to that of Shinno and Ito; however, there are some improvements as follows.

1. Use of design data for classification of the guideway to explicitly consider the jointing method of both units

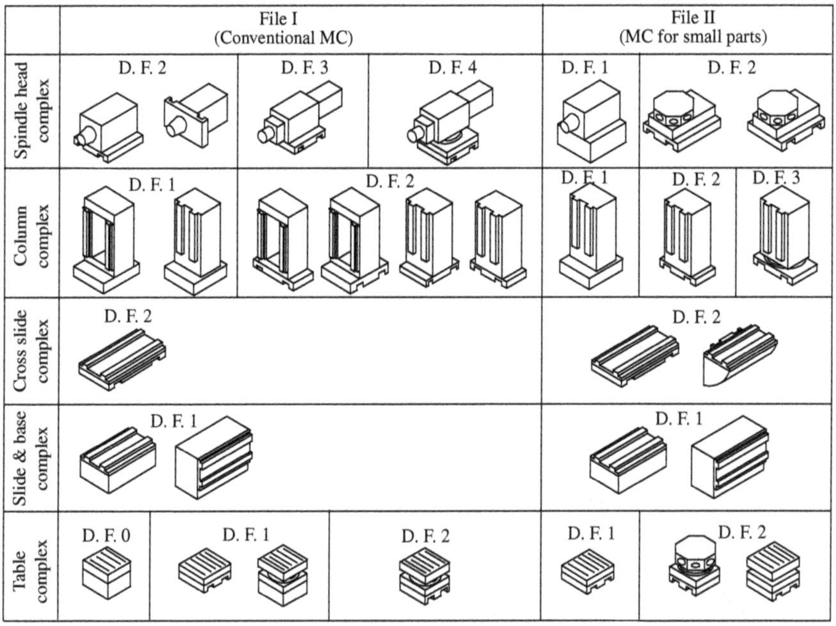

File I (Conventional MC)				File II (MC for small parts)		

D. F. : Degree of freedom in movement function

Figure 4-30   File of module complexes.

2. Use of knowledge representation for the design factors and their standards of the structural component, where the design factor and its standard can be represented by the flame and rule, respectively

4.2.3   Free design for structural configuration

Shinno and Ito once tried to produce the concept drawing with the free design [22]. In this trial, a hierarchical modular system was employed as shown in the design flow in Fig. 4-34, where the basic modules are, in general, the volume elements shown in Fig. 4-35, and the structural configuration can be produced by combining these modules. The volume element can be characterized by its data structure of gravity-origin type. In addition, the idea of a pair of modules is employed, considering furthermore the fixing and supporting functions. In fact, the procedure of this free design is as follows.

1. Determination of a group of modules that are arranged in the form of a pair of modules with relative dimensional specifications.

2. Generation of the initial structure with dimensional information including the FOF and machining space, as shown in Fig. 4-36(a).

S·H : Spindle head
C$_v$: Column
C·S : Cross slide
S·B : Slide & base
T$_v$: Table

(a)

(b)

(c)

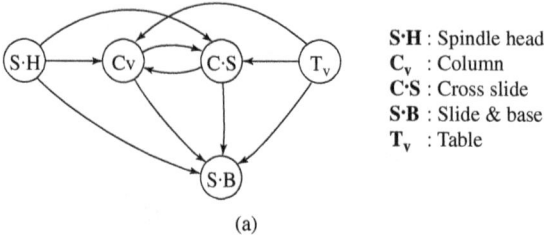

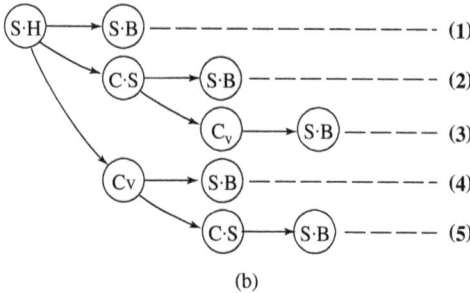

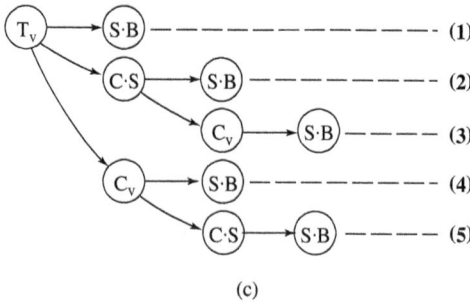

Figure 4-31 Conjunction pattern of module complexes: (a) General conjunction pattern; (b) conjunction pattern for unit complexes in tool branch; and (c) conjunction pattern for unit complexes in workpiece branch.

More specifically, the initial structure is positioned within the (X, Y, Z) coordinate, where the volume element corresponding to the machining space can facilitate correlation of the main FOF to the sub-FOF. In addition, the machining space volume is a virtual one and allocated the origin of (X, Y, Z), corresponding correctly to the leading dimensions of the actual machining space.

3. Modification of the structural configuration by allocating necessary functions to each module, resulting in the determination of the joint surface. In this case, an initial structural candidate can accommodate the following information processing capabilities.

   a. Determination of volume elements (structural modules) with travelling or stationary function

Unit complex for tool branch

Unit complex for workpiece branch

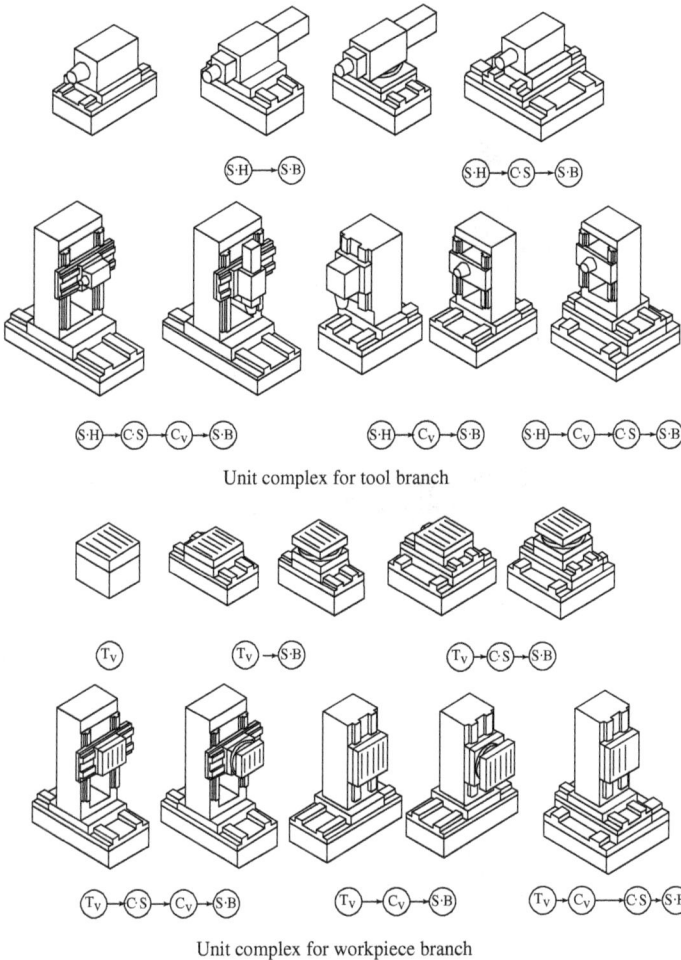

Figure 4-32   Unit complexes generated from design data file I.

   b. Determination of joint surface

   c. Leverage of numbers of the volume element and structural configuration as a whole

4. Dimensional modification of each module.

5. Production of the concept drawing in three-dimensional and plane views along with the structural pattern, as shown in Fig. 4-37.

6. Simulation of the movement function to examine the interference of modules.

It is furthermore worth stating that the free design method of Shinno and Ito was applied to the design of FMC later [23].

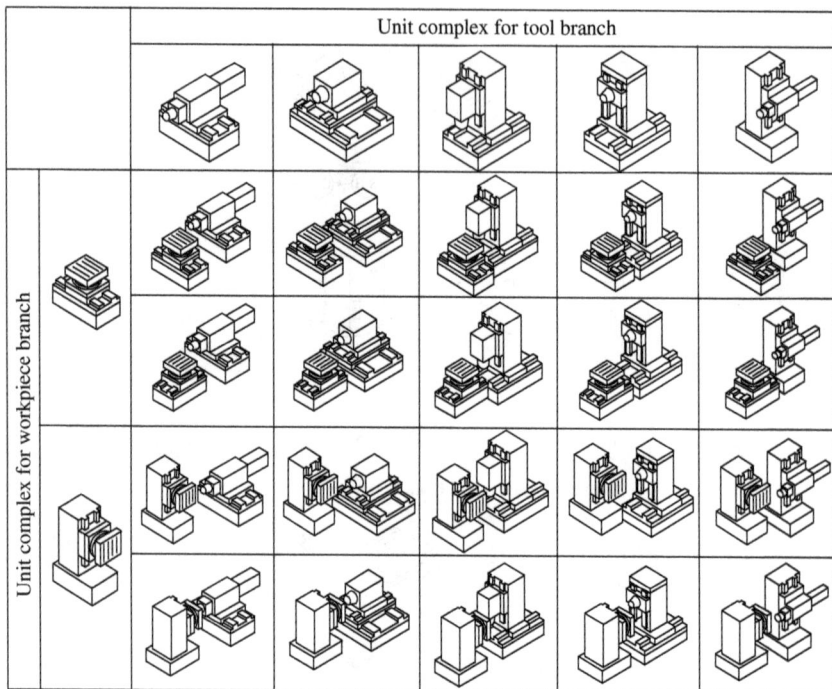

Figure 4-33 Some structural configurations of MC generated from unit complexes.

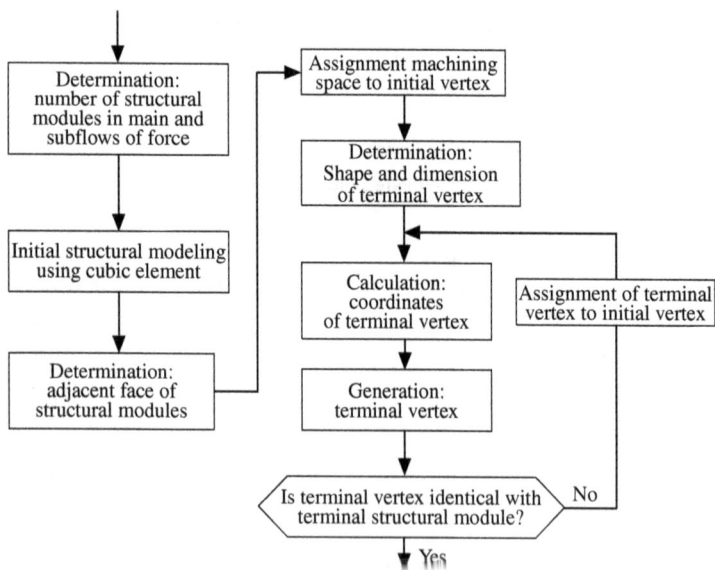

Figure 4-34 Generating procedure for a pair of structural modules.

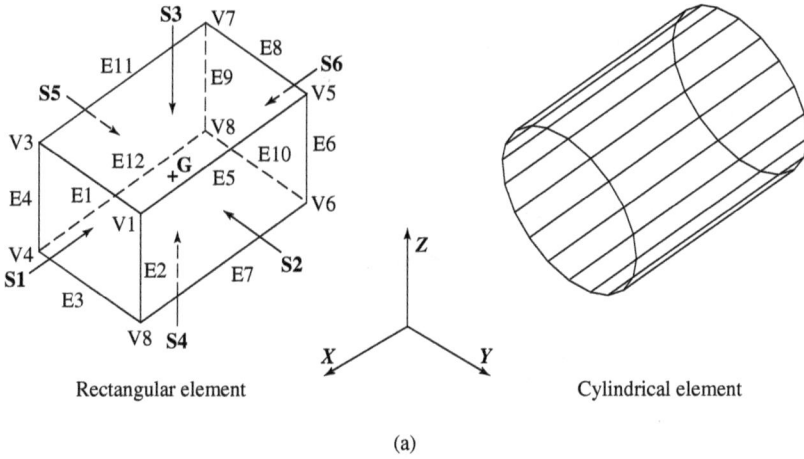

Rectangular element   Cylindrical element

(a)

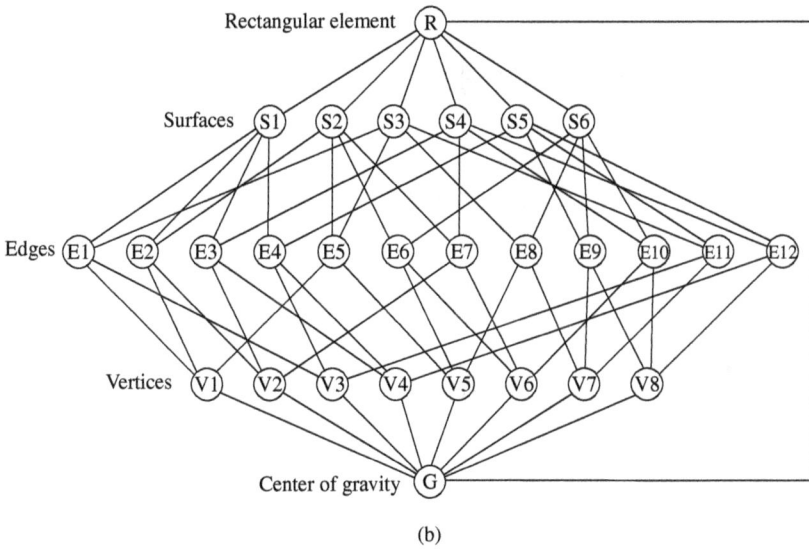

(b)

Figure 4-35 Volume elements and their data structures: (a) Volume elements and (b) data structure of rectangular element.

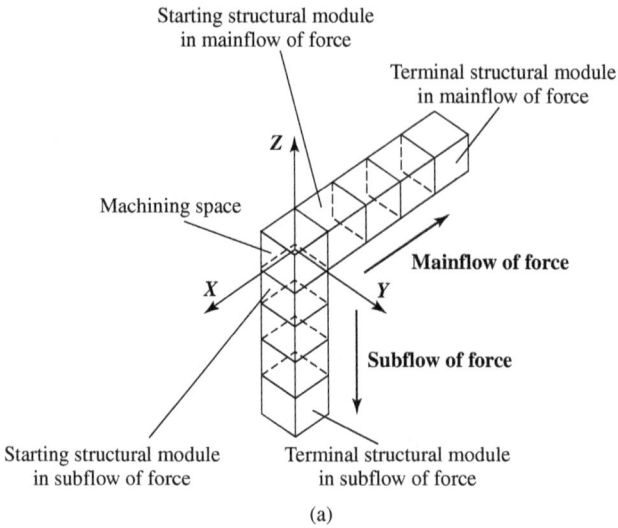

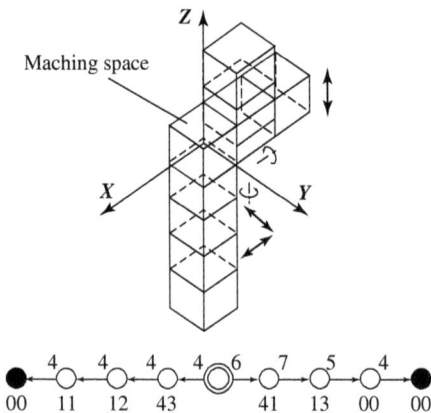

Starting structural module
in mainflow of force

Terminal structural module
in mainflow of force

Machining space

Mainflow of force

Subflow of force

Starting structural module
in subflow of force

Terminal structural module
in subflow of force

(a)

Maching space

00   11   12   43   41   13   00   00

Note:

○ : Structure module

◎ : Machining space

● : Terminal structural module

Upper numeral : GT code to represent joint surface

Lower numeral : GT code to represent movement freedom

(b)

Figure 4-36 Generating procedure for structural configuration using volume elements: (a) Initial structure consisting of volume elements and (b) structural configuration before dimensioning.

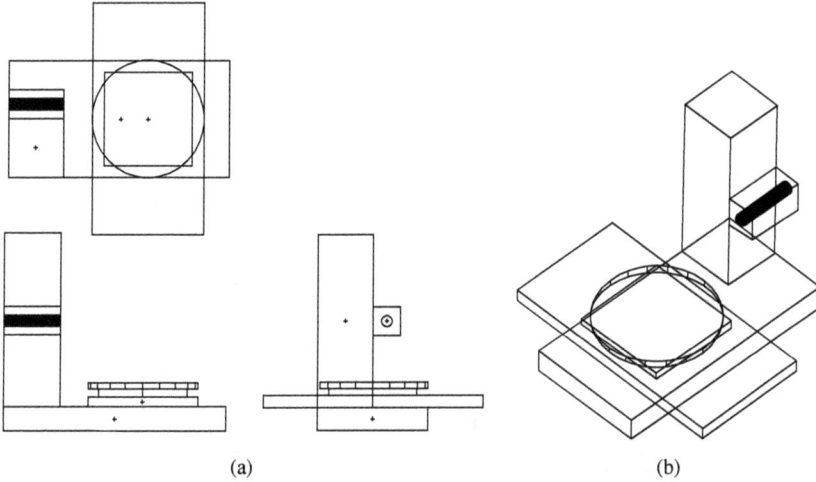

(a)   (b)

Figure 4-37   Outputs of design results: (a) Plane view and (b) three-dimensional view.

References

1. Lee, H. S., H. Shinno, and Y. Ito, "On the Description Method of Machining Centres and Their Structural Configuration Analysis," *J. of JSPE*, 1984, 50(5): 886–890.
2. For example, VDI-Z-Datenbank, *Einspindlige Horizontalbearbeitungszentren Hersteller, Bauform, simultangest. NC-Achsen*, VDI-Verlag, Düsseldorf, 1993.
3. Elb Co. "Modular aufgebaute CD-gerechte flexible Schleifsysteme," in *Schleifen, Läppen, Honen Jahrbuch*, 1984, S. 335-356, Vulkan Verlag.
4. Lee, H. S., E. F. Moritz, and Y. Ito, "Functional Description of Machine Tools and Its Application to Marketability Analyses," *J. of Eng. Des.*, 1996, 7(1): 83–94.
5. Vragov, Yu D., "Structural Analysis of Machine-Tool Layouts," *Mach. and Tool.*, 1972, 18(8): 5–8.
6. Ito, Y., Y. Saito, and Y. Kudama, "Description of Machining Function in FMS and Its Analysis," in B. J. Davies (ed.), *Proc. of 23d Int. MTDR Conf.*, Macmillan, 1983, pp. 503–509.
7. Saito, Y., Y. Ito, and T. Ohtsuka, "Automatisierte Darstellung von Entwurfszeichnungen für Werkzeugmaschinen-Konstruktionen," *ZwF*, 1980, 75(10): 492–495.
8. Saljé, E., and W. Redeker, "Methodisches Planen und Konstruieren spanender Fertigungssysteme," *Werkstatt und Betrieb*, 1974, 107(10): 631–634.
9. Redeker, W., "Systematische Konstruktion spanender Werkzeugmaschinen," Dissertation, Technische Universität Braunschweig, 1979.
10. Saljé, E., "Ein Beitrag zum Entwurf und zur systematische Konstruktion spanender Werkzeugmaschinen," private report, Sept. 17, 1981.
11. Ratchev, T. M., and N. N. Z. Gindy, "Structure and Capability Description of Machine Tools: A Linguistic Approach," in *Proc. of 29th Int. MATADOR Conf.*, Macmillan, 1989, pp. 57–63.
12. Herrmann, J., "Entwurfsoptimierung von Fertigungssystemen am Beispiel der Gestellanordnung numerisch gesteuerter Drehmaschinen," *ZwF*, 1970, 65(7): 323–329.
13. Khomyakov, V. S., and I. I. Davydov, "Prediction of the Accuracy of a Machine Tool at an Early Stage of Its Design, with Layout Factors Taken into Account," *Soviet Eng. Res.*, 7(9): 44–47 [Russian original: *Stanki I Instrument*, 1987, 58(9): 5–7].
14. Moriwaki, T., N. Sugimura, and Y. Miao, "A Model Based Design of Kinematic Accuracy of Machine Tools," in G. J. Olling and F. Kimura (eds.), *Human Aspects in Computer Integrated Manufacturing*, Elsevier Science Publisher, 1992, pp. 673–684.

15. Yokoyama, M., H. Shibuya, and R-K. Park, "A Basic Study of the Automated Generation of Machine Structures" (2d Report, Generation Method of the Functional Structure of Machine), *Trans. of JSME,* 1986, 52(473): 417–422.

16. Ito, Y., and H. Shinno, "Structural Description and Similarity Evaluation of the Structural Configuration in Machine Tools," *Int. J. Mach. Tool Des. Res.,* 1982, 22(2): 97–110.

17. Maeda, J., K. Momose, and T. Sata, "A Proposal for Quantifying Parts Similarity," *J. of JSPE,* 1975, 41(12): 1147–1149.

18. Shinno, H., and Y. Ito, "Generating Method for Structural Configuration of Machine Tools—1st Report, Method of Variant Design Type," *Trans. of JSME(C),* 1984, 50(449): 213–221.

19. Shinno, H., and Y. Ito, "Computer Aided Concept Design for Structural Configuration of Machine Tools—Variant Design Using Directed Graph," *Trans. of ASME J. of Mechanisms, Transmissions and Automation in Design,* 1987, 109: 372–376.

20. Lee, H. S., H. Shinno, and Y. Ito, "Structural Configuration Design of Machining Center—On the Variant Method Using Conjunction Pattern," *J. of JSPE,* 1986, 52(8): 1393–1398.

21. Iwata, K., N. Sugimura, and L. S. Peng, "A Study of the Fundamental Design of Machine Structures for Machining," *Trans of JSME (C),* 1990, 56(523): 803–809.

22. Shinno, H., and Y. Ito, "A Proposed Generating Method for the Structural Configuration of Machine Tools," ASME Paper 84, WA/Prod-22, 1984.

23. Huang, K. J., H. Shinno, and Y. Ito, "Design Methodology for FMC of Compact Type— Spatial Allocation Approach with Functional and Structural Blocks," in *Proc. of 1st Int. Conf. on Automation Technology,* National Chiao Tung University, Taipei, 1990, pp. 165–173.

Engineering Design for Machine Tool Joints— Interfacial Structural Configuration in Modular Design

5

Basic Knowledge of Machine Tool Joints

Professor Kienzle is credited with being the first to conduct preliminary research into the machine tool joint, i.e., comparative research into the damping of the steel welded to cast bed structures, in 1939 [1]. On that occasion, Kienzle implied that the important role of the machine tool joint might be to increase the damping capacity of a machine tool as a whole; however, up to the beginning of the 1970s engineers designed the machine tool by paying special attention to only the body structures.

In fact, there have been many researches into the problem of two elastic bodies in contact, such as Hertz's theory concerning the design of the rolling bearing and stamping problem. Importantly, these earlier works have been based on an assumption: The joint surface has no roughness, waviness, and flatness deviation. This assumption means that the joint is topographically perfect and has an ideal surface, but actually we cannot obtain such a joint surface even when producing it by the utmost leading-edge and ultraprecision processing methods. In short, nearly all the theories of elasticity available at present can deal with the problem of the monolithic elastic body, i.e., elastic body without any joints. Thus, the dire necessity is to establish a theory for the design of machine tool joints, in which the topography of the joint surface has greater influence on the behavior of two elastic bodies in contact.

Other problems of two surfaces in contact are (1) the Mindlin slip theory at microseizure contact points under shear load and (2) the real contact area between two surfaces, proposed by Bowden and Tabor. These are topics in tribology engineering and in certain relation to the machine tool joint. In retrospect, the wear and friction of the guideway, one of the tribology engineering problems, have been a major focus in

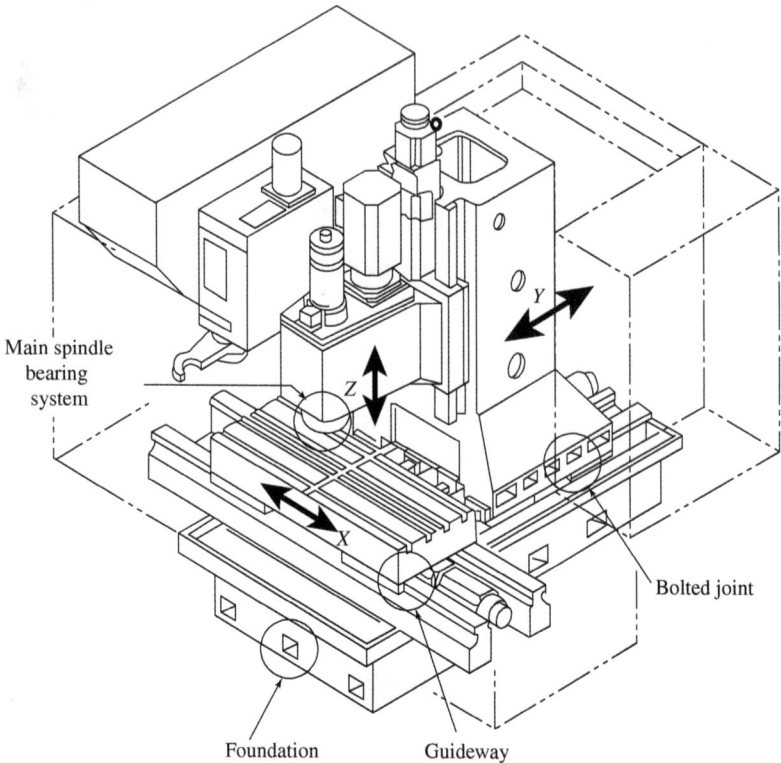

Figure 5-1   Representative joints in a machine tool (courtesy of Okuma).

the design and manufacture of the machine tool. In the late 1950s, a considerable number of publications addressed the wear and friction of the guideway [2–4]. In addition, the MTIRA (*Machine Tool Industry Research Association*) of the United Kingdom[1] often published noteworthy research reports, which were seen only by its members, and within them, we can find of course a report concerning the friction and wear of the slideways [5].

Figure 5-1 delineates how and to what extent a machine tool has joints. In due course, joints govern the overall stiffness of the machine tool as a whole to a large extent; for instance, Píč reported the effects of the joint on the overall stiffness of the bolted column in a horizontal boring and milling machine [6], as shown in Fig. 5-2. As can be readily seen, the static stiffness of a bolted column reduces from 10 to 20 percent compared

---

[1]The successor of the MTIRA is AMTRI (*Advanced Manufacturing Technology Research Institute*)

$$\Delta X\phi_a = \Delta\phi_a \cdot H$$

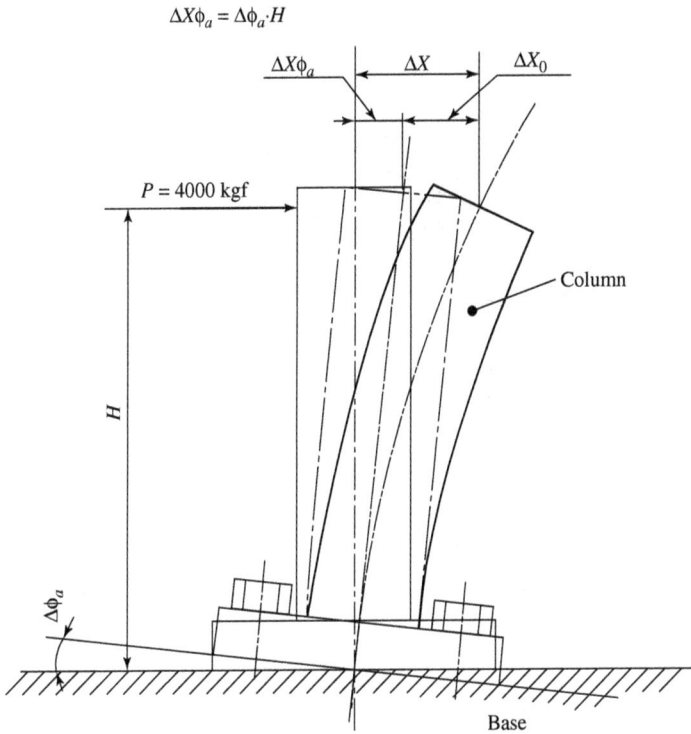

	Overall deflection %	Deflection due to joint %	Deflection of column %	H mm
No joint separation		8	92	1000
	100	15	85	500
With joint separation		12	88	1000
		22	78	500

Figure 5-2 Influencing rate of joint to overall deflection of bolted column for horizontal boring and milling machine (by Píč).

with that of an equivalent solid structure, where *equivalent solid* means the structure having the same shape and dimensions, but no joints. In addition, the designer must be aware that the joint exerts a considerable influence on the total stiffness of a machine tool as a whole, when the joint exists within the FOF (*flow of force, der Kraftfluβ, see Part 1*). In contrast, the damping capacity increases obviously, e.g., in the case of the main spindle of the horizontal boring and milling machine, as

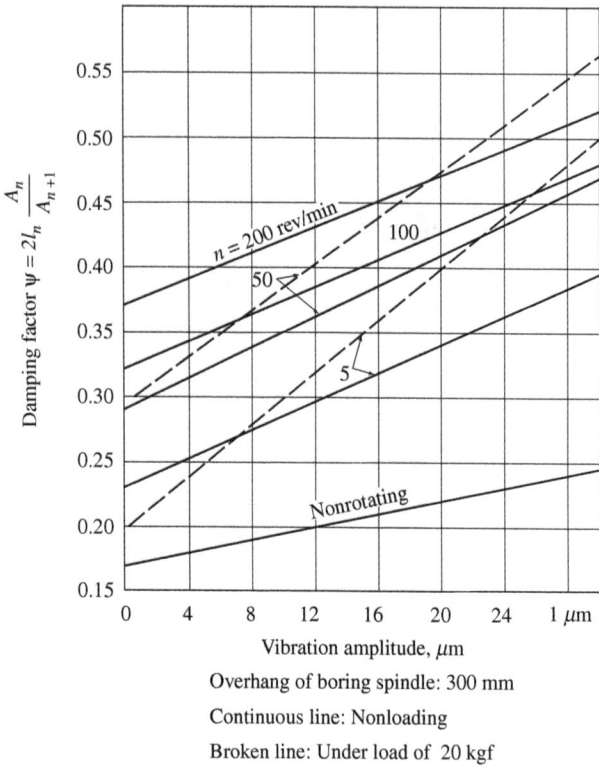

Figure 5-3   Damping capacity of main spindle in horizontal boring machine, measured from decayed free vibration (by Kunin).

shown in Fig. 5-3 [7], where the damping factor is dependent upon both the spindle vibration and the rotational speed. These results show without a doubt the substantial importance of the joint, leading to the conclusion that the joint is also one of the structural body components within a machine tool. In short, the static stiffness reduces and damping capacity increases by providing the joint in the structure, and then the deterioration of the static stiffness is mainly due to the weakness of the joint.

We must, however, provide a certain number of joints within a machine tool in the design procedure. More specifically, the necessities and inevitabilities of providing the joint to the machine tool are as follows.

1. To realize the required form-generating function of a machine tool. The form-generating function is, in principle, the combination of linear movements to the X, Y, and Z axes and rotational movements around these axes when placing the Cartesian coordinates in a

machine tool. The corresponding joints are, e.g., the *slideway* (*guideway*) and *main spindle bearing.*

2. To provide the flexibility in both the functionality and performance to the machine tool, i.e., application of the modular design concept so as to respond to multifarious machining requirements of the customer. In the modular design of unit type, a machine tool structure consists of a group of units, and a root cause of its difficulties lies, in principle, in the connecting method of both units. At present, the modular design prevails especially in the production of conventional MC and TC, in other words, NC machine tools of conventional type (refer to Chap. 1). A corresponding representative is, e.g., the *bolted joint.*

3. To improve the dynamic stiffness of a machine tool. Obviously, it is desirable to provide higher damping while maintaining the higher static stiffness. A representative is the *damping joint* for the welded structure, which was contrived in the United States and widely employed by Japanese machine tool manufacturers [8].

4. For the ease of manufacture, e.g., ease of machining and assembly of units and parts, convenience for component allocation, embodiment of power transmission mechanism, electrical and electronic equipment within a structure, and so on. A representative for fulfilling these requirements is the bolted joint.

5. To simplify the casting procedure, e.g., as a convenience for the separation of wooden pattern, removal of sand from the structural body after casting (fettling), and so on.

6. For the allocation and adjustment of the additional components, e.g., hardened steel strip guideway bolted on the welded or bonded on concrete structure.

7. To solve the transportation troubles in the large-size machine tool.

We must furthermore keep in mind that these requirements from design and manufacturing originally suggested in the era of the traditional machine tool. In due course, such requirements have been modernized with the advent of NC machine tools, aiming at full availability for the machine tool in the year 2000 and beyond. In addition, when we change the viewpoint from hardware to software, at succeeding issue is related to the CAD and CAE (*computer aided engineering*) for the machine tool design. Most machine tool manufacturers worldwide now employ CAE, also known as digital engineering, by emphasizing the computation of the static and dynamic stiffness, and thermal deformation at the stage of the basic layout or embodiment design. In other words, the CAE is now one of the effective tools used to estimate the performance of a machine tool at the design stage along with the assessment

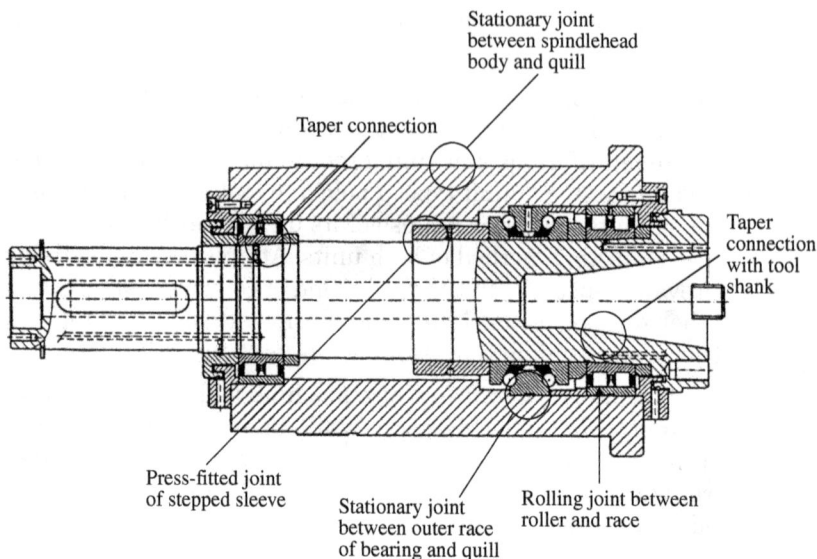

Figure 5-4   Several leading joints in main spindle (modified that of Gebert, courtesy of Carl Hanser).

and acceptance tests.[2] In contrast, a certain number of related machine tool engineers believe even now in the superiority of the well-qualified engineer and skilled worker to the computer.

Whether we rely on the computer or mature human resources, the shortage of authentic and detailed knowledge about the machine tool joint is a root cause of the difficulties in carrying out a machine tool design with higher qualification. Figure 5-4 shows several leading joints within the main spindle system, and the rolling bearing and its surroundings, i.e., a variant of sliding joints, can be replaced with a model, i.e., a couple of variables consisting of a spring (static joint stiffness) and

---

[2]In the beginning of the 1970s, the contact stress in two bodies in contact was a leading issue in the theory of elasticity, and the analysis was carried out using the FEM (*finite element method*). Nearly all those earlier works dealt with the idealized joint surface and contact under the control of Coulomb friction, i.e., with a macroscopic coefficient of friction, both of which are far from the contact condition of the machine tool joint. Despite suffering from such problems in assumptions, the research work in the theory of elasticity provides us with valuable knowledge qualitatively.

Tsuta, N., and S. Yamaji, "Study on Contact Problems with FEM Analysis," *J. of JSME*, 1973, 76(651): 348–358.

Ohte, S., "Analysis of Elastic Contact Stress with FEM," *Trans. JSME* (1), 1972, 38(313): 2210–2216.

Okamoto, N., "Analysis of Nonlinear Contact Problem with FEM," *Trans. JSME*, 1977, 40(870): 0710–3122.

dashpot (damping capacity of joint). In principle, a main spindle system can be replaced with a mathematical model of an elastic beam supported by both the spring and the dashpot at two or three points. In fact, the dynamic performance of the main spindle can be simulated using this mathematical model and the spring and dashpot as variables. In time, the basic necessity is to correctly determine the spring constant and damping coefficient together with the static and dynamic boundary conditions of the spindle system. These variables are, however, still difficult to quantify correctly even when we are using the knowledge obtained so far within the machine tool joint sphere.

In short, the body structure of a machine tool consists of a considerable number of structural body components and their interfaces, i.e., joints. The joint itself can be regarded as one of the structural body components, exerting considerable influence on the static, dynamic, and thermal behavior of a machine tool as a whole. Thus, we should design the functionality and structural configuration of the joint to provide the preferable performance to a machine tool. In addition, we must pay special attention to the joint when machining it and assembling a machine. As a result, we often refer to the design and manufacturing problems related to the joint as the *problems of machine tool joint.* At present, this problem sphere encompasses those problems related to chucking, tooling, jig, fixtures, and so on, although originally the machine tool joint was discussed in the design work on the slideway, bolted joint, and foundation.

Within a machine tool engineering context, it is again worth suggesting that modular design, especially the principle of connection is in the closest relation to the problems of the machine tool joint. A dire necessity is thus to deepen our understanding of the essential features of the machine tool joint, simultaneously crystallizing the keen role of the modular concept in machine tool design.

5.1   Classification of Machine Tool Joints

In a machine tool, there are various kinds of joints, for instance, the bolted joint, guideway, joint between the concrete foundation and the leveling block, taper connection between the cutting tool and the spindle hole, press fitting of the outer race of bearing to the housing, and so on. These joints appear to be very different from one another in their structural configurations; however, these differences depend only upon their required functions to be performed in the machine tool structure. More specifically, the basic structural configuration of various joints is two flat surfaces in contact, i.e., a simple flat joint, although a handful of representative joints can be observed within a machine tool as a whole.

In contrast, it is necessary to classify these various joints as suitable for the structural design of the machine tool. In the classification of the joint, in addition to its apparent features and structural configuration, the following technological aspects should be taken into consideration: in what way and how much the joint to be considered affects the overall behavior of the machine tool. In classifying the joint, moreover, the joint in a machine-tool-work system, e.g., the joint between the cutting tool and the work should be incorporated.

The machine tool joint can thus be classified as shown in Fig. 5-5 depending on the possibility of the relative traveling movement between the two joint surfaces in contact. In short, the stationary, semistationary, and sliding joints are representative, and each joint has a considerable number of variants. In the following, the rudimentary knowledge about each leading joint will quickly be noted (for other joints, refer to Chap. 9 in Part 2).

Stationary joint. In the stationary joint, the macroscopic relative movement cannot be allowed as literally shown, although the microscopic relative displacement of less than micrometer order, i.e., microslip, can be observed. This microslip has, in general, larger effects on the damping capacity of the stationary joint. In this sphere, the representatives are the bolted joint and the foundation.

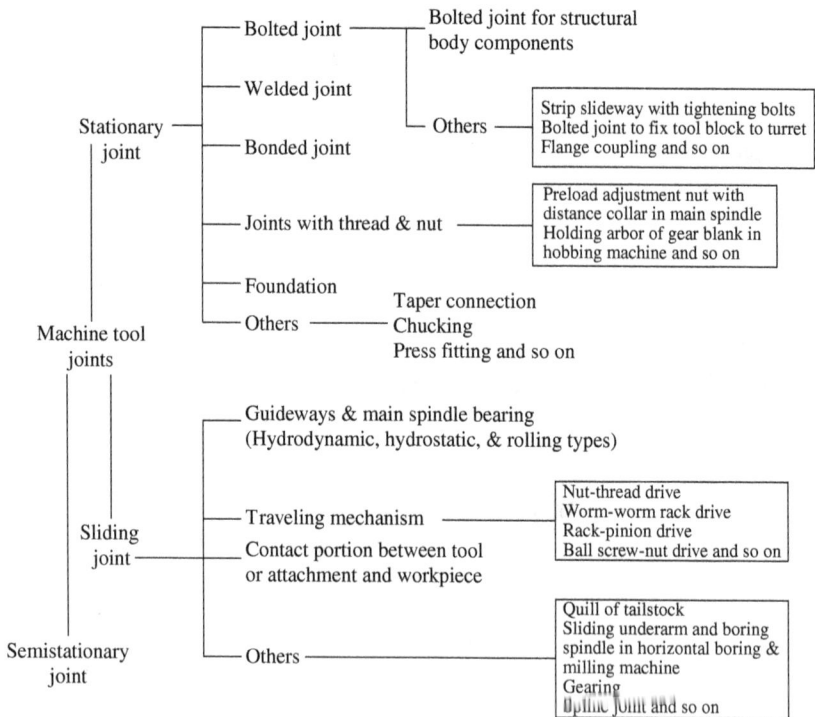

Figure 5-5 Classification of machine tool joints.

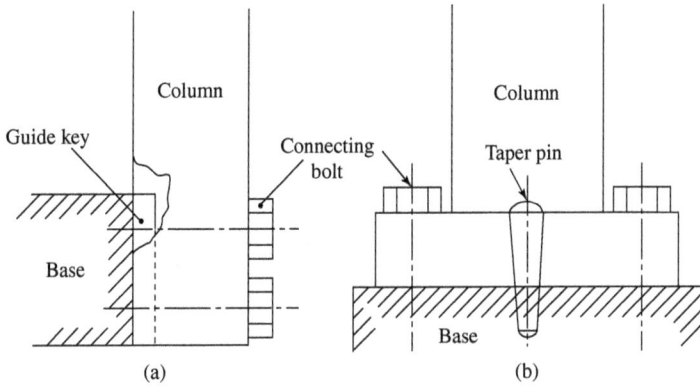

Figure 5-6 Two basic types of bolted joint.

1. *Bolted joint.* There are two kinds of bolted joint in a machine tool structure: one is for the connection of structural body components, such as the base, column, and headstock; the other is for connecting the machine elements, such as the clutch plate, flange, and bearing housing. With respect to the stiffness of the machine tool, the first kind is very important and can generally be classified into the two types shown in Fig. 5-6, depending upon the relative arrangement of the longitudinal axis of the connecting bolt to that of the column. In Fig. 5-7, some examples of the actual bolted joint and their mean interface pressures are shown, where the mean interface pressure has been estimated using Rötscher's proposal, i.e., Rötscher's pressure cone (see Sec. 7.1.1), and the connecting force is assumed to be 0.6 times the yielding stress of the bolt material. As a result, the mean interface pressure of the bolted joint can be estimated to range from 50 to 200 kgf/cm^2 (about 500 to 2000 MPa).

2. *Foundation.* The machine tool is, in general, installed or fixed on the factory floor, and consequently the static, dynamic, and thermal behavior of a machine tool as a whole is largely influenced by the foundation. The representative of the foundation is the leveling block of wedge type, shown in Fig. 5-8. The leveling block has, in principle, several stationary and sliding joints and thus can be called a joint of complex type. It is furthermore notable that the foundation shows very particular behavior under static, dynamic, and thermal loading, which is attributed to the properties of the concrete, grout, and soil.

Sliding joint. The sliding joint has the function of relative traveling movement between the two bodies in contact, so that the necessary form-generating function can be realized with satisfactory machining accuracy. In due course, the representatives are the guideway and main

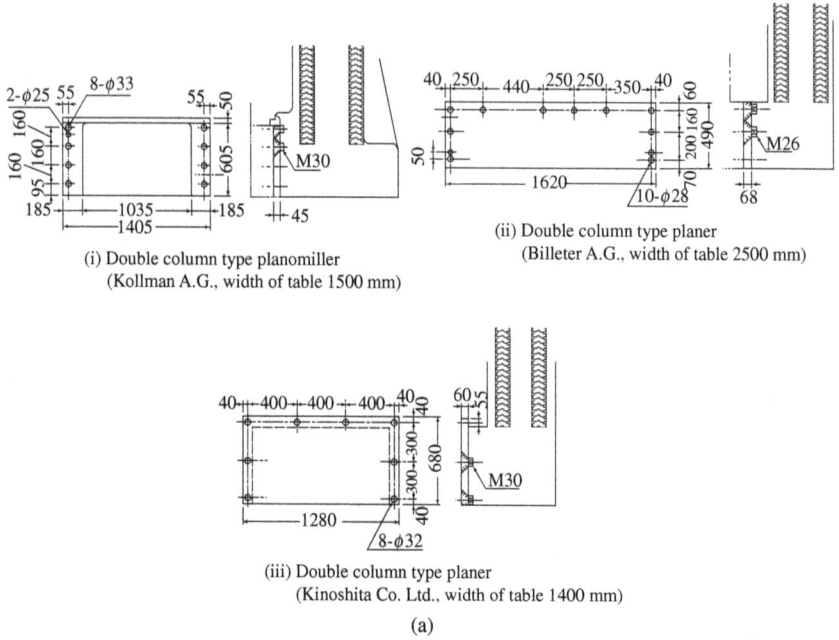

(i) Double column type planomiller
(Kollman A.G., width of table 1500 mm)

(ii) Double column type planer
(Billeter A.G., width of table 2500 mm)

(iii) Double column type planer
(Kinoshita Co. Ltd., width of table 1400 mm)

(a)

	Mean interface pressure p, kgf/mm^2			
	Material of connecting bolt			
	SS41B (JIS)	S45C (JIS) Normalizing	S45C (JIS) Tempering	10 k (DIN)
Planomiller Köllman AG	0.58	0.87	1.25	2.24
Planer Billeter AG	0.30	0.46	0.65	1.17
Planer Kinoshita Co.	0.46	0.71	1.04	1.81

(b)

Figure 5-7   Some examples of bolted joints in practice: (a) Arrangement of bolts and dimension of bolted joints and (b) mean interface pressure estimated by Rötscher's pressure cone (without external applied loads).

spindle-bearing system. Obviously, the anti–wear resistance capability in these joints is additionally of importance with the preferable joint characteristics.

Guideway.   The guideway has a completely different feature from the bolted joint. For instance, a machine tool of mature having bolted joints is capable of being replaced by a structure without bolted joints, i.e., monolithic structural body component, in certain cases of small and

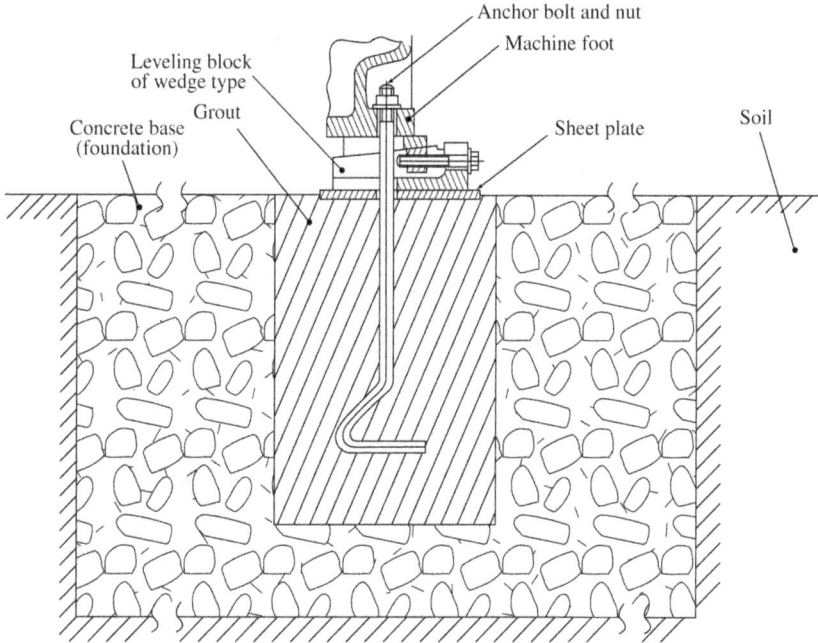

Figure 5-8 Schematic representation of foundation system.

medium-sized machine tools. In contrast, the guideway is necessary and inevitable to provide a machine tool with the basic function, e.g., form-generating movement. The guideway can, in general, be classified into the three types shown in Fig. 5-9, depending upon what the interfacial layers are, i.e., the slideway with hydrodynamic lubrication, the hydrostatic guideway, and the rolling guideway. Of these guideways, the slideway is worth actively investigating, because it is a typical variant of flat joint.

As can be seen from Fig. 5-10, a slideway consists of several flat joints, which are well combined and configured using core parts, e.g., bed (bed slideway), saddle (saddle slideway), gib, and keep plate in the case of the turning machine. In due course, each joint can be replaced with the mathematical model shown in Fig. 5-10. In addition, Table 5-1 summarizes the average magnitude of the interface pressure in the slideway so far reported elsewhere.[3] In many respects, the linear ball bearing and the linear roller bearing have recently prevailed among machine tool manufacturers across the whole world, especially in producing the conventional MC and TC. Importantly, some manufacturers use definitely the slideway, especially

---

[3]For example, see N. S. Atscherkan, *Werkzeugmaschinen—Berechnung und Konstruktion*, Band 1, VEB Verlag, 1961.

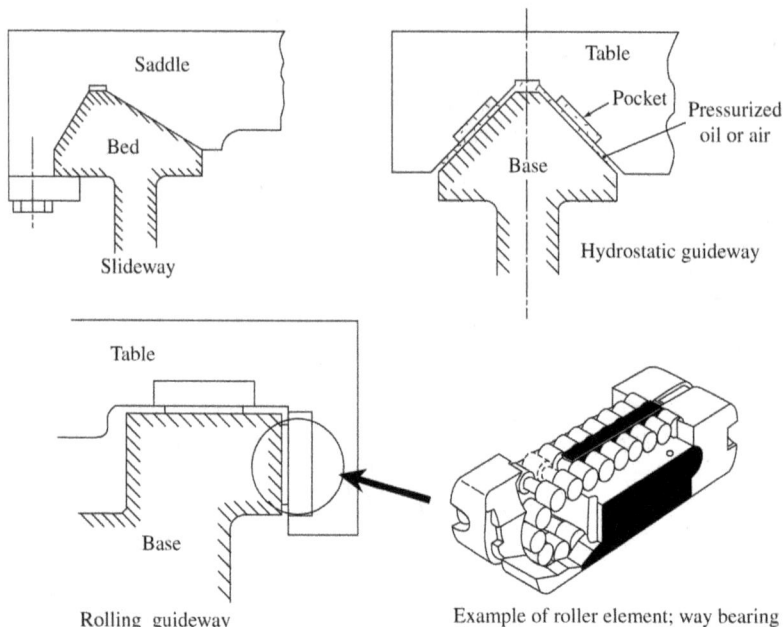

Figure 5-9   Three typical types of guideway.

when producing the MC and TC with specified functions, e.g., for better machining accuracy and heavier cutting.

In fact, the linear rolling guideway facilitates the ease of design and manufacture of the machine tool. Nowadays, the conventional MC and TC can be produced by designing them with the CAD and FEM; by purchasing the servomotor, NC controller, linear bearing, and other necessary parts and units; by in-house machining the structural body component in certain cases; and by conducting in-house assembly. In other words, a certain number of machine tool manufacturers worldwide

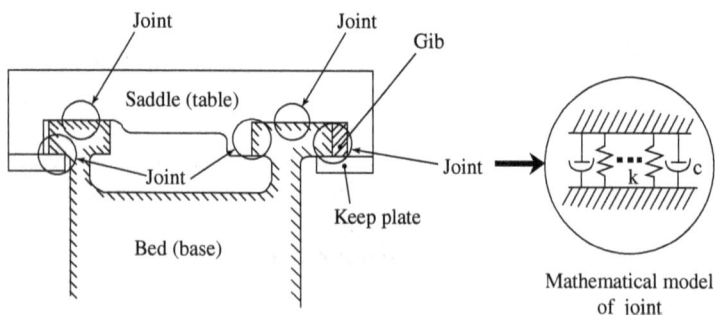

Figure 5-10   Joints in slideway and mathematical model of joint.

TABLE 5-1   Allowable Interface Pressure so Far Publicized

Size of machines	Materials of slideway	Sliding speed	Maximum allowable interface pressure in P_{max}
Small and medium	Cast iron vs. Cast iron	Nearly equal to feed speed (lathe, milling machine)	25–30
		Nearly equal to cutting speed (planer)	8
Large	Cast iron vs. Cast iron	High	4
		Low	10
Always under severe cutting condition Cast iron–steel slideway Steel-steel slideway			$0.75\ P_{max}$ $(1.2–1.3)\ P_{max}$

Source: U.S.S.R. Design Standard H49-2 (by Atscherkan).

Notes:
1. Actual interface pressure of American grinding machine was reported as to be 0.7 kgf/cm^2 by Prof. Atscherkan.
2. For cast iron–cast iron slideway, Prof. Schelesinger recommended that the maximum allowable pressure be from 4 to 6 kgf/cm^2.
3. JSME recommended the following values. In normal conditions, around 1 kgf/cm^2, under good lubrication, 3.5 kgf/cm^2, and slideway especially aiming at precise moving, around 0.3 kgf/cm^2.

become assembly-oriented, and a crucial problem for these manufacturers is to differentiate themselves from the competitors.

In both the large-size and the ultraprecision machine tools, however, the hydrostatic guideway is dominant. It is worth stating that the hydrostatic bearing has much wider application areas, within not only machine tools, but also other industrial machines, and thus we now have a realm of the hydrostatic bearing.[4] This book does not describe the hydrostatic bearing, but gives a quick note as follows.

Within the fundamental research, some representative subjects include (1) the load capacity analysis using FEM, (2) the behavior of the all-metal flexible thrust bearing, (3) the behavior of the aerostatic radial bearing with porous material, and (4) the characteristic features of the servostatic guideway. For example, the review paper of Warnecke [11] offers a firsthand view of the extent to which the related research into the hydrostatic guideway was carried out. In addition, the study of Masuko and Nakahara [12] must be reviewed to understand the far-reaching extent of the research. In their research, the overshoot behavior

[4]Refer to the rudimentary and specified materials such as exemplified by Refs. 9 and 10.

of the hydrostatic guideway in step loading was investigated in consideration of the fluid capacitance in the oil feed line. Obviously, the fluid capacitance is derived from the compressibility of the pressurized oil, elastic deformation of the pipe, and intermixing of air across the whole system. In short, the ratio of the overshoot to the steady displacement of the table increases with increasing fluid capacitance in piping and has a maximum value when the oil film stiffness is a maximum. Within an engineering calculation context, the corresponding mathematical model of the hydrostatic bearing follows that shown in Fig. 5-11 [13].

Main spindle-bearing system.   The main spindle-bearing system including the table system of the vertical turning machine can be regarded as a special case of the sliding joint. In other words, the guideway is for linear movement, whereas the spindle-bearing system is for rotational movement. The spindle-bearing system is of complicated configuration, as already shown in Fig. 5-4. More specifically, the following joints can be observed.

1. Stationary joints between the wall of the headstock and the bearing housing, bearing housing and outer race of bearing, both the side surfaces of the nut and spacer

2. The sliding joint at gearing and the rolling joint between the roller and its races

In these joints, we can also observe some new developments in accordance with the increasing spindle speed, i.e., higher-speed spindle system, as exemplified by the step sleeve, which is press-fitted to the

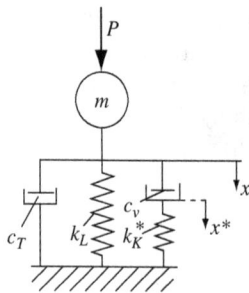

x: Deflection of spring k_L
m: Mass
k_L: Static spring constant of bearing
c_T: Damping coefficient of bearing depending upon dimensional specifications and oil viscosity
c_v: Damping coefficient of bearing caused by oil supplying system

k_k^*: Equivalent spring constant caused by oil compressibility

Figure 5-11   A mathematical model of hydrostatic bearing.

main spindle and can thus render the bearing nut useless to minimize the unbalance. In short, the spindle-bearing system is one of the complex joints, as is the foundation.

Semistationary joint.   This joint can be regarded as a variant of the stationary or sliding joint, and a typical example is a slideway with a clamping mechanism. The slideway behaves like the stationary joint or sliding joint depending on whether the clamping mechanism is in working condition or not.

With the advance of research into the machine tool joint, a rational classification system has been proposed (see Fig. 5-12) based on the essential feature of the joint [14]. The newly proposed classification emphasizes the leading structural configuration of the joint, i.e., that of closed, semiclosed, and open types, to correctly represent the two-dimensional FOF, and the distribution form of the interface pressure. The classification system is thus considered to be applicable to work in the academic sphere rather than in engineering design. In fact, we can analyze the static, dynamic, and thermal behavior in detail by comparing the open- to the closed-type joints, although we do not observe large differences in the engineering measures for both joints.

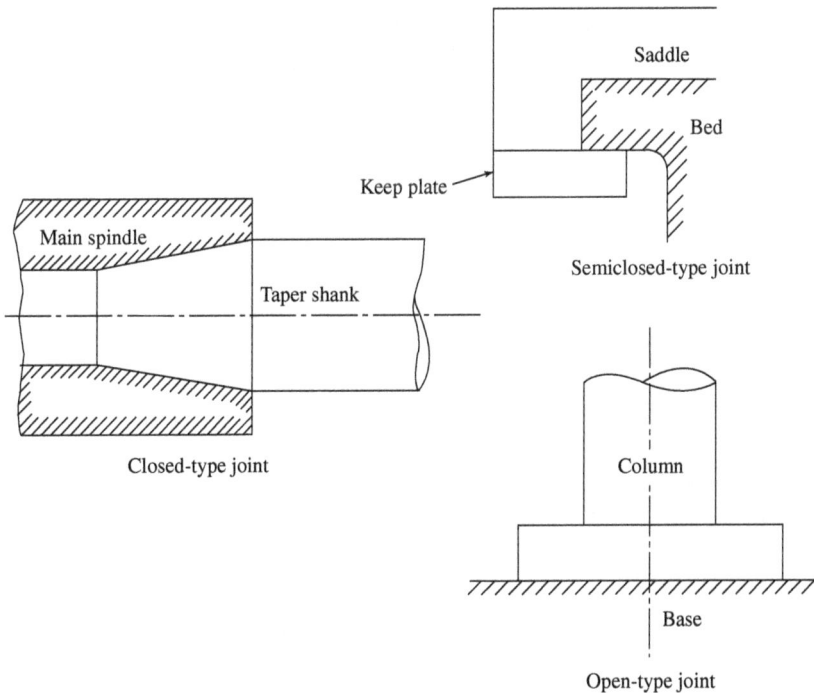

Figure 5-12   Classification of machine tool joints according to configuration aspect.

5.2   Definition of Machine Tool Joint and Representation of Joint Characteristics

Although many studies of the joint have been carried out since 1939, it is difficult to exactly define what the joint is and to determine where the region of joint is. Intuitively, an acceptable definition is that the joint is a portion of roughness and/or flatness deviation of the two surfaces in contact, but this idea is not applicable, because of the evidence obtained from earlier work. As shown in Fig. 5-13 [14], the normal joint deflection, which is given by subtracting the deflection of an equivalent solid from that of the jointed body, is larger than the total amount of surface roughness in contact under certain jointed and loading conditions.

It is thus necessary to propose a new idea for the definition of region of the joint, especially in the academia. In fact, an idea has been proposed in which the joint consists of the following three regions, as shown in Fig. 5-14, depending on the magnitude of the interface pressure [14].

1. The region a corresponds with the sliding joint under very low interface pressure, e.g., the slideway. In this case, the joint deflection is within the surface roughness and may be derived from the deflection of the surface asperities themselves; as a result, the sliding body, such as a table, can be regarded as a stiff body with respect to the base or bed slideway.

2. The regions a and b correspond to the sliding joint under higher interface pressure. In this case, the joint deflection may be determined

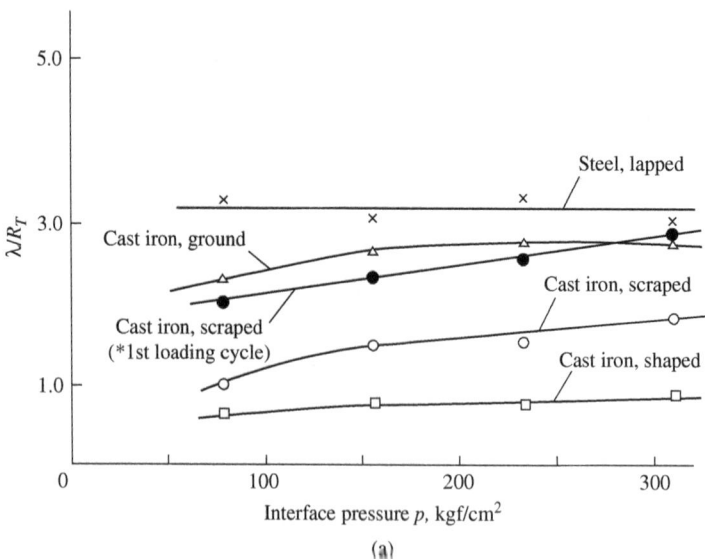

(a)

Figure 5-10   Changes of ratio λ/R_T with increasing interface pressure: (a) Under higher interface pressure and (b) under lower interface pressure.

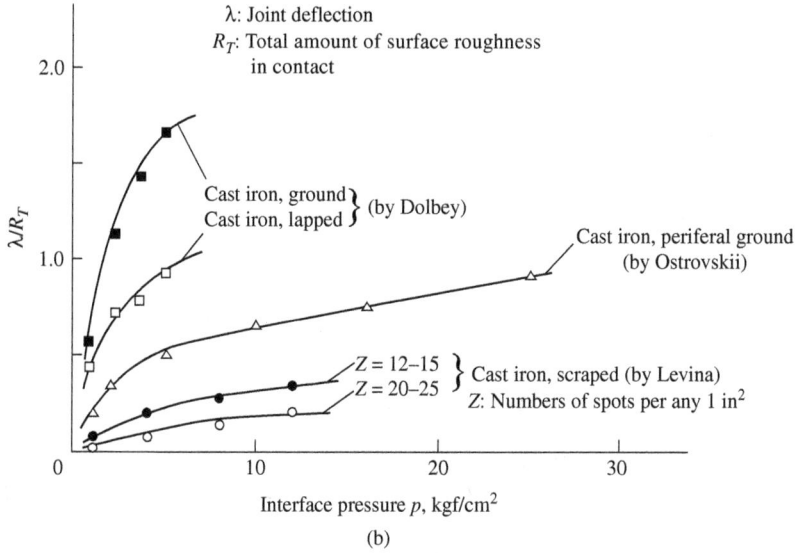

Figure 5-13   (*Continued*)

by the deflection of surface roughness and in part deformation of the surface layer of the body. The slideway showing local deformation in its operating condition falls into this category.

3. The region denoted by a, b, and c corresponds to the bolted joint, because its interface pressure is higher than that of other joints, and the local deformation of the structural body itself, e.g., warping or bedding in the body, is liable to appear.

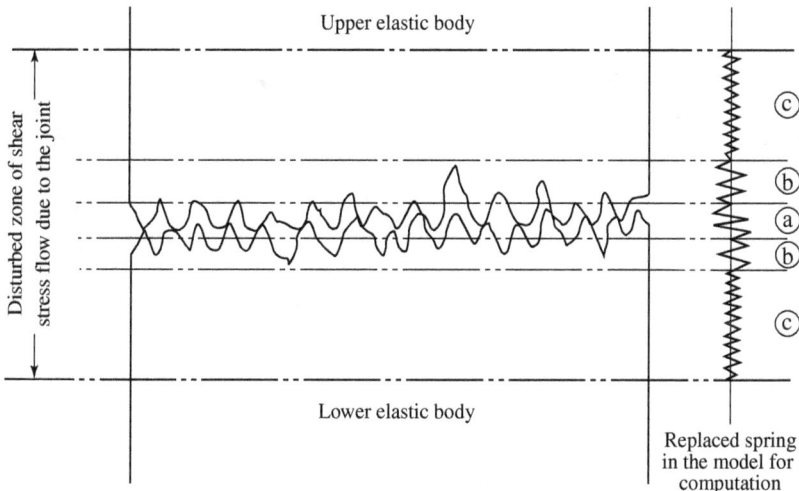

Figure 5-14   The concept of region of joint.

At present, each region cannot be exactly and quantitatively determined; however, the idea described here is considered reasonable and applicable to the analysis, design, and research of the joint. To define the region of the joint and to research the joint, thus, it is of great importance to grasp the magnitude of the interface pressure being acted on the joint. In consideration of some data already reported elsewhere, the magnitudes of the interface pressure can be roughly estimated as follows.

• For the sliding joint without local deformation: from 1 to 50 MPa (about 0.1 to 5 kgf/cm^2)

• For the sliding joint with local deformation: from 50 to 250 MPa (about 5 to 25 kgf/cm^2)

• For the bolted joint: from 500 to 2000 MPa (about 50 to 200 kgf/cm^2)

Although the exact definition of *joint* itself has not been determined yet, it is, at the least, necessary to represent clearly the leading joint characteristics. Based on long-standing experience, there are two representative expressions of the static and dynamic behavior of the structural body component in machine tools, such as in the bolted joint in Fig. 5-15. The expressions for bending stiffness K_{SB} and torsional stiffness K_{ST} are suitable for the engineering design, whereas another expression is aimed at describing directly the joint characteristic itself. The joint characteristic can be defined by using, e.g., the inclination angle $\Delta \phi$ created by the existence of the joint. This inclination angle is very sensitive to change in the essential features of the joint, showing obviously the joint characteristics, and thus the expression using the inclination angle is available for detailed research. Obviously, these two expressions can be converted to one another, as already shown in Fig. 5-15, provided that the flange behaves as a stiff body, resulting in the linear distribution of the interface pressure.

Admitting the validity of the conversion shown in Fig. 5-15, at issue in the machine tool structure is the complex loading condition, and often the flange is prone to show local deformation. Within an engineering design, the stiffnesses K_{SB} and K_{ST} are in fact useful and applicable; however, the equivalent stiffness, which is given by the column without any joints, is required to have the fully quantified evaluation or to carry out an absolute assessment. Figure 5-16 shows such a correlation together with the complex loading condition in the machine tool, for the sake of the ease of understanding. In short, the joint stiffness can be evaluated by comparing, for instance, K_{SB} with K_{OB}, resulting in the expression of relative stiffness.

To enhance the sophisticated knowledge, the emphasis now is on the flat joint, where two characteristics are at issue: one is the normal joint stiffness K_{sn} and the other is the tangential joint stiffness $K_{s\parallel}$, as shown in Fig. 5-17. In not only flat but also bolted joints, we should furthermore

Bending stiffness

$K_{SB} = P/\delta$, kgf/μm

Torsional stiffness

$K_{ST} = T/\psi_T$, kgf · m/rad

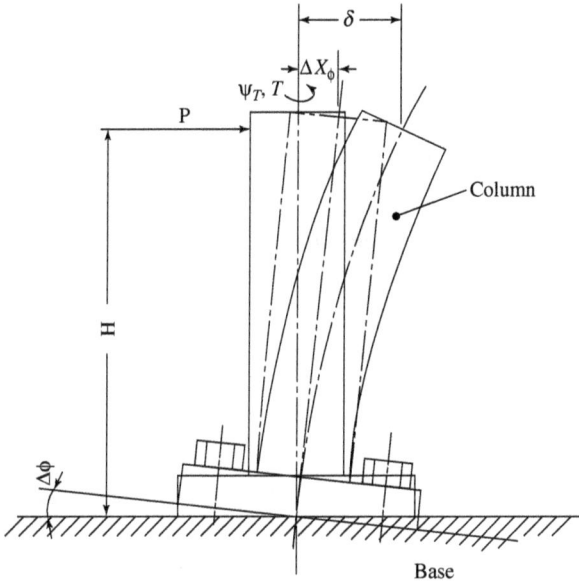

$\Delta X_\phi = H \cdot \Delta\phi$

$\delta = \Delta X_\phi + \dfrac{PH^3}{3EI}$

E: Young's modulus

I: Cross-sectional
  moment of inertia

Figure 5-15   Characteristic expressions for bolted joint.

note that the stiffness per unit length or per unit area is effective along with the stiffness so far used in carrying out the research into and design of the joint.

In contrast to static loading, there are certain kinds of difficulties in representing the characteristics of the joint under dynamic loading. Apart from the joint in real metal-to-metal contact and under normal loading, in general, the joint shows larger damping than the material damping, and thus the phase difference between the dynamic load and the corresponding deflection is important to clarify or to express exactly the dynamic behavior of the joint. In due course, the joint under dynamic loading can be replaced with the spring-dashpot model shown in Fig. 5-18, and using this model, the dynamic joint stiffness K_{dyn} can be written as

$$K_{\text{dyn}} = K_a + jK_b \tag{5-1}$$

where K_a is the in-phase stiffness and K_b is the quadrature stiffness (damping-related stiffness).

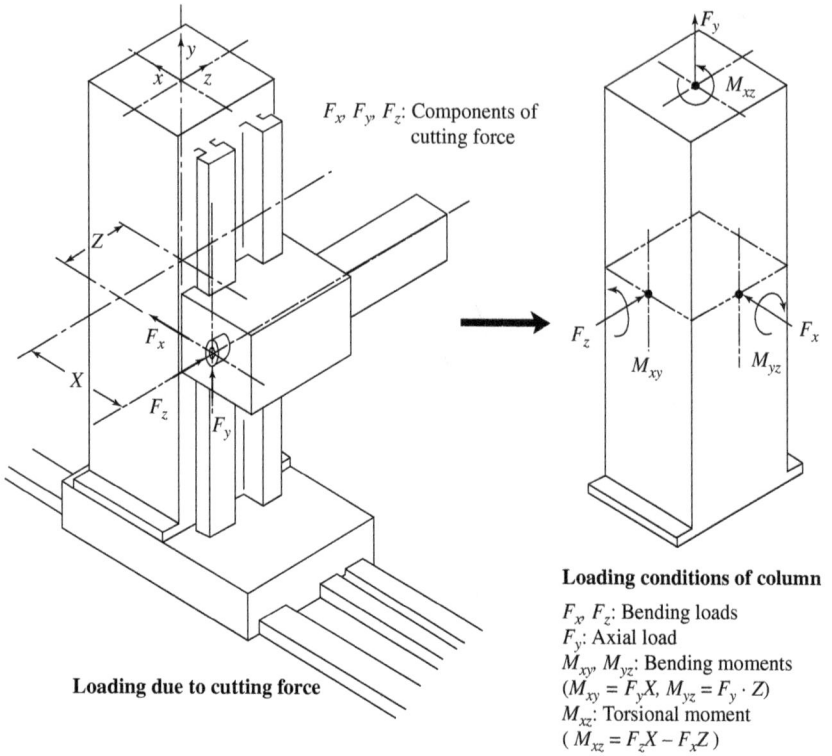

F_x, F_y, F_z: Components of
cutting force

Loading due to cutting force

Loading conditions of column

F_x, F_z: Bending loads
F_y: Axial load
M_{xy}, M_{yz}: Bending moments
($M_{xy} = F_y X$, $M_{yz} = F_y \cdot Z$)
M_{xz}: Torsional moment
($M_{xz} = F_z X - F_x Z$)

(a)

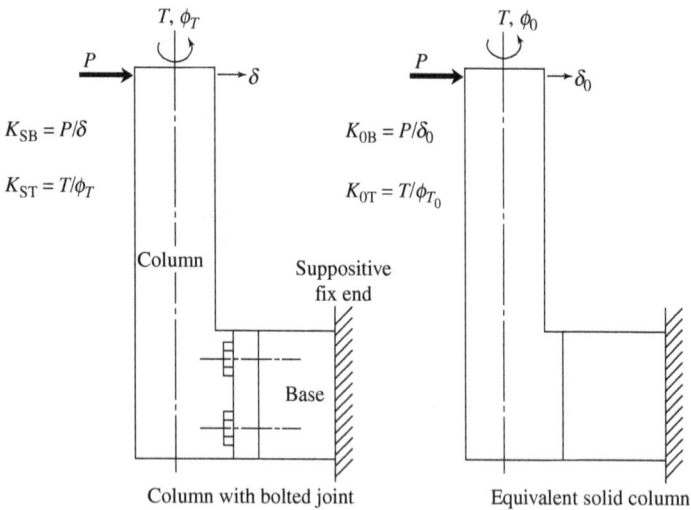

$K_{SB} = P/\delta$

$K_{ST} = T/\phi_T$

$K_{OB} = P/\delta_0$

$K_{OT} = T/\phi_{T_0}$

Column with bolted joint

Equivalent solid column

(b)

Figure 5-16   Stiffness expression for bolted joint in engineering calculation: (a) Complicated loading conditions in a floor type horizontal boring and milling machine and (b) expression for relative joint stiffness.

$K_{sn}, K_{sh} = P/\delta$, kgf/μm

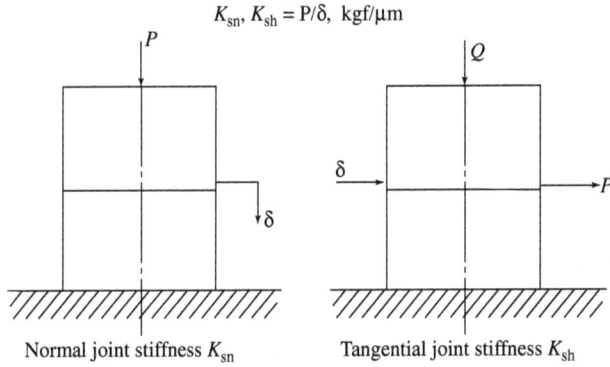

Normal joint stiffness K_{sn} Tangential joint stiffness K_{sh}

Figure 5-17   Stiffness expressions for flat joints under normal and tangential loading.

In fact, Eisele and Corbach [16] and Andrew [17] proposed this expression and verified its validity to a large extent. Assuming that the joint deflection is $a = a_0 e^{j(\omega t - \xi)}$ when the flat joint is subjected to the load $P_{dyn} = P_0 e^{j\omega t}$, both the in-phase and quadrature stiffnesses yield to

$$K_a = k \qquad \text{and} \qquad K_b = \omega c \tag{5-2}$$

In the dry metal joint, the damping capacity in the direction normal to the joint surface is likely to be nil, i.e., $K_b = 0$, and thus the dynamic joint stiffness yields K_a itself.

Within an engineering context, the dynamic behavior of the joint is very complicated compared with the static behavior, and consequently it is not possible to express perfectly the dynamic behavior of the joint using only the dynamic stiffness. In fact, such frequency response and transmitting

$P_{dyn} = P_0 e^{j\omega t}$

$a = a_0 \cdot e^{j(\omega t - \xi)}$

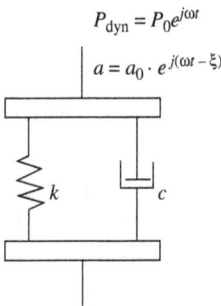

$K_{dyn} = K_a + jk_b = k + j\omega c$

P_{dyn}: Exciting force
 a: Vibration amplitude
 ξ: Phase difference between P_{dyn} and a

Figure 5-18   Expression for dynamic joint stiffness (by Eisele and Corbach).

ability of exciting force should be employed together with the dynamic stiffness, if necessary, to correctly represent the dynamic behavior of the joint. In addition, it is worth suggesting that the concept of *complex damping* might be applicable to an expression of joint behavior [18]. Reportedly, the complex damping is caused by the hysteresis property of the spring, which appears as viscous damping.

5.3   External Applied Loads to Be Considered and Fundamental Factors Governing Joint Characteristics

In the design and manufacture of the structural body component of the machine tool, the bending and torsional loads are of great importance and should be taken into consideration, whereas the axial load has, in general, no significant effects on the static and dynamic behavior of the machine tool. Because the joint is regarded as a structural body component, in nearly all the machine tool joints, the bending and torsional loads are important. In contrast, the axial load becomes dominant as a leading factor for joints comparatively rarely. Actually, when the relative microslip occurs between the two mating joint surfaces, the normal load governs the energy dissipation at the joint, e.g., the damping capacity of the bolted joint. Although some loading conditions for the bolted joint have already been displayed in Fig. 5-16, these examples are not sufficient to understand the various loading conditions of the joint. The further knowledge for this issue will be grasped by referring to the succeeding chapters, in which the actual loading conditions are demonstrated precisely. In this context, note that an optimum design of joint under several different kinds of the load is very difficult at present.

In addition to the loading condition, many factors need to be considered in the designing and production procedures of the joint. Although the importance of each influencing factor varies according to the kind of joint, in general, these factors can be classified into the following three categories.

1. Factors concerned with the applied load, i.e., magnitude, kind, and type of applied load. For the dynamic load, its frequency and static preload should be considered.

2. Factors concerned with the joint surface, i.e., physical properties of joint material, macroscopic and microscopic topographies, machining method, surface roughness, waviness and flatness deviation, interfacial layers (oil, plastic, and metallic shims).

3. Factors concerned with the jointing method, i.e., size and shape of joint surface, number of contact entities across a whole joint surface, machined lay orientation, size and shape of joint surroundings and

their own stiffness. In short, the relative stiffness of joint surroundings to the joint itself is of great importance.

Obviously, by taking the factors summarized above into consideration, we can design the simple flat joint itself, which can be regarded as a basic configuration of all the machine tool joints. Thus for each joint, we should furthermore consider other factors shown, e.g., for bolted and sliding joints, in Table 5-2, and the dominant factors are as follows.

TABLE 5-2 Specified Influencing Factors in Bolted and Sliding Joints

(a) Bolted joint

Resultant factor:
Bearing area and interface pressure distribution
derived from tightening force

Leading factors

Number of connecting bolts Locating position and arrangement of connecting bolts Tightening force of connecting bolt Flange shape and thickness Existence of such a taper pin or guide key

Auxiliary factors

Mechanical properties and diameter of connecting bolt
Kinds of connecting bolts—hexagonal head bolt, hexagonal
                               socket head cap screw and reamer
                               bolt
Interfacial layer
Thread accuracy of bolt and nut

(b) Sliding joint

Resultant factor:
Bearing area and interface pressure distribution
while being still stand and traveling

Leading factors

Number and allocation of guideways Kinds of guideways—slideway, hydrostatic guideway, and rolling guideway Traveling velocity of structural body component Feed force transmitting mechanism including table feeding device coupling Gib and keep plate configuration in slideway

Auxiliary factors

Frictional and wear properties of joint materials Kinds of interfacial layer Viscosity of lubricants Structural configuration—separation of bearing way and guide reference way

1. The bay-type flange is very popular, e.g., in designing the bolted joint for integrating both the structural body components and the connecting mechanism of the ball screw to the table. In due course, the flange variations, e.g., location of the connecting bolt, configuration of the bay-type with or without rib, and bay-type with enclosed bolt pocket, show the larger difference in the joint stiffness.

2. The deformation and vibration modes of the joint surroundings are the leading factors in the determination of the damping capacity.

3. The structural configuration has especially large effects on the interface pressure distribution, resulting in the differing stiffness of the joint.

5.4   Effects of Joint on Static and Dynamic Stiffness, and Thermal Behavior of Machine Tool as a Whole

According to long-standing experiences, the static stiffness of the joint is, in general, very low compared with that of the structural body component itself. As a result, the overall stiffness of a machine tool as a whole deteriorates significantly, as already shown in Fig. 5-2. More specifically, the jointed body structure shows the following characteristic features under static loading.

First, the deterioration of the overall stiffness of a jointed structure is obviously large, and Table 5-3 reproduces some data for the stiffness deterioration due to the joint, which are obtained from reports so far

TABLE 5-3   Rates of Joint to Overall Deflections in Machine Tools

Kinds of machine tools/ units	Objective portion of overall deflection	Dominant joints affecting overall deflection	Influencing rate of joint %
Vertical milling machine	Table–tool	Knee guideways Table guideways	60–70
Vertical boring machine	Bearing surface of ram head–tool	Ram guideways	In axial direction of ram: 25 in max. In perpendicular direction of ram axis: 40 in max.
Engine lathe	Tailstock as a whole work–tool	Barrel guide Saddle guideways	60–70 40 in max.
Gear box	Power transmission systems (in torsion)	Key and spline joints Clutches Bearings	20 30

publicized elsewhere. It is furthermore emphasized that the joint stiffness reduces considerably when the external load or due reaction force at the joint is associated with the shear action.

Second, the relative displacement between the tool and the workpiece due to the joint deflection is duly enlarged by the structural component, when it behaves as a lever, as already shown in Fig. 5-2. In this case, the deflection at the loading point can be obtained by multiplying the inclination angle of the joint by the length of column. This multiplication action of a structural body component is remarkable in most machine tools, having a large influence upon the machining accuracy. Obviously, the data concerning the stiffness deterioration are useful to understand quickly the influence of the joint upon the overall stiffness. In contrast, to observe the joint characteristics in detail, it is necessary to simplify the joint configuration, and thus most of the earlier researches into the joint were carried out with the simple, flat, and small test specimens, which are a model of the actual joint.

In this context, note that the model theory for the machine tool would not be completely applicable for the structure having the joint as reported by Ito and Masuko [19]. In fact, they suggested that the model theory can be applied to the structure with a joint, provided that the length scale factor is more than one-third. Most of the earlier works are thus not directly applicable to the design of the machine tool structure in full-size, but are indirectly applicable.

Third, joint stiffness shows nonlinear behavior; i.e., the load-deflection characteristic of a structure with joints is nonlinear. This nonlinear characteristic of joint stiffness can be observed in all joints notwithstanding their structural configurations. Two typical examples classified by the magnitude of the interface pressure are shown in Figs. 5-19 and 5-20. Figure 5-19 shows the load-deflection characteristic of the joint under higher interface pressure as reported by Thornley et al. [15].

It can be observed that the joint deflection shows, along with nonlinear behavior, hysteresis behavior during loading and unloading. In addition, the residual deformation can obviously be observed after unloading. A further interesting behavior is that the load-deflection characteristics for a given joint under maximum loading cycle can be represented by a curve 0-A-C-E-F, showing also a curve 0-A-C-D under a lower loading. In the lower loading, a curve for reloading and loading far beyond the upper limit of the previous one coincides with that designated by D-C-E-F. Importantly, it is worth suggesting that a linear portion within the unloading curve is identical to that of an equivalent solid, which has the same shape and dimension as that of the flat joint. Without exception, the load-deflection characteristic of the equivalent solid is linear and inversely proportional to Young's modulus, also not showing hysteresis behavior in the loading and unloading cycles.

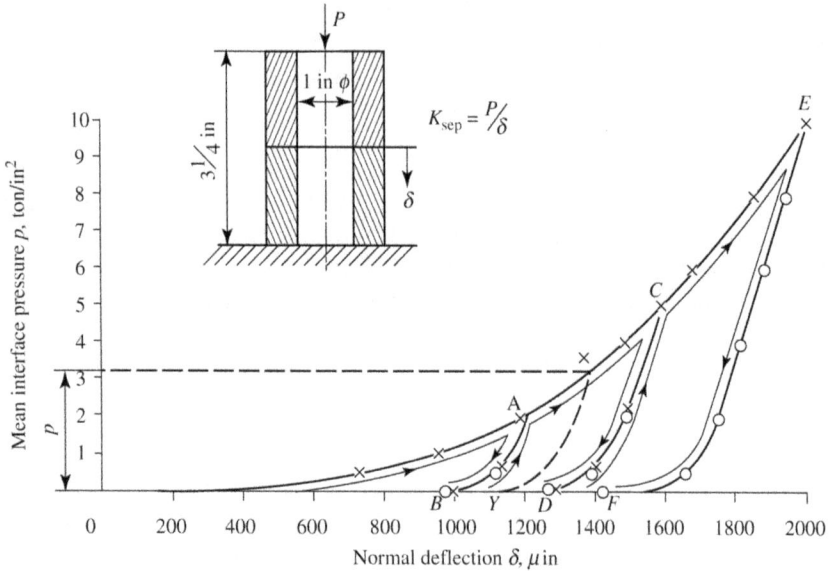

Figure 5-19   Normal deflection in flat joint under higher interface pressure (by Thornley et al. [15]).

	Mean contact pressure, ton/in²	Ratio of stiffness K_{sep}/K_{eq}
Equivalent solid K_{eq}	2.5	1.0
L/L	2.5	0.29
G/L	2.5	0.19
G/G	10.0	0.25
M/M		
M/G	10.0	0.21
M/L		

L: Lapping   G: Grinding
M: Milling
Jointed material: Mild steel

In contrast, Fig. 5-20 demonstrates the load-deflection character-istic of the joint under lower interface pressure [20], and it can be observed that the joint behavior is similar to that shown in Fig. 5-19, although the quantitative values are different, and no residual displace-ment can be observed.

In summary, the static stiffness of the jointed structure is duly detori orated compared to that of an equivalent structure; more specifically, the smoother the joint surface and the less the flatness deviation of the joint

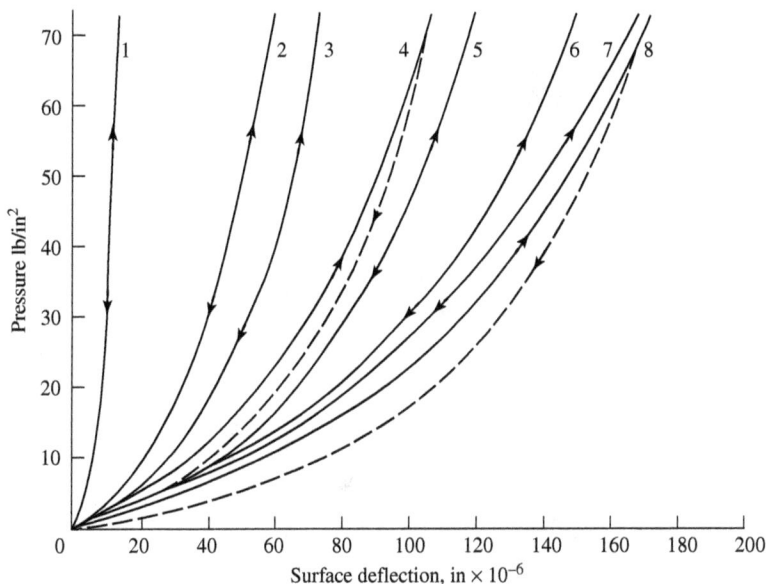

Curve	Upper specimen	Lower specimen
		Material combinations
1	C.I. lapped	C.I. lapped
2	C.I. ground	C.I. ground
3	C.I. ground	C.I. scraped
4	C.I. ground	Ferobestos ground
5	C.I. ground	Glacier DX-plain ground
6	C.I. ground	Glacier DX-dimpled ground
7	C.I. ground	Glacier DU-as received
8	C.I. ground	Tufnol—ground

Note: C.I.: Cast iron

Figure 5-20 Normal deflection in flat joint under lower interface pressure (courtesy of Bell).

surface, the larger is the joint stiffness, such as shown in Fig. 5-21 [21]. In Fig. 5-21, the coefficient b^* is a representative characteristic of the joint stiffness and given by the following expression.

$$\text{Joint stiffness per unit area } dp/d\lambda = b^*p \qquad (5\text{-}3)$$

where p and λ are the mean interface pressure and joint deflection, respectively, and the b^* value is available for the interface pressure up to 10 ton/in^2 · (154 MPa, note: 1 long ton = 2240 lb)

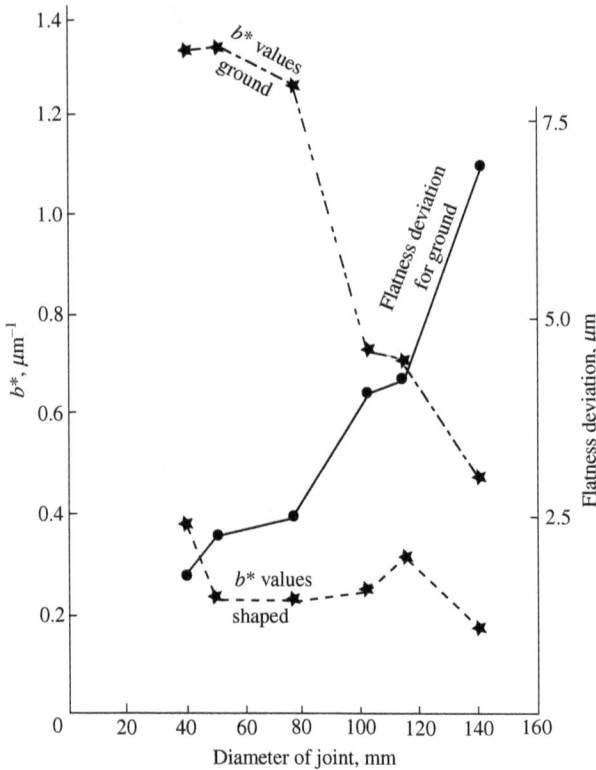

Figure 5-21 Deterioration of joint stiffness by flatness deviation (by Connolly and Thornley).

In general, the flatness deviation and waviness of the finished surface become larger with better surface roughness. As a result, the shaped joint is not sensitive to the flatness deviation, as shown also in Fig. 5-21.

On the basis of the basic research using the model and structure in full-size, the jointed structure has, in general, lower dynamic stiffness than the equivalent solid, resulting in lower natural frequency. Figure 5-22 is a reproduction of the static preload-dynamic stiffness diagram for the flat joint having two-dimensional surface roughness [22], and the dynamic stiffness increases with increasing normal static preload, although it does not approach that of an equivalent solid. It is emphasized that the most remarkable benefit of the joint is the increase of the damping capacity of a machine tool as a whole, and the increasing rate of damping capacity of the machine tool structure having joints depends upon the machining method of the joint surface.

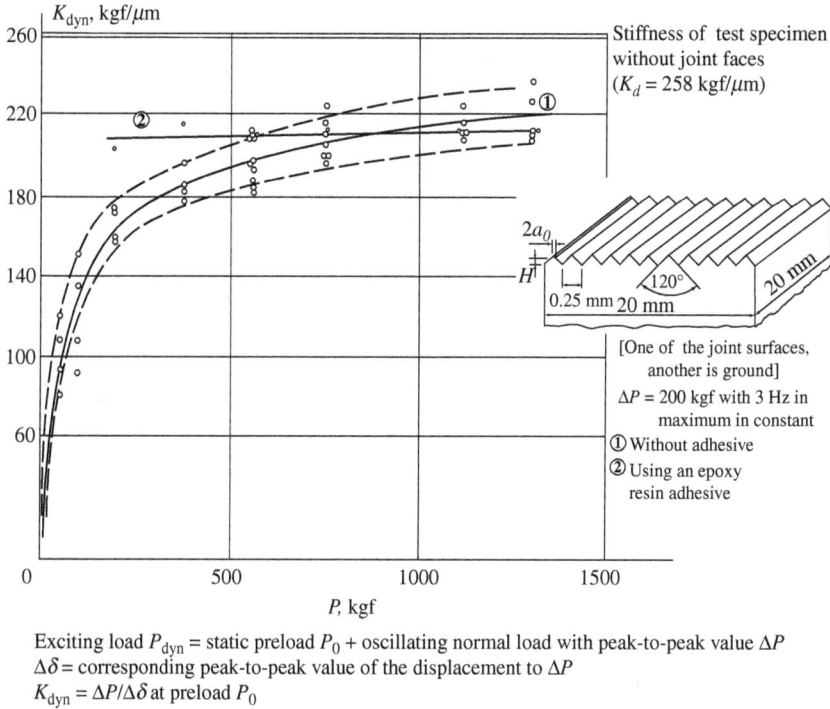

Exciting load P_{dyn} = static preload P_0 + oscillating normal load with peak-to-peak value ΔP
$\Delta \delta$ = corresponding peak-to-peak value of the displacement to ΔP
K_{dyn} = $\Delta P / \Delta \delta$ at preload P_0

Figure 5-22   Dynamic stiffness of flat joint (by Dekoninck).

In the machine tool of full-size, furthermore, oils, dusts, metallic shims, and so on are always found on the joint surface. These interfacial layers have no effect on the static joint stiffness, but have large effect on the dynamic joint stiffness.

Obviously, the joint is also at issue in the thermal deformation of the machine tool, and a crucial problem is the thermal contact resistance. In short, the machine tool shows a certain discontinuity in the temperature distribution when the thermal contact resistance caused by the joint becomes larger, as shown in Fig. 5-23 [23]. In this case, the experiment was carried out to clarify the influence of the interfacial layer between the headstock and base, and we can observe the larger effect of the interfacial layer, resulting in the large vertical displacement of the main spindle. Although the joint is very important even in the thermal deformation of the machine tool, there has not been vigorous research and engineering development, and thus the machine tool designer is requested to refer to the related knowledge in the spheres of the nuclear reactor and space vehicle.

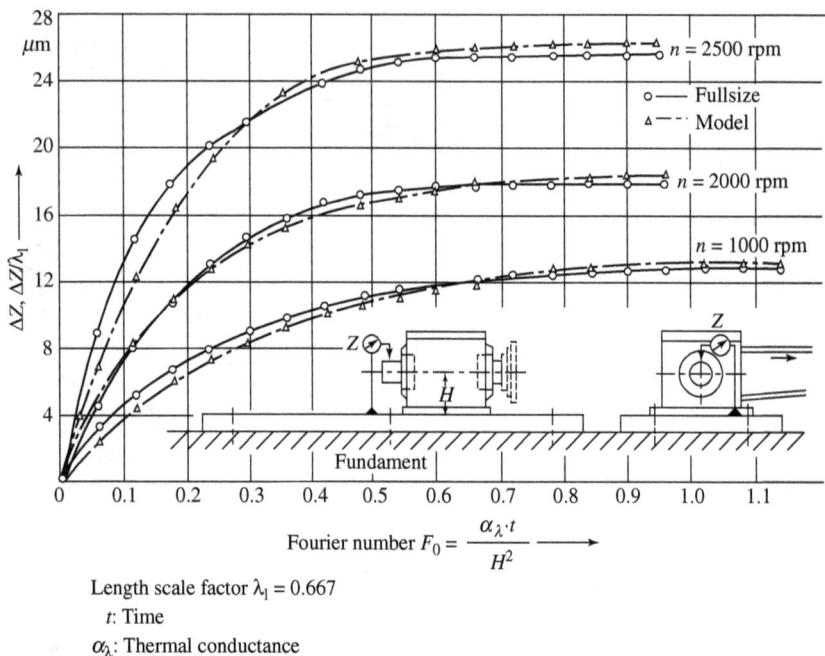

Figure 5-23 Model testing for thermal deformation of main spindle (by Schossig).

5.5 Firsthand View of Research History

The aims and scope of this book lie primarily in the technological aspect; however, knowledge of the history of the research work is somewhat helpful to reinforce the rudimentary knowledge about the machine tool joint. In this section, thus, a firsthand view of research works will be given by emphasizing those from embryonic to first fruition stages before 1975, in the mid-1980s, and in the 1990s. In addition, certain future perspectives will be touched on when necessary.

Research history before 1975. It is very difficult to exactly determine when the first research and engineering work commenced and who conducted them. This is a general rule in the engineering sphere, because each engineering topic is, more or less, subjected to a certain technological inheritance from its predecessors. Despite having such difficulties, Drs. Reshetov and Levina could be credited as the initiators, who explicitly suggested the important role of the joint within the body structure, when realizing the required dimensional and performance specifications of a machine tool from the engineering viewpoint. After carrying out long-standing research, they suggested the necessity of incorporating the joint characteristics into the structural design of the machine tool [24].

In retrospect, there were some symptoms for revealing the importance of the machine tool joint. For instance, Prof. Meyer of Technische Hogeschool in Delft aimed once at the reinforcement of the press-fit portion in the crankshaft for a large-size ship diesel engine without enlarging its size. He conducted a cooperative work with Technische Hochschule Berlin under the control of Prof. Kienzle from 1928 to 1942, and Prof. Meyer clarified the reinforcement effect of the hard particle when applying it to the interface of the single flat joint under normal preload and shear force. In addition, Meyer observed the normal joint deflection by the preload and the nonlinear relationship between the shear force and the corresponding displacement, and duly suggested the dominant role of the joint in the machine tool structure [25]. Following this cooperative work, Kienzle and Kettner [1] showed evidence in 1939 in which the damping capacity of the welded structure was, as already stated, larger than that of equivalent cast structure, and they implied that the welded joint itself is a major cause of this interesting behavior. Thus, we can assert that the first academic research into the joint could be credited to Kienzle in 1939. In fact, the work of Kienzle was carried out to establish a method by which the production time of machine tools could be shortened to respond to the rapidly increasing production volume of the war supplies during World War II.

Figures 5-24 and 5-25 are one of the quick notes to understand the history of research into the machine tool joint in earlier days. In short, Fig. 5-24 shows the ascending and descending flow of the representative research work and subjects in every country, and from it we can understand the diversification trends of the research activities along with implicit mutual influence when the research work was conducted. In contrast, Fig. 5-25 shows explicitly the importance of the human network, information exchange, information transfer by publications, and so on, when the research work was carried out. This is research into the bolted joint and consequent concerns; however, we can expect to have similar results when changing the objective research subject. It is very interesting to know that the world is very small and how the information flies speedily.

Intuitively, it can be seen that the core research works were carried out using mostly the simple flat joint at this stage by choosing the representative joints, e.g., slideway, rolling bearing, and bolted joint. We can thus clarify the basic behavior of the machine tool joint, although not everything has been unveiled. The root cause of some remaining problems at this stage was the lack of effective computational and experimental techniques. In contrast, these earlier works dispelled some old myths. For instance, a famous maxim for the design of the bolted joint, i.e., Rötscher's pressure cone (see Chap. 7), has been verified as applying to only a specified joint; however, a considerable number of designers,

Year	Milestonelike researches	Marked evidence

1939: Performance comparison between cast and welded beds (Kienzle & Kettner, TH Berlin) — Implication for importance of joints

1940

1950 — 1956: Damping in laminated disks, quill structure and bearing (Reshetov & Levina, ENIMS U.S.S.R.) — Initiation of academic researches

1957: Static & dynamic behavior of bolted joint (Schlosser, TH Hannover)

1960

1964: Initiation of researches at MERL of MITI & Tokyo Institute of Technology

1965–1975: Prevailing researches in Belgium, France, Germany, Netherlands, and United Kingdom

Around 1970: CAD in consideration of joints (Back of UMIST & Plock of TH Aachen)

1970

1978: Initiation of researches in North America—Characteristic features of EDM-finished joint (Beards)

1975: Thermal behavior of flat joint (Yoshida, MERL)

End of 1970s: Use of engineering design in consideration of joints Initiation of researches in Egypt and P. R. of China Symptom of new research deployment

1980: Influences of surface waviness (Kops, Univ. of McGill)

1980

Around 1980: Deployment to researches into chuck & modular tooling

1990

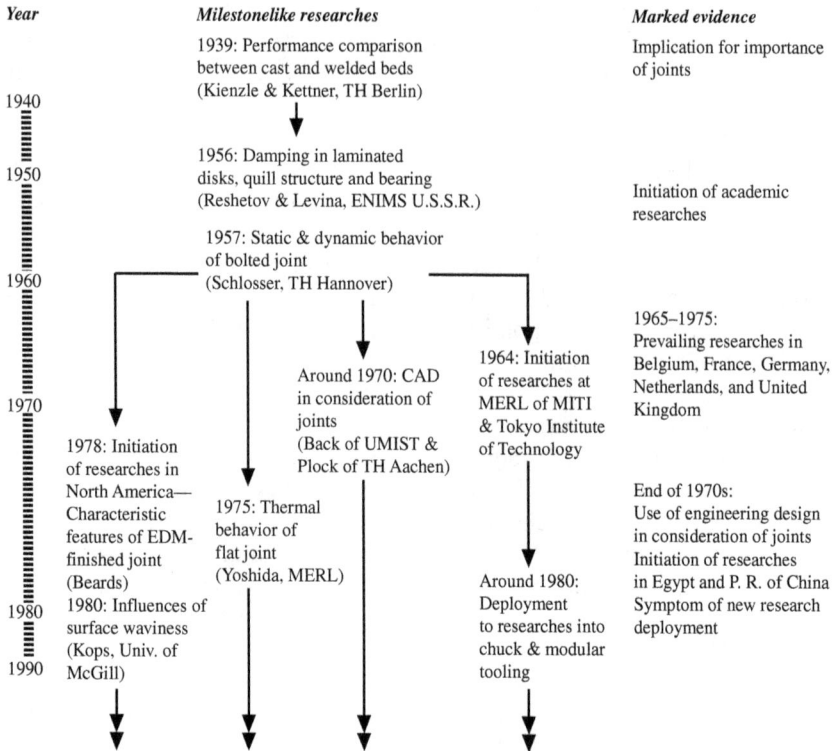

Figure 5-24 Firsthand view of research history in machine tool joint.

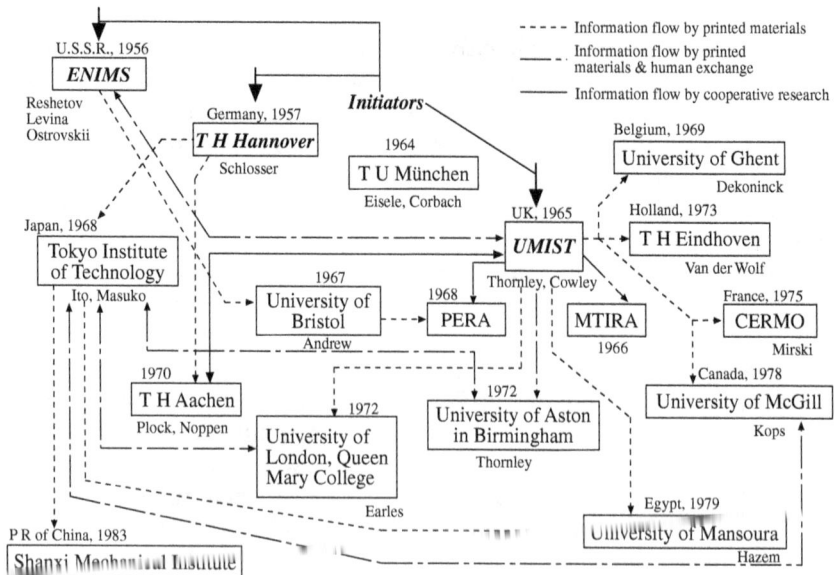

Figure 5-25 Space-time continuumlike research map for machine tool joints.

especially those in civil engineering, believe in the wider availability of Rötscher's pressure cone even in the 2000s.

Research trends between 1975 and 1985. The static, dynamic, and thermal behavior of the joint has been clarified to a larger and various extent by the research works carried out up to 1975. In due course, the designer is capable of conducting the design of the body structure, if necessary, in consideration of the joint characteristics. In addition, it is possible to estimate the behavior of other joints that have not yet been investigated by bringing together the knowledge about the representative joints so far investigated in detail.

In this second phase, researchers and engineers extended their focus to the joints in the accessories and peripheral equipment, whereas the research subject of the sliding joint was duly ramified in accordance with the performance enhancement of the machine tool. In fact, at burning issue was the taper connection between the main spindle and the tool shank, as well as the connecting mechanism between the table and its driving system. To continue with the former subject, the focus is, as will be stated later, the HSK (*hollow shank tool*) for MC and TC, and it becomes a pronounced area of interest in the 1990s. In contrast, the latter is one of the key technologies for the ultraprecision machine tool. Figure 5-26 is a time series–like firsthand view of the research and engineering development subjects of the taper connection. From these, it can be readily seen that the research objective has shifted to complex ones more than ever

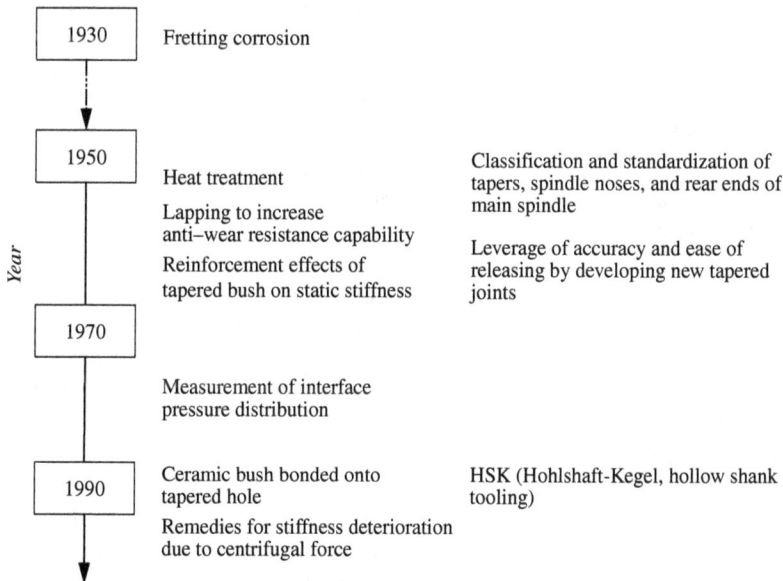

Year		
1930	Fretting corrosion	
1950	Heat treatment	Classification and standardization of tapers, spindle noses, and rear ends of main spindle
	Lapping to increase anti–wear resistance capability	
	Reinforcement effects of tapered bush on static stiffness	Leverage of accuracy and ease of releasing by developing new tapered joints
1970		
	Measurement of interface pressure distribution	
1990	Ceramic bush bonded onto tapered hole	HSK (Hohlshaft-Kegel, hollow shank tooling)
	Remedies for stiffness deterioration due to centrifugal force	

Figure 5-26   Time-series visualization for technology development in taper connection.

before with the enhancement of the machine tool technology and the advent of new technologies available for the related engineering activities.

For further understanding, now let us look at an example of the problem of the taper connection. As already shown in the research history in Fig. 5-26, the first facing problem was fretting corrosion, and the engineer endeavored to clarify its cause. In this regard, it is very interesting that the machine tool even in the 1990s suffered from fretting corrosion at the tapered hole of the main spindle. Actually, some of the representative MCs have employed the ceramic tapered hole as one of the remedies. Next, the stiffness of the taper connection was investigated including the reinforcement effect of the tapered bush press-fitted on the main spindle [26]. On that occasion, the fitting tolerance of the taper connection became crucial. To improve the machining accuracy, it is necessary to use tight fitting by having the bearing area extended across the whole joint surface; however, this leads to greater difficulty in the release of taper shank. For instance, the two-stepped taper was widely employed for jig borers. The MC was furthermore requested to enhance its cutting performance in considering that the tapered connection between the main spindle and the end mill is the weakest within the machine-tool-work system, during machining the plastic injection mold. Within an engineering design context of the tapered connection, one remedy is to produce smooth and full contact around the larger end diameter of the tapered hole.

Research activities in the 1990s.   In the late 1990s, research into the machine tool joint became less active compared to the first and second phases. This observation is acceptable in part; however, we have to pay the special attention to the changing trend of the structural design. Importantly, an emerging trend is the wider application of higher-speed cutting, for instance, face milling of mission casing made of Mg alloy and end milling of plastic injection molding die in the industrial sector. The joint design is thus forced to change, and new research subjects are suggested according to new requirements based, e.g., on the fast-growing employment of the linear motor to the table traveling system, linear roller guideway, and HSK.

The higher-speed table system consists of a combination of (1) ball screw-nut and linear roller guideway, (2) linear motor and roller guideway, or (3) linear motor and hydrostatic guideway. As is widely said, the linear motor and rolling guideway have less damping capacity than the ball screw-nut driving system and slideway, and thus at issue is how to increase the damping capacity of table traveling system. In retrospect, the damping problem of the joint was vigorously investigated in the first phase, and the connecting mechanism of the table and its driving system was investigated in the second phase. In this context, some grass root like subjects are being investigated such as the following.

1. Use of logarithmic curved roller to the roller bearing, in order to realize the desirable uniform distribution of the interface pressure between the roller and the race. This remedy may increase the bearing stiffness [27], and following to this contrivance, an innovative method for chamfering the roller edge is being developed using water-jet processing.

2. Use of water hydrostatic guideway to dissipate the heat generated by higher-speed traveling of the table in the grinding machine [28].[5] In this case, the guideway is a weld-in-place type to the base, which has a T-shaped layout and lightweight structure consisting of viscoelastic-covered steel tubes, i.e., a variant of two-layered plate. More specifically, four tubes are integrated into one component to form the central and wing base units. A similar idea was previously employed in the cross rail (10,000 mm in length) of the vertical turning machine of Skoda make in 1956, the maximum allowable machining diameter of which is 12,500 mm. Importantly, the base has a number of welded and bonded joints.

The higher-speed cutting has furthermore unveiled a serious problem, i.e., expansion of the tapered hole at the larger end diameter and of the outer diameter of the spindle nose due to the centrifugal force. To cope with this new problem especially evident in the case of 7/24 taper, a new tapered connection, i.e., HSK, was proposed, and the corresponding research and development has been very active up to now.

Given that the main spindle driving system consists of a built-in-motor to some extent, another problem is the joint between the rotor of the motor and the main spindle. In this joint, the question is how to reduce the magnetic vibration—and to what extent—by realizing the sure fixation between the rotor and the main spindle. With the advance of higher-speed cutting, the built-in-motor technology faces new challenges, i.e., the need to reduce the centrifugal force and dissipate the generated heat as well as to clarify the effect of the rear bearing of movable type on the spindle stiffness. In addition, an ever-growing problem is the tooling system, i.e., the interface between the tool block and the turret head, which is compatible with higher-speed cutting.

To this end, Fig. 5-27 shows some representative research and development subjects related to the machine tool in general as well as specific

---

[5]With the reevaluation of the water lubrication, not only in the machine tool, but also in the stepper in LSI (*large scale integrated circuit*) manufacturing, the natural wood material, i.e., lignum vitae, is in resurgence, and its modern alternative material is the phenol resin. For example, in deep hole drilling by the BTA (*British Tool Association*) method, the pad made of lignum vitae was employed by Uni Sig Co. in 1982. Lignum vitae was furthermore used at the bearing of the hydroelectric power generator and propelling system of the submarine.

Improvement of economic aspect →	Bearing with turn-finished race	

Improvement of performances →	Advanced light-weighted structure consisting of sandwich plate with porous metal or Truss™ Grid (composite material)	Application of ceramic resin concrete to structural body components to reduce thermal deformation
	Rolling bearing—roller with logarithmic curve	Preloading mechanism for rolling bearing with KMT-C nut
	Linear motor drive with MRF (magneto-rheological fluid) to increase damping	Sure fixation of motor rotor to main spindle
	Honing finishing for traveling quill instead of manual lapping	Linear roller guide with larger stiffness and damping
		Water hydrostatic guideway

Development of multiple-function integrated elements →	Table driving system with directional orientation in stiffness	Innovative driving system of pinion-rack type
		CARB bearing of SKF make
	Rolling bearing with larger stiffness, larger damping, and lower frictional force by improving race configuration	Tool block clamping with multiple function type

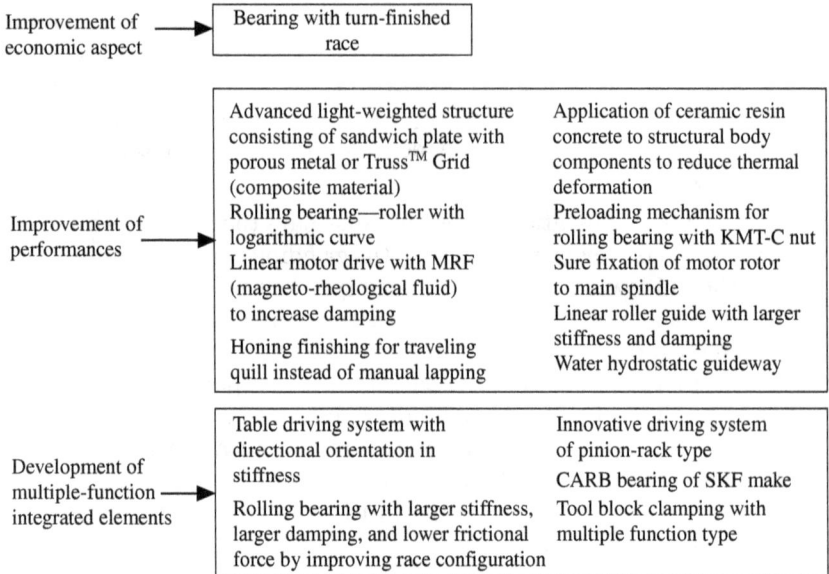

Figure 5-27   Some representative engineering subjects around the year 2000.

to the joint itself in the year 2000 and beyond. Although the primary concerns are to contrive more effective remedies for both the self-exciting chatter and the thermal deformation, main stress is placed on the performance enhancement for the main bearing and traveling mechanism. Some interesting newcomers are as follows.

1. *Multiple-function integrated joint*, such as the CARB bearing developed by SKF. In short, the CARB bearing has a beneficial feature obtained from the combination of the cylindrical roller, spherical roller, and needle roller bearings [29].

2. *Joint with directional orientation function.* A representative is a connection mechanism of the table-feed driving system having only the higher stiffness to the table traveling direction [30]. In addition, research into the engineering calculation of damping capacity and improvement of interface pressure measurement using the ultrasonic waves method are seen as increasingly important.

References

1. Kienzle, Kettner H., "Das Schwingungsverhalten eines gußeisernen und eines stählernen Drehbankbettes," *Werkstattstechnik und Werksleiter*, 1 Mai 1939, 33(9): 229–237.
2. Raymond, L. W., "Can Chrome Plating Solve Machine Way Problems," *American Machinist*, Aug. 1957, pp. 124–127.
3. Angus, H. T., "The Wear of Cast Iron Machine Tool Slides, Shears and Guideways," *Wear*, 1957/58: 1: 40–57

4. Birchall, T. M., and A. I. W. Moore, "Friction and Lubrication of Machine Tool Slideways," *Machinery*, Oct. 1958, 93: 824–831.
5. Research Report No. 47, "Friction and Wear of Materials for Machine Tool Slideways," The MTIRA, Macclesfield, Nov. 1972.
6. Píč, J., "Die Starrheit der Schraubenverbindung," *Konstruktion*, 1967, 19(1): 7–12.
7. Kunin, E. A., and V. M. Faingauz, "Damping in a Horizontal Boring Machine Spindle Unit," *Machines and Tooling*, 1967, 38(1): 16.
8. Kronenberg, M., P. Maker, and E. Dix, "Practical Design Techniques for Controlling Vibration in Welded Machines," *Machine Design*, July 12, 1956, pp. 103–109.
9. Rippel, H. C., "Design of Hydrostatic Bearings, Parts 1–10," *Machine Design*, Aug. 1, 1963–Dec. 5, 1963.
10. Report No. 134, "Hydrostatic Bearing System Design and Its Application to Machine Tools," Production Engineering Research Association, Jan. 1965.
11. Warnecke, H. J., "Konstruktion und Eigenschaften aerostatischen Larger und Führungen," *Annals of CIRP*, 1972, 21(2): 191–204.
12. Masuko, M., and T. Nakahara, "The Influences of the Fluid Capacitance in the Oil Feed Line System on the Transient Response of Hydrostatic Guideways," *Int. J. Mach. Tool Des. Res.*, 1974, 14: 233–244.
13. Mießen, W., "Berechnung des dynamischen Verhaltens hydrostatischer Spindel-Larger-Systeme auf Digitalrechneranlagen," *Konstruktion*, 1976, 28(7): 275–281.
14. Ito, Y., and M. Tsutsumi,"Determination of Mathematical Models in Structural Analysis of Machine Tools—Second Report, Determination of Mathematical Models for Normal Static Stiffness of Joints," *Bulletin of JSME*, 1981, 24(198): 2234–2239.
15. Thornley, R H., et al., "The Effect of Surface Topography upon the Static Stiffness of Machine Tool Joints," *Int. J. Mach. Tool Des. Res.*, 1965, 5(1/2): 57–74.
16. Eisele, F., and K. Corbach, "Dynamische Steifigkeit von Führungen und Fugenverbindungen an Werkzeugmaschinen," *Maschinenmarkt*, 1964, 70(89): 88–93.
17. Andrew, C., "Metal Surfaces in Contact under Normal Forces: Some Dynamic Stiffness and Damping Characteristics," in *Proc. IMechE*, 1967–68, 182-Pt3K: 92–100.
18. Myklestad, N. O., and U. Ill, "The Concept of Complex Damping," *J. Appl. Mechanics*, Sept. 1952, pp. 284–286.
19. Ito, Y., and M. Masuko, "Influence of Bolted Joint on the Model Testing of Machine Tool Construction," *Trans. of JSME*, 1970, 36(284): 649–654.
20. Dolbey, M. P., and R. Bell, "The Contact Stiffness of Joints at Low Apparent Interface Pressures," *Annals of CIRP*, 1971, 19: 67–79.
21. Thornley, R H., R. Connolly, and F. Koenigsberger, "The Effect of Flatness of Joint Faces upon the Static Stiffness of Machine Tool Joints," in *Proc. IMechE*, 1967–68, vol. 182, pt.1.
22. Dekoninck, C., "Experimental Investigation of the Normal Dynamic Stiffness of Metal Joints," *Int. J. Mach. Tool Des. Res.*, 1969, 9: 279–292.
23. Opitz, H., and J. Schunck, "Untersuchungen über den Einfluß thermisch bedingter Verformungen auf die Arbeitsgenauigkeit von Werkzeugmaschinen," *Forschungsberichte des Landes Nordrhein-Westfalen Nr. 1781*, Westdeutscher Verlag, 1966.
24. Reshetov, D. N., and Z. M. Levina, "Damping of Oscillations in the Couplings of Components of Machines," *Vestnik Mashinostroyeniya*, 1956, no. 12: 3–13 (translated into English by Production Engineering Research Association of Great Britain [PERA]).
25. Meyer, P., "Erhöhung des Haftbeiwertes von Preßpassungen durch eine Körnerlage in der Fuge," *Werkstatttechnik und Maschinenbau*, 1949, 39(11/12): 321–325.
26. Opitz, H., et al., "Untersuchungen an Werkzeugmaschinenspindeln, Wälzlagern und hydrostatischen Lagerungen," *Forschungsberichte des landes Nordrhein-westfalen Nr. 1331*, Westdeutscher Verlag, 1964.
27. Gebert, K., "Zylinderrollenlager für schnelldrehende Spindelsysteme," *Werkstatt und Betrieb*, 1997, 130(9): 702–708.
28. Owen, J. V., "Materials Drive Grinding," *Manufacturing Engineering*, 1998, 120(2): 42–50.
29. Brown, J., "Three-in-One Bearing Does More," *Power Transmission Des.*, June 1997, pp. 23–25.
30. Ito, Y., and K. Höft, "A System Concept for Culture- and Mindset-Harmonized Manufacturing Systems and Its Core Machining Function," in F.-L. Krause, and E. Uhlmann (eds.), *Innovative Produktionstechnik*, Carl Hanser Verlag, 1998, pp. 175–186.

6

Fundamentals of Engineering Design and Characteristics of the Single Flat Joint

In a machine tool, there are, as already shown in Chap. 5, a handful of representative joints; however, the basic structural configuration of these joints is two flat surfaces in contact, i.e., a single flat joint. To understand clearly the fundamental and complicated behavior of all the machine tool joints, thus, the basic necessity is to have authentic and correct knowledge about the flat joint. In due course, the single flat joint is a fundamental entity in establishing the engineering calculation and computation for the machine tool joint with various configurations. More specifically, primary concerns in engineering design are to determine the mathematical model, to arrange the database to make it available for carrying out the calculation and computation, and to verify the validity of the design results. Of these, the mathematical model of the joint that is widely acceptable is the spring (static joint stiffness) dashpot (damping capacity of joint) coupling, as already shown in Fig. 5-16, and in due course the database must contain the calculation formula and concrete design data for the spring constant and the damping capacity.

Importantly, the design database for the spring constant and damping capacity so far has been given for the single flat joint without local deformation, and in fact there have been a considerable number of related proposals. By reason of the simplicity of single flat joint, these proposals have, from one viewpoint, greater possibilities to apply to the practical cases, e.g., bolted joint with bay-type flange and the slideway with gib and keep plate. Intuitively, the difficulties lie in how to consider the characteristic factors within each practical joint, when applying those of a single flat joint.

In the following, thus, first the quick notes for the single flat joint will be stated, and then the design formula and some characteristic features for each variant of the single flat joint will be summarized. In addition, the thermal behavior of the single flat joint will be touched on.

6.1   Quick Notes for Single Flat Joint, Determination of Mathematical Model, and Fundamental Knowledge about Engineering Design Formulas

Although the flat joint appears to be very simple, the amount of related knowledge is huge, even when we limit our discussion of the formulation of the structural design data in terms of the static and dynamic stiffness also including thermal deformation. In addition, the single flat joint has, in principle, myriad influencing factors in relation to the joint stiffness, e.g., magnitude and direction of the external and internal loads, finished condition of joint surfaces, and the relative dimensional difference between the upper and lower joint surroundings.

Importantly, the systematic classification of the single flat joint can be, in principle, made by using the following three dominant facets, which have been clarified through the long-standing experience in the research and engineering development.

1. In consideration of larger effects of the direction of external load on the joint behavior, the flat joint is required to branch into those (a) under normal loading, (b) under tangential loading with normal preload, and (c) under moment with normal preload.

2. The joint can be, in principle, classified into the two types depending on the magnitude of the interface pressure. The flat joint under lower interface pressure corresponds with the slideway, whereas the flat joint under higher interface pressure is an idealized model of bolted joint. In fact, the bolted joint is prone to present a slight local deformation, resulting in the nonuniform interface pressure distribution. In other words, the bolted joint in the structural body shows, in nearly all cases, bedding-in and warping of clamped component, flairlike deformation at the bay-type flange, and so on (see Chap. 7).

3. More specifically, at issue is the relative stiffness of joint surroundings to the joint stiffness itself, resulting in the apparent difference in joint deformation. As can be readily seen, the joint deformation is subject to the magnitude and distribution of the interface pressure.

   a. In the case of $K_0 > K_j$, where K_0 and K_j are the stiffnesses of joint surroundings and of the joint itself, respectively, the joint surface does not separate from itself and may deform uniformly across the whole joint surface, when the normal load is applied. Consequently, the interface pressure is in uniform or linear distribution, and we can observe this kind at the slideway.

b. In the case of $K_0 < K_j$, the joint surface is, in general, liable to separate, and consequently the joint doesn't show any linear deformation across the whole joint surface, resulting in the nonuniform or nonlinear interface pressure distribution. As can be seen, we can observe this kind at the bolted joint.

On the basis of these dominant facets, the flat joint should be classified into several representative variants, as shown in Fig. 6-1, which can be regarded as the basic model of some representative machine tool joints. As can be seen, Fig. 6-1 may be associated with the structural design of the machine tool and its joints to a larger extent; however, the three variants, i.e., V_A, V_B, and V_C types in Fig. 6-1, are not in reality in the structural body component of full-size. Summarizing, the single flat joint can be characterized by the dominant factors, i.e., correlation between the magnitude of the interface pressure and the relative stiffness of joint surroundings to the joint itself, and also the direction of the external applied load.

In consideration of the characteristic feature of joint surroundings, proposed is a classification system of machine tool joints shown in Fig. 6-2, which can be considered as suitable for the determination of the mathematical model [1]. More specifically, first the machine tool joint should be classified from the viewpoint of its structural configuration, i.e., open, semiclosed, or closed type. Then considering the magnitude of interface pressure, the joint must be detailed, and finally the mathematical model should be determined in consideration of the correlation between the joint

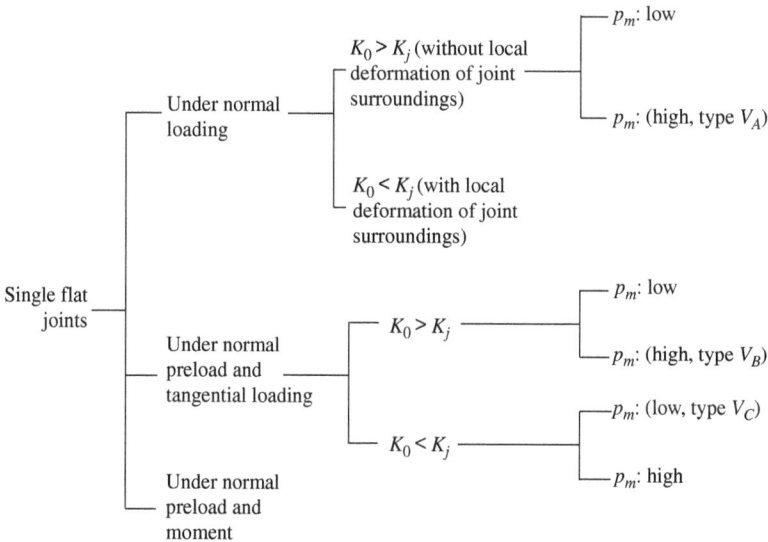

K_0: Stiffness of joint surroundings K_j: Joint stiffness p_m: Mean interface pressure

Figure 6-1   Classification of single flat joint.

Interface pressure p_m and relationship between joint stiffness K_j and joint surrounding's stiffness K_0		A: Example of applicable joints	B: Mathematical models for computation of stiffness in A
p_m: low (0–20 kgf/cm²)	$K_0 >> K_j$	Guideways (linear interface pressure distribution)	Rigid body on elastic foundation
	$K_0 < K_j$	Guideways (local deformation)	Plates or beams on elastic foundation
p_m: medium (<35 kgf/cm²)	$K_0 > K_j$	Tapered joints	Beams on elastic foundation (supported by independent springs)
p_m: high (<50 kgf/cm²)	$K_0 < K_j$	Bolted joints	Beams or plates on elastic foundation (springs having cross receptance effect)

Joints in machine tools

Open type

Semiclosed type

Closed type

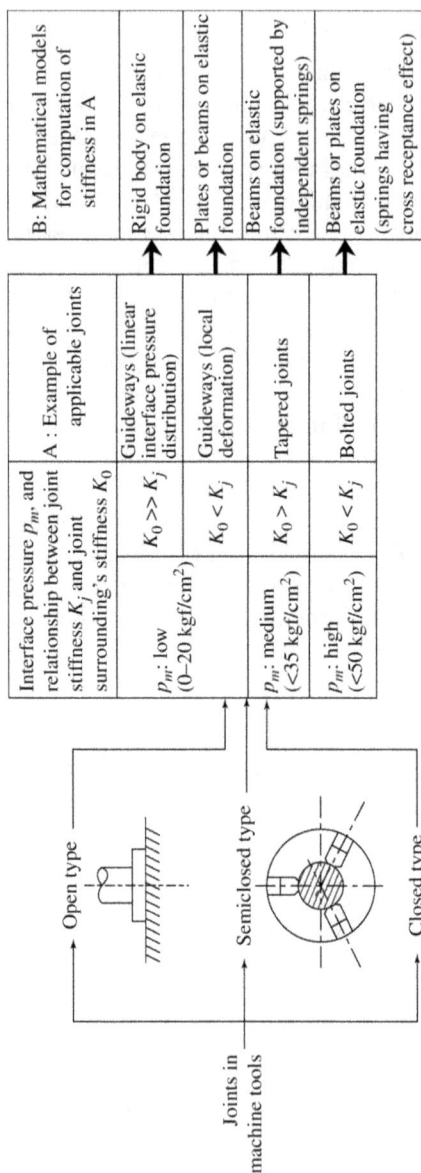

Figure 6-2 Classification of machine tool joints to determine mathematical models.

stiffness and the stiffness of the joint surroundings. In fact, the proposed classification system is very convenient when we apply the design database for the spring constant and damping capacity of the single flat joint to the practical structural design. In addition, it is notable that nearly all the machine tool joints belong to one of these joint types, as mentioned already in Chap. 5. Conceptually, Fig. 6-2 may assist the understanding of the analytical procedure in the engineering calculation with special respect to what a mathematical model is, although nowadays the computation method is dominant. In the computation method, the FEM model has been employed without exception, and the joint can be also replaced with the model consisting of the spring-dashpot couple.[1,2]

Given that the joint can be represented with the spring-dashpot model and characterized by the state of interface pressure distribution as mentioned above, a primary concern is first how to determine the spring constant and damping capacity within the engineering design formula. As will be shown later, there have been a considerable number of expressions relating to the normal and tangential joint stiffness under static loading, and also to the damping capacity.[3] In due course, another crucial issue is the applicability of these expressions to the engineering design. Within the expression context, only the expression for the normal joint stiffness proposed by Ostrovskii has, in the wider scope of engineering calculation, proved its validity without revealing any serious problems by Kaminskaya, Back, Nakahara, and PERA[4] to a large extent. In other words, we can, under satisfactory conditions, conduct the engineering design of the structure with the joint, e.g., slideways of flat and dovetail types, taper connection, and bolted joint under static normal loading.

Reportedly, the model theory is can be applied to the structure with the joint, provided that certain prerequisites are satisfied (refer to App. 2) [2], and thus these expressions facilitate, in principle, the engineering design of the joint. It is furthermore recommended that the constants in the expression be varied, if possible, in consideration of the actual condition of the joint to be designed.

---

[1]Engineers benefit by the analytical method. Typically, the influencing factors governing the machine tool performance and rates of their effects can be grasped without any difficulties by investigating only the final expression of the analytical solution.

[2]The mathematical model can be determined in full consideration of the (1) structural configuration, (2) capability of available program, and (3) ability of engineer who may determine the mathematical model.

[3]Hijink and van der Wolf reported once a firsthand view of the joint stiffness and damping in the beginning of 1970s.

Hijink, J. A. W., and A. C. H. van der Wolf, "Survey on Stiffness and Damping of Machine Tool Elements," *Annals of CIRP*, 1972, 22(1): 123–124.

[4]PERA (*Production Engineering Research Association of Great Britain*) Report Nos. 180 and 198, "Machine Tool Joints, Part 1 and Part 2," late 1960s.

6.2   Design Formulas for Normal Joint Stiffness and Related Research

6.2.1   Expressions for static normal joint stiffness

The single flat joint under normal loading can be characterized by the following two facets.

1. The joint deflection is in nonlinear relation to the applied load.
2. The load-deflection curve shows, in general, the hysteresis behavior. In addition, the single flat joint of $K_0 > K_j$ shows the uniform interface pressure across the whole joint surface.

In full consideration of these characteristic features, the expressions for the normal joint deflection have been proposed as shown in Table 6-1. Of these, the expression of Ostrovskii, as widely accepted in the engineering sphere, can be considered to be applicable to the engineering design to a large extent.

More specifically, the expression of Ostrovskii can be written as

$$\lambda = Cp^m \tag{6-1}$$

where λ = joint deflection in normal direction
$\quad p$ = interface pressure
$\quad C, m$ = constants

TABLE 6-1   Expressions for Normal Joint Deflection λ

	Expression	Conditions Obtained Expressions			Examples of joints to be applicable expressions
		p, kgf/cm^2	Relationships between K_j and K_0	Shape and size of joint surface	
Levina [3]	$\lambda = C_0 p$	< 4	$K_0 \gg K_j$	Slideways of machines in full-size	Slideway (including locally deformed condition)
Ostrovskii [4]	$\lambda = Cp^m$	0–25	$K_0 > K_j$ — $K_0 < K_j$	Circular type Area: 16 cm^2	Slideway (including locally deformed condition)
Connolly & Thornley [5]	$p = ae^{b*\lambda}$	8–500	$K_0 > K_j$ — $K_0 < K_j$	Annular ring type Area: around 13 cm^2	Bolted joint

λ: Normal joint deflection p: Interface pressure

a, b^*, C_0, C and m: Constants

Note: Numbers in brackets indicate references

As can be readily seen, the joint deflection λ is in exponential proportion to the interface pressure p. In addition, C and m are the constants depending mainly on the joint material, machining method and roughness of joint surface, machined lay orientation, flatness deviation, and size of joint area. The joint stiffness per unit area is thus given by

$$dp/d\lambda = [1/(Cm)] \cdot p^{(1-m)} \qquad (6\text{-}2)$$

In the engineering design, it is necessary to first determine the values C and m in consideration of the dimensional and performance specifications of the objective joint. In general, the constant C is to be the lower value for the joint made of high tensile strength material, having small mating area, with high stiffness of joint surroundings, and with mating surface of higher quality.

Table 6-2 shows some representative values for C and m available for the joint, where the joint of 16 cm^2 contact area is made of cast iron and the interface pressure is less than 25 kgf/cm^2. These values were shown by Back et al. [6], after arrangement of the experimental data reported by Levina [3, 7], and Ostrovskii [4]. Although it has some limitations, the expression can obviously unveil the essential feature of the joint, clearly showing that m is, in general cases, equal to 0.5. In addition, it is obvious from Eq. (6-2) that the joint stiffness is dependent upon the interface pressure, simultaneously showing nonlinearity.

TABLE 6-2   Values of C and m (Available up to $p = 25$ kgf/cm^2) (Arranged by and courtesy of Back)

	Finishing methods		Constants	
	Points in any 1 in^2 of bearing area	Depth of scraping or surface roughness, (μm)	C	m
Hand-scraped/ hand-scraped	20–25	3–5	0.3	0.5
		6–8	0.5	0.5
	15–18		0.8–1.0	0.5
	10–12		1.3–1.5	0.5
	5–12		1.5–2.0	0.5
Hand-scraped/ ground	15–18	1.0 R_{CLA}	0.8–1.0	0.4
Peripheral ground/ peripheral ground		1.0 R_{CLA}	0.6–0.7	0.4–0.5
Finish planning/ finish planning			0.6	0.5

Cast iron joint

Note: Values of C and m are available when λ and p are in μm and kgf/cm^2, respectively.

TABLE 6-3   Values of *C* and *m* (Courtesy of Bell)

Material combination		m (dimensionless)	C	C (per Table 6-2)
Cast iron	Cast iron	0.50	7.25	—
	Ferobestos	0.32	43.6	3.98
	Tufnol	0.39	26.0	2.36
	Glacier DU (as received)	0.50	19.5	—
	Glaxier DX	0.43	19.2	—
	Glacier DX (dimpled)	0.41	25.2	—

Ground-to-ground joint surfaces

Note: Ferobestos: asbestos reinforced plastic with colloidal graphite
Tufnol: resin-impregnated paper.
Glacier DU: PTFE and lead-impregnated bronze.
Glacier DX: acetal resin polymer on porous bronze.
Values of C and m are available when λ and p are in μin and lb/in^2, respectively.

As mentioned above, the validity of the expression was already verified to a great extent by many researchers from both the experimental
and theoretical aspects, including a considerable number of applications
to the engineering calculation. Thus there have been some trials to expand
the availability of Ostrovskii's expression from both the joint material and
the magnitude of interface pressure aspects. Tables 6-3 and 6-4 are for the
values of C and m especially focusing on the nonmetallic materials for
the slideway, which were reported by Dolbey and Bell [8],[5] and by Ito
after arranging the experimental results of Eisele and Corbach [9].[6] As

---

[5]In their paper, the unit of C is given by μin (microinches). This appears to be a mis-
print, and in Table 6-3, the unit is deleted. In contrast, the values of C reported by Back
are added.

[6]In the U.S.S.R., the vibratory burnishing was tried to apply it to the slideway. The vibratory attachment is of planetary movement type and uses a diamond ball as a cutting tool.
The stiffness of vibratory burnished (vibratory burnishing after grinding, then grinding)
and scraped flat joints under repeat loading is 1.5 and 0.83 (kgf/mm^2) μm, respectively,
where the joint material is cast iron. The vibratory burnishing after grinding gives 35 to
46 contact spots per 25 × 25 mm^2, whereas scraping gives 24 to 36 points in any 1 in^2 of
bearing area.

Ryzhov, E. V., et al., "Increasing Contact Stiffness by Vibratory Burnishing," *Machines
and Tooling*, 1972, 43(1): 59–60.

Shneider, Yu G., et al., "Vibratory Burnishing of Machine Tool Slideways," *Machines and
Tooling*, 1972, 43(11): 51–52.

TABLE 6-4   Values of C and m for Nonmetallic Materials (in Part, Calculated from Data of Eisele and Corbach)

Type of surfaces in contact		C	m	Remarks
Cast iron (GG26), scraped	Bronze (SnBz8), scraped	0.3	0.65	
	MoS$_2$ compressed	0.52	0.85	
	Polyamide	8.8	0.35	
	Backlite laminated woven clothes	11.0	0.5	Valid range of p: 2.0–7.5 kgf/cm^2
Cast iron, ground ($R_{max} = 2\ \mu m$)	Turcite, scraped ($R_{max} = 30\ \mu m$)	2.0	0.6 –	Reported by Furukawa, Tokyo Metropolitan Univ., elsewhere
	Turcite, ground ($R_{max} = 6\ \mu m$)	1.4	0.7	

Note: Values of C and m are available up to 6 kgf/cm^2 and when λ and p are in μm and kgf/cm^2, respectively.

can be seen, the value of m is not 0.5, but is less than 0.5 in the case of the plastic material and more than 0.5 in the cases of bronze and MoS$_2$. In addition, it is noticeable that MoS$_2$ compressed material and bakelite of woven clothes laminated type show some peculiar characteristics, although both have relatively low joint stiffness: the former shows a large value of m, but the latter shows a large value of C compared with those of other joint materials.

In addition, further noteworthy behavior can be observed as follows.

1. Apart from certain kinds of nonmetallic joint and joints of laminated type, the interface pressure–joint deflection curve does not show any hysteresis, even when the loading and unloading procedures are repeated.

2. The joint stiffness is in proportion to the modulus of elasticity of the joint material. With the increase of the modulus of elasticity, the joint stiffness becomes higher.

3. With improving surface roughness, the joint stiffness increases, provided that the joint surface has no flatness deviation and/or waviness.

4. In the cast iron joint, its surface roughness has no effects on the joint stiffness.

In many respects, it is very desirable to apply the expression of Ostrovskii to the joint under higher interface pressure, and from such a viewpoint, Taniguchi et al. [10] investigated the further availability

TABLE 6-5   Values of C and m for Joints under Higher Interface Pressure

(i) Ground surface $R_{max} = 1.0\ \mu m$			(ii) Lapped surface $R_{max} = 1.4\ \mu m$		
D/h	m	C	D/h	m	C
100/40	0.40	0.089	100/40	0.54	0.051
80/40	0.36	0.046	80/40	0.60	0.039
60/40	0.75	0.017	60/40	0.60	0.032
50/40	0.58	0.008	50/40	—	—
100/24	0.39	0.090	100/24	0.56	0.045
80/24	0.58	0.053	80/24	0.43	0.074
60/24	0.61	0.022	60/24	0.55	0.027
50/24	0.21	0.048	50/24	—	—
Average	0.48	0.046	Average	0.55	0.045

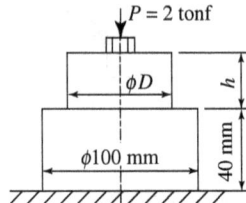

Joint material: semihard steel
(S45C of JIS)

Note: Values of C and m are available when λ and p are in micrometers and megapascals, respectively.

of Ostrovskii's expression in the case of steel-to-steel joint, and Ito and Tsutsumi [11] reported the interesting behavior as follows.

1. The Ostrovskii expression can be used for the joint, where the interface pressure is up to 100 kgf/cm^2. Table 6-5 shows the values for C and m.

2. The value of m is around 0.5 as reported by Back et al. for the joint with lower interface pressure, although the values of C differs largely from those reported by Back et al.

3. In the joint with local deformation such as the bolted joint of the column to the table guideway of planomiller, as already reported by Kaminskaya, the joint stiffness should be determined to be smaller than that calculated from Eq. (6-2).

Within this context, Connolly and Thornley proposed another expression, as already shown in Table 6-1.[7] They emphasized that within a machine tool design context, a root cause of the difficulties lies in the

---

[†]They proposed later a modified expression to clarify the effects of the surface roughness together with considering the waviness and flatness deviation [16].

uncertainty in quantitatively determining the magnitude of the flatness deviation, although the surface roughness can clearly be indicated on the drawing. This induces another problem—the test specimen with quantified flatness deviation cannot be produced. In consideration of such unfavorable influences of the flatness deviation on the joint stiffness, they reported the value b^* for the single flat joint under higher normal loading and not showing any local deformation, such as shown in Table 6-6.

Importantly, a further problem in the expression of the joint stiffness under higher interface pressure is to establish a modified expression with special respect to the bolted joint, which is preferably based on that of Ostrovskii and takes into consideration an effect of cross receptance, i.e., mutual spring action of nonlinear type [12]. With the increase of the interface pressure, the cross receptance in the joint stiffness could become generally strong; however, the details have not yet been clarified.

To this end, the wider applicability of the expression of Ostrovskii will be stated. In accordance with the expression of Levina, the value of m for the slideway under lower interface pressure can be regarded as unit, and thus the joint stiffness per unit area is equal to C as already shown in Table 6-1. In due course, Levina suggested the value of C_0 for such a slideway shown in Table 6-7, and verified its validity in the engineering calculation. Figure 6-3 is two examples of the comparison between the theoretical and experimental values, and as can be seen, good agreement between both values can be observed. In the slideway, furthermore, at issue is the flat joint subjected to complex loading, i.e., normal loading with

TABLE 6-6 Value of b^* for Expression of Connolly and Thornley (Courtesy of Thornley)

Joint surfaces		b^* 10^{-4} in^{-1}			Valid range of p ton/in^2
Machined finish	Average surface roughness μ in CLA	High	Low	Average	
Shaped or planed	188	8.85	0.81	4.01	$0.05 < p < 2.5$
Turned	117	13.10	2.23	5.37	$0.05 < p < 3$
Milled	81	4.10	0.62	1.85	$0.05 < p < 3$
Ground	18.5	14.85	2.12	7.19	$0.05 < p < 1$

Joint material: Mild steel

Notes:
1. $p = ae^{b^*\lambda}$, and then $dp/d\lambda = ab^*e^{b^*\lambda} = b^*p$
2. Value of b^* is available when λ and p are in μ in and ton/in^2, respectively.

TABLE 6-7   Recommended Values of C_0 for Horizontal Slideways (by Levina)

Mean interface pressure p_m, kgf/cm^2	Width of slideway mm	C_0, $\mu m \cdot$ cm^2/kgf
< 3.0	< 50	0.5–0.7
	< 100	1.0
	< 200	2.0–2.5
	< 300	3.2
	< 400	4.0
> 3.0–4.0	(40–50)% lower than above values	

Notes: 1. In the case of slideway with local deformation, the recommended values of C_0 are (50–70)% higher than those of slidways under p_m < 3.0 kgf/cm^2.

2. In the case of vertical slideway, the values of C_0 should be (30–40)% higher than those for horizontal slideways.

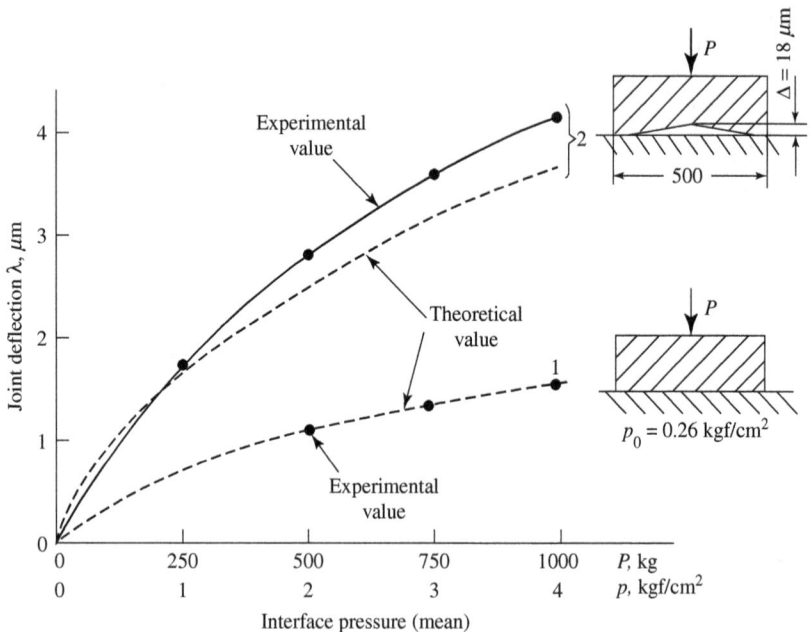

Note: Theoretical value was calculated on the basis of the expression of $\lambda = Cp^m$.

Figure 6-3   Applicability of Ostrovskii's expression to engineering calculation (by Levina).

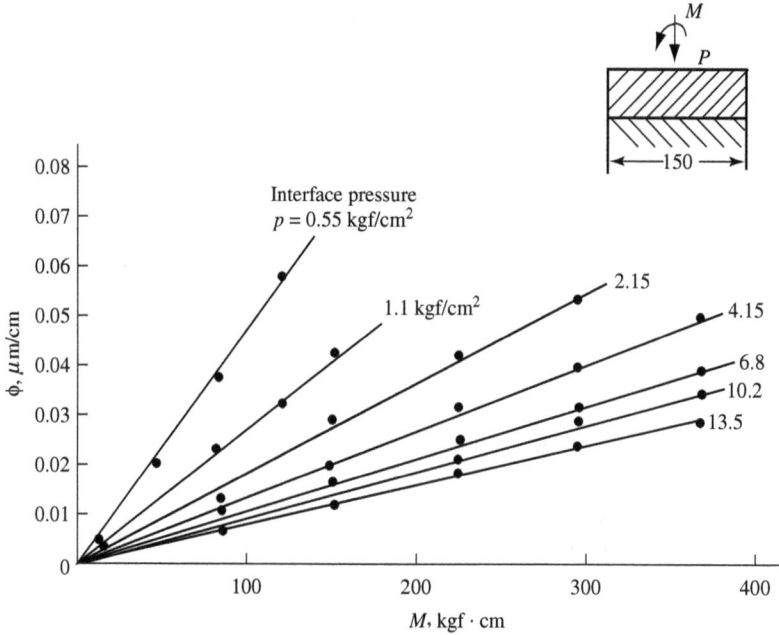

Note: The joint has a rectangular shape, both faces are hand-scraped, both elements are made of cast iron.

Figure 6-4   Moment–angular deflection relationships of flat joint under complex loading (by Levina).

moment, which should be considered as a variant of the fundamental flat joints. In this case, the joint stiffness can be represented by the inclination angle ϕ, and according to the report of Levina [7], the relationship between the moment M and the inclination angle ϕ is always linear, as shown in Fig. 6-4, when the mean preinterface pressure distributes uniformly across the whole joint surface. In addition, the joint stiffness under the moment is liable to reduce by the flatness deviation or waviness. For instance, Tenner [13] reported that the stiffness of the table slideway under the moment around the vertical axis is within the range of 220 to 560 kgf/μm/m, when it is measured on seven single-column jig borers of the same production batch. He pointed out that this stiffness variation can only be attributed to the fitting errors in the slideway.

6.2.2   Representative researches into behavior of the single flat joint under normal loading

Figure 6-5 shows a firsthand view of representative research activities on the single flat joint under normal static loading, and as can be easily understood, nearly all the representative research activities were carried out in

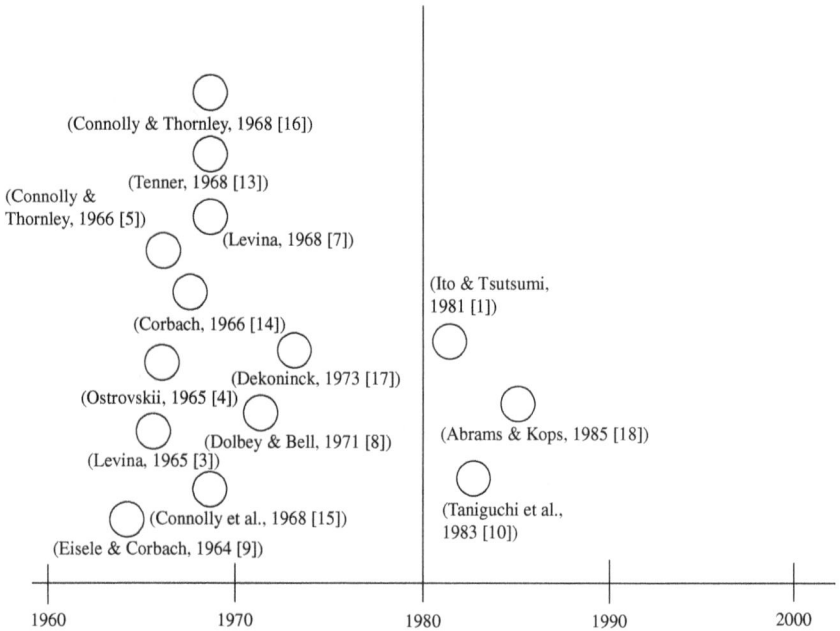

Figure 6-5   Firsthand view for research into single flat joints without local deformation.

the 1960s. In other words, the static behavior of the single flat joint without local deformation was clarified already in the 1960s to a large extent.

On the basis of such evidence, two representative researches and concerns will be discussed in the following. Figure 6-6 is the schematic view of Ostrovskii's test rig and the specimen made of cast iron, its contact area being 16 cm^2, where the joint surfaces are scraped, fine-planed, and ground in full consideration of the slideway of full-size. Figure 6-7 shows some of the measured results for the joint deflection when the interface pressure is varied as well as the machining method of the joint surface. Although the test rig is very simple, using, for instance, the lever loading mechanism and dial gauge for the measurement of joint deflection, the general static behavior of the flat joint can be obviously observed, as shown already in Fig. 6-7. With the improvement of the surface roughness even using the same machining method, e.g., from coarse scraping to fine scraping, the joint deflection decreases largely, and the machining method of the joint surface has greater effects on the joint deflection. In short, the static stiffness increases with increasing interface pressure, approaching a certain constant value, and largely depends upon the finishing method of the joint surface. In addition, the interface pressure–joint deflection curve does not show any hysteresis even when the loading and unloading cycles are repeated, provided that the joint surface is not made of certain kinds of nonmetallic and laminated

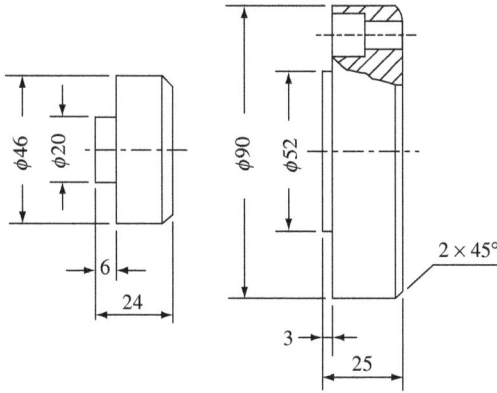

Test piece

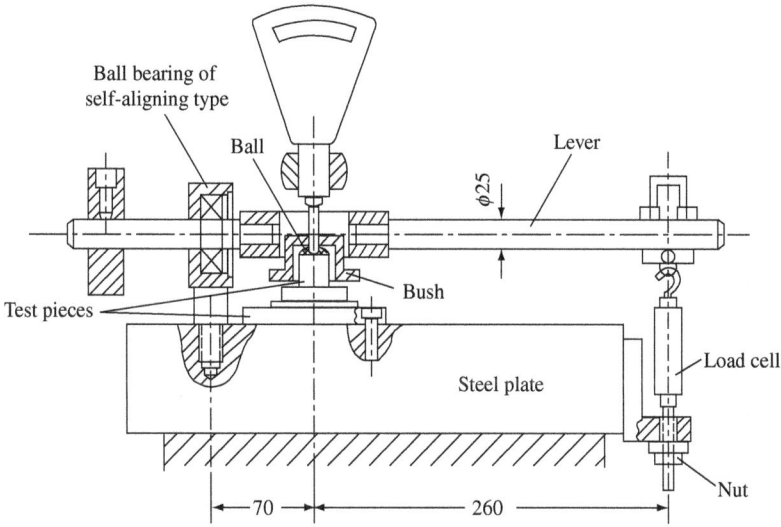

Test rig

Figure 6-6   Test rig and test pieces used by Ostrovskii.

materials. It is worth pointing out that Eisele and Corbach [9] also reported the same evidence as Ostrovskii at the same time. Figure 6-8 shows some interface pressure–joint deflection curves to understand what is different from cast iron to nonmetallic joints.

Following that the research of Ostrovskii, Dolbey and Bell [8] conducted a further investigation into flat joints including those made of new slideway materials using the rectangular specimen of 3×3 in^2. In this experiment, the specimens used were made of gray cast iron (BS Grade 14),

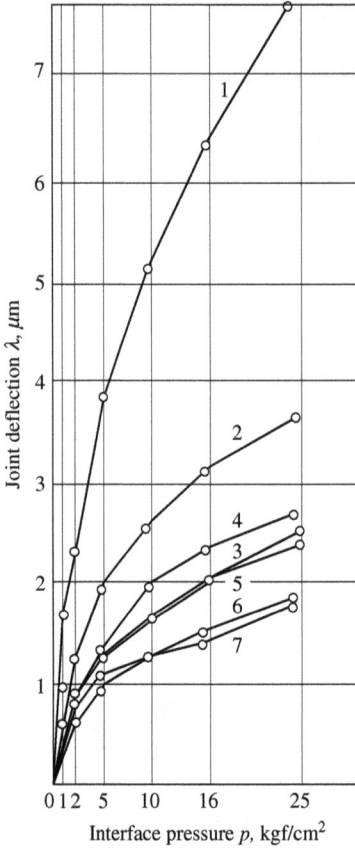

1—Coarse scraping 25×25 mm^2, number of spots per
  inch^2 $z = 5$ to 10
2—Conventional scraping $z = 12$ to 18
3—Fine scraping $z = 24$ to 36
4—Finish planing
5—Grinding with wheel face
6—Grinding with wheel periphery
7—Scraping with flat broad scrapers

Figure 6-7 Effects of interface
pressure and machining method
on joint deflection (by Ostrovskii).

Ferobestos (asbestos reinforced plastics with colloidal graphite), Tufnol
(resin impregnated paper), Glacier DU bearing material (PTFE and lead
impregnated bronze), and Glacier DX bearing material (acetal resin polymer on porous bronze). They showed the typical interface pressure–joint
deflection curve as already shown in Fig. 5-20. In the flat joint made of
certain kinds of thick plastic material, i.e., Ferobestos and Tufnol, a typical hysteresis behavior can be observed; however, any residual joint
deflections are not observed after unloading, although some hysteresis
behavior appears during the loading and unloading cycle.

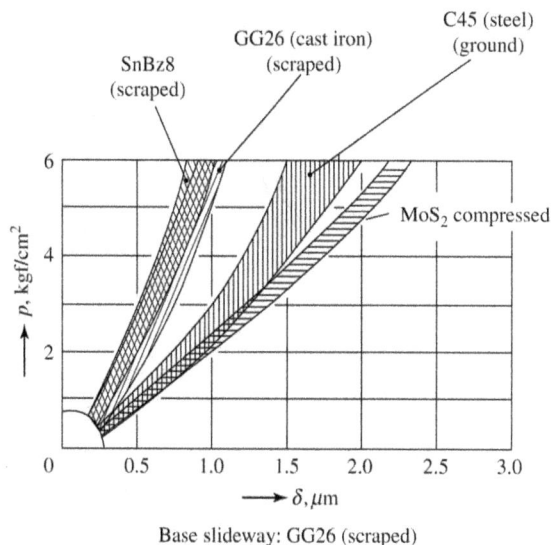

Base slideway: GG26 (scraped)

Figure 6-8 Normal joint deflection under lower interface pressure (by Eisele and Corbach).

In contrast, Thornley and coworkers conducted a series of famous researches into behavior of the single flat joint under higher interface pressure in the late 1960s. The single flat joint under higher magnitude and linear distribution of the interface pressure is, as already stated, not the case of the actual joint, but a model of the bolted joint. As stated already in Chap. 5, the bolted joint can often be observed within the structural body component, and thus the flat joint under higher interface pressure is worth investigating to understand the essential features of the bolted joint, although the joint is idealized. Following the research of Thornley, Dekoninck conducted some further interesting researches.

In accordance with the results obtained from these earlier works, the flat joint under higher interface pressure can primarily be characterized by the appearance of the apparent residual displacement in its interface pressure-joint deflection curve for the first loading and unloading cycle, as already shown in Fig. 5-19. In addition, after the first loading cycle, the joint deflection shows good repeatability, provided that the applied load does not exceed its maximum in the previous loading cycle. This fact implies that the joint deflection of the flat joint under higher interface pressure consists of the elastic and plastic components, where the plastic one is derived from the due deformation of surface asperities.

The interesting results of Thornley et al. were obtained using the test specimens of hollow cylinder form, and made of mild steel, cast iron,

Directions of
machining relative to each other

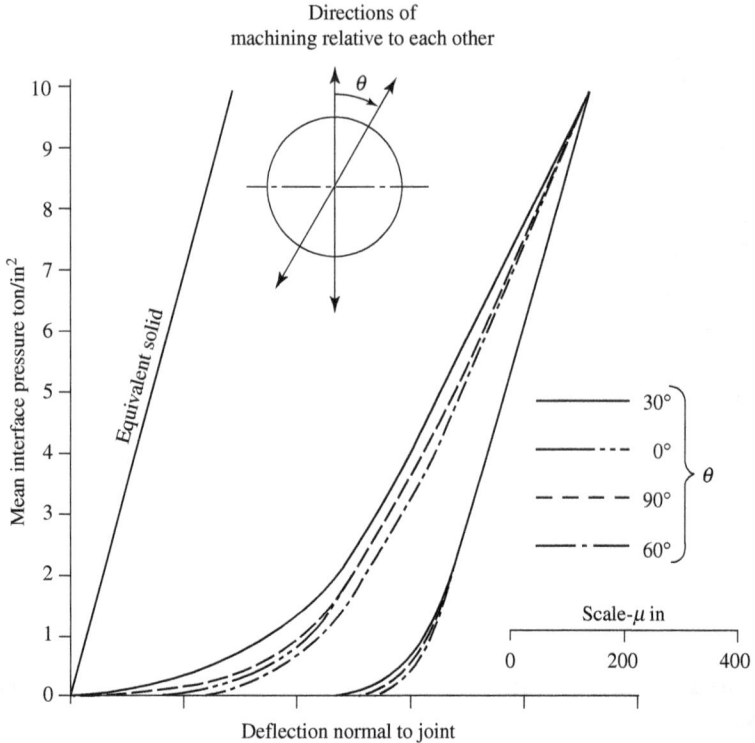

Figure 6-9   Effects of machined lay orientation on joint deflection—shaped mild steel joint (by Thornley and coworkers).

brass and Al alloy, the size of specimen being 1 in bore, 15/8 in height, and 2 in^2 cross-sectional area.[8] Although they do suffer from some limitations, the works of Thornley et al. involve much interesting evidence and thus in the following, some noteworthy results will be stated.

Machined lay orientation.   Thornley is credited with being the first researcher to suggest the importance of the machined lay orientation on the joint deflection. Although the shaped flat joint made of mild steel does not show obviously the effect of the machined lay orientation, as seen in Fig. 6-9, the machined lay orientation is dominant in certain joints, and thus it should often be taken into consideration.

---

[8]It is worth suggesting the following report for the ease of understanding of the achievements of Thornley and coworkers, although the report itself is confidential to members of MTIRA.

Connolly, R. and R. H. Thornley, Research Report No. 13, "The Static Stiffness of Joints between Machined Surfaces," The MTIRA, March 1966.

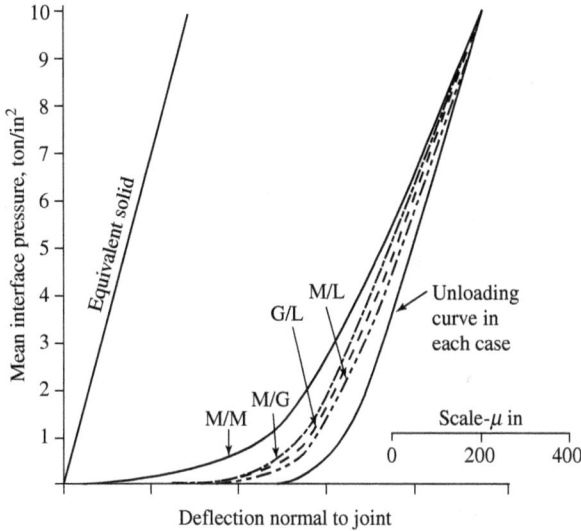

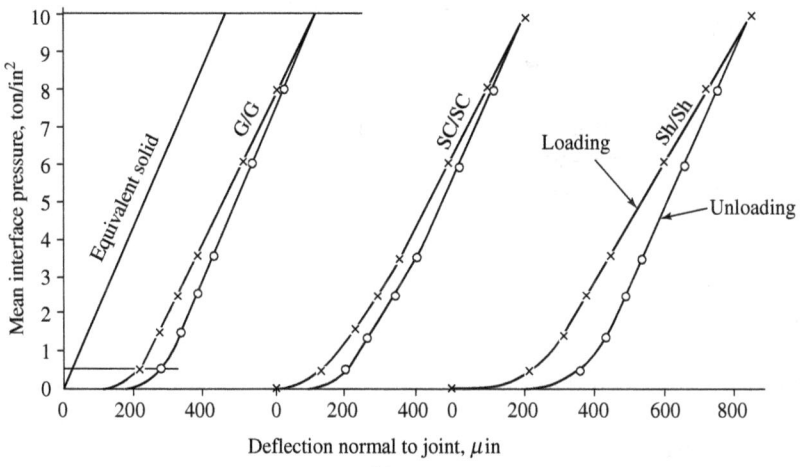

G: Ground   L: Lapped   M: Milled

SC: Scraped   Sh: Shaped

Figure 6-10 Effects of surface finishes on joint deflection: (a) Mild steel joint and (b) cast iron joint (by Thornley and coworkers).

Type of surface finish.   Figure 6-10 shows the effects of the type of surface finish on the joint deflection. It is obvious that the surface finish has a considerable effect on the joint deflection during the first loading, but no effect on those of unloading procedures at all.

Surface roughness.   Notwithstanding the machining method of the joint surface, the higher joint stiffness can be realized when the joint surface is smoother, provided that the joint surface has no waviness and/or flatness deviation.

Oil and grease as interfacial layer.   In the actual machine tool joint, it is difficult to realize the pure dried condition of the joint surface, but the oil and grease exist always between the joint surfaces. In the flat joint under static loading, however, these interfacial layers have no effect on the joint deflection except for the lapped joint.

Hardness of joint surface.   In general, Young's modulus of the material is independent of the hardness, and then the joint stiffness after first loading appears not to have the influence of hardness. In accordance with the measured results for the shaped joint made of EN 9 steel, where the hardness of the joint surface was varied using direct hardening and tempering or direct hardening and annealing, as reported by Thornley and coworkers, the joint stiffness corresponding to unloading is, contrary to the suggestion stated elsewhere, somewhat dependent on the hardness of the joint surface. In short, it is very interesting that the joint stiffness increases with hardening of the joint surface, simultaneously showing less residual deflection.

6.3   Design Formulas for Tangential Joint Stiffness, Related Researches, and Peculiar Behavior of Microslip

6.3.1   Expressions for static tangential joint stiffness

The machine tool joint is often subjected to tangential loading together with the normal preload, resulting in the occurrence of shear at the interface. This shear deteriorates, for instance, the positioning accuracy of the carriage of the engine lathe, where the positive stopper can be used. In contrast, the shear induces duly the residual displacement or microslip, by which the damping capacity at the jointed surface can be determined. The joint under tangential loading can be represented with a model that is a simple flat joint under the normal preload and tangential loading and, in due course, is worth investigating as well as that under only normal loading to understand deeply the characteristics of the machine tool joint (see Fig. 5-17).

Importantly, Kirsanova [19] is credited to the first researcher for the tangential joint stiffness in 1967. In due course, Kirsanova represented the tangential joint deflection with an empirical expression to assist the design procedure of the slideway. Table 6-8 summarizes the expressions for the

TABLE 6-8   Expressions for Tangential Joint Deflection δ and Stiffness K_s

	Expression	Conditions obtained expressions				Examples of joints to be applicable expressions
		p kgf/cm^2	p_τ or τ	Relationships between K_j and K_0	Shape and size of joint surface	
Kirsanova [19]	$\delta = K_\tau \tau$	0.9–3.6	< 2.0 kgf/cm^2	—	Rectangular type Area: 225 cm^2	Slideways
		1.8–15			Circular type Area: 51 cm^2	
Koizumi et al. [20]	$\delta = C\,(\tau/p)$	0–100	< 20 kgf	$K_0 > K_j$ $K_0 < K_j$	Annular ring type Area: 2–26 cm^2	Bolted joint
Back et al. [21]	$K_s = (1/R)\,p^S$	<60 (presumed)	< 7.0 kgf/cm^2 (presumed)	—	—	—

p_τ: tangential load, τ: shear stress, p: normal interface pressure, C, K_τ, R and S: constants.

Note: For the constant in the expression of Koizumi, please refer to Table 6-9.

tangential joint deflection so far proposed, although not guaranteeing their reliabilities as well as that for damping, because no other investigations were carried out after then by other researchers and engineers. In short, the expression of Kirsanova can be written as

$$\delta = K_\tau P_\tau \qquad (6\text{-}3)$$

where δ = elastic tangential deflection, μm
 K_τ = coefficient of contact shear compliance, μm · cm^2/kgf
 P_τ = tangential load, kgf/cm^2

The coefficient K_τ is a function, in which the normal preload and surface finish of the joint are variables, as shown in Fig. 6-11. For those of Koizumi et al. and Back et al., the values for the constants are, in due course, given as shown in Tables 6-9 and 6-10, respectively. It is very interesting that the constant S in the expression of Back et al. is 0.5, the same as that in the expression of Ostrovskii.

6.3.2   Representative researches into behavior of the static tangential joint stiffness and the microslip

Owing to the complexity of the characteristic features, the tangential stiffness of the flat joint is not fully clarified yet, although interesting behavior was already observed and reported elsewhere. More specifically,

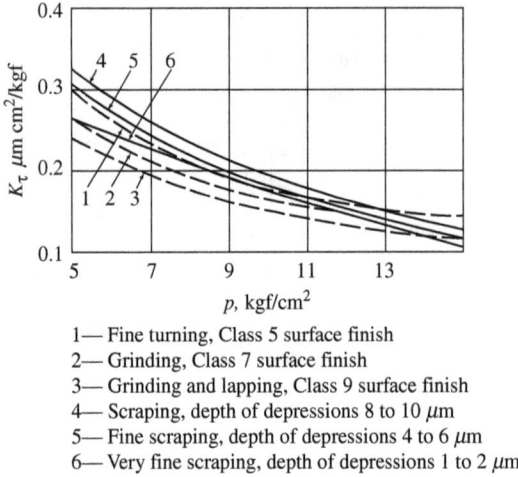

1— Fine turning, Class 5 surface finish
2— Grinding, Class 7 surface finish
3— Grinding and lapping, Class 9 surface finish
4— Scraping, depth of depressions 8 to 10 μm
5— Fine scraping, depth of depressions 4 to 6 μm
6— Very fine scraping, depth of depressions 1 to 2 μm

Figure 6-11   Values of K_τ in connection with p (by Kirsanova).

Kirsanova showed a typical result as seen in Fig. 6-12. The tangential displacement reduces with increasing normal interface pressure; and under constant pressure, the tangential displacement increases with the tangential load, where the joint stiffness is always constant and only the residual displacement component, i.e., microslip, increases oppositely. In addition, the fundamental characteristic feature in the load-displacement curve is not affected by the lubricated condition of the joint surface, and by the lapse of time after applying the preload. It is furthermore notable that the maintaining time after jointing has an effect on the increase in the displacement. Figure 6-13 reproduces the

TABLE 6-9   Values of C in Expression of Koizumi

Materials and concerns	C	Materials and concerns		C
S45C	1.0	Furnace cooling	$R_{max} = 0.4$	0.93
FC25	0.52		$= 1.3$	2.9
BsBM2	0.58		$= 3.0$	0.77
A2017BE	1.5	SK 3	Air cooling $R_{max} = 0.4$	2.0
S45C D = $\begin{cases} = 30 \\ = 40 \\ = 50 \\ = 60 \end{cases}$	3.6		Tempering $R_{max} = 0.4$	1.2
	2.4		$R_{max} = 0.4$	0.51
	1.5	Oil quenching	$= 1.3$	2.4
	2.7		$= 3.0$	0.85

D: diameter of test piece mm; R_{max}: surface roughness, μm

TABLE 6-10   Values of C_τ and S (Courtesy of Back)

	Finishing methods		Constants	
	Points in any 1 in² of bearing area	Depth of scraping or surface roughness, mm	C_τ	m
Hand-scraped/ hand-scraped	20–25	3–5	0.39	0.5
		6–8	0.65	0.5
	15–18		1.0–1.3	0.5
	10–12		1.7–2.0	0.5
	5–12		2.0–2.6	0.5
Hand-scraped/ ground	15–18	$1.0R_{CLA}$	1.0–1.3	0.5
Peripheral ground/ peripheral ground		$1.0R_{CLA}$	0.8–0.9	0.5
Finish planning/ finish planning			0.78	0.5

Cast iron joint

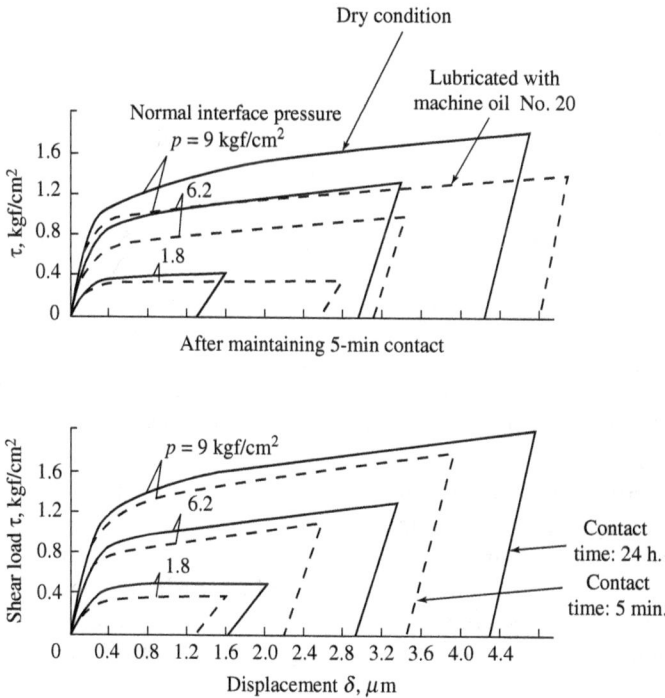

Figure 6-12   Load-deflection curve at first loading cycle (linear type, by Kirsanova).

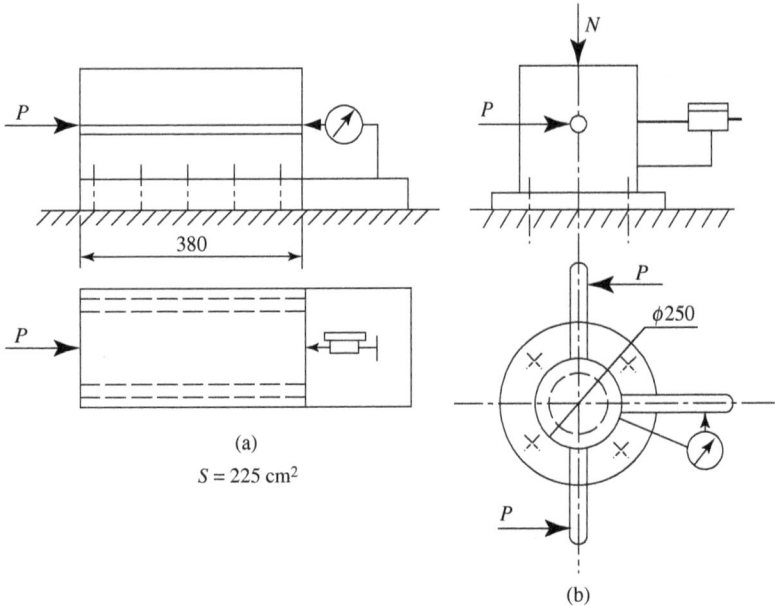

Figure 6-13 Test rigs of Kirsanova: (a) Linear type and (b) circular type.

schematic view of the test rig used by Kirsanova. In this test rig, the joint material is the gray cast iron, and both joint surfaces are ground and scraped (16–20 spots/in^2). In addition, the normal preload can be varied by the dead weight on the rectangular slideways.[9]

In fact, Kirsanova provides us with much noteworthy knowledge about the flat joint under tangential loading, although using a very simple test rig.

In general, the joint is subjected to the repeated loading across the whole machine tool life, and it is furthermore necessary to investigate the effects of repeated loading. Within this context, it is natural to recall a maxim that the friction characteristic of the slideway may be changed with the running time, called the *maturity of sliding surface*, and thus what happens at the flat joint when the repeated tangential load is applied is very interesting. Intuitively, the microslip can be furthermore considered as a major cause of the large damping capacity of the joint.

Having in mind such an implication, Masuko and coworkers investigated the behavior of the tangential stiffness and microslip of the single bolt-flange assembly to crystallize their ideas that the damping capacity of the bolted joint will show certain time dependence, simultaneously

---

[9] For reasons of some difficulties in measuring the smaller deflection, Kirsanova used also the test rig of circular type, its contact area being 51 cm^2.

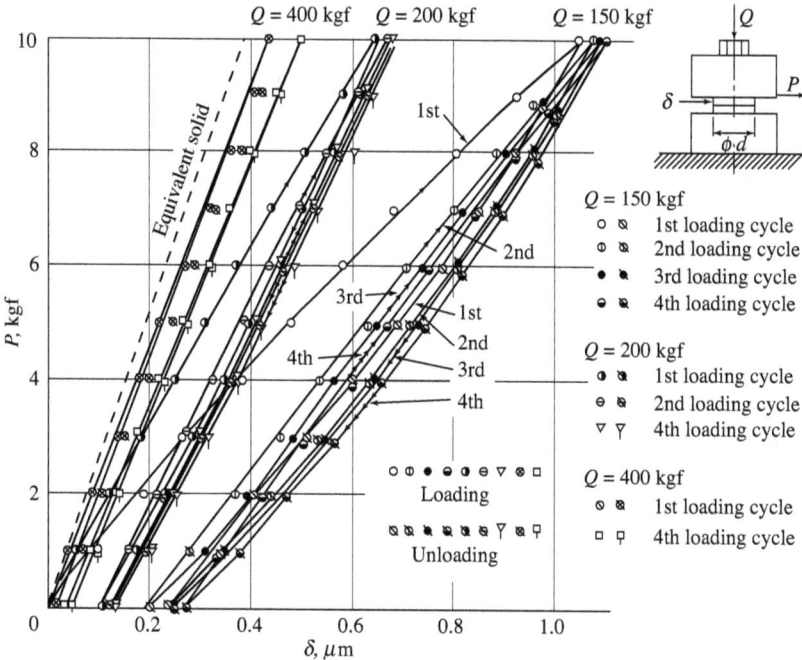

Figure 6-14 Tangential joint deflection under repeated loading cycles—with higher preinterface pressure.

aiming at the establishment of a calculation method of damping capacity of the bolted joint [22, 23].[10] Figure 6-14 is a typical load-deflection curve in tangential loading, and as can be readily seen, there appears a considerable residual displacement in the first loading; however, in the succeeding loading cycles, in which the maximum load is maintained to be within that of first loading, the load-deflection curve repeats nearly the same behavior, showing the constant hysteresis loop and no residual deflection. More specifically, in both the ground joints made of mild steel and brass, the hysteresis loop remains in constant in the second loading cycle and beyond, i.e., steady-state loop type. In contrast, the ground joint made of cast iron shows the gradually progressing hysteresis loops in

---

[10]Although some influences are caused by the connecting bolt, the single bolt-flange assembly with uniform interface pressure distribution is, from one aspect, convenient to investigate the basic behavior of the single flat joint under higher interface pressure. Obviously, such a single bolt-flange assembly can be regarded as a basic entity of the bolted joint. In the single flat joint under higher interface pressure, the deflection to be measured is very small, even in maximum only on the order of 1 μm, together with showing the time dependence under constant loading. All the experiments were thus carried out in the temperature-controlled room.

the second loading cycle and beyond, i.e., progressive loop type. Importantly, Masuko and his coworkers suggested that the hysteresis loop is caused by the microslip at the contact asperities under the elastic and plastic deformation. As a result, damping at the single bolted joint can be characterized by its viscous—and dry frictionlike property (see Chap. 7).

In the immediate previous research, Masuko and coworkers investigated the joint behavior at first loading to observe the essential features of the tangential joint stiffness [22]. Figure 6-15 shows the tangential

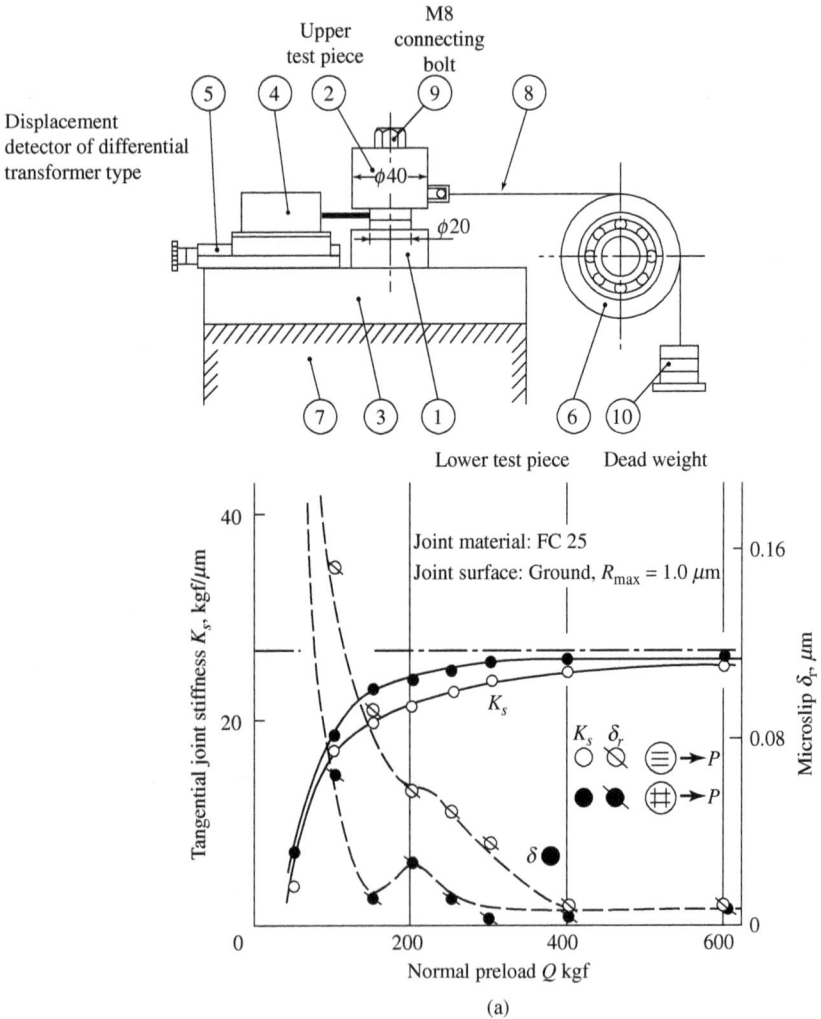

Figure 6-15 Tangential joint stiffness and microslip in varying normal preload: (a) For cast iron joint and (b) for brass joint.

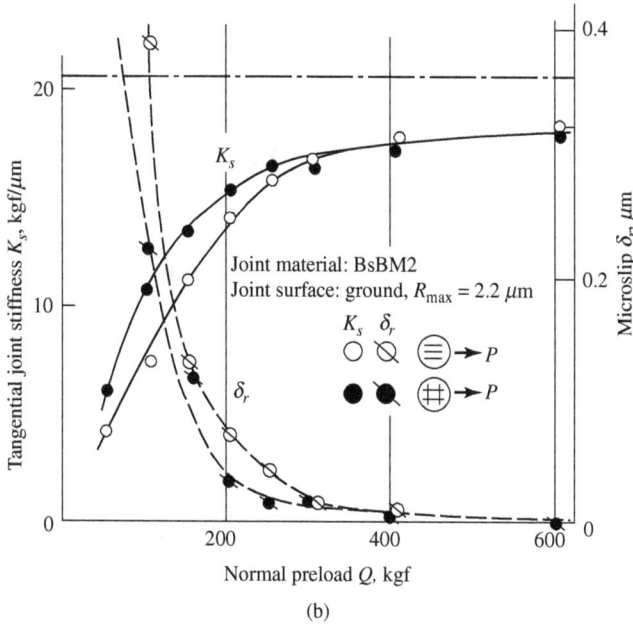

Figure 6-15 (*Continued*)

joint stiffness and microslip with increasing normal preload, when the joint material and finishing method of the joint are varied. In addition to the observation reported by Kirsanov, they unveiled further interesting behavior as follows.

1. The joint material and surface finishing have considerable effect on the tangential joint stiffness. In this context, there is a desirable surface roughness, at which the joint stiffness shows a maximum value. Figure 6-16 shows such a characteristic in the case of a scraped joint, and in fact, the joint stiffness is maximum when the number of contact spots in any 1 in^2 is around 20.

2. The machined lay orientation has also considerable effect on the joint stiffness, and in general, the perpendicular layout shows larger stiffness than the parallel layout.

Following those studies of Masuko et al., Boothroyd and coworkers investigated the single flat joint of annular ring type [24, 25] to analyze the essential feature of structural damping in the wheelhead of a grinding machine. In addition, Burdekin et al. conducted some related studies on the single flat joint of laminated type [26, 27]. Figure 6-17 is a firsthand view of a research map regarding the tangential deflection and microslip of the single flat joint, and summarizing all the observations

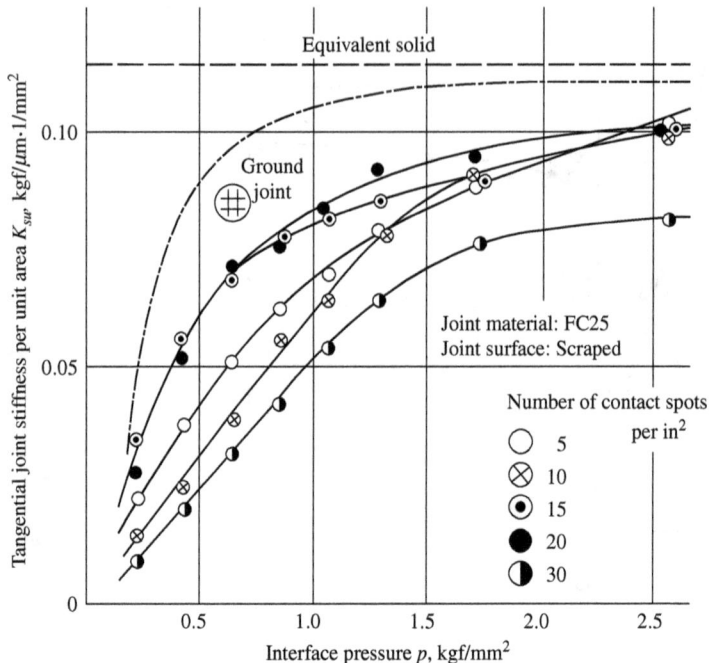

Figure 6-16   Effects of surface finishing quality on joint stiffness.

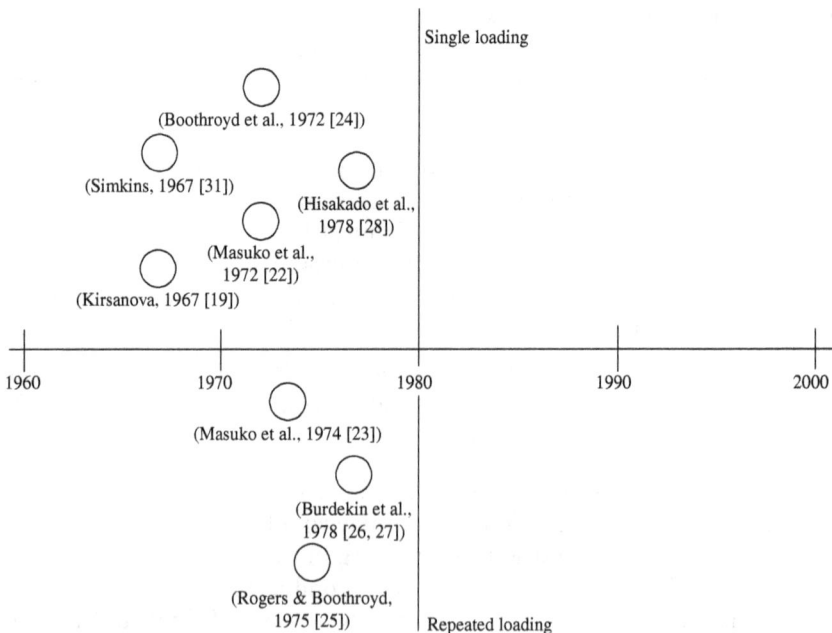

Figure 6-17   Firsthand view for research into static joint stiffness under tangential loading and normal preload.

obtained from the related researches, e.g., those of Kirsanova, Boothroyd et al., and Burdekin et al., the following characteristic behavior can be revealed.

1. The shear stiffness increases with normal preload and also the improvement of the surface finish.

2. The tangential deflection consists of the elastic deflection and microslip, showing a considerable residual displacement after unloading of the first loading cycle.

3. The hysteresis loop decreases its area with the increase of normal preload and repeated number of loading cycles.

4. The hysteresis loop encompasses gradually wider area with increasing tangential force under constant normal preload, as shown in Fig. 6-18, simultaneously maintaining constant slope of loops.

5. The joint deflection in tangential direction is, in most cases, comparable with that in normal direction. In addition, the load-deflection curve shows an opposite trend to that subjected to normal loading: the incremental tangential stiffness is maximum at the commencement of loading, whereas the incremental normal stiffness increases continuously with loading.

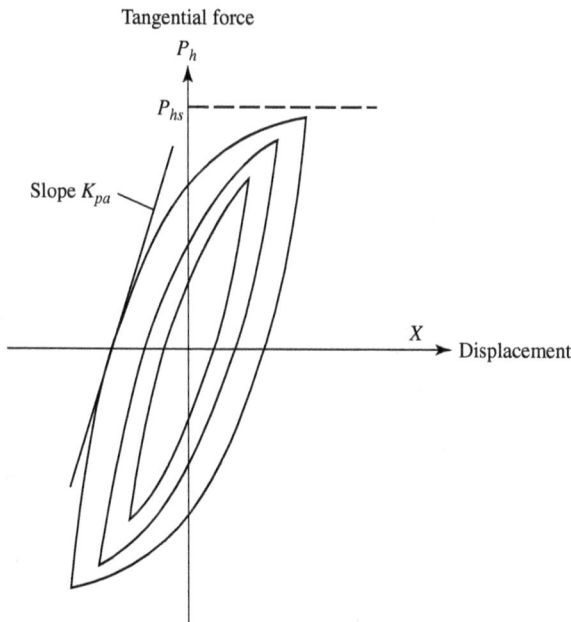

Figure 6-18   Hysteresis loops with increasing tangential loads under constant normal preload (courtesy of Boothroyd).

6. The rate of loading has an effect on the microslip, resulting in the reduction of the tangential joint stiffness. According to the results of Kirsanova, the tangential stiffness reduces 20% when the rate of loading increases 2.5 times.

To this end, Fig. 6-19 shows the test rig used by Boothroyed and coworkers. This test rig can be characterized by its smart ideas as follows.

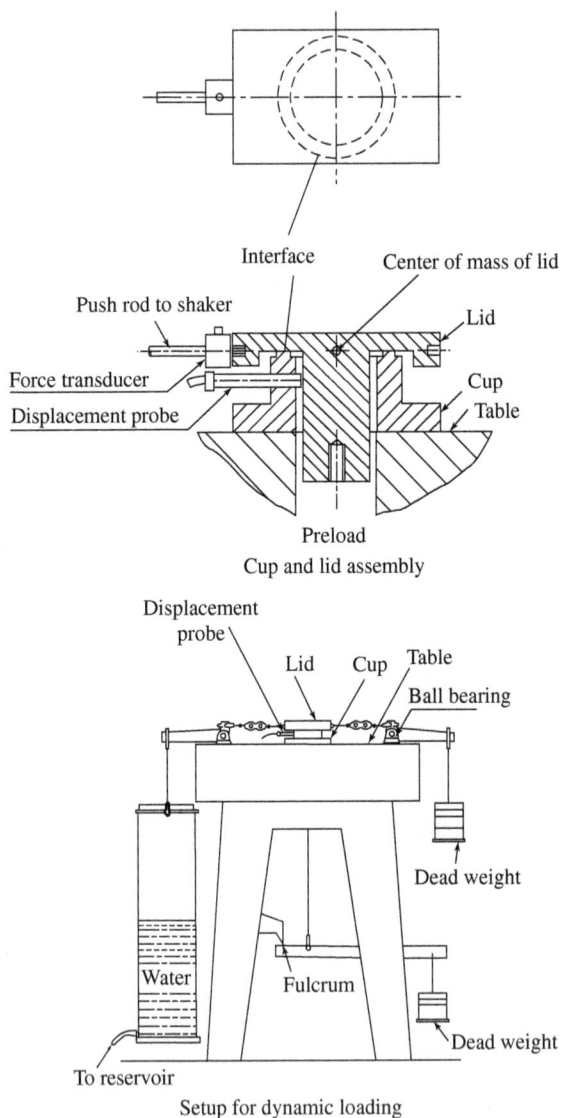

Interface   Center of mass of lid

Push rod to shaker   Lid

Force transducer   Cup

Displacement probe   Table

Preload

Cup and lid assembly

Displacement probe

Lid   Cup   Table

Ball bearing

Dead weight

Water   Fulcrum

Dead weight

To reservoir

Setup for dynamic loading

Figure 6-19 Test rig for dynamic tangential behavior of flat joint (courtesy of Boothroyd).

- Realization of the line action of each force on the body. The lid and cup configuration can facilitate preferable loading passing through the center of mass, i.e., center of gravity of the body.
- Applying normal preload with flexible cable and dead weight.

These remedies are favorable to ensure the accuracy of the experiment, although they give rise to some difficulties in machining the joint surface of the lid.

6.3.3   Peculiar behavior of microslip

In relation to the flat joint under preload and tangential loading, at further issue is the deflection- (displacement-) dependent characteristic of the microslip. In accordance with the general sense, the macroscopic-slip (gross slip) occurs when the following condition is fulfilled, i.e., the rule of Coulomb friction.

$$P_h > \mu\, Q \qquad\qquad (6\text{-}4)$$

where P_h = tangential load
Q = normal preload
μ = macroscopic coefficient of friction

This famous principle can also be accepted at the machine tool joint; however, we must be aware that the microslip is allowed even when the external applied load is less than the friction force, i.e., $P_h < \mu Q$. Actually, a microslip occurring prior to the start of the macroscopic slip, which obeys Eq. (6-4), is one of the most characteristic features of the machine tool joint, and determines definitely the damping capacity of the joint. To distinguish this microslip, obviously it is better to use the term *tangential force ratio* μ_T instead of the *coefficient of friction* μ under $P_h < \mu Q$.[11,12]

In short, the tangential force ratio is equivalent to the coefficient of friction in the condition of the microslip, and its utmost characteristic feature is of *displacement dependence*, as reported first by Courtney-Pratt and Eisner [29]. Figure 6-20 shows a relationship between the microslip

---

[11]Kirsanova [19] reported that the tangential force ratio is, in general, around one-half of the static coefficient of friction. For example, in the joint finished by very fine scraping, the tangential force ratio and static coefficient of friction are 0.14 and 0.28 in dry condition, and furthermore 0.12 and 0.24 in lubricated condition, respectively.

[12]Although we don't have the relevant definition of the microslip, we have the term *slip damping*, which can be observed even in the press fit portion of the turbine blade. A marked suggestion in it is the existence of the optimum pressure, at which the damping capacity is maximum. In fact, damping of the two-layered beam has been investigated (see Chap. 7).

Goodman, L. E., and J. H. Klumpp, "Analysis of Slip Damping with Reference to Turbine-Blade Vibration," *J. Appl. Mech. ASME*, Sept. 1956, pp. 421–429.

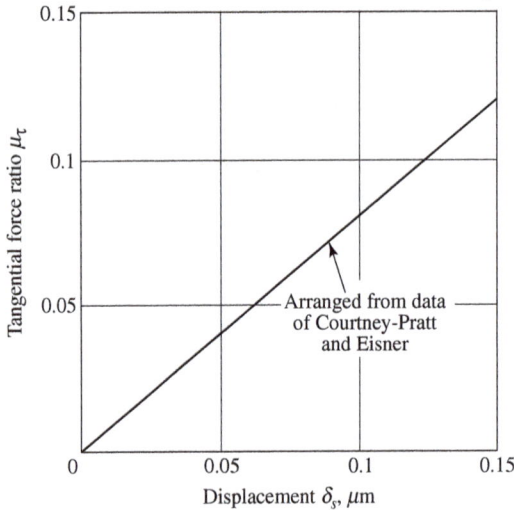

Figure 6-20   Displacement dependence of tangential force ratio in steel joint.

and the tangential force ratio, which was arranged by Masuko et al. [30] on the basis of the data obtained by Courtney-Pratt and Eisner, for the ease of understanding and in order to associate such a relationship with the engineering calculation of the damping capacity of the two-layered beam. Figure 6-21 is a reproduction of the data of Courtney-Pratt and Eisner, where they investigated the metallic joint of sphere-to-flat surface form in small size and made of gold, platinum, tin, indium, and mild steel.

Following that of Courtney-Pratt and Eisner, Simkins [31] also investigated the displacement dependence of the tangential force ratio and typified the microslip by its stepwise-like movement. In fact, Simkins used a smart apparatus as shown in Fig. 6-22, where the displacement detector is of fiber-optic type and capable of resolving 10^{-7} μin, a steel rectangular slider weighing 653 gr can travel on the parallel-piped guide, and also the two surfaces in contact are of 63 μin rms in roughness. When the shear force is applied by the water, the slider shows clearly a stepwise-like movement within the range $P_h < \mu Q$, as shown in Fig. 6-22, and at the point B_{cr}, where $P_h = \mu Q$, the microslip develops rapidly into a gross slip. In general, the number of the microslips that occur depends upon the joint surface quality and loading rate: It reduces with the improvement of the surface quality and speed-up of the loading rate. As a result, it can be said that the tangential force ratio increases monotonically and finally approaches the value of the coefficient

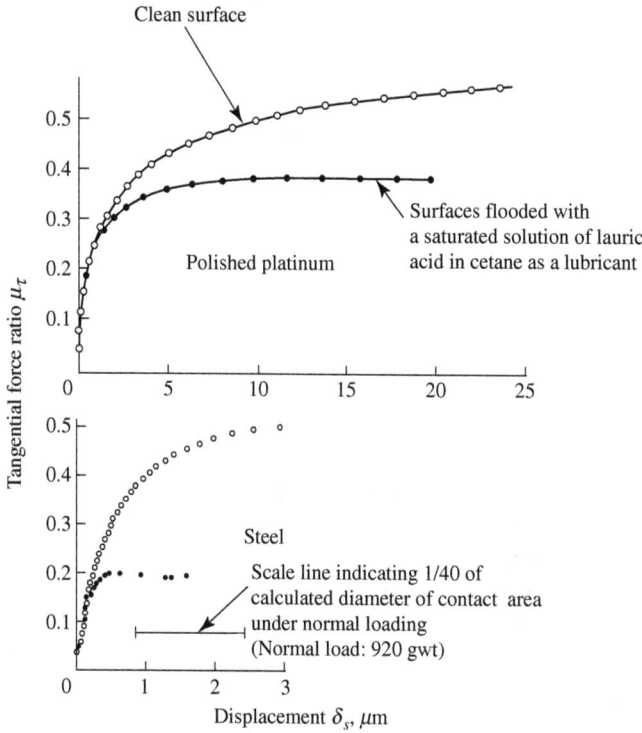

Figure 6-21   Original data for displacement dependence of tangential force ratio (by Courtney-Pratt and Eisner).

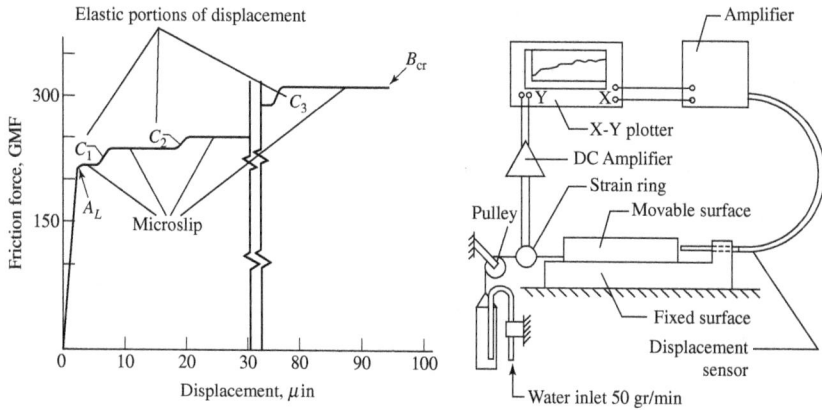

Figure 6-22   Microslips and stick-slip-like movement prior to start of macrosliding (by Simkins).

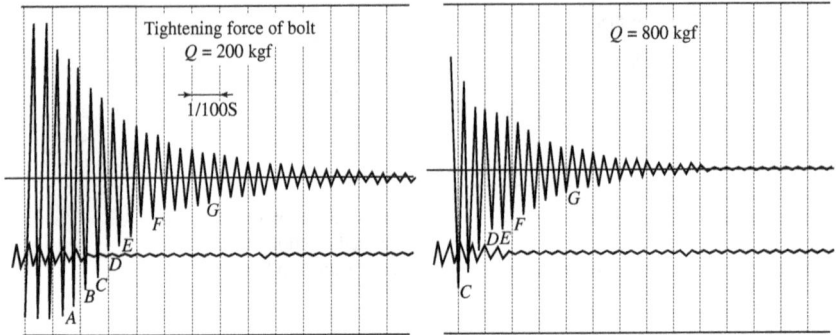

Note: Measurement was carried out with the bolted beam of cantilever form under bending vibration (see that of Ito in Chap. 3).

Figure 6-23   Records of stick-slip-like amplitude changes in damped decay-free vibration.

of friction, always showing proportionality to the tangential displacement. Within this context, Boothroyd and coworkers suggested also the same result, when the displacement is less than 1 μm [25].

Importantly, Ito and coworkers suggested that the hysteresis loop is caused by the microslip at the contact asperities under the elastic and plastic deformation. As a result, damping at the single bolted joint can be characterized by its viscous and dry friction-like property. This suggestion may be ascertained by scrutinizing the experimental results reported by Simkins, and in fact the decayed free vibration curve of a bolted beam shows a stick-slip-like change in vibration amplitude as seen in Fig. 6-23, where the portions D-E and E-F appear to correspond with the stick and slip, respectively. In addition, the portions A-C and F-G appear to be dry friction-like and viscouslike damped vibration.

When we investigate and discuss the marked characteristics in the single flat joint under normal preload and tangential loading, e.g., hysteresis loop in load-deflection curve, stick-slip-like movement of test piece, and appearance of microslip, the research in the tribology sphere is somewhat useful, although the test rig and piece may be designed to be suitable for the wear and friction problem. Figure 6-24 shows thus a firsthand view of the research in the tribology sphere carried out so far, and reportedly these are an extension of those related to Hertz' and Mindlin's theories [37].

6.4   Design Formulas for Damping Capacity and Related Researches

In the flat joint, the dynamic behavior is of course one of the important engineering problems as well as the static behavior. In general, the

Static loading

(Courtney-Pratt &
Eisner, 1957 [29])
Spherically ended cone
bearing on a plane
(Seireg & Weiter, 1966 [33])
Ball-to-plane contact
(Johnson, 1955 [32])
Ball-to-plane contact

(Fujimoto et al., 1998 [34])
Annual plane-to-annual
plane contact

1960   1970   1980   1990   2000

(Seireg & Weiter, 1962 [35])
Ball-to-pin contact

(Goodman & Brown, 1962 [36])
Sphere-to-plates contact

Dynamic loading, microslip,
and damping capacity

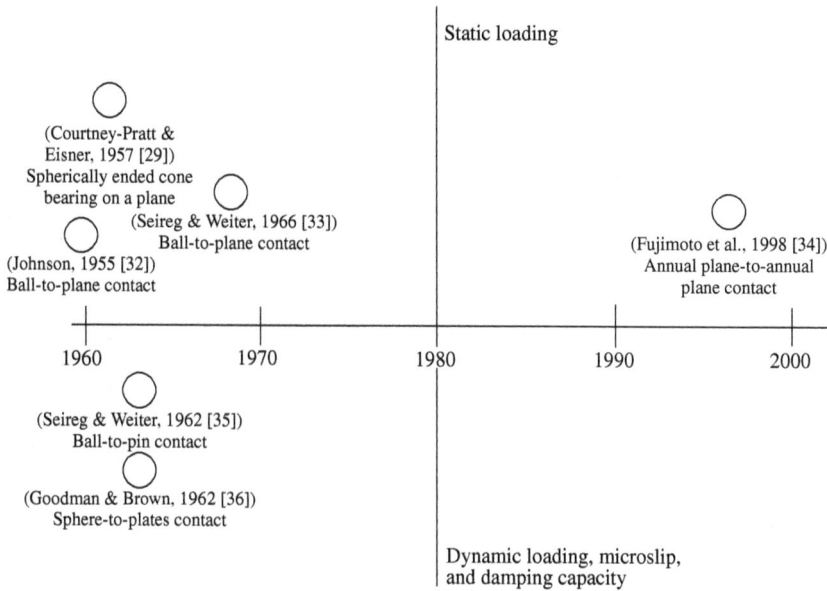

Figure 6-24   Firsthand view for research into two bodies in contact under tangential loading within tribology sphere.

dynamic behavior can be determined by the static stiffness, damping capacity, and self-weight of the objective itself. In relation to the static behavior, we can use the knowledge and database mentioned in Sec. 6.3, and thus at issue is the damping capacity, i.e., energy dissipation at the joint, when the dynamic behavior of the single flat joint is discussed. In general, we must remember the following maxim: *The damping capacity varies inversely with the static stiffness and is derived from the microslip mentioned in Sec. 6.3.3.*

6.4.1   Expressions for damping capacity

The energy dissipation at the joint is likely due to a friction loss, although the microslip is dominant in the machine tool joint rather than the gross slip observed widely in other machines, resulting in the appearance of the viscous damping-like property. Actually, the gross slip is subject to the rule of Coulomb friction, resulting in the appearance of the decayed free vibration curve with linearly damped amplitude.

Table 6-11 summarizes the expressions for the damping capacity proposed so far, and Tables 6-12 and 6-13 show the values of constants for those expressions of Groth and Dekoninck. It is regrettable that the

TABLE 6-11   Expressions for Damping Capacity

	Expression	Conditions obtained expressions			Examples of joints to be applicable expressions	Remarks
		p kgf/cm^2	Relationships between K_j and K_0	Shape and size of joint surface		
Reshetov and Levina [38]	$\psi = A_C l^{/3} \sqrt{p}$	<15		Annular ring type		Test piece of multiple-laminated type Interface layer: Oil
Groth [39]	$\zeta = \beta p^{-\gamma}$	<100	$K_0 < K_j$	Rectangular type Area: 210×210 mm^2	Bolted joint	Test piece of flanged column type
Dekoninck [40]	$D = C\tau^m$	<523	$K_0 \gg K_j$	Rectangular type Area: 40 mm^2		Tangential load per area $100 < \tau < 300$ kgf/cm^2
Tsutsumi and Ito [41]	$\psi = C_a \theta_p^{\chi}$	<12.7	$K_0 > K_j$	Annular ring type Area: around 11.4 cm^2		Torsional loading (exciting frequency < 100 Hz) θ_p: Peak angular displacement

ψ: dissipating energy, ζ: damping factor, D: damping energy,
p: normal interface pressure, A_C, C, C_a, m, β, γ and χ: constants

Note: Figures in the column of proposer designate the number of references.

validities of these expressions are not verified as yet, even in the beginning of 2000.[13]

In the following, some suggestions and recommendations will be stated from the viewpoint of the engineering calculation.

1. In the studies of Reshetov and Levina, the constant A_C is 1.35 and 0.9 for industrial oil 45 and industrial oil 12, respectively. In addition, Reshetov and Levina reported that the dissipating energy is proportional to the square of vibrational amplitude in the case of the laminated joint. Moreover, they applied their expression to calculate the damping capacity of the rolling bearing, key connection, and spline shaft.

2. The expression of Dekoninck was obtained for the joint made of mild steel and for the normal contact pressure up to 500 kgf/cm^2. In addition, the joint consists of the ground surface with roughness of about 2 μm in

---

[13]A root cause of difficulties in the verification lies in the measurement of the damping capacity. In the utmost preferable case, the larger load excitation and due response signal detections of noncontact type are needed to ensure the measurement of the damping capacity with higher accuracy. Figure 6A-1 reproduces the measured damping capacities of the vertical turning machine reported by Groth [39]. The measurement was carried out for five similar machines, and as can be readily seen, there is a considerable scatter of the measured values. This scatter may be caused by something related to the roller guideways at the traveling ram and headstock.

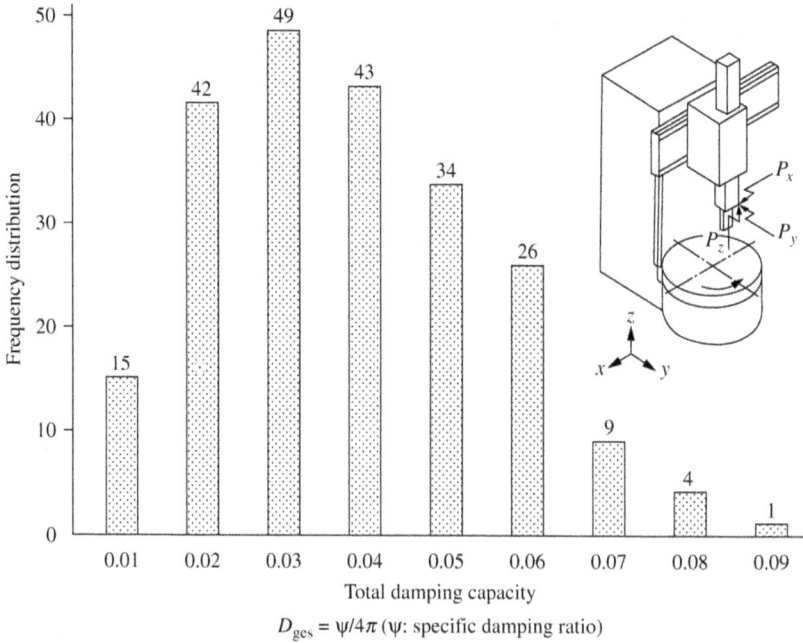

$$D_{ges} = \psi/4\pi \; (\psi: \text{specific damping ratio})$$

Figure 6A-1   Scatter of measured damping capacity in practical cases (courtesy of Groth).

TABLE 6-12   Values of β and γ (Courtesy of Groth)

Type of contact surface		β	γ	Valid range of interface pressure p kgf/cm²
Scraped/ scraped	$A = 122$	0.017	1.16	
	$A = 122$, lubricated			
	$v = 21$	0.044		
	$v = 33$	0.036	0.33	
	$v = 117$	0.021		
Scraped/ planed	$A = 122$	0.00415		0.3–100
	$= 208$	0.0080	0.66	
	$= 260$	0.012		
	Lubricated ($v = 117$) Machined lay orientation: Long	0.018	0.91	
	Short	0.010		
Scraped/ ground	$A = 244$	0.0165	0.82	0.3–40

A: Area of joint surface cm²; v: Oil viscosity at 50°C, cSt.

TABLE 6-13   Values of Constants C and m in Dekoninck's Expression

	Normal contact pressure p, kgf/mm^2	$C \times 10^6$	m	Valid for P_τ larger than (kgf/mm^2)
Prismatic roughness peaks	4.80–5.23			
$s = 0.5$ mm, $2 \times \beta_0 = 90°$, pressed against mild steel		27.9	6.85	1.5
$s = 2$ mm, $2 \times \beta_0 = 90°$, pressed against mild steel		115.0	4.65	1.5
$s = 2$ mm, $2 \times \beta_0 = 160°$, pressed against mild steel		65.0	4.86	1.8
$s = 2$ mm, $2 \times \beta_0 = 160°$, pressed against mild steel; use of an adhesive		166.0	4.90	1.0
Idem as before but adhesive layer was broken before experiment		67.0	6.56	1.5
Shaped roughness peaks				
$s = 0.6$ mm, pressed against mild steel		263.0	4.05	1.3
Test specimen without contact surfaces		201.0	1.99	

R_a and milled surface with two-dimensional prismatic roughness peak. As will be clear from the above, the expression may be used for rough estimation of damping, because the joints investigated are far from the actual ones. In addition, the dependence of displacement in the microslip appears not to be considered rationally.

3. When we consider the availability of the model theory for the jointed structure (see App. 2), the expression of Groth appears to be directly applicable to the machine tool structure of full-size.

For the flat joint under dynamic tangential loading, the most interesting subject to understand is the characteristic of the hysteresis loop, which is in closer relation to the damping capacity derived from the joint. Following to the investigation into the static tangential loading, Rogers and Boothroyd proposed an expression to represent the hysteresis loop on the basis of its schematic representation, already shown in Fig. 6-18.[14]

$$P_h = P_{hs}(1 - e^{-K_{pa} \delta_s / P_{hs}}) \qquad (6-5)$$

where P_h = tangential load

δ_s = tangential displacement

K_{pa} = stiffness parameter. This value is of great interest in understanding the dynamic behavior of the flat joint subjected to the exciting force; however, its nature is not fully clarified yet.

P_{hs} = asymptotic value of tangential load (see Fig. 6-18)

In the hysteresis loop context, the marked observations are that the energy loss per cycle is in inverse proportion to the value K_{pa}, and that the maximum energy loss can be obtained by approaching the tangential force ratio to be close to the coefficient of friction and using the joint surface with a low value of K_{pa} and high coefficient of friction.

Obviously, Boothroyd et al. and Dekoninck provide us with the informative data, and thus we summarize the further major findings as follows.

1. Under dynamic tangential loading, the joint shows apparently the nonlinear hysteresis loop in conjunction with the load-deflection curve, and this loop has a slight time dependence: The damping is larger in the commencement of loading.

2. The nonlinear hysteresis curve is derived from the interfacial microslip, which shows linearity and nonlinearity at lower and larger loads, respectively.

3. The energy loss at the joint is significant and changes considerably depending on the applied load in the previous loading cycle.

4. The energy loss per cycle is independent of the excitation frequency ranging from 5 to 200 Hz.

5. It is notable that larger values of D in Dekoninck's expression can be obtained for the joint with the adhesive, the same as the flat joint subjected to the excitation force in the direction normal to the joint surface.

---

[14]Koizumi et al. proposed an empirical expression for the tangential microdeflection of the single bolt-flange assembly under tangential loading. According to their proposal, the tangential microdeflection is a function of the tangential force ratio, consisting of the elastic deflection and microslip.

Koizumu, T., Y. Ito, and M. Masuko, "Experimantal Expression of the Tangential Microdisplacement between Joint Surfaces," *Trans. of JSME*, 1978, 44(384): 2861–2870.

Expressions

(Groth, 1972 [39])

(Tsutsumi & Ito, 1980 [41])

(Reshetov & Levina, 1956 [38])

(Dekoninck, 1972 [40])

|———————|———————|———————|———————|———
1960   1970   1980   1990   2000

(Rogers & Boothroyd, 1975 [25]) Hysteresis loop

(Beards & Neroutsopoulos, 1980 [43]) Loss factor of EDM joint

(Hashimoto & Kume, 1972 [42]) Vibration transmittability

Hysteresis loop, vibration transmittability, loss factor and so on

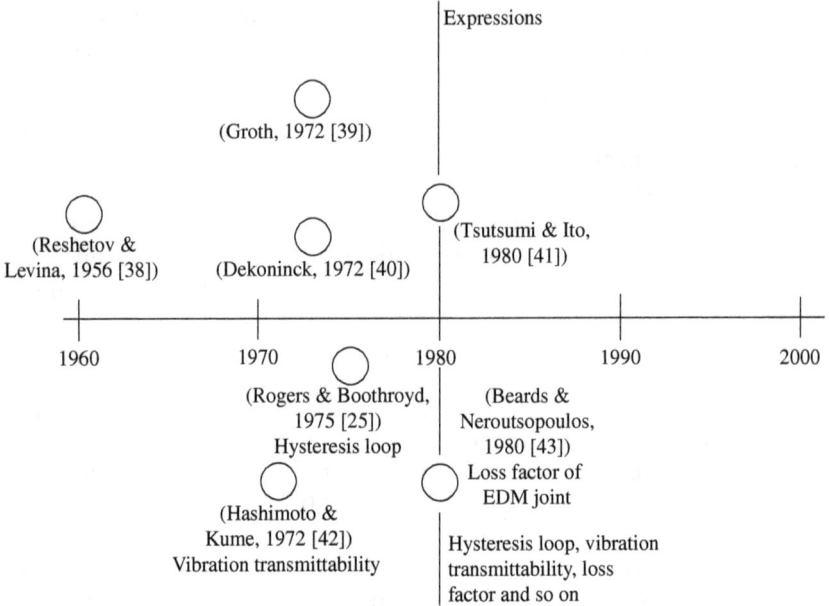

Figure 6-25 Firsthand view for research into expressions for damping.

To this end, a firsthand view of research into the expression of damping is given in Fig. 6-25. Within this view, we can find the interesting report. For instance, Hashimoto and Kume investigated the transmission characteristics of the torsional vibration through the lubricated conical joint, i.e., tapered cylinder-to-bush joint, which is made of mild steel and with surface roughness of 6 μm in R_{max}. In accordance with their report, the response waves change to the rectangular form from sinusoidal form, when occurring the gross slip, and it depends upon the interface pressure and amplitude of the exciting vibration. As is easily understood, the rectangular response disappears with increasing interface pressure.

6.4.2 Representative research into dynamic behavior

Figure 6-26 is a firsthand view of the representative researches into the dynamic behavior of the machine tool joint (see alternately Fig. 6-25). Except for those of Schaible and Burdekin, the research activities were carried out using the joint under normal dynamic loading, and they were two-pronged: one is for the single flat type, and the other is for multiple-laminated flat type. Of these, the multiple-laminated joint consisting of nine stacked hollow disks is first employed by Reshetov and

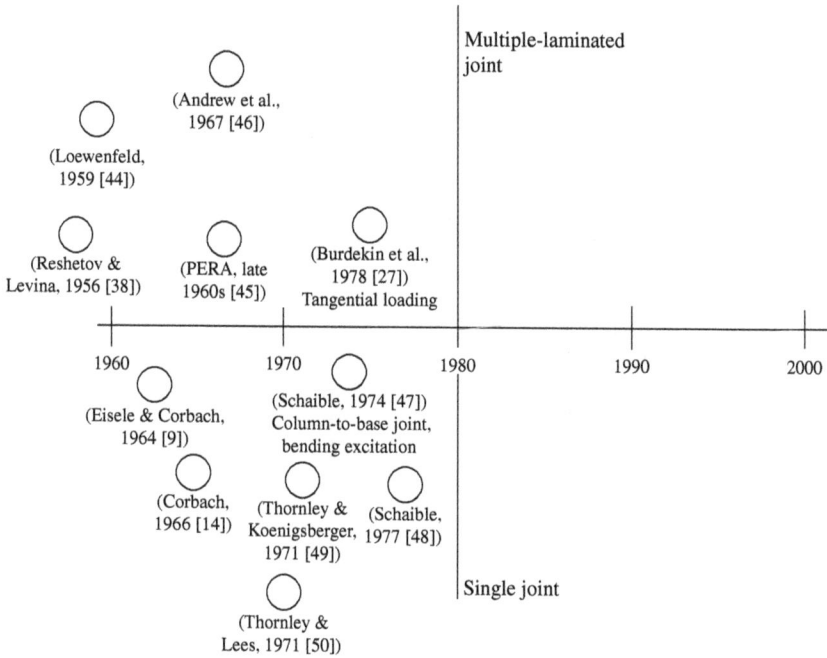

Figure 6-26 Firsthand view for research into dynamic stiffness.

Levina so as to facilitate the experiment by magnifying the damping capacity to be measured. Following their research, Loewenfeld [44], PERA [45], and Andrew et al. [46] conducted investigations using the test rig similar to that of Reshetov and Levina. In retrospect, the first noteworthy research into the damping capacity of the machine tool joint is credited to Reshetov in 1956, simultaneously evoking the importance of the machine tool joint in the structural design, although the flat joint of multiple-laminated type can be considered to be far from the joints observed in the machine tool of full-size.

In the flat joint under the normal preload and tangential loading, the damping mechanism even in the dry joint condition can be understood without difficulty; however, it is hard to imagine the increase of damping at the dry flat joint under only normal dynamic loading, when it is compared with that of an equivalent solid. In accordance with the earlier work, some people assert the increasing effects of the damping due to the joint even under only normal loading, and at the same time other people deny it. Given such disputations, we quickly touch on those of Corbach and Thornley in the following. In fact, they investigated the single flat joint under dynamic normal loading and with lubricants.

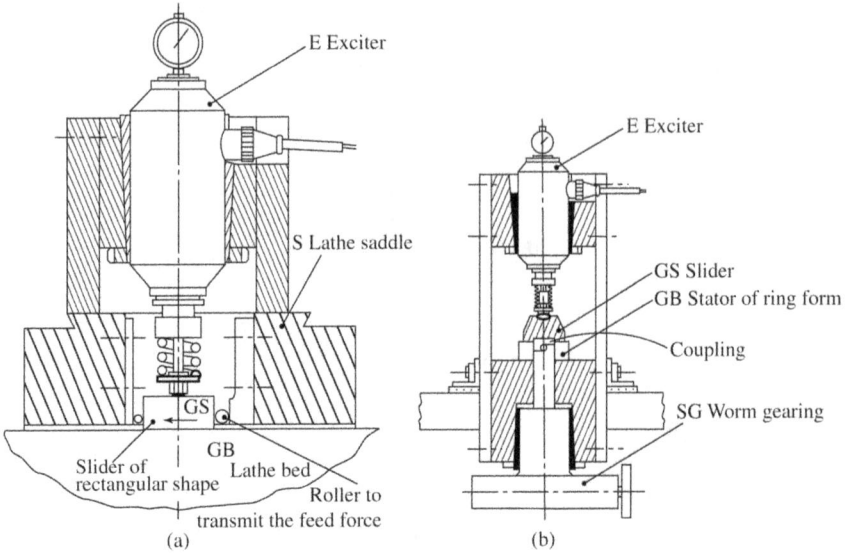

Figure 6-27   Test rigs used by Eisele and Corbach: (a) Linear type and (b) rotary type.

Single flat joint under lower preload and dynamic normal loading—that of Eisele and Corbach [9, 14].   Although suggesting no expressions available for the engineering calculation, Eisele and Corbach conducted a series of investigations using the test rig of linear or rotary motion type, intending to clarify the dynamic behavior of the sliding joint. As will be shown below, they unveiled important basic knowledge about the dynamic behavior available for the design of the slideway. Importantly, they mainly used such a test rig of rotary motion type shown in Fig. 6-27 by reason of difficulty in the measurement of the relative displacement. In this test rig, the lower test piece GB of ring form is fixed and the upper test piece GS is driven by the worm gearing through the coupling, which can transmit only the driving torque, but not the axial force, to the upper test piece. The sinusoidal exciting force is applied to the upper test piece, simultaneously measuring it by the load cell of piezoelectric type. The corresponding joint displacement is measured by the displacement detector of capacitance type, and the joint characteristics are displayed using the in-phase component K_a and loss factor η (K_b/K_a, see Chap. 5). In short, K_a increases with the static preload P_{st}, and the loss factor η shows the opposite behavior. These are the general characteristic features of machine tool joints, and without exception, the behavior of the dynamic joint stiffness is determined definitely by its in-phase component. In addition, the oil as an interfacial medium has excellent effects on both the in-phase and quadrature components, showing, e.g., the considerable increase of the in-phase component and maintaining the same tendency to the static preload.

In the slideway, it is crucial to clarify the effect of the sliding velocity, and thus Fig. 6-28 shows a typical result when the excitation frequency is varied. In this case, the sliding joints investigated are a pair of gray cast irons, and also a pair of gray cast iron (fixed branch) and cloth laminated bakelite, where the joint surface made of gray cast iron is scraped. Although the dynamic stiffness K_a in still condition ($s = 0$) is larger than that under $s \neq 0$, the sliding velocity of more than 250 mm/min has no apparent effect on the joint stiffness. In addition, as well as the excitation frequency, the oil viscosity has considerable effect on both the in-phase and quadrature components. For the damping capacity represented here with the loss factor η, the same tendency mentioned above can be

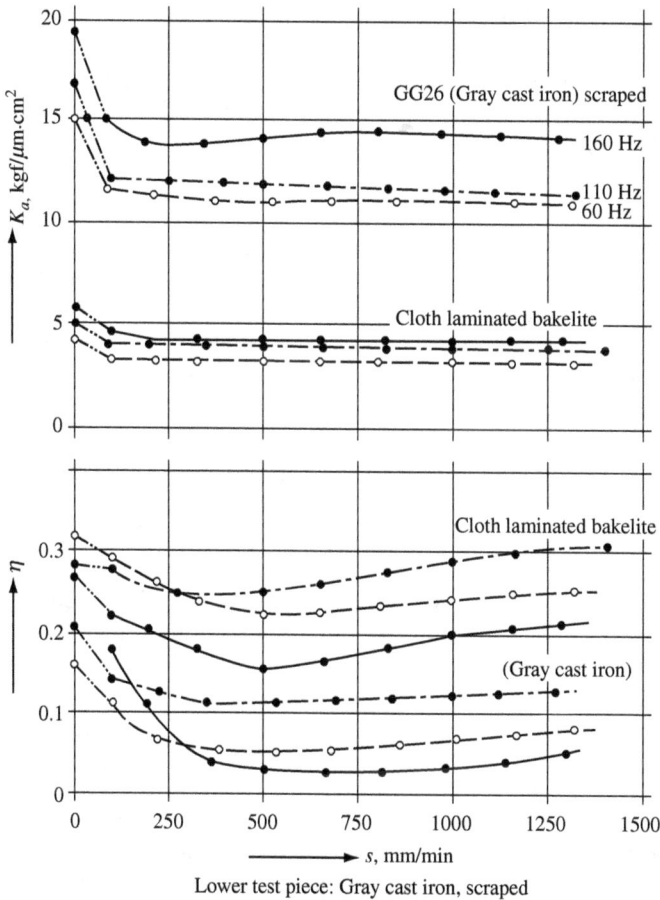

Figure 6-28 Effects of sliding velocity on joint stiffness and damping when excitation frequency is varied (by Corbach).

observed, but it is more complicated than that of dynamic stiffness. More specifically, the following should be stated.

1. The larger the in-phase component, the lower the quadrature component; and both components increase by providing the oil at the joint in the case of scraped cast iron joint.

2. The in-phase component increases and decreases with the excitation frequency and oil viscosity, respectively, in the joint made of gray cast iron. In contrast, the quadrature component decreases and increases with the excitation frequency and oil viscosity, respectively.

3. In the joint made of cloth laminated bakelite, the oil viscosity shows the opposite behavior to that for the joint made of gray cast iron.

In accordance with the reports of PERA, there is higher possibility of improving the static and dynamic joint stiffness as well as damping when the liquid polyisobutylene or antithixotropic polymer solution is applied to the multiple-laminated joint. In this context, furthermore, Groth [39] observed, by contrast, the larger increase of damping under sliding up to 3000 mm/min in nearly all the bending and rocking vibration modes.

Single flat joint under higher preload and dynamic normal loading— researches of Thornley and Koenigsberger [49] and Thornley and Lees [50]. Thornley and coworkers are thought to have conducted the first experimental work regarding the dynamic behavior of the single flat joint. They employed the test rig and measuring system shown in Fig. 6-29, and the noteworthy feature of their test rig is a mechanism to apply uniformly the higher preload across the whole joint area. For this purpose, the static preload is applied using the hydrostatic oil thrust bearing with three pads. Following those studies of Thornley and coworkers, Dekoninck [40] conducted certain experiments to expand the needed knowledge using similar test rigs and experimental techniques. The theoretical and experimental evidence reported by these earlier researches indicates obviously that the dynamic behavior of the single flat joint under dynamic load normal to the joint surface is, in nearly all cases, identical to that of the multiple-laminated joint.

In the following, some representative behavior of the single flat joint will be shown, focusing on the relevant effects of major jointing factors.

1. *Static preload.* Figure 6-30 shows the effect of the static preload on the dynamic stiffness. Although not shown here, the dynamic stiffness of the joint under higher preload increases steeply with the preload,

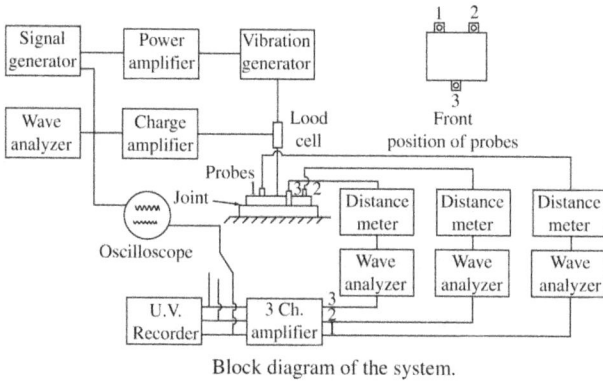

Block diagram of the system.

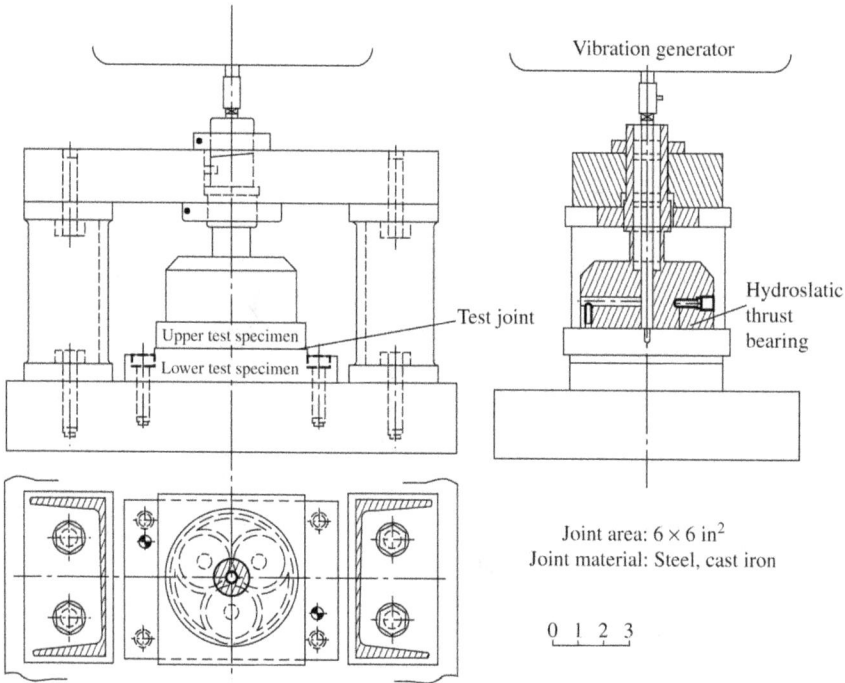

Joint area: 6×6 in^2
Joint material: Steel, cast iron

Dynamic test rig for normal loading of joint faces

Figure 6-29   Test rig and measuring system for single flat joint (courtesy of Thornley).

finally flattening up to that of an equivalent solid, as already shown in Fig. 5-22. In short, the dynamic joint stiffness increases duly with the preload, showing some dependence on the excitation frequency. This behavior is, as easily understood, identical to that observed in the static stiffness of the single flat joint and also in the dynamic stiffness of the multiple-laminated joint.

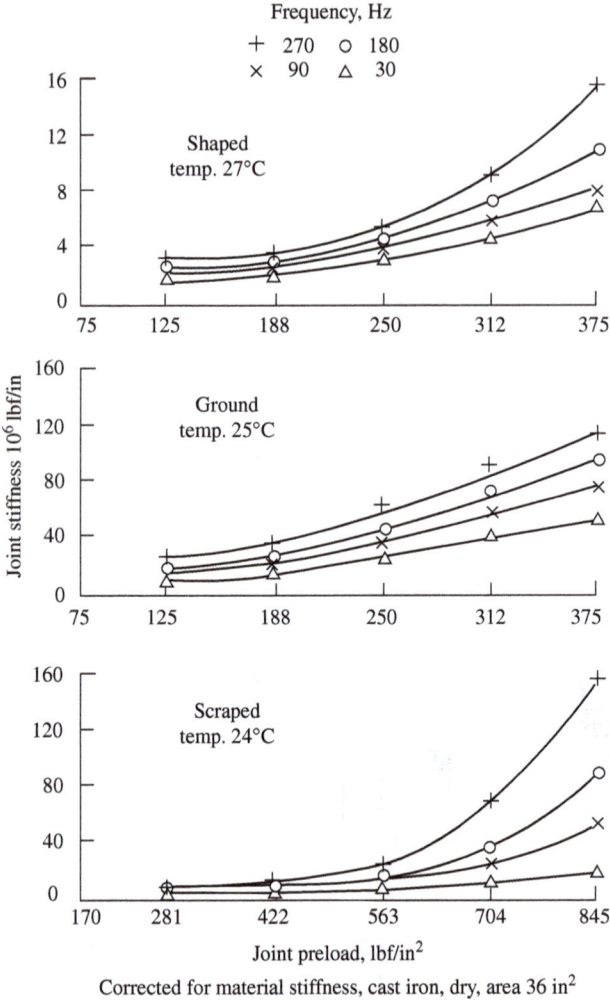

Figure 6-30 Relationships between joint stiffness and static preload when excitation frequency is varied (courtesy of Thornley).

2. *Lubricants.* When there is oil between the joint surfaces, the in-phase and quadrature components increase in all the cases, although their rate of increase depends upon the surface topography, oil viscosity, joint material pairings, static preload, excitation frequency, and hardness of joint material. Figure 6-31 shows the marked effect of the oil on the dynamic joint stiffness. The higher the oil viscosity, the higher stiffness. It is, however, difficult to have the clear tendency for the effect of each kind of oil.

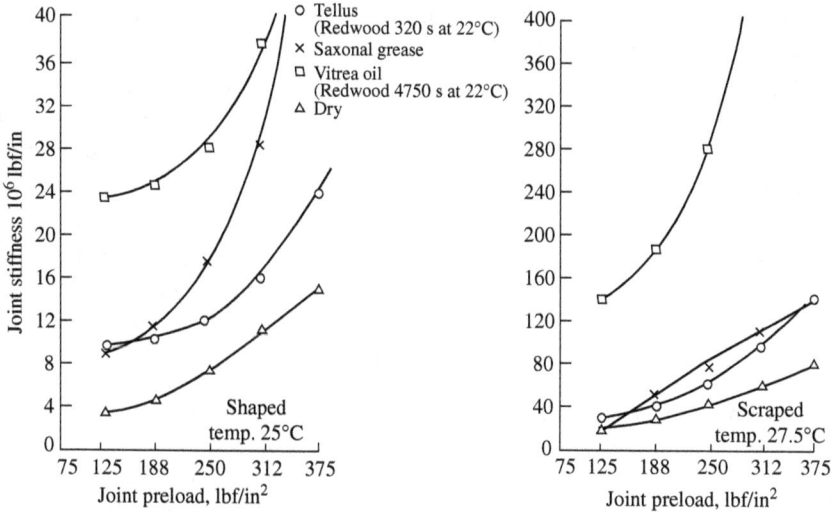

Note: Change in magnitude of joint stiffness axis. Mild steel, frequency 90 Hz, area 36 in^2.

Figure 6-31 Effects of lubricants on joint stiffness (courtesy of Thornley).

3. *Surface topography.* The dry rough joint has, in general, lower in-phase and higher quadrature components compared with those of the smooth joint. As a result, the smooth joint is stiffer than the rough joint, provided that the joint surfaces have no flatness deviation.

4. *Apparent area of contact.* Apart from some special cases, i.e., the joint under higher preload, the apparent area of contact has no effect on the dynamic joint stiffness.

5. *Effects of planform shape.* When the apparent joint area is kept constant and the mild steel specimen with shaped dry surface is used, the planform shape has very little influence on the stiffness, whereas the surface topography has considerable influence.

Remembering that the magnitude of the interface pressure in these earlier research activities ranges to various extents, the dynamic behavior of the single flat joint can be summarized as follows.

1. When the static preload increases, the dynamic joint stiffness increases whereas damping decreases.

2. The interface layers, such as oil and grease, increase both the dynamic stiffness and the damping capacity. The oil viscosity shows extreme effects on the increase of stiffness: the more viscous the oil, the greater the increase of stiffness.

3. The in-phase and quadrature components show complicated behavior regarding the excitation frequency of the applied load.

4. Although a rough surface shows slightly better damping than a smooth surface, the rough joint surface has less stiffness than a smooth surface does, provided that the joint has no flatness deviation and/or waviness.

5. The bonded joint contributes considerably to increase the dynamic joint stiffness. The bonding enables the flatness deviation of the joint to be improved, because it fills up the valley of the wavy surface. In addition, such a lead foil can improve the quality of joint surface, resulting in an effect similar to that of the bonded joint.

To this end, we must be aware that the basic knowledge is of great importance, but is not useful without establishing the model theory, and that the facing problem is especially serious in the case of dynamic behavior. For instance, the vibration mode is dominant in relation to the dynamic behavior of the joint to a large extent, although it is not obvious in the basic investigation using the small test piece. Schaible suggested such effects of the vibration mode [48], when the bolted joint exists between the two box-like cast beams (GG30 of DIN and 600 × 300 × 250 mm^3) with 20 mm in wall thickness. Obviously, the relative displacement at the joint can clearly be observed in the first vibration mode, i.e., the bending mode (400 Hz), resulting in the increase of damping due to the joint. In fact, the magnitude of the damping coefficient ranges from around 1.0×10^{-2} to 0.6×10^{-2} for milled joint surface with increasing interface pressure from 0.5 to 4.5 N/mm^2. In contrast, no relative displacements are observed at the joint in the second mode, i.e., the drum-like mode (900 Hz); as a result, the box-like structure can accommodate only the material damping, i.e., around 0.3×10^{-2} in damping coefficient notwithstanding the magnitude of the interface pressure.

6.5 Thermal Behavior of Single Flat Joint

In the design procedure of the machine tool, the thermal behavior is one of the most important factors,[15] and as well as static and dynamic behavior, the thermal behavior is subjected to the joint characteristics within

---

[15]In the late 1990s, at burning issue was how to reduce the thermal deformation of the machine tool. The machine tool with small or no thermal deformation should be realized with the growing requirements for higher machining accuracy even under heavy cutting with higher cutting speed. For instance, Fig. 6A-2 summarizes some well-known remedies used so far. In addition, it is worth suggesting that the modernized design for the panel cover on the new horizon has two aspects: One is an advance of industrial design by providing the machine with an amenity-oriented appearance, and the other is the reduction of thermal deformation caused by the heat dissipated from the machine body and accumulated between the body structure and the panel cover.

Various remedies for thermal deformation

Compensation of thermal deformation

Allocation of additional heat sources for counterbalance
— Providing additional thermal deformation for counterbalance
— Compensation of NC information, e.g., in accordance with output signal of temperature sensor

Isolation of heat sources from structural boby

Isolation of heat sources, such as lubrication oil tank, hydraulic unit, main motor, and so on, from structural body
— Insert of heat isolation materials

Positive use of cooling effects of machine elements

e.g., increase of diameter of V-pulley and fan effects of ribs on rear surface of 4-jaw chuck

Unification of temperature distribution within structural body

e.g., circulating system of warm wind with constant temperature in structural body

Cooling remedies for heat sources/prompt elimination of generated heat

Electric cooling devices
Oil shower cooling for inside wall of structural body
Heat pipe
Forced air or oil cooling

Circulating method within main bearing housing

Air circulating within structural body
Air cooling for rolling element within main bearing

Remedies from structural materials

Increase of heat dissipation by enamel coating (double magnitude as compared with that of cast surface)
Use of raw material with lower thermal expansion coefficient, e.g., invar
Use of raw material with larger thermal inertia, e.g., concrete

Remedies from structural design

To be identical thermal deformation vector with cutting speed direction: e.g., turret lathe of drum type
Shortening scale dimension related to thermal deformation
Jointing method of unit, e.g., fixing method of headstock

Nonconstraint structure for thermal deformation
Nonsensitive structure for thermal deformation
Structure minimizing thermal deformation

Thermally symmetric structure, e.g., housing structure and twin-spindle

Reduction of generated heat

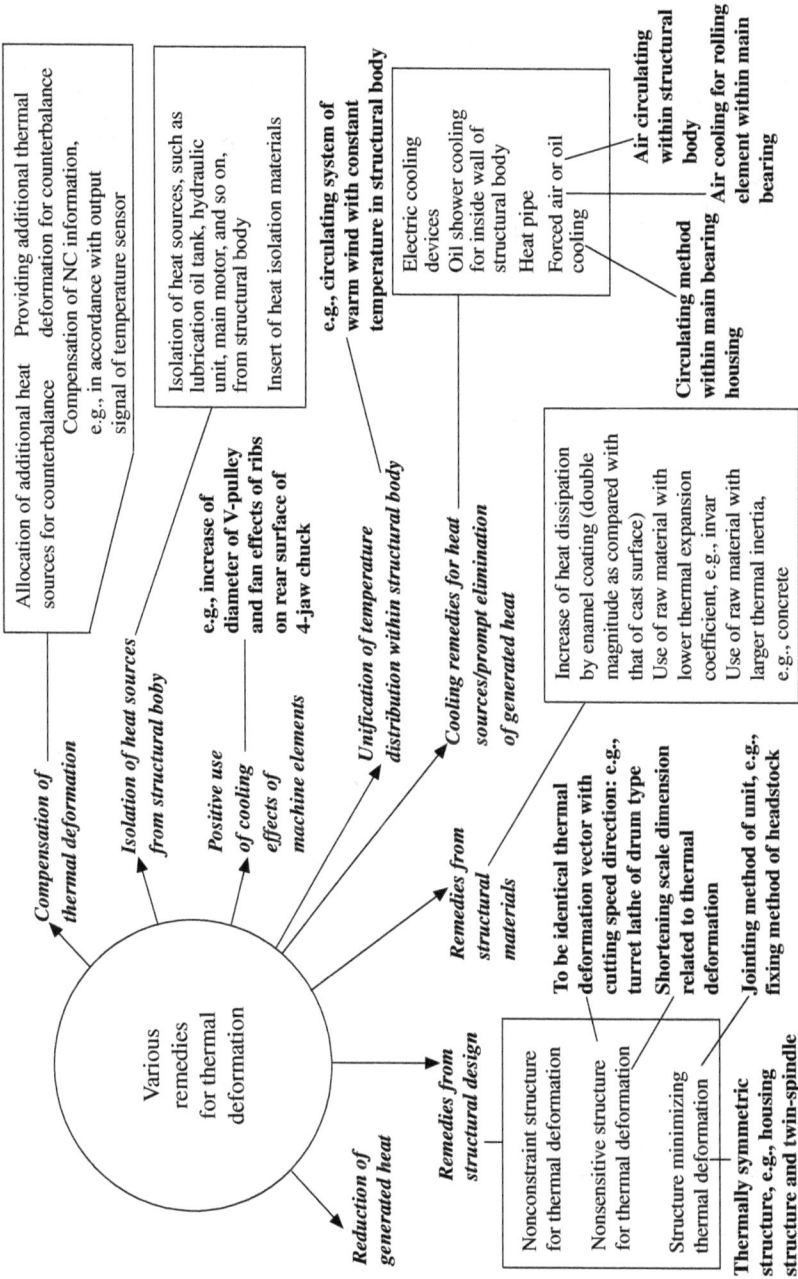

Note: The twin-spindle is very effective at preventing skewing of the spindle.

Figure 6A-2   Firsthand view of remedies for thermal deformation.

a machine tool as a whole to a large extent, where the utmost representative characteristic is the thermal contact resistance. In this context, it is crucially necessary to understand that the research into the thermal behavior of the joint is in the embryonic stage, even as of the beginning of 2000, although the thermal contact resistance is one of the well-established engineering problems especially in the heat transfer sphere from the old days.[16] In contrast to the machine tool joint, however, there have been no ideas of closed-loop effects within the heat transfer sphere, nor has there been consideration of the thermal deformation of the joint surroundings.

In general, the thermal contact resistance can be determined by (1) the contact surface condition, (2) the joint material and its hardness, (3) the interface pressure, and (4) the interfacial layer, and quickly the following three are noted as the representative behavior.

1. The thermal contact resistance decreases with the increase of the interface pressure.

2. The thermal deformation of the joint is both space- and time-dependent.

3. The thermal deformation of the joint is subjected considerably to the interface pressure and its distribution together depending on the mechanical constraint condition.

More specifically, as is well known, the thermal contact resistance is caused by shrinkage of the heat flow at the joint surface, and partly is derived from a certain effect of the oxide film and interface layer, such as shown in Fig. 6-32. In considering the case where two cylinders are in contact with each other at their side flat surfaces, the thermal contact resistance r_c can be written as

$$r_c = \Delta T / q$$

---

[16]For example, there are the following materials.

- Yovanovich, M. M., and W. M. Rohsenow, "Influence of Surface Roughness and Waviness upon Thermal Contact Resistance," Technical Report No. 6361-48 (sponsored by NASA), June 1967, Department of Mechanical Engineering, Massachusetts Institute of Technology. In this report, the contact pattern was measured using the X-ray.

- Fontenot, J. E., Jr., "The Thermal Conductance of Bolted Joint," Dissertation of Louisiana State University, February 1968.

- Yovanovich, M. M., A. H. Hegazy, and J. DeVaal, "Surface Hardness Distribution Effects upon Contact, Gap and Joint Conductances," AIAA/ASME 3d Joint Thermophysics, Fluids, Plasma and Heat Transfer Conf., 1982, Paper No. AIAA-82-0877.

- Yovanovich, M. M., A. Hegazy, and V. W. Antonetti, "Experimental Verification of Contact Conductance Models Based upon Distributed Surface Micro-hardness," AIAA 21st Aerospace Sciences Meeting, 1983, Paper No. AIAA-83-0532.

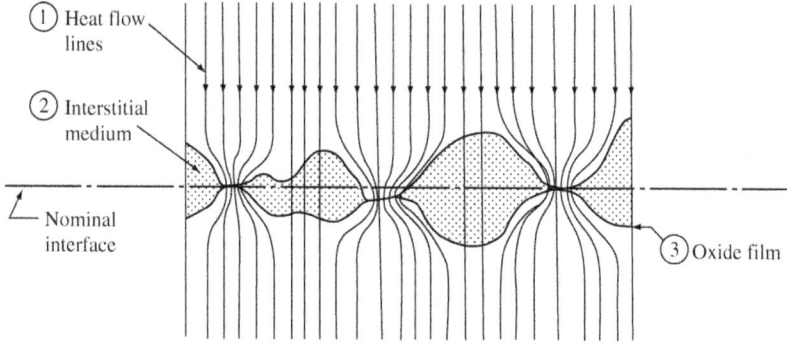

Figure 6-32   Causal sources of thermal contact resistance.

where q and ΔT are, as shown in Fig. 6-33, the heat flux and the temperature difference at the contact surface, respectively.

If the temperature gradient is known, the heat flux is given by

$$q = -\alpha_\lambda \, (dT/dx)$$

where α_λ is the thermal conductivity of the solid body.

In addition, the temperature distribution becomes linear within each cylinder, except a zone near the joint surface, and thus the equivalent contact length Δl and the thermal contact coefficient r_θ can be used to represent the thermal contact resistance. In this case, r_θ can be defined by the following expression.[17]

$$r_\theta = \Theta_{20}/\Theta_{10}$$

In short, the nonlinear temperature distribution at the joint surface is dominant in the structural design, and the designer is duly requested to pay special attention to the thermal contact resistance, in other words, thermal deformation of the joint itself, but not requested to consider the convection and radiation heat transfer through an interface in nearly all cases of the machine tool joint. In addition, the thermal deformation of the joint, the interface pressure distribution, and the mechanical constraint condition are in close relation to one another, i.e., with these three factors being in a closed-loop interaction. More specifically, the closed-loop interaction is effected as follows.

---

[17]In a vacuum environment, the thermal conductance of the two surfaces in contact can be represented by the ultrasonic transmission with higher sensitivity.

Wolf, L., Jr., and C. Kostenko, "Ultrasonic Measurement of the Thermal Conductance of Joints in Vacuum," in *Proc. 7th Conf. on Thermal Conductivity* (NBS Special Publication No.302), 1968.

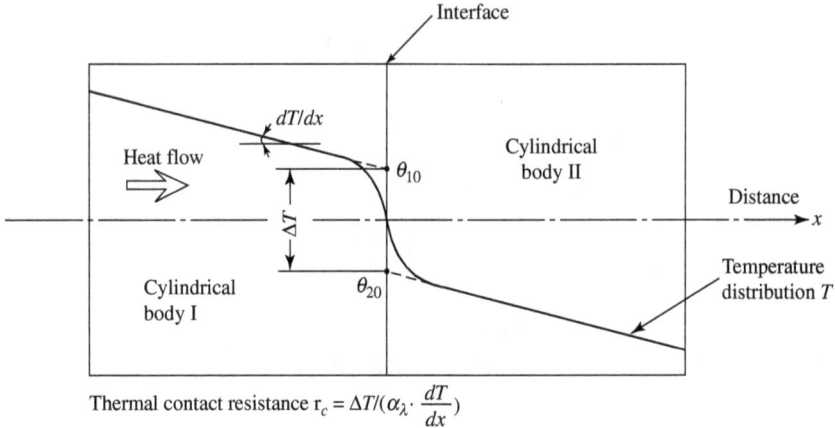

Thermal contact resistance $r_c = \Delta T/(\alpha_\lambda \cdot \dfrac{dT}{dx})$

Figure 6-33 Definition of thermal contact resistance.

1. The total interface pressure across the whole joint, which is derived from mechanical loading, results in the change of the thermal contact resistance, depending upon the interface configuration.

2. Both the thermal contact resistance and external thermal loading can determine the thermal field within the structural components.

3. In accordance with the thermal field, the thermal deformation of the structural components can be given in consideration of the mechanical constraints.

4. In due course, the thermal stresses across the whole joint can be obtained, resulting in change in the interface pressure.

In short, the mechanical constraint has a considerable effect on the thermal deformation of the structural body component, resulting in further effects on the interface pressure distribution. The closed-loop concept proposed by Attia and Kops [51] is considered the most suitable for interpreting the thermal behavior at the joint to a certain extent.[18]

Admitting that there have been very few research activities on the thermal contact resistance from the viewpoint of the machine tool joint, two examples are given in the following.

Yoshida [52] conducted a series of experimental investigations, trying to present the necessary design data. Figure 6-34 is the schematic view of his experimental setup, where the test piece is 30 mm in diameter and 50 mm in length. In the experimental setup, it is very important to always keep the thermal boundary condition constant, and to ensure such the condition, Yoshida conducted all the experiments in a

---

[18]Attia and Kops did not suggest the repeated number of the closed loop, and later Ito and coworkers revealed that the closed loop is not repeated so often (see Chap. 7).

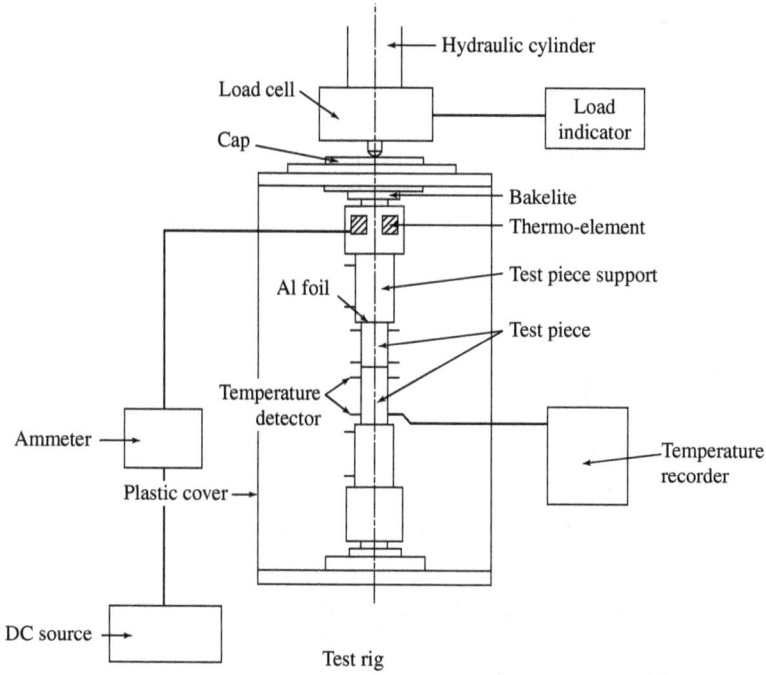

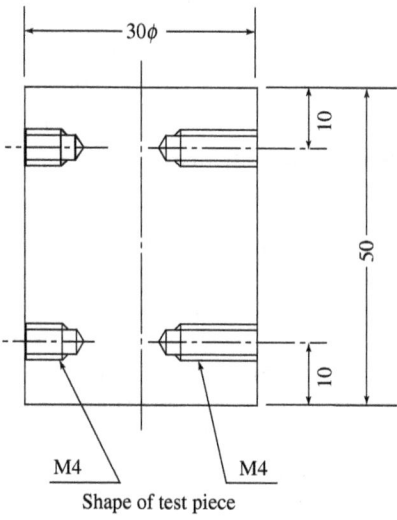

Figure 6-34 Setup to measure thermal contact resistance (courtesy of Yoshida).

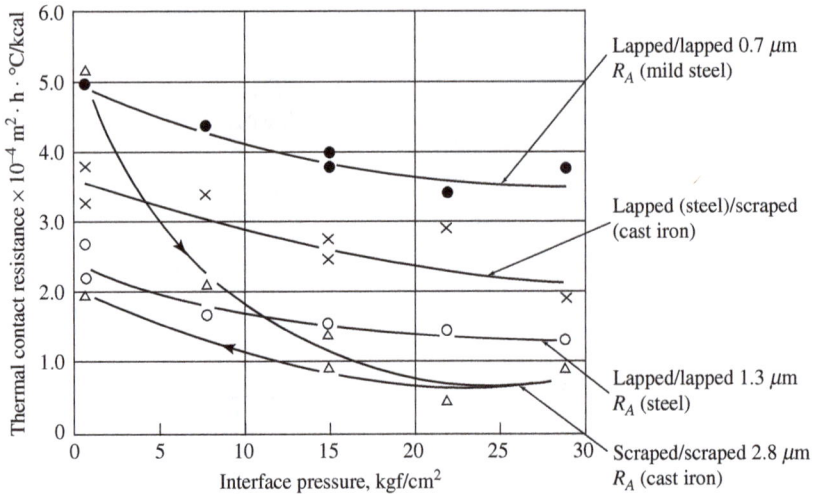

Figure 6-35   Thermal contact resistance in dry joint (courtesy of Yoshida).

temperature-controlled room, with a constant temperature of $20 \pm 0.5°C$. Figure 6-35 is one of the measured results for the thermal contact resistance of the steel and cast iron joints in dry condition. In addition to the basic behavior already mentioned, we can anticipate the following.

1. The cast iron joint shows apparent hysteresis behavior, corresponding to the loading and unloading cycles. This is considered to be due to the elasticity of the joint surface.

2. With the improvement of the surface quality, the thermal contact resistance increases.

The machine tool joint has, in general, the lubricant oil as an interfacial layer, and this oil has considerable effect on the thermal contact resistance. As shown in Fig. 6-36, the thermal contact resistance becomes independent of the interface pressure and temperature of the test piece by applying the oil (turbine oil No. 90) to the interface. In this case, the coefficient of thermal contact resistance r_c ranges from 0.95 to 0.97. According to the results of Yoshida, the joint has less effect on the thermal behavior of a machine tool as a whole.

Following the study of Yoshida, Saito and Nishiwaki [53] conducted a series of investigations on the effect of the interfacial layers, using a similar test rig to that of Yoshida. More specifically, they investigated both the thermal contact resistance and the thermal resistance between two noncontact surfaces (maximum gap being less than 100 μm), when the surface roughness and clearance at the joint were varied. In fact, they intended to apply the research result to the loose and light tightening

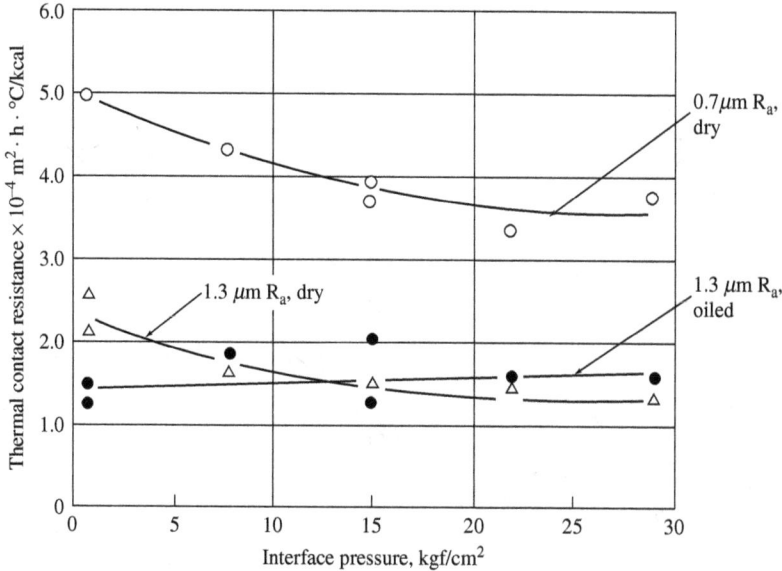

Figure 6-36 Thermal contact resistance in lapped joint with oil (courtesy of Yoshida).

fits in the machine tool structure, e.g., that for traveling spindle and ram (see Chap. 9). Figure 6-37 shows a result ranging from the contact to noncontact.[19] In the contact condition, they obtained an interesting result—the thermal contact resistance decreases with the improvement of the surface roughness. This is opposite to that of Yoshida, implying the important influence of the machining method. In fact, Yoshida used the lapped joint, which is liable to produce the waviness. In the non-contact condition, furthermore, the thermal resistance increases with increasing the surface roughness and clearance, and with decreasing thermal conductivity of the fluid in the clearance, resulting in the need to determine the equivalent clearance.

6.6 Forerunning Research into Single Flat Joint with Local Deformation

When the joint stiffness K_j is larger than the stiffness of the joint surroundings K_o, the flat joint shows a local deformation of joint surroundings under normal loading, e.g., bedding in or warping, resulting

---

[19]In Fig. 6-37, the solid line indicates the theoretical values obtained from the computation by the finite difference method. In the computation procedure, it is assumed that the thermal resistance is only derived from the heat conduction between the solid and the solid, and furthermore, the joint surface is replaced with the simple geometric model, i.e., two-dimensional model with triangular-form surface asperities.

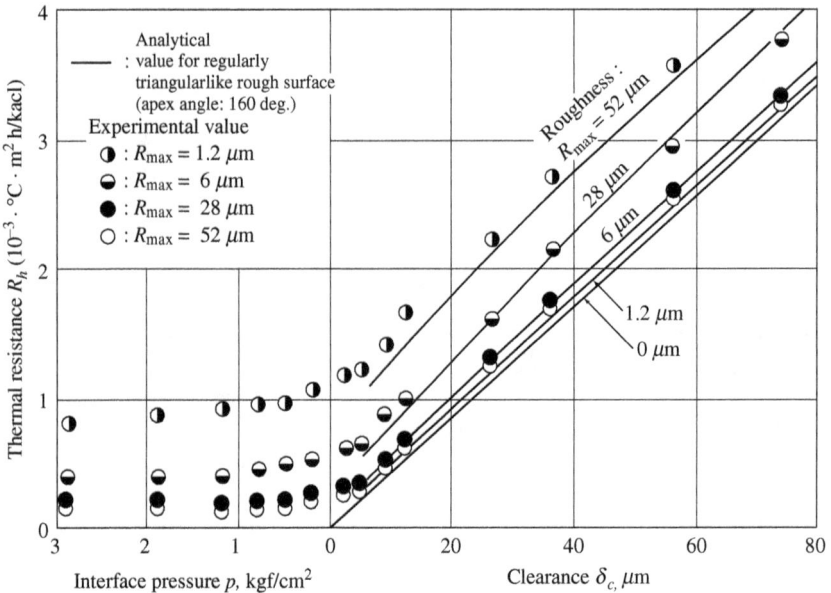

Figure 6-37   Thermal resistance of joints with positive and negative gaps (courtesy of Saito and Nishiwaki).

in the nonuniform interface pressure distribution. Warping of the joint can often be observed at many portions of the machine tool, for instance, those at the table of a large-size machine tool, such as planer, planomiller, and guideway grinder; bay-type flange of the structural body component; gib and keep plate in the slideway. Although the joint with warping is not preferable from the viewpoint of machine tool design, the machine tool structure involves such joints within itself by nature. In fact, the designer needs sufficient knowledge about warping of the joint when carrying out, for instance, the following engineering calculation.

- Calculation of the table deformation of large-size machine tool.
- The flairlike local deformation pattern of the bay-type flange in the bolted joint.

Although the flat joint with local deformation is a basic model of joint dominant in the body structure, the corresponding research has not been vigorously carried out even as of the beginning of 2000, except that for the static normal stiffness. In short, Levina [7], Masuko and Ito [54], and Shin and Ito [55] conducted some investigations in order to establish the design method of the slideway under relatively high static loading, aiming actually at the determination of a mathematical model applicable to the engineering design.

In their researches, the corresponding joint can be represented by a model of an elastic beam or plate on elastic foundation, and duly at issue is how to determine the modulus of foundation, i.e., joint stiffness per unit length or area. In this context, Levina first investigated the applicability of a simplified expression of Ostrovskii, as already mentioned in Sec. 6.1, and she verified that the simplified expression is fully applicable for calculating even the nonuniform joint deflection, provided that the mean interface pressure is less than 4 kgf/cm^2. In addition, she suggested that most of the sliding or traveling components in machine tools, for instance, ram, table, quill, barrel in tailstock, and so on, should be considered as a beam on elastic foundation with linear spring characteristic, although such a foundation has, by nature, a spring of nonlinear characteristic. In her case, the joint consists of a cast iron beam, 500 mm in length and 120 or 200 mm in height, and rigid cast iron plate; furthermore the joint surface is scraped or ground.

Assuming also the simplified expression of Ostrovskii to be applicable, Nakahara et al. conducted a further investigation into the effects of C and m on the deformation [56]. They chose a finite elastic beam under line loading at its center position, and for the ease of computation, Ostrovskii's expression was converted as follows.

$$p = K_i y^i \tag{6-6}$$

where p = interface pressure, kgf/mm^2, and y = joint deformation, mm. When we consider the differences in the unit used in both the expressions,

$$i = 1/m$$

$$K_i = 10^{(3i-2)}/C^i \quad \text{kgf/mm}^{(i+2)}$$

In general, $m = 0.5$, i.e., $i = 2$, and thus in Fig. 6-38 some deformation curves are shown when the values of i are varied, as well as K_2 from 0.69×10^4 to 6.25×10^4 kgf/mm^4 (C: from 0.4 to 1.2). In due course, the following conclusions can be drawn from the results.

1. The value C has a larger effect than m, and the depth of bedding-in and mating area of beam increases with the increase of C.

2. The effects of C and m appear apparently when the bending rigidity of beam is lower.

Importantly, with reference to the work of Levina, it appears that the joint with local deflection can be solved easily; however, the root cause

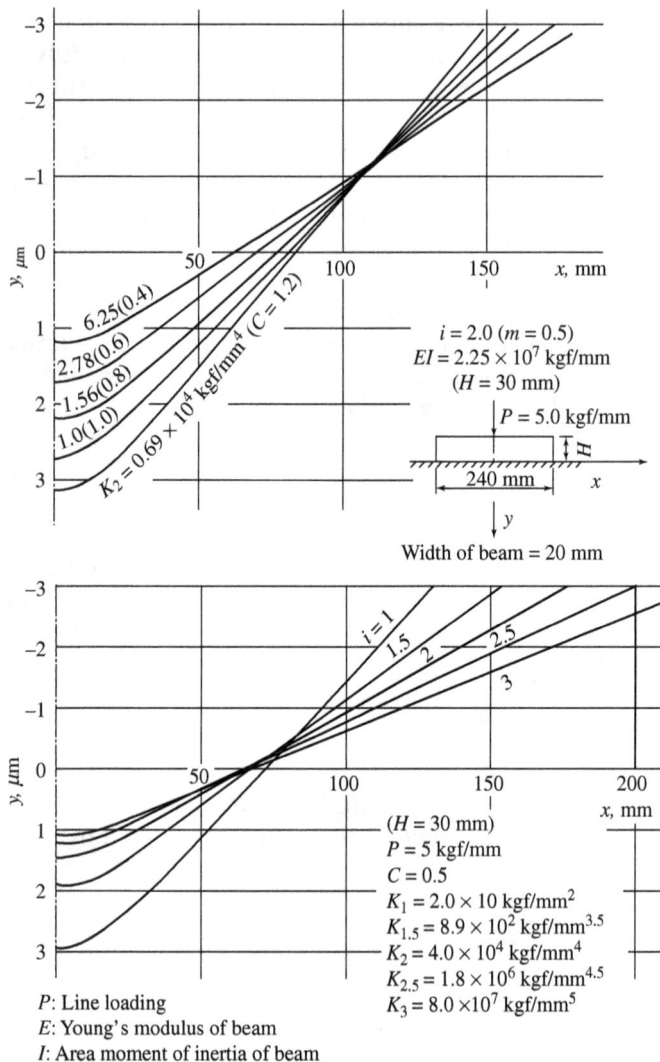

Figure 6-38 Effects of C and m on deformation pattern of cast iron beam.

of the difficulties in this engineering problem lies substantially in the following.

1. The measurement of the joint deflection is not easy, because it changes at every point across the whole joint.

2. It is necessary to measure both the joint deflection and the corresponding local pressure at the same point.

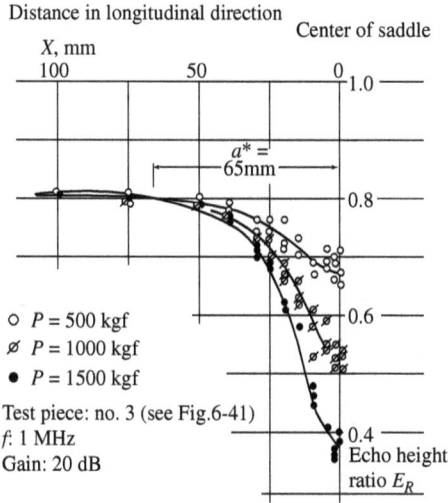

Figure 6-39 Contact pattern of saddle on slideway.

In this regard, Ito proposed an application of contact pattern measurement by means of ultrasonic waves as one of the effective experimental techniques (see App. 1) and suggested that the theory for an elastic beam on an elastic foundation should be applied to the joint with warping or bedding-in. By assuming the modulus of foundation to be linear within a certain range of the interface pressure, a beam on an elastic foundation has been analyzed by emphasizing its deformation curve. In accordance with the theory of elasticity, the beam deforms showing a constant point, at which all the deformation curves corresponding to different magnitudes of the load intersect one another, such as shown in Fig. 6-39. In Fig. 6-39, the deformation curve of the elastic beam was replaced by the contact pattern, where the contact pattern is proportional to the deformation curve. These contact patterns were measured using the experimental setup shown in Fig. 6-40. Thus, by denoting the distance between the loading and this constant points as a^*, the modulus of foundation is given by

$$K_e = 4EI \cdot (2.356/a^*)^4 \qquad (6\text{-}6)$$

where E = Young's modulus of the beam and I = cross-sectional second moment of the beam.

It can be seen from the measured result that the value a^* shows the following behavior.

- The value a^* decreases with the applied load, when the thickness of the beam, i.e., stiffness of joint surroundings, is constant.
- The value a^* increases with the thickness of the beam.

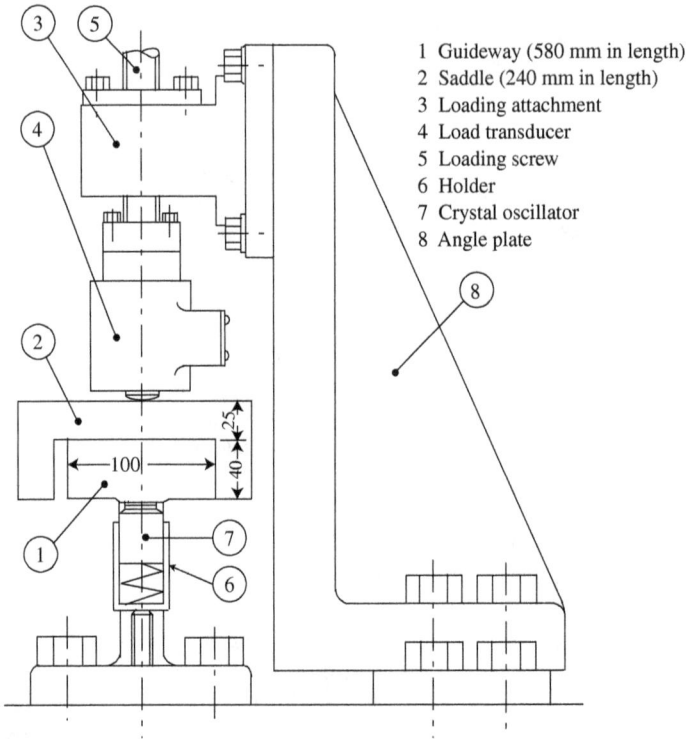

1 Guideway (580 mm in length)
2 Saddle (240 mm in length)
3 Loading attachment
4 Load transducer
5 Loading screw
6 Holder
7 Crystal oscillator
8 Angle plate

Figure 6-40 Experimental setup for measuring joint stiffness by means of ultrasonic waves.

- The value a^* becomes constant, notwithstanding the change in the applied load, when the thickness of the beam is very thin.

On the basis of the experimental results applicable to the design of slideway, Fig. 6-41 summarizes the joint stiffness K_e and the critical interface pressure p_{cr}, where p_{cr} is the maximum allowable interface pressure within which the deformation of the flat joint is uniform.

In the flat joint with local deformation, furthermore, we can observe the larger effects of the friction between both joint surfaces. Back et al. suggested already the importance of friction; however, they did not show any quantitative results. Keeping that in mind, Shimizu [57] investigated later such a problem theoretically by taking a single bolt-flange assembly as an example. According to his results, the joint deflection reduces to some extent with the coefficient of friction, as shown in Fig. 6-42. This fact implies that there is a possibility of appearance of microslip, which is derived from the relative tangential displacement between both joint surfaces even though no external tangential forces act on the joint. Figure 6-43 is one of the measured hysteresis loops for the flat joint with

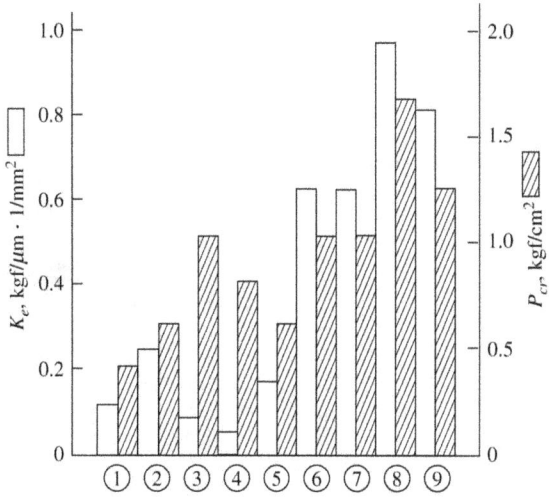

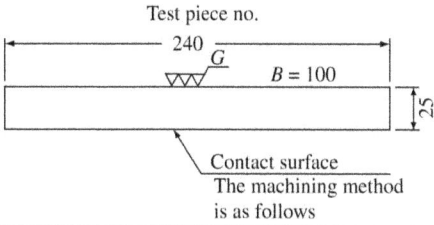

No.	Machining method	Material	Surface roughness R_{max} μm	Surface treatment
1	Scraped	FC35	—	Numbers of contact point 30/1 in2
2			—	Numbers of contact point 60/1 in2
3	Ground	FC25	1.5	
4			4.0	
5	Ground	FC25		Chrome plated thickness; 0.5 mm
6	Ground	FC25		Flame hardening Depth: 1 mm, H_B: 285
7			3.0	Flame hardening Depth: 1 mm, H_B: 375
8	Ground	SNCM1		Quenching H_{RC}: 53
9	Ground	SNC21		Case hardening Depth: 0.5 mm, H_{RC}: 59

※ Guideway: Lapped, flame hardening; depth: 1 mm
H_B: 450B
R_{max}: 20 μm, material: FC25 (cast iron)

Figure 6-41   Values of K_e and p_{cr}.

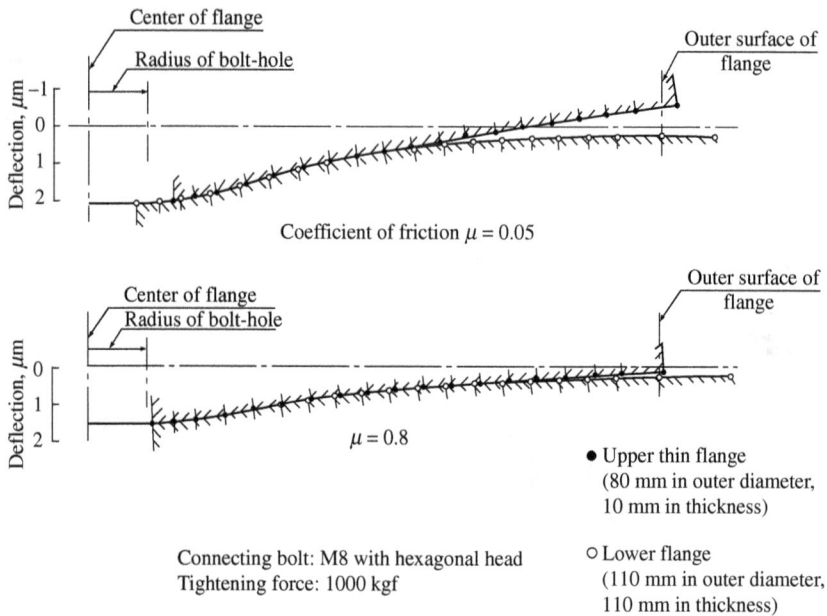

Figure 6-42 Constraint effect of friction on joint deflection (courtesy of Shimizu).

Coefficient of friction $\mu = 0.05$

$\mu = 0.8$

- Upper thin flange
  (80 mm in outer diameter,
  10 mm in thickness)

Connecting bolt: M8 with hexagonal head
Tightening force: 1000 kgf

○ Lower flange
  (110 mm in outer diameter,
  110 mm in thickness)

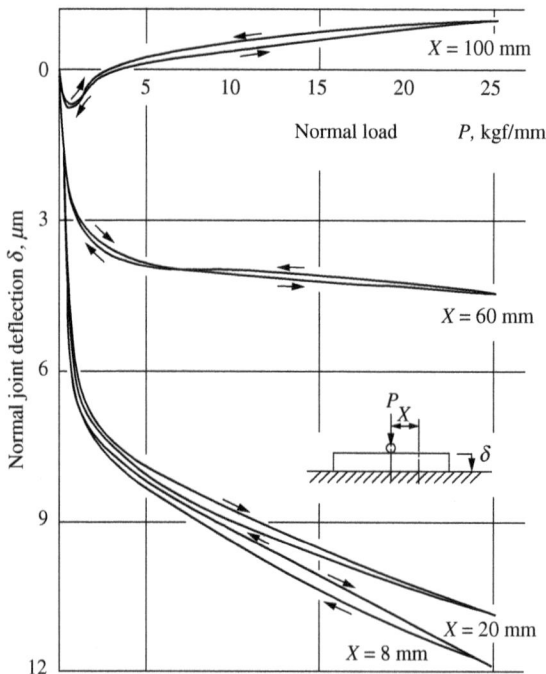

Beam: Mild steel, joint surface roughness 3 μm in R_{max}
  (240 × 20 × 20 mm)

Figure 6-43 Hysteresis loop in flat joint with local deformation.

Normal joint deflection
Flat joint with local deformation

Normal joint deflection   Tangential displacement
Flat joint without local deformation

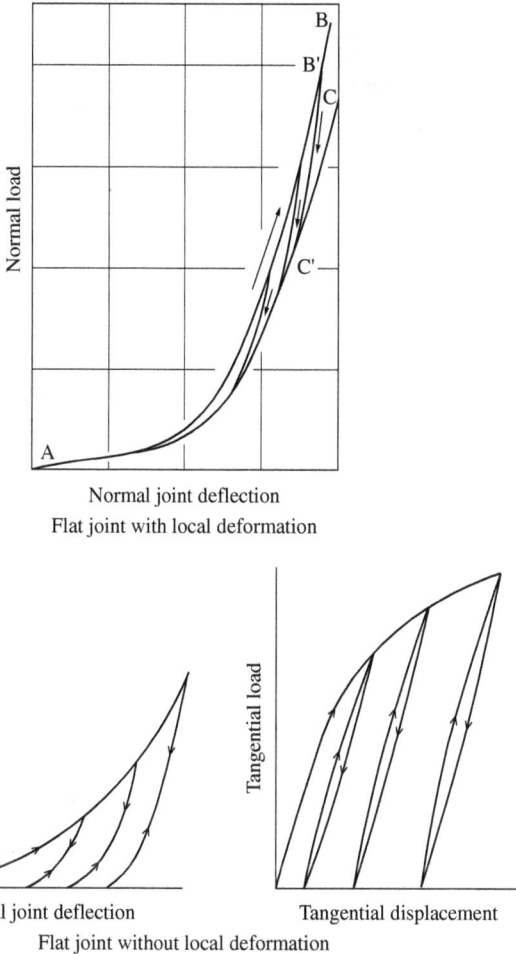

Figure 6-44   Comparison of load-deflection curves for representative flat joints.

local deformation and under line loading [58]. The clear hysteresis occurs due to the difference of deformations between the loading and unloading procedures. This hysteresis curve is similar to that of flat joint under tangential loading, but not the same: it does not show any residual displacement as comparatively and schematically depicted in Fig. 6-44. More specifically, the hysteresis loop for the joint with local deformation shows the following interesting behavior.

1. The joint material has considerable effects on the area of the hysteresis loop.

2. The area of the hysteresis loop is smaller and larger at the center and tailoff of the deformation curve, respectively.

3. With the increase of the beam depth and deterioration of surface roughness, the area near the center becomes larger.

References

1. Ito, Y., and M. Tsutsumi, "Determination of Mathematical Models in Structural Analysis of Machine Tools—2nd Report, Determination of Mathematical Models for Normal Static Stiffness of Joints," *Bull. of JSME*, 1981, 24(198): 2234–2239.
2. Ito, Y., and M. Masuko, "Influence of Bolted Joint on the Model Testing of Machine Tool Construction," *Trans. of JSME*, 1970, 36(284): 649–654.
3. Levina, Z. M., "Calculation of Contact Deformations in Slideways," *Machines and Tooling*, 1965, 36(1): 8–17.
4. Ostrovskii, V. I., "The Influence of Machining Methods on Slideway Contact Stiffness," *Machines and Tooling*, 1965, 36(1): 17–19.
5. Connolly, R., and R. H. Thornley, "The Significance of Joints on the Overall Deflection of Machine Tool Structures," in S. A. Tobias and F. Koenigsberger (eds.), *Proc. of 6th Int. MTDR*, Pergamon, 1966, pp. 139–156.
6. Back, N., M. Burdekin, and A. Cowley, "Review of the Research on Fixed and Sliding Joints," in S. A. Tobias and F. Koenigsberger (eds.), *Proc. of 13th Int. MTDR Conf.*, Pergamon, 1973, pp. 87–97.
7. Levina, Z M., "Research on the Static Stiffness of Joints in Machine Tools," in S. A. Tobias and F. Koenigsberger (eds.), *Proc. of 8th MTDR Conf.*, Pergamon, 1968, pp. 737–758.
8. Dolbey, M. P., and R. Bell, "The Contact Stiffness of Joints at Low Apparent Interface Pressures," *Annals of CIRP*, 1971, 19: 67–79.
9. Eisele, F., and K. Corbach, "Dynamische Steifigkeit von Führungen und Fugenverbindungen an Werkzeugmaschinen," *Maschinenmarkt*, 1964, 70(89): 88–93.
10. Taniguchi, A., M. Tsutsumi, and Y. Ito, "Treatment of Contact Stiffness in Structural Analysis—1st Report, Mathematical Model of Contact Stiffness and Its Applications," *Trans. of JSME*, 1983, 49(443): 1282–1288.
11. Ito, Y., and M. Tsutsumi, "Design Data for Machine Tools with Modular Construction System," Report of Cooperative Research Committee of JSME RC-SC42, 1978.
12. Ito, Y., M. Koizumi, and M. Masuko, "One Proposal to the Computing Procedure of CAD Considering a Bolted Joint—Study on the CAD for Machine Tool Structures, Part 2," *Trans. of JSME (C)*, 1977, 43(367): 1123–1131.
13. Tenner, O. G., "Contact Stiffness of Friction Slideways," *Machines and Tooling*, 1968, 39(3): 3–6.
14. Corbach, K., "Die dynamische Steifigkeit ruhender und beweglicher Verbindungen an Werkzeugmaschinen," *Maschinenmarkt*, 1966, 72(79): 19–29.
15. Connolly, R., R. E. Schofield, and R. H. Thornley, "The Approach of Machined Surfaces with Particular Reference to Their Hardness," in S. A. Tobias and F. Koenigsberger (eds.), *Proc. of 8th Int. MTDR Conf.*, Pergamon, 1968, pp. 759–775.
16. Connolly, R., and R. H. Thornley, "Determining the Normal Stiffness of Joint Faces," *Trans. of ASME, J. Eng. for Ind.*, 1968, pp. 97–106.
17. Dekoninck, C., "Experimental Study of the Normal Static Stiffness of Metallic Contact Surfaces of Joints," in S. A. Tobias and F. Koenigsberger (eds.), *Proc. of 13th Int. MTDR Conf.*, Macmillan, 1973, pp. 61–66.
18. Abrams, D. M., and L. Kops, "Effect of Waviness on Normal Contact Stiffness of Machine Tool Joints," *Annals of CIRP*, 1985, 34(1): 327–330.
19. Kirsanova, V. N., "The Shear Compliance of Flat Joints," *Machines and Tooling*, 1967, 38(7): 30–34.
20. Koizumi, T., Y. Ito, and M. Masuko, "Experimental Expression of the Tangential Micro-displacement between Joint Surfaces," *Trans. of JSME*, 1978, 44(384): 2861–2870.

21. Back, N., M. Burdekin, and A. Cowley,. "Analysis of Machine Tool Joints by the Finite Element Method," in S. A. Tobias and F. Koenigsberger (eds.), *Proc. of 14th Int. MTDR Conf.*, Macmillan, 1974, pp. 529–537.
22. Masuko, M., Y. Ito, and C. Fujimoto, "Behaviour of the Horizontal Stiffness and the Micro-sliding on the Bolted Joint under the Normal Preload," in F. Koenigsberger and S. A. Tobias (eds.), *Proc. of 12th Int. MTDR Conf.*, Macmillan, 1972, pp. 81–88.
23. Masuko, M., Y. Ito, and T. Koizumi, "Horizontal Stiffness and Micro-slip on a Bolted Joint Subjected to Repeated Tangential Static Loads," *Trans. of JSME*, 1974, 40(331): 855–861.
24. Boothroyd, G., C. Poli, and I. P. Migliozzi, "Damping in a Preloaded Metallic Interface," SME Technical Paper MR72-229, 1972.
25. Rogers, P. F., and G. Boothroyd, "Damping at Metallic Interfaces Subjected to Oscillating Tangential Loads," *Trans. of ASME, J. Eng. for Ind.*, August 1975, pp. 1087–1093.
26. Burdekin, M., A. Cowley, and N. Back, "An Elastic Mechanism for the Micro-sliding Characteristics between Contacting Machined Surfaces," *J. of Mech. Eng. Sci. (IMechE)*, 1978, 20(3): 121–127.
27. Burdekin, M., N. Back, and A. Cowley, "Experimental Study of Normal and Shear Characteristics of Machined Surfaces in Contact," *J. of Mech. Eng. Sci. (IMechE)*, 1978, 20(3): 129–132.
28. Hisakado, T., et al., "Deformation Mechanism of Solid Bodies in Contact and under Tangential Loading," *Trans. of JSME*, 1978, 44(382): 2080–2087.
29. Courtney-Pratt, J. S., and E. Eisner, "The Effect of a Tangential Force on the Contact of Metallic Bodies," in *Proc. Roy. Soc., Series A*, 1957, 238: 529–550 (Plate 22).
30. Masuko, M., Y. Ito, and K. Yoshida, "Theoretical Analysis for a Damping Ratio of a Jointed Cantibeam," *Trans. of JSME* (Part 3), 1973, 39(317): 382–392.
31. Simkins, T. E., "The Mutuality of Static and Kinetic Friction," *Lub. Eng.*, January 1967, pp. 26–31.
32. Johnson, K. L., "Surface Interaction Between Elastically Loaded Bodies under Tangential Forces," in *Proc. Roy. Soc. A*, 1955, 230: 531–548 (Plate 13).
33. Seireg, A., and E. J. Weiter, "Viscoelastic Behaviour of Frictional Hertzian Contacts under Ram-type Loads," *Proc. Inst. Mech. Engrs*, 1966–1967, 181(Pt. 3): 200–206.
34. Fujimoto, T., et al., "Micro-displacement Characteristics under Tangential Force between Surfaces in Contact (Part 4)," *Tribologist*, 1998, 43(6): 485–492.
35. Seireg, A., and E. J. Weiter, "Frictional Interface Behavior under Dynamic Excitation," *Wear*, 1963, 6: 66–77.
36. Goodman, L. E., and C. B. Brown, "Energy Dissipation in Contact Friction: Constant Normal and Cyclic Tangential Loading," *Trans. of ASME, J. Appl. Mech.*, March 1962, pp. 17–22.
37. Mindlin, R. D., "Compliance of Elastic Bodies in Contact," *J. Appl. Mech.*, Sept. 1949, pp. 259–268.
38. Reshetov, D. N., and Z. M. Levina, "Damping of Oscillations in the Couplings of Components of Machine," *Vestnik Mashinostroyeniya*, 1956, No. 12: 3–13. (Translated into English by PERA.)
39. Groth, W. H., "Die Dämpfung in verspannten Fugen und Arbeitsführungen von Werkzeugmaschinen," Dissertation RWTH Aachen, 1972.
40. Dekoninck, C., "Deformation Properties of Metallic Contact Surfaces of Joint under the Influence of Dynamic Tangential Loads," *Int. J. Mach. Tool Des. Res.*, 1972, 12: 193–199.
41. Tsutsumi, M., and Y. Ito, "Damping Mechanism of a Bolted Joint in Machine Tools," in F. Koenigsberger and S. A. Tobias (eds.), *Proc. of 20th Int. MTDR Conf.*, Macmillan, 1980, pp. 443–448.
42. Hashimoto, F., and Y. Kume, "Critical Condition of Rectangular Wave Response for Coupling Model—Fundamental Studies on Dynamic Characteristics of Contact Surface in Machine Tool Structure (1st Report)," *J. of JSPE*, 1972, 38(10): 844–849.
43. Beards, C. F., and A. A. Neroutsopoulos, "The Control of Structural Vibration by Frictional Damping in Electro-Discharge Machined Joints," *Trans. of ASME, J. of Mech. Des.*, 1980, 102: 54–57.
44. Loewenfeld, K., "Zusatzdämpfung von Werkzeugmaschinen durch lamellenpakete," *Maschinenmarkt*, März 1959, Nr. 19: 28–35.

45. PERA Report Nos. 180 and 198, "Machine Tool Joints, Part 1—Effect of Intermediate Viscous Films on Stiffness and Damping/Machine Tool Joints, Part 2—Effect of Intermediate Viscous Films on Stiffness and Damping of Cast Iron Joints," PERA, August 1969.
46. Andrew, C., J. A. Cockburn, and A. E. Waring, "Metal Surfaces in Contact under Normal Forces: Some Dynamic Stiffness and Damping Characteristics," in *Proc. Inst. Mech. Engrs.* 182 Pt 3K, *IMechE,* 1967–1968, Paper 22.
47. Schaible, B., "Dämpfung in Fugenverbindungen," *wt-Z. ind. Fertig.,* 1974, 64(2): 81–86.
48. Schaible, B., "Dämpfung in Fugenverbindungen—Zweiter Teil: Dämpfungskennwerte," *wt-Z. ind. Fertig.* 1977, 67(5): 301–305.
49. Thornley, R. H., and F. Koenigsberger, "Dynamic Characteristics of Machined Joints Loaded and Excited to the Joint Face," *Annals of CIRP,* 1971, 19: 459–469.
50. Thornley, R. H., and K. Lees, "The Effect of Planform Shape on the Normal Dynamic Characteristics of Metal to Metal Joints," in *Proc. of Tribology Convention, IMechE,* 1971, pp. 71–79 (Paper C62/71).
51. Attia, M. H., and L. Kops, "Nonlinear Thermoelastic Behaviour of Structural Joints—Solution to a Missing Link for Prediction of Thermal Deformation of Machine Tools," *Trans. of ASME, J. of Eng. Ind.,* 1979, 101: 348–354.
52. Yoshida, Y., "Research on Thermal Deformation of a Vertical Milling Machine," Technical Report of MEL (*Mechanical Engineering Laboratory, Ministry of International Trade and Industry, Japan*), 1975, No. 82.
53. Saito, Y., and N. Nishiwaki, "Thermal Resistance between Two Non-Contact Surfaces," in *1st Int. Conf. on Metrology and Properties of Engineering Surfaces,* April 1979, Leicester Polytechnic.
54. Masuko, M., and Y. Ito, "Distribution of Contact Pressure on Machine Tool Slideways," in S. A. Tobias and F. Koenigsberger (eds.), *Proc. of 10th Int. MTDR Conf.,* Pergamon, 1970, pp. 641–650.
55. Shin, B. S., and Y. Ito, "Joint Stiffness at a Metal Contact under Local Deformation Such as Warping," *J. of JSLE,* 1974, 19(8): 570–576.
56. Nakahara, T., T. Endo, and Y. Ito, "Analysis for a Local Deformation of Two Flat Surfaces in Contact," *J. of JSLE,* 1976, 21(11): 764–771.
57. Shimizu, S., "Study on the Deformation and Interface Pressure Distribution of Bolted Joints," Dr. Dissertation, Sophia University, Tokyo, Japan, 1981.
58. Koizumi, T., T. Ohtsuka, and Y. Ito, "Hysteresis Phenomena Arising from a Local Deformation of Two Flat Surfaces in Contact," *J. of JSLE,* 1978, 23(9): 678–684.

Supplement: Theoretical Proof of Ostrovskii's Expression

In the single flat joint under uniform interface pressure, the contact theory established in the tribology sphere is applicable to the calculation of the joint deflection resulting from the deformation of surface asperities. The theoretical proof of the expression of Ostrovskii can thus be carried out, as reported by Dolbey and Bell [8], on the basis of contact theory.

In the theoretical analysis of Dolbey and Bell, the following assumptions are made.

- The real contact area A_r is proportional to the applied load P, i.e.,

$$P = K_r A_r \quad (K_r \text{ constant})$$

- The real contact area is proportional to the number of contact points N, and the average area of each contact point A_i is constant, i.e.,

$$A_r = NA_i$$

When a load P is applied to the joint, the joint surface approaches a distance λ so as to support the load by the sum of the contact points formed as the surface approaches. This relationship can be written as

$$N = \eta_{TN}A_a \int \phi(z)\, dz \qquad (\int \text{ is from 0 to } \lambda)$$

where A_a = apparent contact area
η_{TN} = total number of asperities per unit area
$\phi(z)$ = distribution function of contact formation (probability of contact being formed by asperity of height z)

Thereby,

$$P = GA_a \int \phi(z)\, dz \qquad (\int \text{ is from 0 to } \lambda)$$

where $G = K_r \eta_{TN}A_i$.

The mean interface pressure yields to $p = P/A_a$, and then

$$p = G \int \phi(z)\, dz \qquad (\int \text{ is from 0 to } \lambda)$$

By assuming $\phi(z)$ as to be a power law distribution, i.e., $\phi(z) = bz^{(1-m)/m}$

$$P = Gbm\lambda^{1/m} \qquad \text{or} \qquad \lambda = Cp^m$$

where $C = 1/(Gbm)$.

In the theoretical analysis of the expression of the static joint stiffness, we must be aware of the importance of how to assume $\phi(z)$. For instance, Thornley and coworkers assumed $\phi(z)$ to be a function of exponential form.

Design Guides, Practices, and Firsthand View of Engineering Developments—Stationary Joints

As already stated in Chap. 5, a machine tool as a whole has many stationary joints, ranging from the foundation and bolted joint connecting both the structural body components, through screw-nut fixation for ball screw and spline connection, to stacked blank fixation in the hobbing machine. Of these, the bolted joint and foundation are primary concerns from the structural design of the machine tool. In consequence, we deal with the bolted joint and foundation in this chapter, emphasizing their engineering design aspects.

7.1 Bolted Joint

The bolted joint is the most popular method used to connect machine components, not only in machine tools, but also in industrial machines. In fact, there have been myriad activities to clarify the behavior of the bolted joint under static and dynamic loading and often under nonuniform temperature distribution with complex boundary condition. In retrospect, these earlier activities were concentrated on those in connection with the relaxation mechanism of the tightening force, fatigue strength of a bolt-flange assembly, control of tightening force, reliability of bolted connection and so on, such as summarized in Fig. 7-1. As can be readily seen, the engineering development and research concerned with the bolted joint have been aimed at ensuring the strength of a bolt-flange assembly. In contrast, in the case of the machine tool structure, the stiffness of the bolted joint is of great importance instead of its strength,

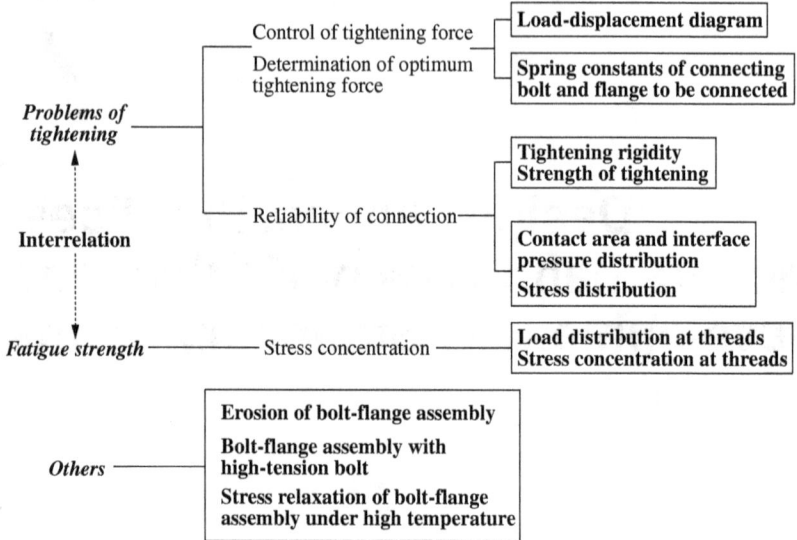

Figure 7-1 Research and engineering development subjects of bolt-flange assembly.

resulting from the design principle of machine tools, i.e., *allowable deflection-based design.*

We must furthermore remember that the bolted joint is of particular importance when the machine tool is designed using the modular principle. In fact, the modular design can be facilitated with the bolted joint to connect both modules, especially both structural body components. Importantly, the bolted joint can guarantee the higher joint stiffness and assembly accuracy as well as enable the ease of assembly and disintegration. Figure 7-2 shows thus a firsthand view of the engineering developments and researches for the bolted joint within the machine tool. Although considerable activities have been carried out, there remain still many problems to be solved. A root cause of such unsolved problems lies in the configuration complexity of the bolted joint. Obviously, the bolted joint consists of, even in the simple case, the connecting bolt and washer, bay-type flange, stiffening rib or bolt pocket and aperture, although its basic configuration is of a flat joint.

More specifically, the distinct differences between the bolted joint and the flat joint are as follows.

1. The interfacial pressure distribution given by the tightening force is not uniform across the whole joint surface.

2. Such a flange portion of column to be connected is liable to show warping or bedding-in, resulting in a nonuniform deformation of

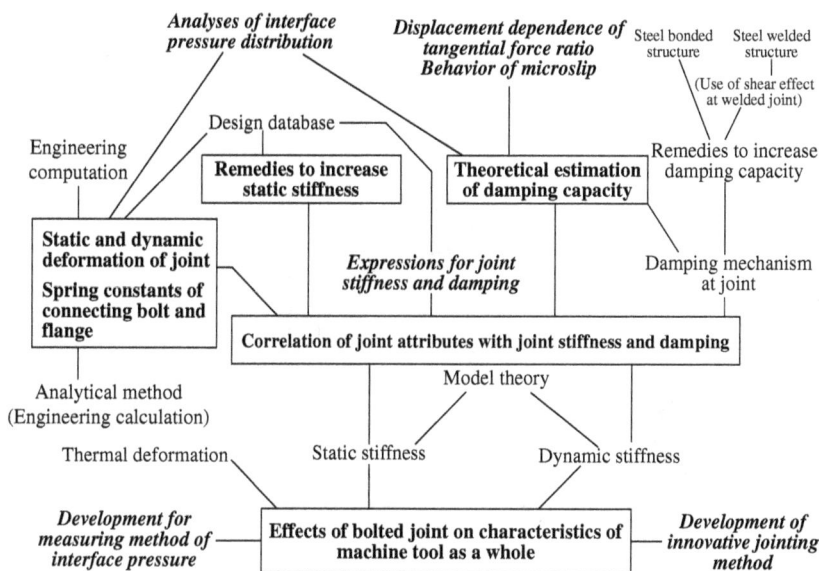

Figure 7-2   Research and engineering development subjects of bolted joint in machine tools.

joint. Actually, the relative stiffness of a flange is not so large compared with the joint stiffness.

3. The damping derived from the contact surface of a bolt head to a flange or at threads cannot be disregarded, when we assess the damping capacity of the bolt-flange assembly.

In addition, in normal loading, it is necessary to understand that the tightening force of the connecting bolt can be regarded as the preload in the flat joint.

For ease of understanding, Fig. 7-3 shows the load and tightening force dependence of the joint stiffness under normal bending loading. This can be considered one of the representative characteristics of the bolted joint. Similar to the flat joint, the stiffness of the bolted joint shows the nonlinear characteristic. In general, the joint stiffness increases with the tightening force, finally flattening to a certain constant value, which is, even in the preferable case, lower than that of equivalent solid. Furthermore, the joint stiffness decreases with the applied load especially when the joint is in partly separated condition under loading, whereas the joint stiffness increases with the applied load such as shown in Fig. 7-4, when the joint surface does not separate under loading. In short, the load- and tightening force-joint stiffness characteristics are largely dependent upon (1) the correlation of the stiffness of the clamped

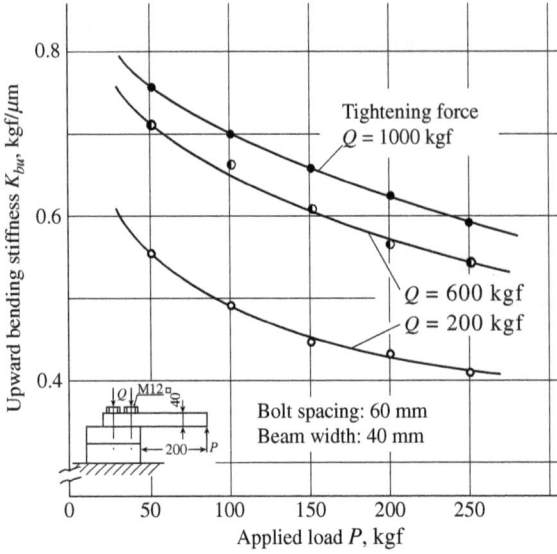

Figure 7-3 Load and tightening force dependence of joint stiffness—separation in joint.

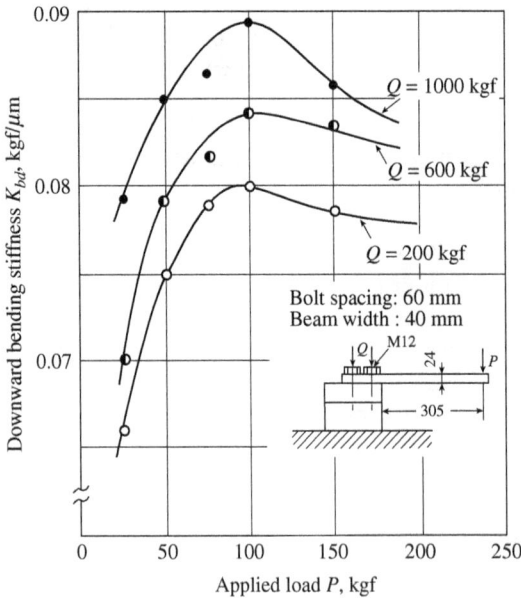

Figure 7-4 Load and tightening force dependence of joint stiffness—nonseparation in joint.

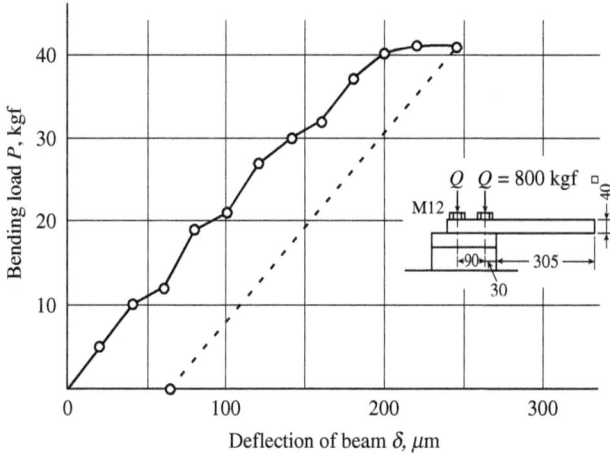

Figure 7-5 Microscopic stick-slip observed at bolted joint under tangential bending load.

component, i.e., joint surroundings to the diameter of the connecting bolt, (2) correlation of the joint stiffness to the stiffness of joint surroundings, and (3) the number of connecting bolts.

Figure 7-5 shows a load-deflection relationship of bolted joint under tangential bending loading, and as similar to that of the single flat joint, the stepwise deflection can be observed with the applied load. In the practical structure, the bolted joint has often the taper pin or guidekey to ensure, maintain, and reproduce the positioning accuracy of the structural body components to be integrated with each other. This is because the assembly and disintegration are inevitable in the production of the machine tool. In due course, these machine elements benefit to reinforce the joint stiffness under tangential bending loading to some extent (see Sec. 7.1.4). In addition, the bolted joint shown already in Fig. 7-5 (type A) can be regarded as a variant of the bolted joint of type B under torsional loading, implying higher possibilities for the interchangeability of the engineering knowledge.

Figure 7-6 shows the change of the damping ratio with the tightening force reported by Groth [1], and as a rule of the machine tool joint, the damping ratio decreases with the tightening force. Importantly, the damping capacity of the bolted joint shows the peak value with special respect to the interface pressure in certain joint conditions such as shown in Fig. 7-7 as reported by Ito and Masuko [2]. In addition, Ito suggested an interesting idea for the damping mechanism of the bolted joint. Figure 7-8 reproduces his idea, and there are the two possibilities for the maximum damping in relation to the tightening force.

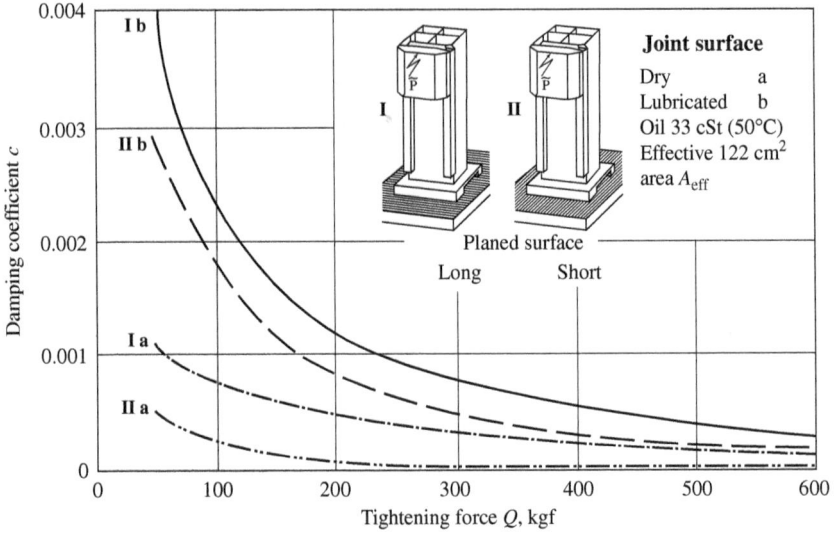

Figure 7-6 Effect of tightening force on damping capacity in consideration of machined lay orientation (courtesy of Groth).

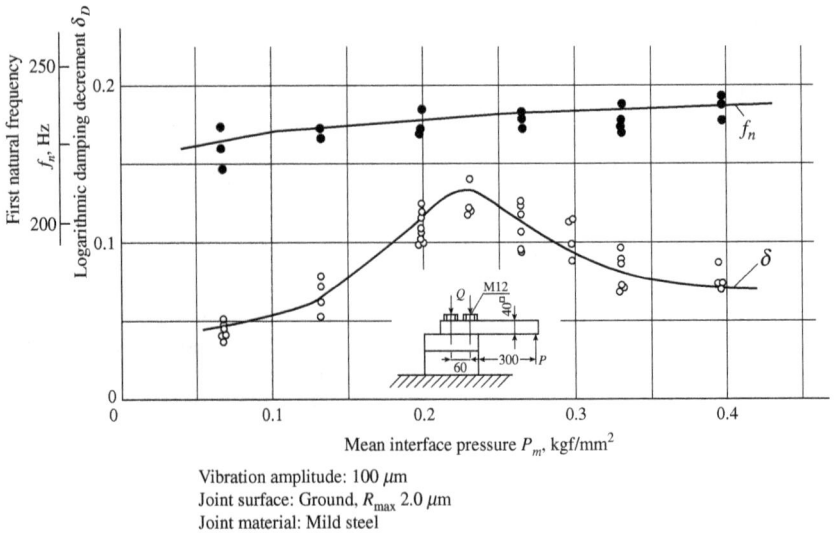

Figure 7-7 Bolted joint showing maximum damping capacity with respect to interface pressure.

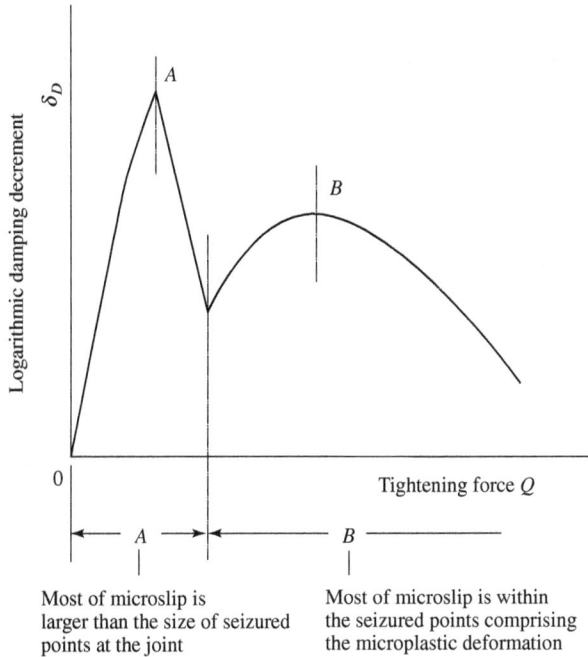

Figure 7-8 Qualitative relationships between damping capacity and tightening force.

Regarding the thermal behavior of the bolted joint, at issue is the differing thermal inertia of each component consisting of the bolt-flange assembly even made of the same material, as will be stated in Sec. 7.1.6.[1]

Keeping in mind these observations mentioned above and concerns, in the following first we note some knowledge available for the engineering design, and then state the firsthand view of the related researches and engineering developments. In addition, some marked researches will be viewed to deepen the understanding for the bolted joint. To this end, it is worth suggesting that the following three subjects are even now of the utmost importance; however, the due research activities are not vigorous.

---

[1]Nowadays, we can, without any difficulties, conduct the engineering calculation and computation for the structural design in consideration of the joint to some extent using the software package on the market. There is, at least, no need to state the computational method for the static behavior of the machine tool as a whole, and thus, some rudimentary knowledge about the computation will be stated in Supplement 1 at the end of this chapter together with the research history.

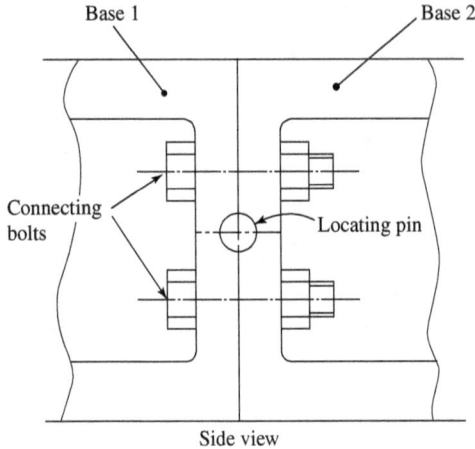

Figure 7-9 Bolted joint for producing long-length base—case of large-size NC horizontal boring and milling machine of Skoda make, 2004.

1. Quantitative estimation of damping capacity.

2. Clarification for the nonlinearity of the joint stiffness and its cross receptance effect. In general, the joint stiffness, i.e., spring constant, at any local points across the whole joint is not affected by that of other local points; however, in the case of the bolted joint, the joint stiffness at a point might be determined in consideration of the cross-effect derived from those other points.

3. Application of knowledge so far obtained to the design of a variant, which can be observed at the joint between both beds to produce the longer-length bed as shown in Fig. 7-9.

7.1.1   Design guides and knowledge—pressure cone and reinforcement remedies from structural configuration

Within a bolted joint context, one of the basic necessities is to understand which of its features differ from those of the single flat joint. In Table 5-2, thus, the factors having considerable influence on the behavior of the bolted joint were already shown after the corresponding factors were classified into those related to the flat joint and to the bolted joint itself. As can be seen, there are many and various leading factors, and of these factors the engineering problems in the stiffness of the connecting bolt have been solved to a large extent. Dare to say, at issue is how to consider the nonlinearity of the stiffness derived from the meshing portion of threads such as schematically shown in Fig. 7-10 [3].

Although there are myriad influencing factors within the bolted joint, they can be totally represented by the magnitude of the interface pressure and its distribution form in the analysis for, research into and

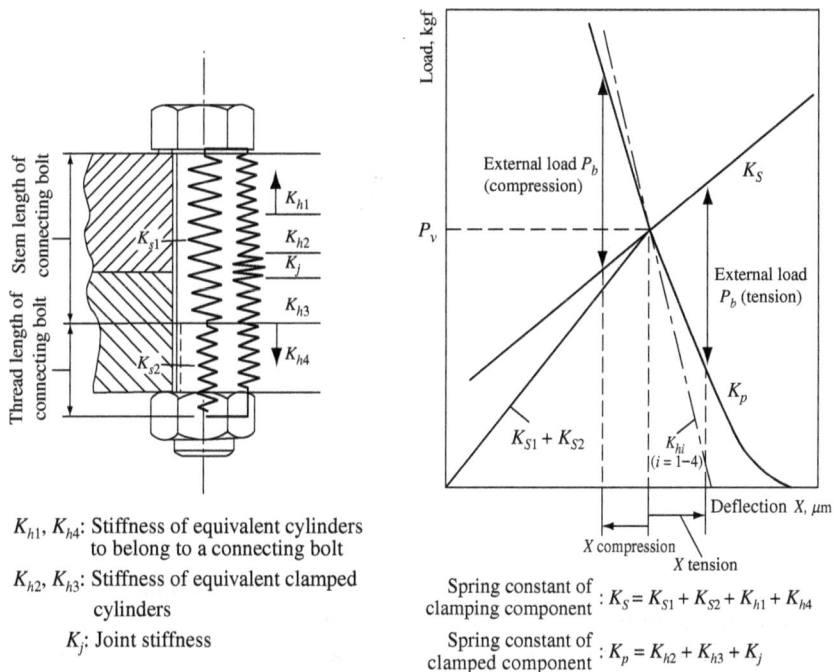

K_{h1}, K_{h4}: Stiffness of equivalent cylinders
   to belong to a connecting bolt

K_{h2}, K_{h3}: Stiffness of equivalent clamped
   cylinders

K_j: Joint stiffness

Spring constant of clamping component: $K_S = K_{S1} + K_{S2} + K_{h1} + K_{h4}$

Spring constant of clamped component: $K_p = K_{h2} + K_{h3} + K_j$

Figure 7-10   Load-deflection diagram of bolt-flange assembly in consideration of nonlinearity in spring constant (by Plock, courtesy of Industrie-Anzeiger).

engineering development of the bolted joint. Consequently, there are two leading engineering design data: one is the half angle of pressure cone, which is a representative index of the interface pressure distribution, and the other is the reinforcement remedies from the aspect of structural configuration.

Half angle of pressure cone.   As already delineated in Chaps. 5 and 6, the mean interface pressure of the bolted joint is considerably higher than that of other joints, and in addition the joint surroundings are liable to deform. In the engineering design, thus, the interface pressure distribution and its spreading area are the leading attributes to estimate the static stiffness, damping capacity, and thermal deformation of the bolted joint.

On the basis of the achievements obtained from the earlier theoretical and experimental works, we can summarize the rudimentary knowledge about the interface pressure distribution in the bolted joint as follows, provided that the joint surface has no considerable flatness deviation and/or waviness.

1. The effective area of the interface pressure distribution does not change considerably with the tightening force.

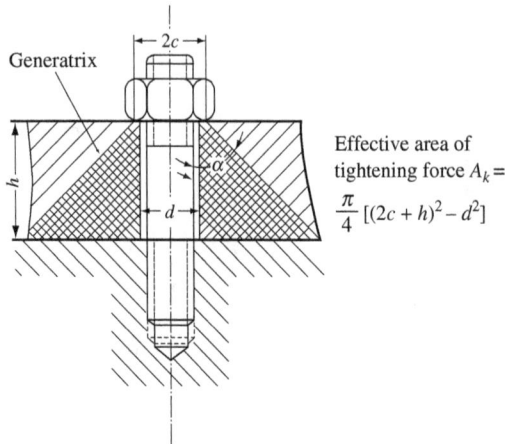

Generatrix

Effective area of
tightening force $A_k =$

$$\frac{\pi}{4}[(2c+h)^2 - d^2]$$

Figure 7-11   Concept of pressure cone proposed by
Rötscher.

2. The ratio of the flange thickness to bolt diameter, the material of the clamped flange, and the joint surface topography have large effects on the form and effective area of the interface pressure distribution.

3. The interface pressure distribution is of truncated conical form in the case of the single bolt-flange assembly and of the multiple-bolt-flange assembly with thinner flange.

   Keeping in mind the engineering knowledge in general mentioned above, we now discuss the pressure cone in detail. The pressure cone is one of the engineering guides to estimate only the effective area of an interface pressure derived from the tightening force,[2] and it was first proposed by Rötscher. We used to call it *Rötscher's pressure cone*. In the proposal, shown in Fig. 7-11, the tightening force in a clamped component is constrained it's influence only within the truncated cone having a vertical angle of 90° with the axis of the bolt, i.e., the half angle of pressure cone being 45°, and thus the spring constant of a clamped component can be determined by the elastic deformation of an equivalent cylinder, with a diameter of it passing through the center of the cone's generatrices.

   In consequence, the effective area of a tightening force at the joint surface is within the base circle of the truncated cone. Obviously, the concept of the pressure cone is very simple and useful, provided that the

---

[2]There is a belief, by which Rötscher proposed the concept of the pressure cone within his book entitled *Die Maschinenelemente*. The book was published in 1927 by Springer-Verlag.

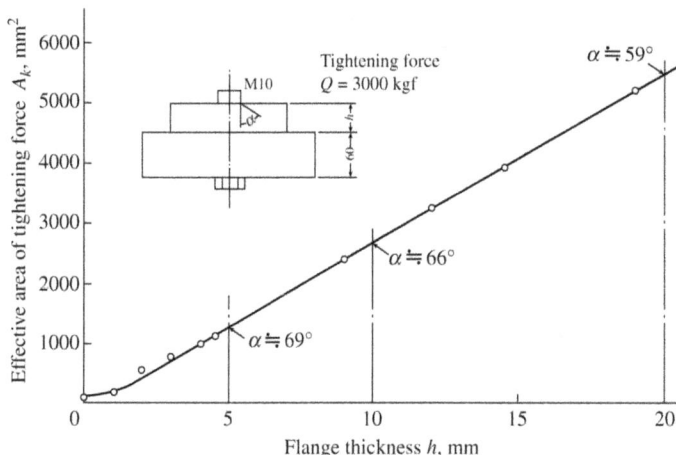

Figure 7-12 Effective area of tightening force (by Plock, courtesy of Industrie-Anzeiger).

half angle is modified in accordance with the condition of the objective joint. In other words, the vertical angle of the pressure cone is one of the fundamental design data.

Figure 7-12 shows thus the effective areas of the tightening force reported by Plock [3, 4], where the measurement was carried out by means of pressure sensitive paper. On this paper the intensity of color changes in proportion to the interface pressure. As can be readily seen, the vertical angle of pressure cone in actuality is from 60° to 70° depending on the thickness of flange, although there are some uncertainties by varying the joint characteristics due to the inclusion of the foreign interfacial layer.[3]

Following Plock, Ito et al. [5] and Ito [6] have publicized similar results by measuring the effective area of the tightening force with ultrasonic waves, which is one of the nondestructive methods giving no changes in the characteristics of the joint surfaces. As shown the effective area of the tightening force in Table 7-1, the vertical angle of pressure cone depends to a large extent on the flange material, also decreasing its value with the increase of the flange thickness [5]. Table 7-2 summarizes the half angles of the pressure cone obtained from the flat joint with local deflection [6].[4]

---

[3]As stated in App. 1, the vertical angle of pressure cone measured is prone to represent a relatively large value, because of the inclusion of the soft interfacial layer.

[4]The experiment was carried out using the same test rig shown in Fig. 6-40, but changing the upper test piece to that of flat bar type.

TABLE 7-1   Measured Values of Half Angle of Pressure Cone

h, mm	α, deg		
	Flange material		
	S45C	Bs BM1	Al B1
8	55	73	73
16	36	59	59
24	30	47	47
32	24	39	39

Admitting that the ratio of the flange thickness to the diameter of the connecting bolt widely used is less than 2 to 2.5, it can be concluded that the angle of the pressure cone is, in general, larger than that of Rötscher as suggested by Plock. It is furthermore interesting that the angle α for a bolt-flange assembly with smooth joint surfaces is smaller than that for rough joint surfaces. This noteworthy feature is protrudent in the bolt-flange assembly with thinner flange, and Fig. 7-13 shows the effect

TABLE 7-2   Effects of Joint Surface Qualities on Half Angle of Pressure Cone

Joint Surfaces						Half angle of pressure cone α, deg
Lower specimen			Upper specimen			
Material	Machining method	Surface properties	Material	Machining method	Surface properties	
Cast iron (FC 25)	Lapped	H_B: 450–470 R_{max}: 2.0 μm FH (Depth: 1 mm)	Cast iron (FC 35)	Scraped	30/in 60/in^2	67 63
			Cast iron (FC 25)	Ground	R_{max}: 1.5 μm R_{max}: 4.0 μm R_{max}: 2.8 μm H_{RC}: 40 (FH)	69 72 58
				Lapped	R_{max}: 1.7 μm	63
			Case hardening steel	Ground	R_{max}: 2.5 μm H_{RC}: 59	61
Cast iron (FC 35)	Scraped	30/in^2	Cast iron (FC 35)	Scraped	30/in^2 60/in^2	67 63
			Cast iron (FC 25)	Ground	R_{max}: 2.0 μm R_{max}: 3.2 μm H_{RC}: 30 (FH)	70 63
				Lapped	R_{max}: 1.7 μm	63
Mild steel (SS41B)	Ground	R_{max}: 2.0 μm	Mild steel (SS41B)	Ground	R_{max}: 2.0 μm	73

Note: FC 25, FC 35, and SS41B: material descriptions per JIS. FH: flame hardening.

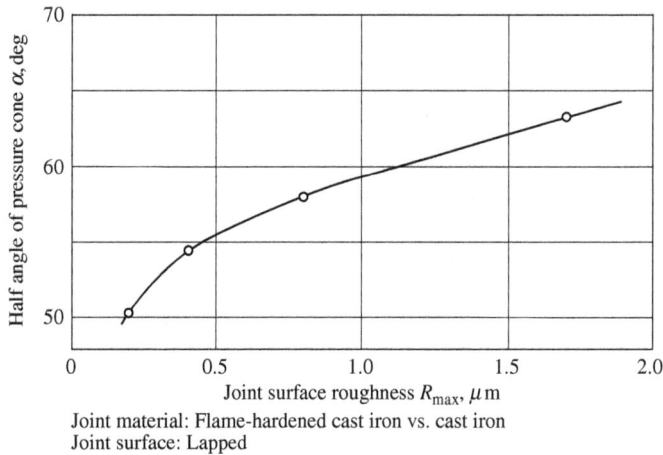

Joint material: Flame-hardened cast iron vs. cast iron
Joint surface: Lapped

Figure 7-13   Changes of half angle of pressure cone with joint surface roughness.

of the joint surface quality upon the value of α reported by Ito [6].[5] To deepen the understanding, furthermore, Fig. 7-14 reproduces the effects of the tightening force on the interface pressure distribution, and as can be seen, the tightening force has no apparent effect on the interface pressure distribution so far clarified elsewhere. However, these new findings have not reduced the valuableness of the concept of Rötscher's pressure cone, but from the engineering calculation point of view, his proposal is very effective. This is because by varying only the vertical angle of pressure cone, his proposal can facilitate understanding of the behavior of the bolted joint to a larger degree.

In short, it is envisaged that the verification of Rötscher's pressure cone from a practical viewpoint is credited to Plock in 1971. Importantly, through several experimental researches afterward, it can be concluded that the half angle of pressure cone is considerably larger than that proposed by Rötscher apart from a special case. In addition, the interface pressure concentrates around the center of the single bolt-flange assembly, when the apparent joint surface is smaller than the effective area of the interface pressure. In this case, the interface pressure distribution becomes a form, which is superimposed the tail-off part of the

---

[5]To improve or to enhance the bearing condition, the small recess has been machined at the joint surface from the old days. According to the experiment conducted by Itoh et al. [11], the annular recess, i.e., shape pattern of bearing surface, in a single bolt-flange assembly has greater influence on the interface pressure distribution. In fact, the distribution shape and region of the interface pressure can be determined by the allocation of the recess and deformation of the clamped component.

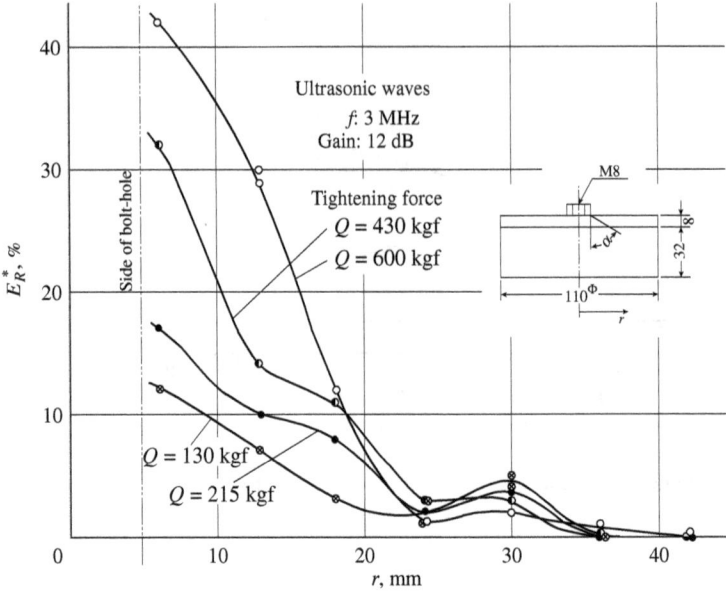

Figure 7-14 Qualitative interface pressure distribution when tightening force is varied.

distribution onto that around the bolt-hole, indicating the applicability of the *moment image method* [11]. In fact, these researches based on the practical viewpoint render the earlier proposals for the calculating procedures of the spring constant of the clamped component useless, where the validity of Rötscher's pressure cone, i.e., pressure cone with half angle of 45° was believed.

Because of the importance of some new findings in the pressure cone, a quick note about such earlier researches will be stated in the following, and Fig. 7-15 shows first a Puttick grid in order to understand the state of earlier researches into the interface pressure distribution, to verify the validity of Rötscher's pressure cone, and to propose a reliable calculating method for the spring constant of the clamped component in a bolt-flange assembly.

For shortage of effective measuring technologies, the earlier work was first carried out from the theoretical aspect as exemplified in that of Fernlund [7]. Such works can be two-fold by the model of a bolt-flange assembly: One is the infinite plate with a hole, and the other is the finite hollow cylinder, both of which have no joint surfaces. In these works, thus, the distribution of normal stress σ_z on the midplane ($z = 0$) and along the r direction of the model is regarded as the interface pressure distribution derived from the tightening force. In addition, the stress

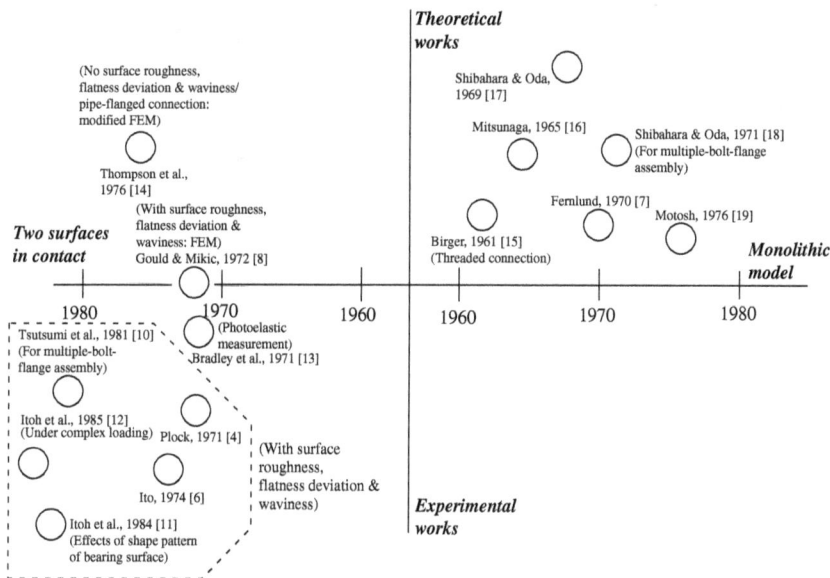

Figure 7-15   Researches into interface pressure distribution of bolt-flange assembly.

distribution was measured using the frozen pattern photoelastic methods; however, this method cannot correctly model the bolt-flange assembly from the material aspects. In due course, there were considerable discrepancies between the theoretical and Rötscher's pressure cones, and thus the theoretical spring constant calculated from the pressure cone or pressure barrel is not in good agreement with the experimental one.

With the advent of the FEM, the bolt-flange assembly was dealt with as a problem of two surfaces in contact as exemplified by Gould and Mikic [8]; however, the idealized joint surfaces, i.e., flat and smooth surfaces, were assumed. More specifically, the interface pressure distribution and radii of contact zone were computed using the FEM model consisting of the annular ring element, and the bolt-flange assembly was treated as a three-dimensional problem with mixed boundary condition in the theory of elasticity. Figure 7-16 shows one of the computed results using the FEM for the bolt-flange assembly with ideal joint surfaces (two plate analysis) and compares it with those not considered the joint surfaces (single plate analysis). From this comparison, it can be observed that the two-plate model yields somewhat different stress distribution from that approximated from the single-plate model.

In consideration of such apparent differences mentioned above, Fig. 7-17 reproduces a comparison between the computed results of Gould and Mikic and the measured interface pressure distributions reported by Ito et al. [5], where the measurement was carried out by means of the

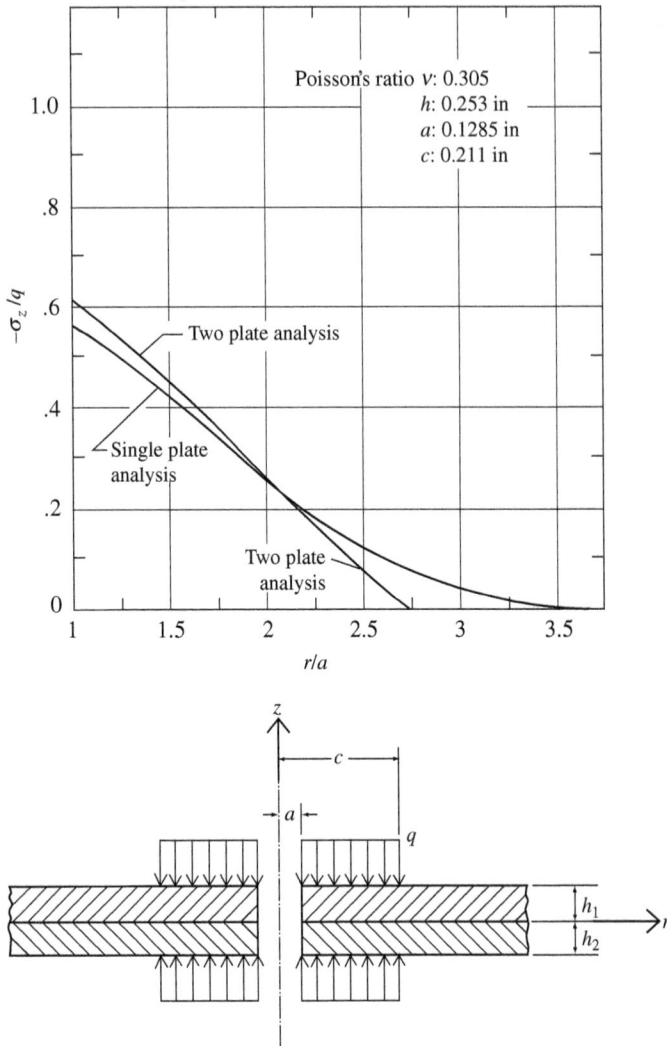

Figure 7-16 Finite element analysis for bolt-flange assembly of 1/4-in plate pair (by Gould and Mikic, courtesy of ASME).

ultrasonic waves method, although the objective was the bolt-flange-threaded hole connection. Obviously, there are considerable disagreements between the theoretical and measured interface pressure distributions as follows.

1. The measured pressure distribution is very much wider than the theoretical one in both the bolt flange assemblies with lapped and ground-to-lapped joints.

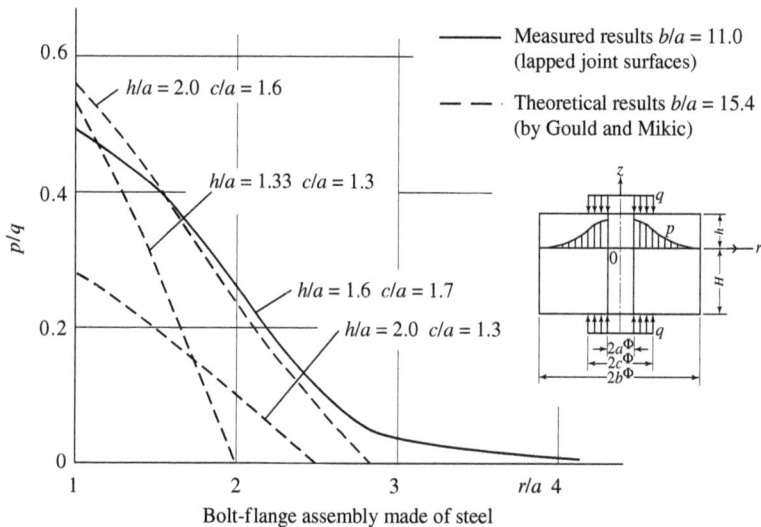

Figure 7-17   Comparison between theoretical and experimental interface pressure distribution.

2. In the case of a thin upper flange, for instance, the measured pressure is lower than the theoretical one around the bolt-hole and higher than at the skirt of the distribution area together with showing the longer tail-off.

3. The theoretical and measured results show qualitatively similar behavior with respect to the effect of flange thickness on the interface pressure distribution and on its effective area.

4. In the theoretical results, the dimensions of the bolt-flange assembly have large effects upon the pressure distribution; however, in the measured pressure distribution, the flange dimensions have less pronounced influence.

These disagreements may be substantially attributed to the disregard of the topography in the actual joint surface. In addition, we must be aware of the ease of warping in the actual bolt-flange assembly, even when the nondimensional values of both the bolted joints in theoretical and experimental works are identical.

As can be readily seen, we can conclude that the earlier researches shown in Fig. 7-15, except those of Plock, Ito, and Tsutsumi, have not dealt with the actual bolt-flange assembly from both the theoretical and experimental aspects. In short, in the future we need to figure out how to incorporate adequately the surface topography of the joint surface in the computing procedure for the interface pressure distribution

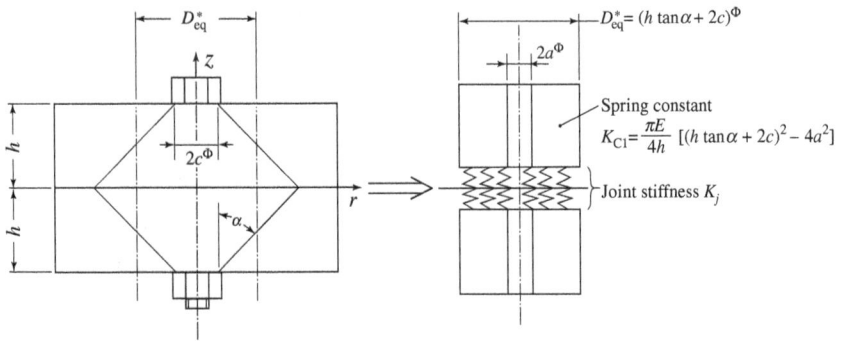

Figure 7-18 Mathematical model for calculating spring constant of clamped component based on pressure cone.

of the bolt-flange assembly. Figure 7-18 is a mathematical model for calculating the spring constant of clamped component in consideration of the joint proposed by Ito, where the spring constant of the bolt-flange assembly can be determined by the stiffness of the joint and equivalent hollow cylinders. In this context, Itoh et al. tried to calculate the spring constant and showed good agreement between the theoretical and measured spring constants. In the trial, they employed the pressure cylinder determined from the interface pressure distribution by using the moment image method [9].

More importantly, Gould and Mikic investigated the radii of separation in a bolt-flange assembly, using both the autoradiographic technique and the mechanical polishing method to verify the validity of their computed result. In the latter case, the radii of separation can be measured as the footprint resulting from the polished area around the bolt-hole of the clamped components, which is derived from sliding under load within the contact zone. In their investigation, the thin stainless steel plates with better surface quality, i.e., R_{rms} and flatness deviation being better than 0.15 and 0.3 μm, respectively, were used as clamped components, and thus warping is prone to appear. As a result, there are certain difficulties concerning whether the footprint indicates exactly the effective area of the tightening force. Despite such uncertainties, the half angle of pressure cone can, from the data for the radii of separation, be estimated to be from 42° to 56° according to the increase of the clamped plate thickness, and these results are in good agreement with those of Plock and Ito.

Reinforcement remedies from structural configuration.   In the machine tool structure, the bolted joint with flange of bay-type is the most popular configuration, and in general, its stiffness becomes larger with increasing column stiffness; however, the deterioration rate of the stiffness due to the joint is, in inverse, larger. In addition, the joint stiffness shows the considerable deterioration where a separation of joint surface can be observed under certain connecting and loading conditions. Importantly, the apparent local deformation in the bolted column appears near the bolt-hole and intersecting portion of the flange to the column wall, and this local deformation causes the deterioration of the joint stiffness as verified by Tsutsumi et al. [20].

In consequence, there are two-fold remedies, i.e., improvement of the flange configuration and realization of full contact condition across the whole joint surface under expected loading.

More specifically, these remedies can be detailed as follows; however, we must be aware that these remedies show often the mutual cross-effects among one another.

1. Optimization for the ratio of the flange thickness h to the diameter of connecting bolt d. In the preferable case, the ratio should be chosen so as to produce the uniform distribution of the interface pressure.

2. Preferable configuration design for bolt pocket in consideration of the allocation of the connecting bolt.

3. Reinforcement of the surroundings around the bolt-hole using the stiffening rib to reduce the bending stress in the flange and connecting bolt.

4. Integrated arrangement of the connecting bolt with sufficient tightening force to realize the oiltight-like contact zone in the bolted joint.

5. Employment of the bay-type flange together with allocating the connecting bolt as near the column wall as possible to maintain the complete contact condition across the whole joint surface.

6. Finishing of the joint surface so that it has no flatness deviation and/or waviness. As suggested by Connolly and Thornley [21], the waviness shows more dominant influence on the joint stiffness than the surface roughness.

In the following, some engineering knowledge about item 1 will be quickly detailed.

Optimum ratio of flange thickness h to diameter of connecting bolt d. In the bolted joint, the bay-type flange is very popular in realizing the sufficient joint stiffness as well as maintaining the ease of assembly and disintegration. Intuitively, the leading design attributes of the bay-type flange are the ratio h/d and relative dimension of flange thickness to thickness of the column wall, which show the maximum stiffness in the bolted column.

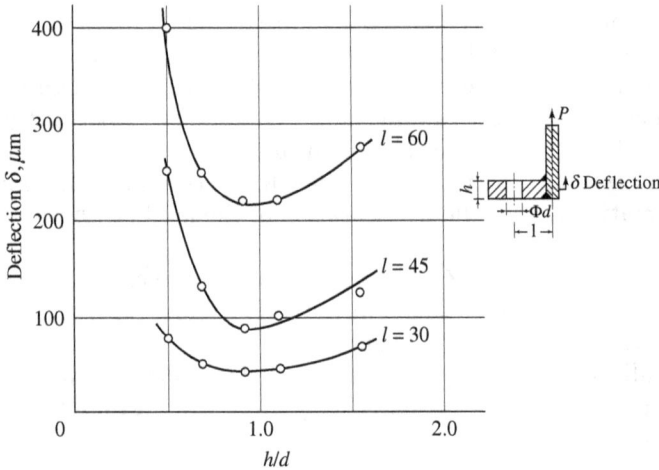

Figure 7-19   Optimum flange thickness of L-shaped bolted flange (by Opitz and Bielefeld).

Although Opitz and Bielefeld reported the optimum flange thickness, in general, there are no optimum values in the ratio of h/d. More specifically, Fig. 7-19 is one of the results presented by Opitz and Bielefeld [22], when carrying out the investigation into the simple L-shaped bolted joint. As can be seen, the joint stiffness is maximum when the ratio h/d is from 0.9 to 1.1. In accordance with reports of Plock [3], Schlosser [23], and Tsutusmi et al. [20], however, we cannot find any obvious optimum values such as shown in Fig. 7-20, although they carried out the due investigations using the models having closer features to the actual bolted joint. As a result, it appears reasonable to determine the optimum value of h/d when the stiffness of the jointed column is around 90% of that of the equivalent column, simultaneously in considering that the connecting bolt shows lower stiffness with the increase of flange thickness. In due course, the optimum value of h/d is from 2.0 to 2.5.

7.1.2   Engineering design for practices—
suitable configuration of bolt pocket
and arrangement of connecting bolts

When the validity of a proposal has been verified by many people to a various extent, such a proposal can be considered as reference material for the design guide. In contrast, although a proposal may be considered very valuable, it must be dealt with during a prestage of the design guide, when the proposal's validity has not been verified to a certain extent yet. In the following, such design knowledge will be stated.

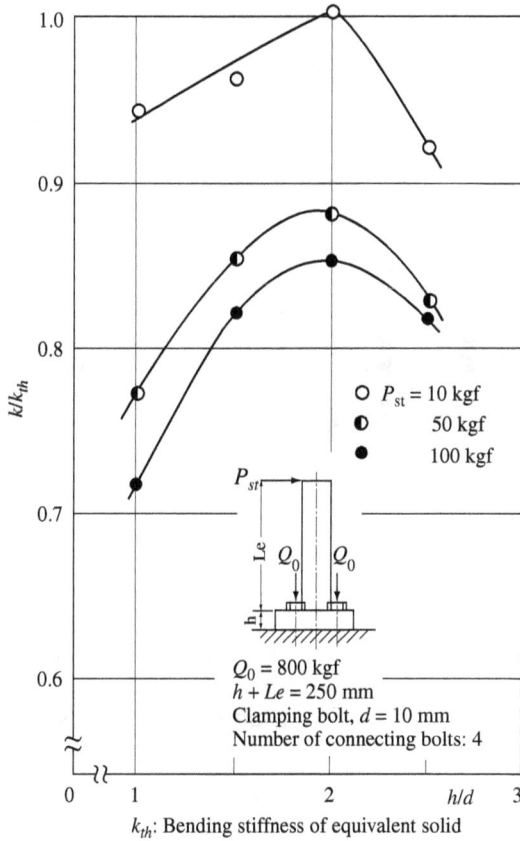

Figure 7-20   Optimum ratio h/d in bolted joint of B type.

Suitable configuration of bolt pocket—Reinforcement around bolt-hole. In the bolted joint, the bolt-hole is one of the weak portions, and thus there are several remedies, e.g., affixing the stiffening rib and employment of the bolt pocket, to reinforce the joint stiffness. Within this context, Opitz, Plock, Thornley, and Ito conducted, as mentioned above, the researches and on the basis of their achievements, the due design knowledge can be enumerated as follows.

1. In general, the deterioration of the joint stiffness is prominent when the bolted joint under certain connecting and loading shows a considerable separation of jointed surface.

2. The connecting bolt must be allocated closer to the side wall of the structural body component, so that only the tensile load acts on

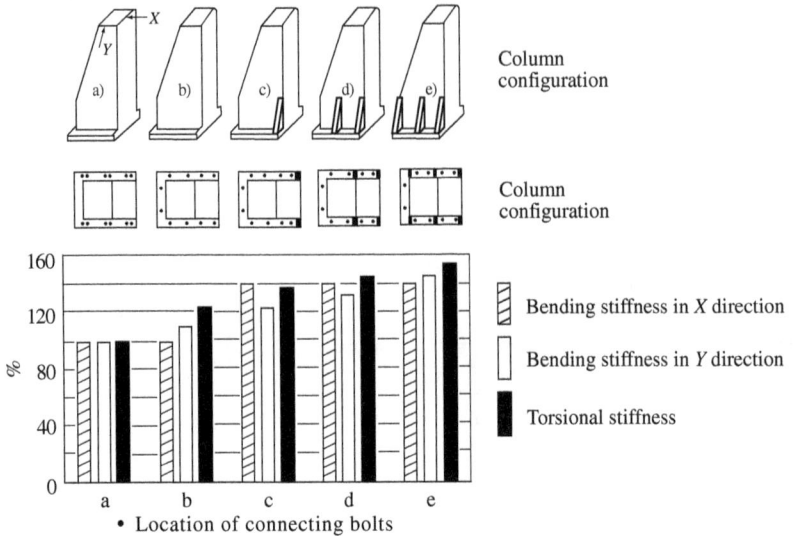

Figure 7-21 Effects of stiffening ribs on stiffness of bolted column with bay-type flange (by Opitz and Bielefeld).

the connecting bolt. According to the report of Opitz and Noppen [24],[6] the deflection of the L-shaped bolt-flange assembly reduces to 24% of that with offset bolted type by allocating the two connecting bolts at the pockets within the side wall.

3. The bolt pocket must be of closed type, although the production cost increases considerably.

In general, the stiffening rib is the most popular remedy used so far, with the expectation of the improvement of the joint stiffness with cheaper production cost. Figure 7-21 shows also qualitatively the reinforcement effect of the stiffening rib reported by Opitz and Bielefeld [22], where they varied the number of the ribs and connecting bolts in the flange-bolted column. As can be readily seen, the stiffness of the column increases with the number of stiffening ribs. To continue these earlier activities, Thornley conducted a comparative research into the reinforcement effects of the various configurations near the bolt-hole, and showed the same results as those of Opitz and Plock. In that of Thornley, the experiment was carried out using the comparatively large specimens, i.e., 18(length) × 4 · 1/2(width) × 4 · (3/4)(height) in with three

---

[6]Using a model of L shape made of plastics, Yasui et al. investigated the effects of the tightening force, number and diameter of connecting bolts, allocation of the connecting bolt, flange thickness, and rib on the joint stiffness.

Masui, T., et al., "The Rigidities of the Jointed Parts of Machine Tools (1)," Laboratory Research Reports of MEL within MITI, 1968, 22(3): 1–10.

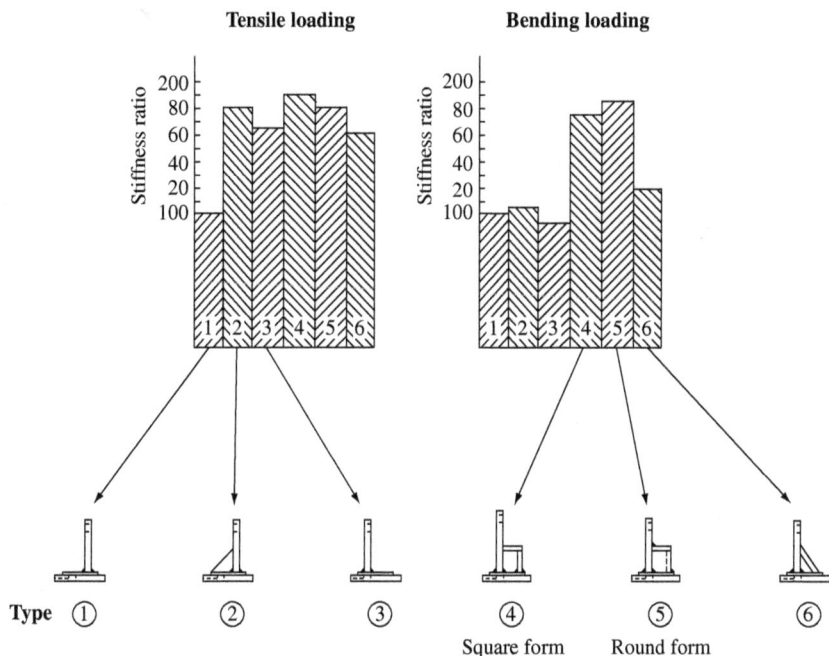

Figure 7-22   Comparison of various remedies for reinforcement (courtesy of Thornley).

connecting bolts, and he ascertained the better effect of the bolt pocket of round- or square-enclosed type than the stiffening rib, as shown in Fig. 7-22 [25]. In addition, Plock verified experimentally the validity of research results of Opitz and Bielefeld and later showed the importance of the allocation of the connecting bolt, as shown in Fig. 7-23 [3]. More specifically, the bending stiffness of the bolted column with bay-type flange can be reinforced by providing the stiffening rib; however, its stiffening effect is smaller than that obtained by placing the connecting bolt at the side wall.

Since the enclosed bolt pocket shows larger reinforcement effect than those given by other representative remedies, a further necessity is to unveil what is the essential role of the bolt pocket. Thus Ito and coworkers conducted an interesting research into the variation of the interface pressure distribution when the bolt pocket configuration is changed [26]. They measured the two-dimensional interface pressure distribution, using the ultrasonic waves method of focus type transducer (oscillating frequency = 5 MHz) and setup with automatized scanning function. Figure 7-24 is a reproduction of typical measured result under bending, and it can be seen that the bolt pocket can, in short, accommodate the directional orientation effect, resulting in the larger reinforcement of the joint stiffness. In detail, the bolt pocket can facilitate

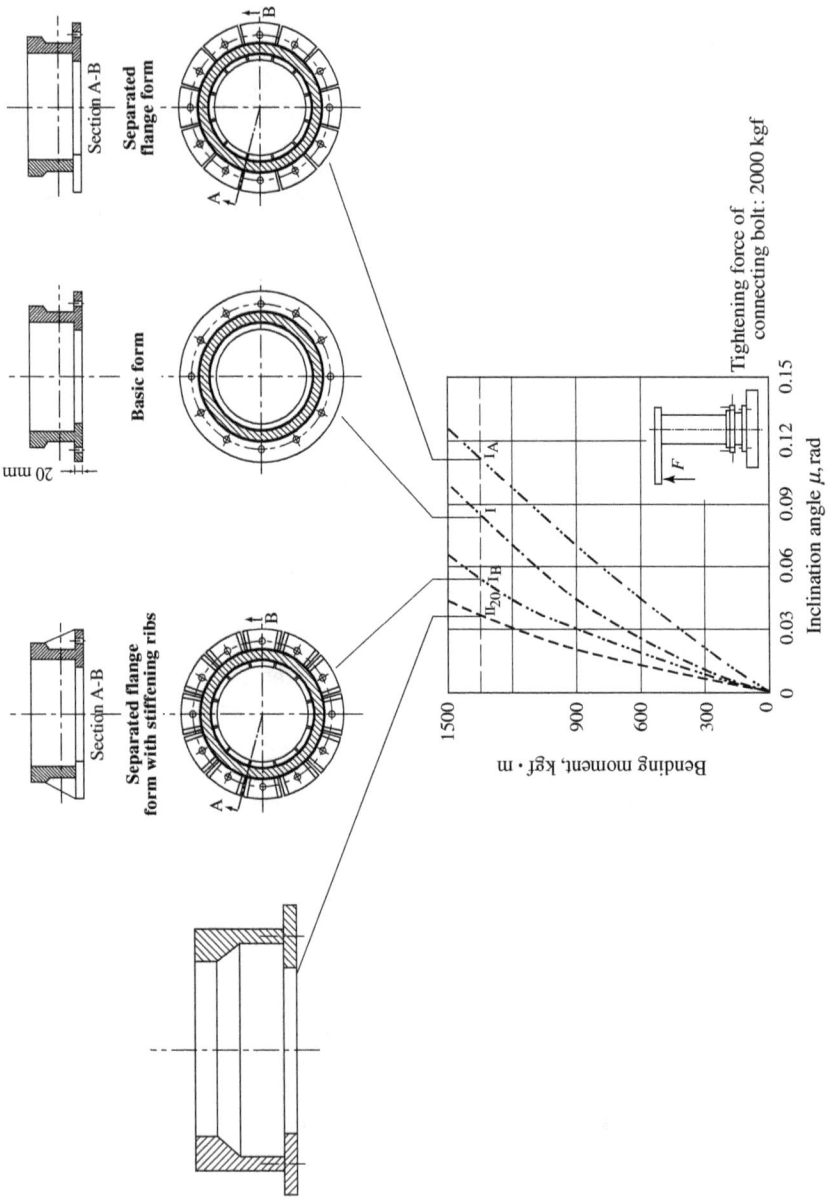

Figure 7-23 Effects of allocation of connecting bolts and flange configurations (by Plock, courtesy of Industrie-Anzeiger).

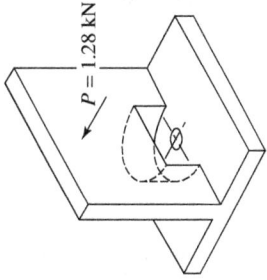

Connecting bolt: M8
Flange thickness: 16 mm
Thickness of side wall: 10 mm
Inner diameter of pocket: 50 mm
Thickness of pocket wall: 10 mm

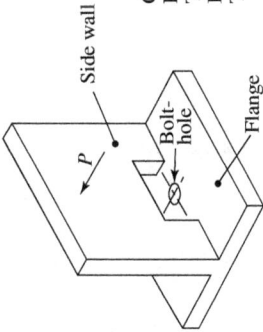

Tightening force $Q = 5$ kN, 0.1 $E_R^* = 5.5$ MPa

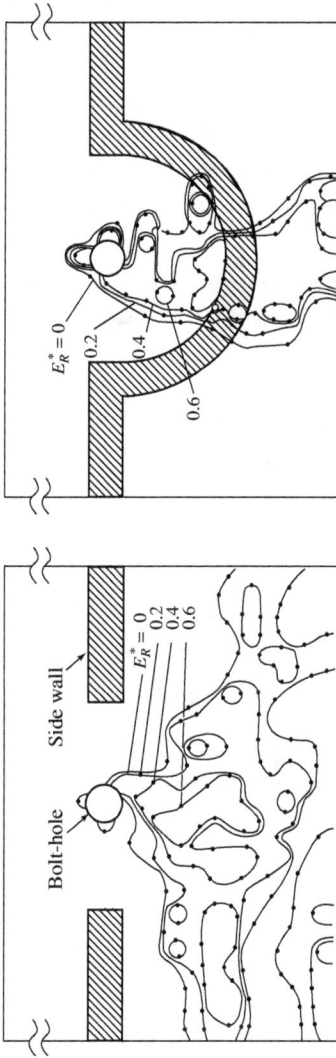

Figure 7-24 Directional orientation effect of bolt pocket—interface pressure distribution of bolted joint under normal loading.

the efficient reaction to the applied load by varying the widely spread interface pressure to bandlike form, the centerline of which coincides with the acting direction of bending loading. In contrast, Fig. 7-25 shows the interface pressure distribution in the case of bending along the side wall, and there is no directional orientation effect, although the bandlike form of the interface pressure distribution can be observed. In the tightening condition without bending, the bolt pocket has no significant effects on the distribution form of the interface, showing the concentriclike form within the bolt pocket. In addition, the distribution area becomes smaller as the stiffness of joint surroundings reduces.

Number and arrangement of connecting bolts. In general, increasing the number of connecting bolts results in the considerable improvement of

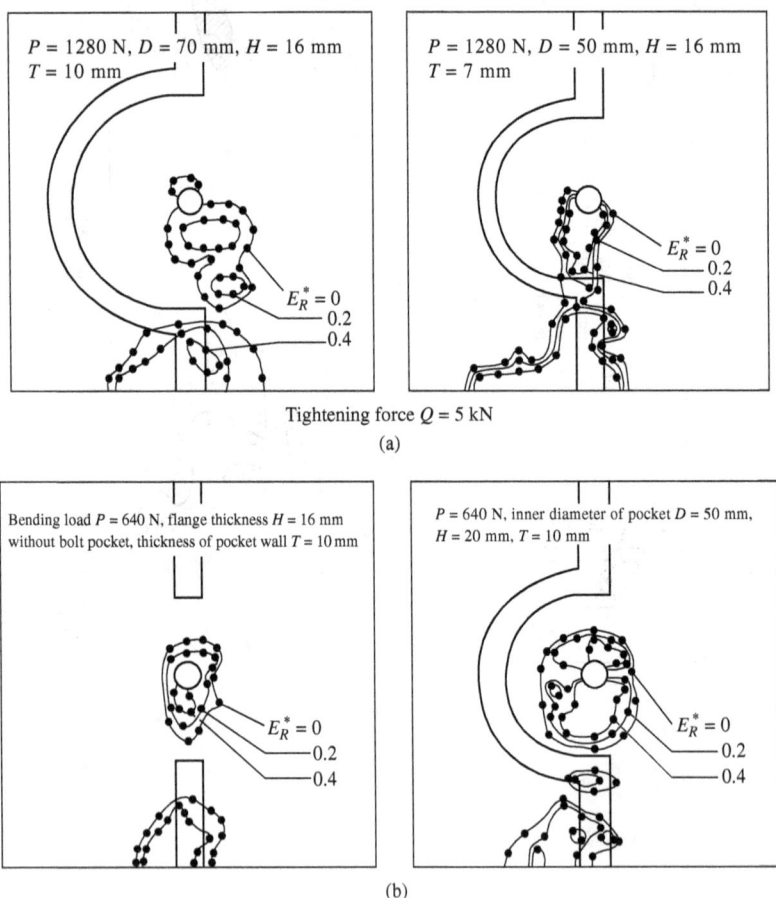

Tightening force $Q = 5$ kN

(a)

(b)

Figure 7-25 Interface pressure distribution in parallel loading: (a) Varying the pocket configuration and (b) effect of the bolt pocket.

the joint stiffness. At issue is thus to determine the optimum number of connecting bolts, if possible, in consideration of both the technological and economic aspects. Within this context, however, it is very difficult to produce a desirable instruction because the earlier researches reported obviously the following evidence.

1. By Schlosser [23, 27], the effective number of connecting bolts is around 5, as shown in Fig. 7-26, whereas Ito and Masuko [28] observed

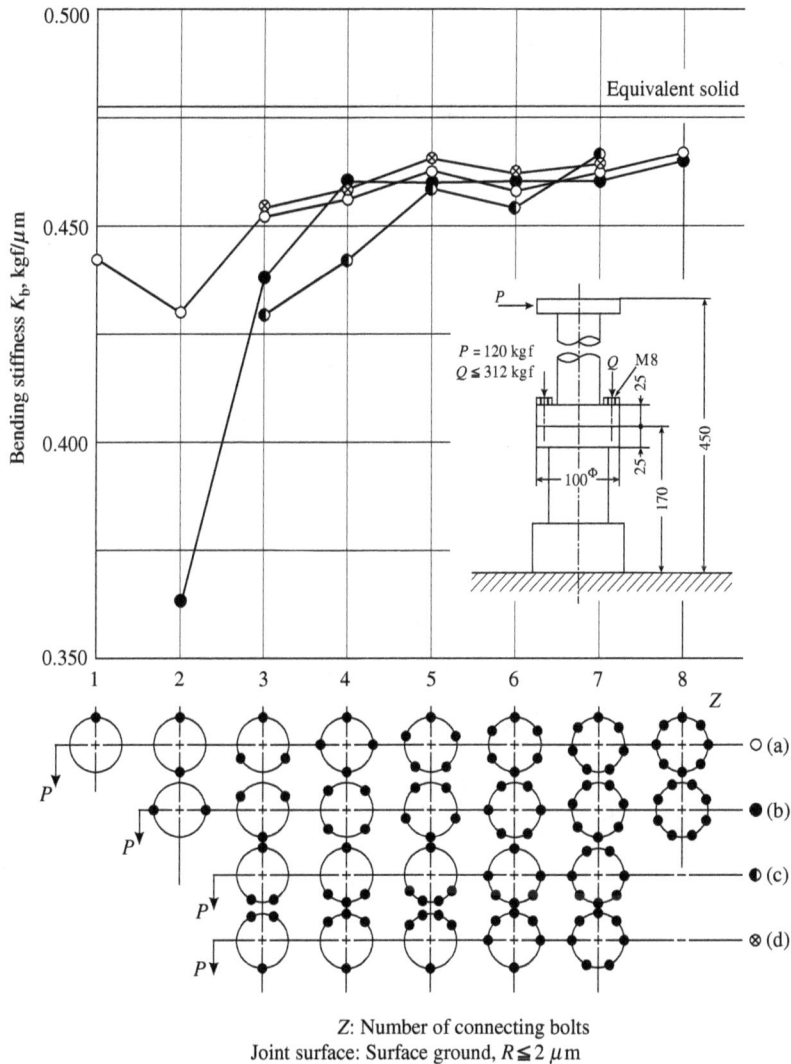

Z: Number of connecting bolts
Joint surface: Surface ground, $R \leq 2\,\mu$m

Figure 7-26   Effects of arrangement and number of connecting bolts (by Schlosser).

the stepwise-like increase of the joint stiffness with the increasing number of connecting bolts, as shown in Fig. 7-27. In Fig. 7-27, the elongation of the connecting bolts is also shown, implying the importance of the full joint contact to maintain the higher joint stiffness. As can be seen, furthermore, the nonsymmetry of the elongation in the front connecting bolt can be observed, resulting in the

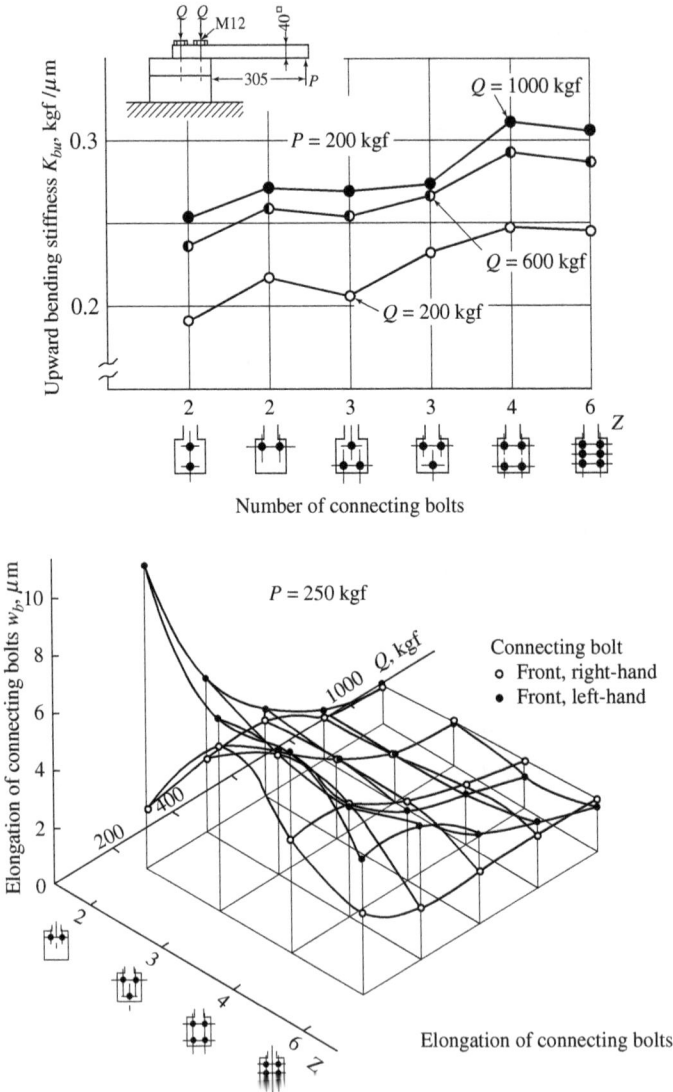

Figure 7-27   Effects of arrangement and number of connecting bolts.

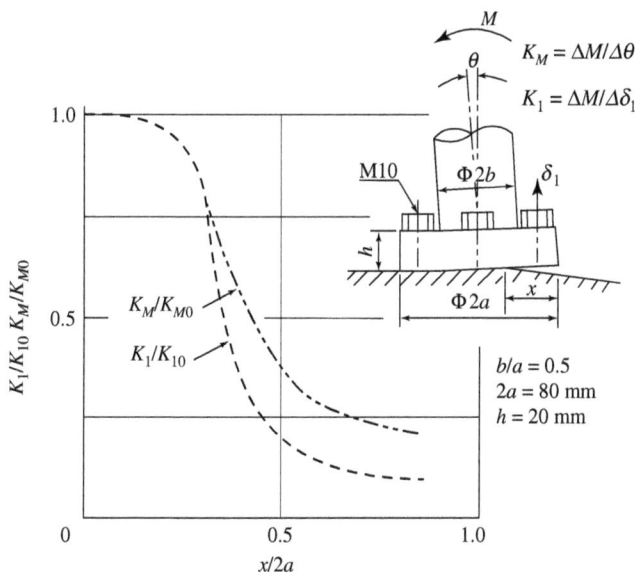

Figure 7-28   Influences of joint separation on deterioration of joint stiffness.

complicated behavior of the bolted joint. This nonsymmetry is due to the geometric difference of each connecting bolt and arbitrary location of the hexagonal bolt head.

2. Figure 7-28 shows the deterioration of the joint stiffness when the joint surface occurs the separation [29].

3. Plock suggested the greater importance of both the allocation of the connecting bolts and the location of the connecting bolt in respect to the side wall of the column than the number of connecting bolts, as shown in Fig. 7-29, and showed uncertain effects of the number of connecting bolts [3]. In fact, the joint stiffness increases when the bending neutral axis moves in the opposite direction to the bending load.

As a result, some recommendations are available for the design:

1. The close arrangement of connecting bolts so as to produce the contact pattern of enclosed type at the joint surface, or along the pitch circle of connecting bolts, i.e., realization of the overlap of the pressure cone of each connecting bolt.

2. The connecting bolts must be allocated so that the joint does not show any separation under external loading.

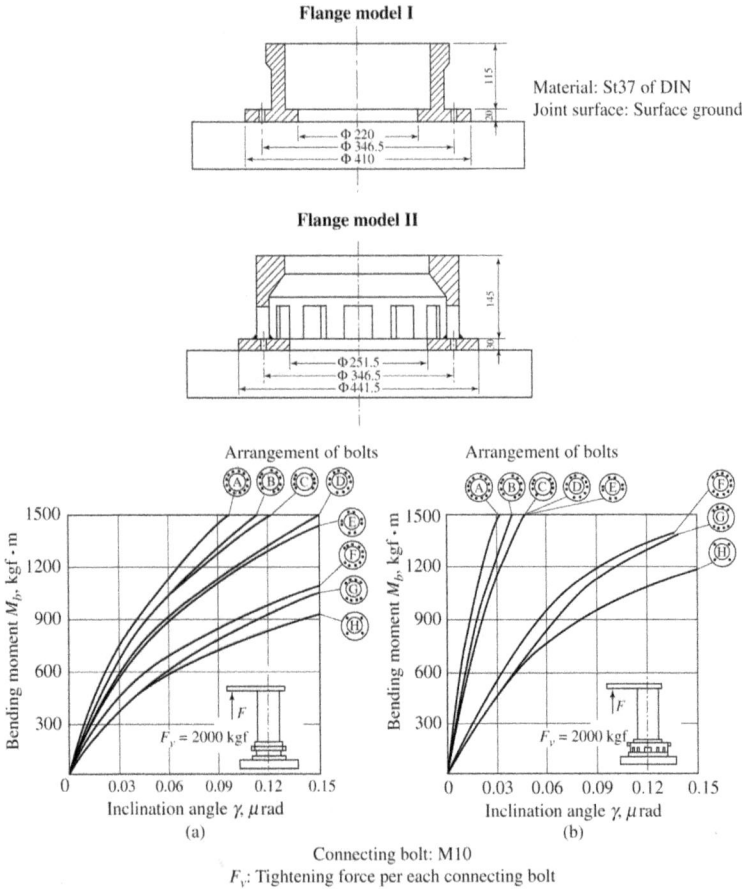

Figure 7-29   Effects of arrangement and number of connecting bolts for a large model: (a) Flange model I and (b) flange model II (by Plock, courtesy of Industrie-Anzeiger).

To this end, the bolt spacing must be discussed; however, it has no noticeable influence on the joint stiffness, except that under low tightening force, provided that the joint area and the number of the connecting bolts are kept constant, as reported elsewhere [30].[7,8]

---

[7]There is another report on the effect of bolt spacing as follows.

Meck, H. R., "Analysis of Bolt Spacing for Flange Sealing," Technical Briefs in *Trans. of ASME*, Feb. 1969, pp. 290–292.

[8]Although not showing noticeable influence on the joint stiffness, the bolt spacing is very important to bolt the hardened strip onto the slideway with less waviness. This is another design subject of the machine tool joint, and thus a quick note will be stated in Supplement 2, at the end of the chapter.

7.1.3 Engineering calculation for damping capacity

In the machine tool joint, at burning issue is to estimate the damping capacity derived from the joint and to evaluate its contribution to the total damping capacity of the machine tool as a whole. To gain certain clues to solve these problems, Fig. 7-30 shows a suggestion for the damping mechanism at the joint [31]. As will be clear from Fig. 7-30, the initial contact points of the base A and beam B move to the points A' and B', respectively, when the beam vibrates on the base without the separation of the joint. Thereby, a relative microdisplacement u, which is nearly equal to a line segment $A'B'$, is produced at the joint surface, and this microdisplacement may be considered one of the leading causes of the energy dissipation at the joint. Assuming that $p(x)$ and k are constant along the width of beam, the instantaneous

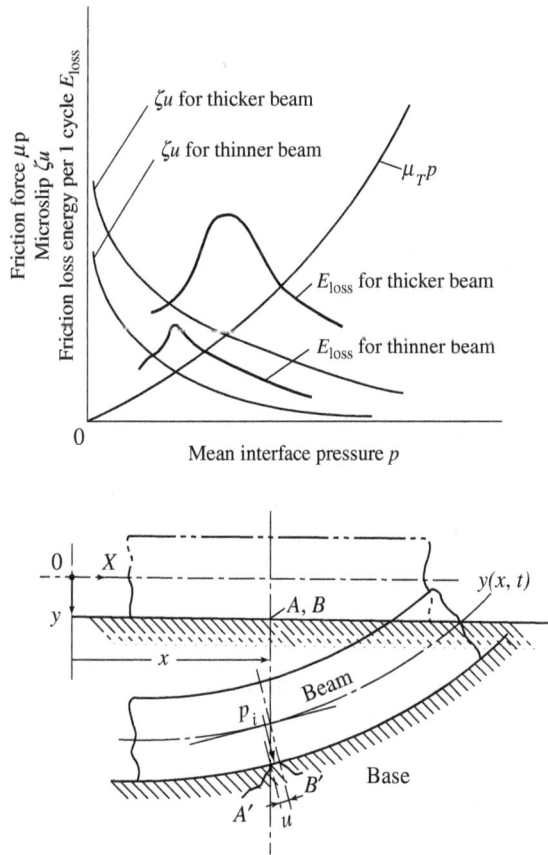

Figure 7-30   Damping mechanism of bolted joint of type A.

frictional force being acted on at the distance dx in coordinate x can be first written as

$$\mu_T p(x) \cdot b \, dx = \mu_T (p_i + ky)b \, dx \qquad (7\text{-}1)$$

where b = width of beam
$p(x)$ = interface pressure at x
p_i = initial interface pressure due to tightening force of connecting bolt
k = normal stiffness per unit area of base
ky = reaction pressure from the base
μ_T = tangential force ratio (equivalent coefficient of friction), which is function of microdisplacement u

Importantly, a part of microdisplacement can be recovered elastically during the half-cycle of vibration. Then the microslip u_s between both joint surfaces yields to

$$u_s \doteqdot \zeta(h/2)[\partial y(x,t)/\partial x] \qquad (7\text{-}2)$$

where h = height of beam
$y(x,t)$ = bending vibration mode of beam
t = time
ζ = coefficient less than unity and determined by roughness and machining method of joint surface, and concerns

In consideration of the relative microslip velocity at dx, the energy loss per cycle E_{loss} can be given by

$$E_{\text{loss}} = \iint \mu_T (p_i + ky)b \, \zeta \, (h/2)[d/dt(\partial y(x,t)/\partial x)] \, dx \, dt$$
$$(\textstyle\int: 0 \to V_T, \int: 0 \to l) \qquad (7\text{-}3)$$

where l = effective length of interface pressure and V_T = period of vibration.

As a result, the qualitative relationships between the interface pressure and the energy loss can be obtained such as shown together in Fig. 7-30, implying the possibility of existence of an optimum pressure to have maximum damping at the joint. As can be readily seen, the frictional force increases and in contrast microslip decreases with the interface pressure, and as a result the energy loss may have a maximum value at a certain interface pressure.

Reportedly Tsutsumi et al. also proposed a damping mechanism for flange-bolted column [32]. Although it is a similar mechanism to that shown in Fig. 7-30 (see line segment $C'D'$), the further microslip depicted as Δu_s should be considered, where Δu_s is derived from the bending load acting parallel to the joint, as shown in Fig. 7-31.

Considering that the root causes of difficulties in the estimation of damping caused by the joint lie in the determination of the absolute

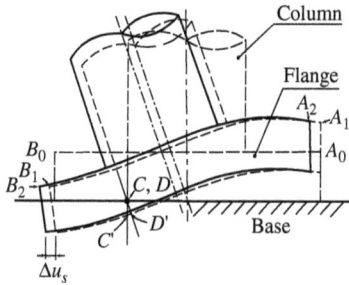

Figure 7-31 Damping mechanism of bolted joint of type B.

value of microslip, tangential force ratio, and interface pressure distribution in the actual bolted joint, a calculating procedure of the damping capacity will be stated in the following by taking the two-layered beam of cantilever configuration as an objective. Within this context, there have been many trials, and an exemplification is that of Ockert [33]; however, these earlier trials considered the frictional energy loss as the Coulomb friction. Reportedly, the utmost marked feature of damping at the bolted joint is that of viscouslike damping including the Coulomb friction-based damping in part.

Figure 7-32 shows a two-layered beam of cantilever configuration. In this beam, the two plates with same thickness are being clamped to each other using n jointing elements, and thus the interface pressure is distributed discontinuously. Assuming that each plate shows no extension of its neutral axis and no distortion in its cross section, a relative displacement $u(x,t)(\Delta u_1 + \Delta u_2)$ appears at the interface, as shown in Fig. 7-32, when the jointed beam vibrates freely [34].

By defining the X-Y coordinates as shown in Fig. 7-32, the relative displacement $u(x,t)$ at $X = x$ yields approximately to

$$u(x,t) \doteqdot \Delta u_1 + \Delta u_2 = 2h \tan[\partial y(x,t)/\partial x] \tag{7-4}$$

In consideration of the damping mechanism mentioned above, the microslip $u_s(x,t)$ can be written as

$$u_s(x,t) = \zeta u(x,t) \tag{7-5}$$

where $y(x,t)$ is the bending deflection of the jointed beam in the Y direction and t is the time.

In contrast, the frictional force F_r at dx is given by

$$F_r = \mu_T p\,(x)b\,dx \tag{7-6}$$

where b = width of beam
$p(x)$ = interface pressure at $X = x$
μ_T = tangential force ratio

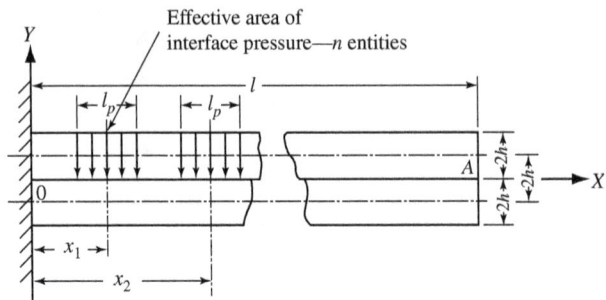

Mathematical model

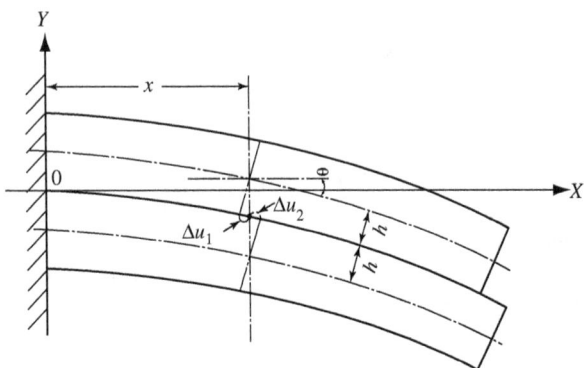

Mechanism for relative microdisplacement

Figure 7-32 Mathematical model of and damping mechanism in two-layered beam in cantilever configuration.

After passing an arbitrary time from the initial time, the loss energy E_{loss} and elastic recovered energy E_{ne} per half cycle can be obtained by referring to Fig. 7-33, and thus the damping ratio D is given by

$$D = E_{\text{loss}}/(E_{\text{loss}} + E_{\text{ne}}) = 1/[1 + (1/\varphi_m)] \tag{7-7}$$

where E_{ne} = elastic recovered energy and in correspondence with the area ABH in Fig. 7-33 and E_{loss} = loss energy dissipated by the microslip and in correspondence with area OAB in Fig. 7-33.

$$\varphi_j = E_{\text{loss}}/E_{\text{ne}} = \{4\Sigma \iint \zeta \mu_T \, bhp(x)$$
$$\partial/\partial t \, [\tan(\partial y(x,t)/\partial x)]dx \cdot dt\}/\{(3EI/l^3)y^2(1, j\pi/2\omega) \tag{7-8}$$

$(\Sigma: i = 1 - n)$

$(\int: j\pi/2\omega - (j + 1)\pi/2\omega)$

$(\int: x_i - l_p/2 - x_i + l_p/2)$

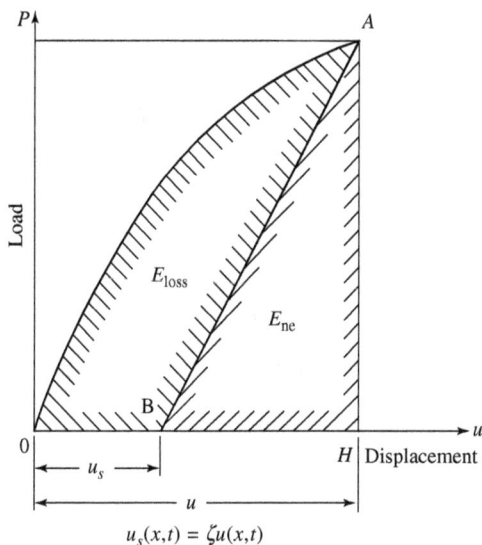

Figure 7-33 Definition of damping ratio D.

$$u_s(x,t) = \zeta u(x,t)$$

By assuming that the stored energy corresponding to the amplitude of vibration a_n is equal to Ca_n^2 (C constant), we can obtain the following relationship between the damping ratio and the logarithmic damping decrement δ_D.

$$\delta_D = \ln(E_n/E_{n+1})^{1/2} = (1/2) \cdot \ln[1/(1 - D)] \tag{7-9}$$

In the case of $D \ll 1$, the Maclaurin expansion for the above-mentioned expression yields the following, when we ignore the higher-order terms more than D^3.

$$\delta_D = (1/2) \cdot [D + (1/2)D^2] \qquad \text{or} \qquad \delta_D = (1/2) \cdot \ln(\varphi_m + 1) \tag{7-10}$$

As can be readily seen, the slip ratio ζ has a larger influence on the damping ratio, and on the basis of the earlier work, ζ may be represented by

$$\zeta = u_s/u = Ge^{-w} \qquad w = a^* p^\rho \tag{7-11}$$

where G, a^*, and ρ are constants determined by the interfacial condition, and furthermore the value of ρ is, in general, unity.

In the simplified case, where the interface pressure is uniformly spread across the whole contact area bl, $p(x)$ yields to p. Besides, by converting $\tan[\partial y(x,t)/\partial x]$ into $\partial(x,t)/\partial x$ and putting $j = 0$, Eq. (7-8) yields to

$$\varphi_0 = \{(4\,\mu_T\,bhp\zeta)/[(3EI/l^3)y^2(1,0)]\} \iint [\partial^2 y(x,t)/\partial x\,\partial t]dx\,dt \qquad (7\text{-}12)$$

$(\int: 0 - \pi/(2\omega))$

$(\int: 0 - 1)$

In consideration of the boundary and initial conditions, i.e.,

$$y(1, 0) = y_0 \qquad \partial y\,(1,0)/\partial t = 0$$

the bending deflection of beam being vibrated $y(x, t)$ can be written as

$$y(x, t) = Y(x)[y_0/Y(l)]\cos \omega t \qquad (7\text{-}13)$$

where $Y(x)$ is an eigenfunction given by

$$Y(x) = (\sinh \lambda_1 + \sin \lambda_1)(\cosh \lambda_1 x/l - \cos \lambda_1 x/l)$$
$$- (\cosh \lambda_1 + \cos \lambda_1)(\sinh \lambda_1 x/l - \sin \lambda_1 x/l) \qquad \lambda_1 = 1.875$$

Thus,

$$D = 1/[1 + k/(4\mu_T y^* pGe^{-w})] \qquad w = a^* p^\rho \qquad (7\text{-}14)$$

where

$$k = 3EI/(l^3 b) \qquad y^* = h/y(1,0)$$

In Fig. 7-34, the relationships between the damping ratio and the interface pressure are shown when the constants G, a^*, and ρ are varied. In these calculations, $k = 3.33 \times 10^{-2}$ kgf/mm^2 and $y^* = 20$, corresponding to the jointed beam shown in Fig. 7-35.[9]

In summary,

1. The value G has large effects on the magnitude of the damping ratio, and in turn the damping ratio is in proportion to the value G.

2. The values a^* and ρ have small effects on the magnitude of the damping ratio, but have large effects on the behavior of the damping ratio at the higher interface pressure.

In short, the damping ratio shows obviously a peak and larger magnitude, when the microslip decreases steeply with increasing interface pressure. Such a behavior of the microslip has been called the *negative derivative characteristic*, and, e.g., the joint with rough surface and

---

[9]The measurement of damping was carried out for the cantilever configuration and also under decayed free vibration, where the length of cantilever is 360 mm.

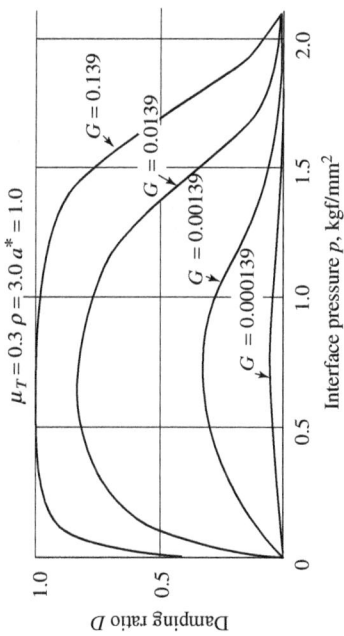

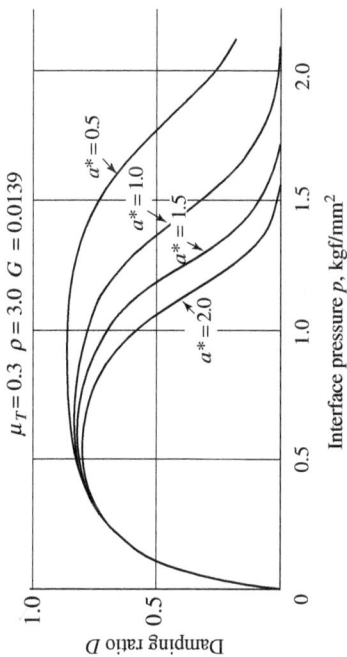

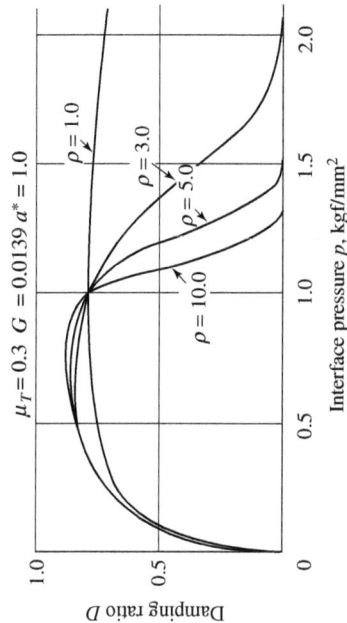

Figure 7-34 Effects of constants G, a^*, and ρ on damping ratio D.

317

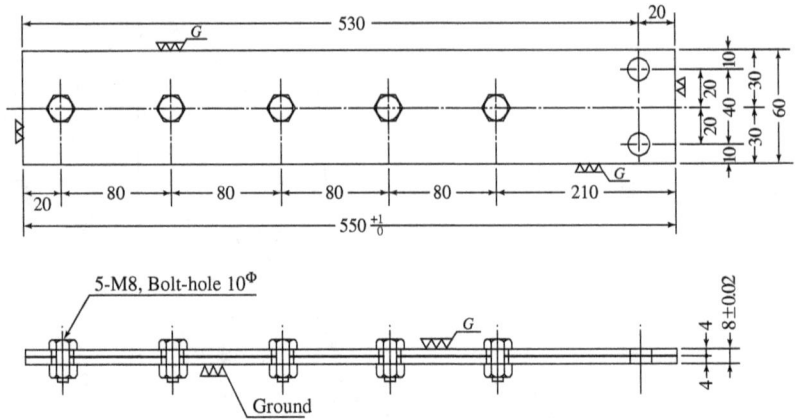

Figure 7-35 Two-layered beam for experiments.

made of material of low flow pressure is liable to indicate the negative derivative characteristic.

Figure 7-36 demonstrates a comparison between the theoretical and experimental values for the jointed beam shown already in Fig. 7-35. As can be seen, both values are in good agreement, although the theoretical

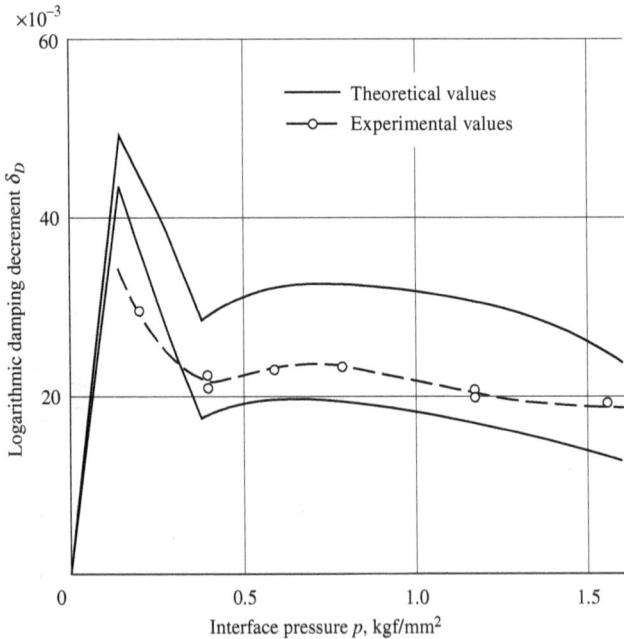

Figure 7-36 Comparison between theoretical and experimental damping capacities.

values are given as the bandlike curve owing to the characteristic features of the microslip. In the calculation, furthermore, special attention was paid to the following points.

Determination of characteristics in microslip. The general behavior of the microslip was already discussed in Chap. 6 and thus we now need to quantify the constants G, a^*, and ρ. In due course, these are determined from the data for tangential deflection in the single bolted joint, which is identical to a bolted entity within the jointed beam. As shown in Fig. 7-37, the microslip is in nonlinear relation to the interface pressure and consists of the three regions, i.e., I, II, and III, and shows certain scatter. In consideration of such scatter, it is necessary to represent the theoretical values with bandlike curve.

Tangential force ratio. The equivalent coefficient of friction was already defined as the *tangential force ratio* in Chap. 6. In consideration of the displacement-dependence characteristic, the tangential force ratio is

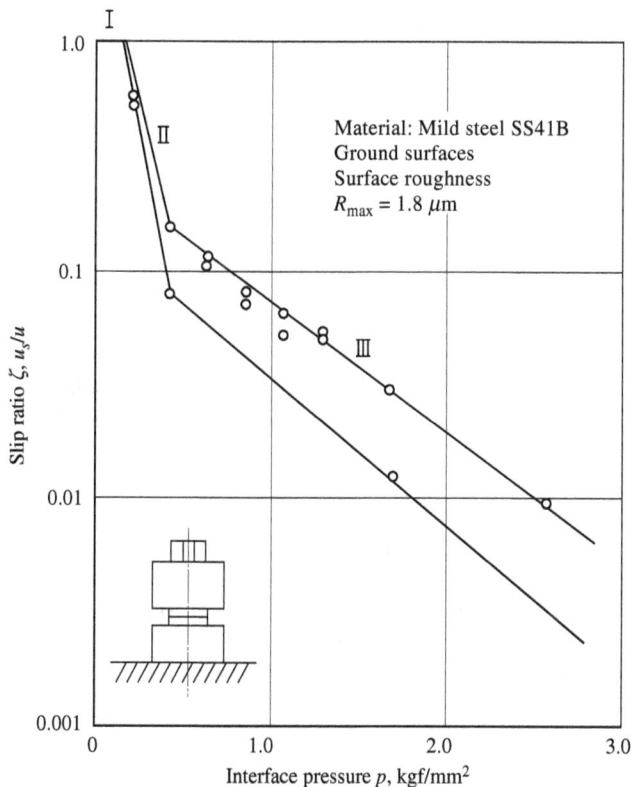

Figure 7-37 Characteristics of microslip in single bolted joint.

determined here, provided that the microslip in the jointed beam examined ranges from 0.01 to 0.03 μm at the tip of the jointed beam when the total vibration amplitude at the tip is 100 μm. In fact, μ_{Tst} is assumed to be 0.02, and furthermore, its dynamic value is assumed to be 0.009 in consideration that the macroscopic coefficient of friction in static condition is 2 to 2.5 times larger than that in dynamic condition.

7.1.4   Representative researches and their noteworthy achievements—static behavior

The static behavior of the bolted joint has been unveiled to a various and larger extent through the most collaborated work conducted by Schlosser for type B [23, 27] and by Ito et al. for type A [28, 30, 35–37], and Schlosser is credited with being the first to commence a series of researches in 1957. In that of Schlosser, some marked features in the experiment are as follows.

1. The test pieces made of St 50 (steel) and GG22 (cast iron) of DIN were produced from the melting condition of the same charge to maintain the homogeneity of the matrix structure.

2. The experiment was carried out after maintaining the test rig and test piece up to 1.5 h in the temperature-controlled room with $20 \pm 0.2°$, so that the accurate value of elastic deflection less than 1.0 μm can be detected.

3. The elongation of the connecting bolt was measured using the detector of strain gauge type.[10]

Importantly, nearly all research results reported by them have been ascertained later by many other researchers and also through practical experience. To understand what are the characteristic features of the bolted joint, thus, a firsthand view for representative behavior will be stated in the following in addition to those already shown beforehand.

Size of and contact entities pattern in joint surface under normal loading. Figure 7-38 shows the effects of the apparent contact area and shape pattern of contact entities on the bending stiffness of the bolted column of type B. In Fig. 7-38(a), the apparent contact area is marked with the relative ratio, where the largest area of about 7260 mm^2 is regarded as 100% after subtracting the area of the bolt-holes. As can be

---

[10]In those of Ito and coworkers, the strain gauges bonded onto the stem of the connecting bolt can facilitate the accurate measurement of the tightening force, although the available clearance to bond the strain gauges is very small.

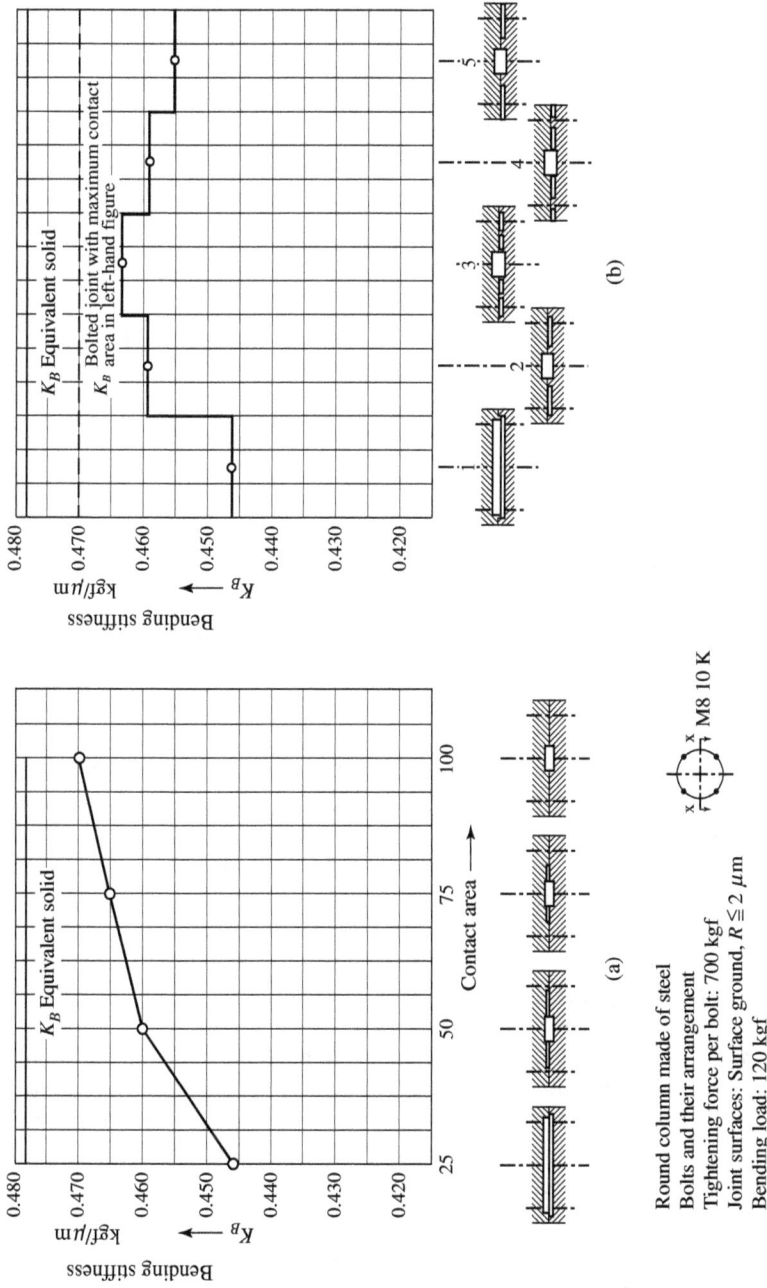

Figure 7-38 Effects of (a) Contact area and (b) contact entities pattern on joint stiffness (by Schlosser).

321

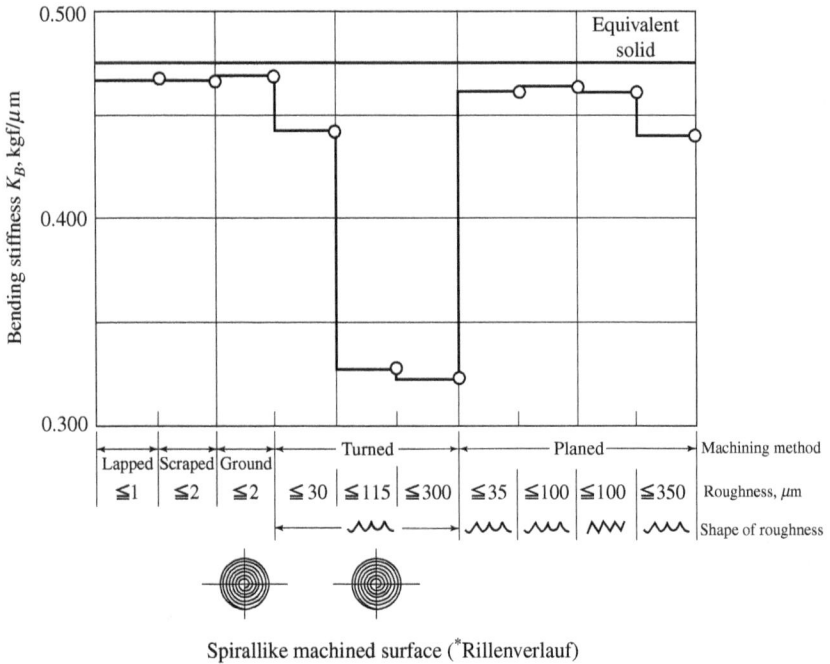

Spirallike machined surface (*Rillenverlauf)

Note: The test piece and loading conditions are as same as those shown in Fig. 7-26.

Figure 7-39   Effects of surface roughness on joint stiffness (by Schlosser).

easily imagined, the joint stiffness increases with the contact area, although the increasing rate is around 10% maximum. Plock also publicized the same result when using the large test piece, and it is very interesting that the increasing rate becomes larger together with showing certain size effect. In addition, the joint stiffness increases to some extent by varying the contact entities pattern such as shown in Fig. 7-38(b), where the better contact quality is achieved by providing smaller recesses across the whole joint surface.

Roughness of joint surfaces under normal loading. Figure 7-39 shows one example of the effects of the surface roughness upon the bending stiffness of the bolted joint. Apart from the planed surface, the joint stiffness increases considerably with improving the quality of the joint surface. In the bolted joint in full-size, the ground or scraped surface is widely employed, and then the deterioration of the joint stiffness due to the surface roughness appears not to be large, provided that the joint surface has no flatness deviation and/or waviness. Figure 7-40 reproduces the other experimental results to show the effects of the various scraped surfaces on the joint stiffness, and as can be seen, the contact

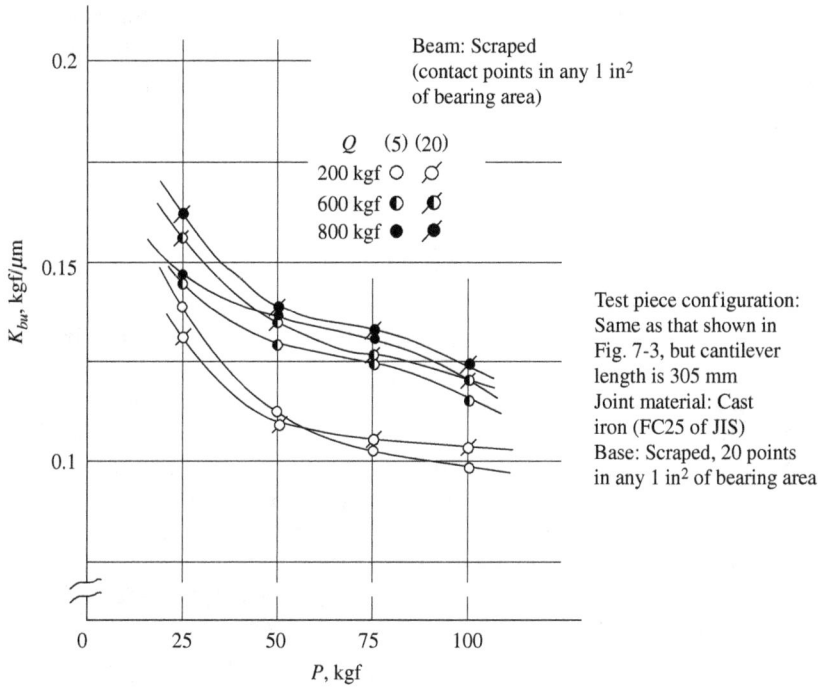

Figure 7-40   Effects of scraping quality on joint stiffness.

points in any 1 in^2 of bearing area has no apparent influence on the joint stiffness. In Fig. 7-39, furthermore, the planed joint shows fairly higher stiffness, i.e., joint stiffness nearly equal to that of the scraped surface. Schlosser deduced that this interesting behavior might be derived from the micromeshing mechanism (die Mikroverzahnung) at the joint, i.e., a variant of directional orientation effect proposed later by Thornley (see Chap. 6).

Interfacial layer under normal loading.   Through a series of experiments, Schlosser confirmed that the interface layer has no effect on the joint stiffness. In the experiment, he investigated the SiC powder, metal foil, and plastic foil as the interfacial layer [23]. Following Schlosser, Thornley et al. [38] investigated the effects of the grease and oil, and reported that these interfacial layers have no effect on the joint stiffness apart from the lapped joint.

Taper pin and guidekey under tangential loading.   As already shown in Fig. 7-5, the bolted joint under tangential loading or torsional loading parallel to the joint surface is, in general, not the design objective.[11] In fact, the joint stiffness under tangential loading appears to be equal to that under bending loading; however, as shown in Fig. 7-41, the joint is prone to occur the macroscopic slip, i.e., sliding, under tangential loading, and the critical load becomes dominant [37]. More specifically, the tangential deflection is in linear relation to the applied load, when the applied load is within the friction force from a macroscopic viewpoint. After exceeding the friction force, i.e., critical load, the bolted joint loses its restoration ability. Obviously, the critical load is in proportion to the tightening force, and then the tangential stiffness increases slightly with the tightening force. In this context, it is worth emphasizing that Ito and coworkers suggested the shear deflection at seizure points in the real contact area as the leading cause of the spring action under tangential loading.

Prior to start of the macroscopic slip, there is a microslip, and thus from the damping capacity point of view, the behavior of the bolted joint under tangential loading should be clarified. However, because of the very low value of the critical load, the bolted joint must have certain remedies for the structural design aspect to enlarge the critical load. In general, such remedies are those of placing "locating elements," e.g., taper pin and guidekey, and in special case, the reamer bolt or guide bolt should be employed notwithstanding the raise of production cost.

Figure 7-42 shows the tangential deflection of the bolted joint with the taper pin, and as can be seen, the load-deflection curve consists of the three sections as follows.

1. In the section $0A$, the tightening force can control the tangential deflection.

2. In the transient section AB, the applied load exceeds the critical load, which can be determined by the frictional force due to the tightening force, and the taper pin starts to function, showing also the apparent microslip.

3. In section BC, the tangential deflection can be determined by the stiffness of the taper pin.

---

[11]Schlosser conducted also a series of researches into the torsional stiffness of the bolted joint of type B; however, the emphasis is laid on the bending stiffness in his report. The noteworthy results are as follows.

1. As same as the bending stiffness, the machining method and roughness of the joint surface have considerably large effects on the joint stiffness.

2. The size and shape of joint surface have no influence on the torsional stiffness.

3. Differing from the bending stiffness, the flange thickness has no effect on the torsional stiffness.

4. The interfacial layer has absolutely no effect on the joint stiffness.

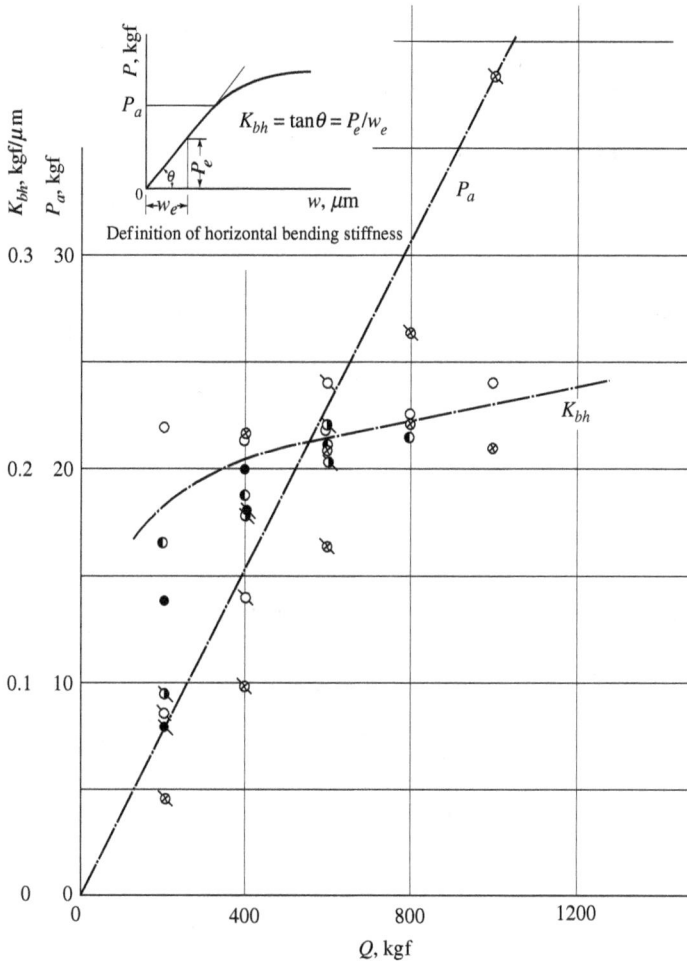

Figure 7-41   Critical bending load and tangential stiffness in bolted joint of A type.

Connecting bolts: $2 \times M12$ Beam size: 40 mm in height, 40 mm in width
Bolt spacing: 90 mm   Length of cantilever: 305 mm

These phenomena may be due to the fitting quality, i.e., locating accuracy, between the taper pin and the taper hole. In other words, the taper pin induces a particular behavior depending on the fitting tolerance, i.e., additional torsion of the bolted cantilever as shown in Fig. 7-42. In fact, the front connecting bolt elongates upward with the tangential loading, and thus it may be recommended that the taper pin be arranged far beyond the effective area of the tightening force of the connecting bolt. In accordance with long-standing experience, this remedy can assist the improvement of assembly accuracy.

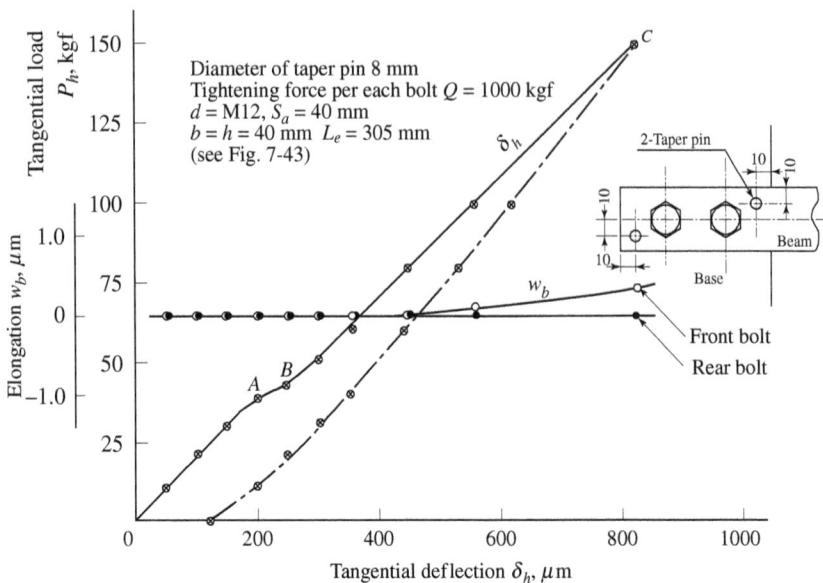

Figure 7-42   Tangential deflection of bolted joint with taper pin.

Similar behavior can be observed in the bolted joint with the guidekey as shown in Fig. 7-43. As can be readily seen, a typical differing feature is the longer section AB compared with that observed in the bolted joint with taper pin. In addition, the distance AB far exceeds the fitting tolerance between the guidekey and the keyway. This may be attributed

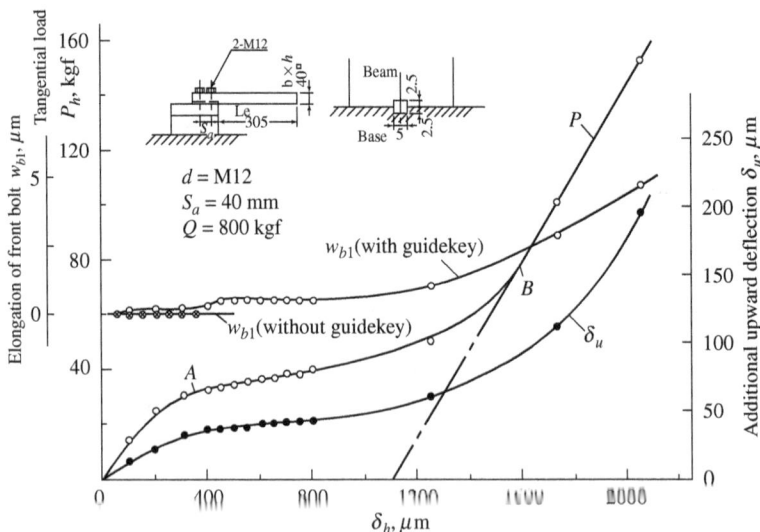

Figure 7-43   Tangential deflection of bolted joint with guidekey.

to the rotation of the guidekey within the keyway, because the tangential load is applied to the upper side of the joint surface. In due course, the upward deflection of the beam occurs geminately with the tangential deflection when the tangential load is increasing.

To this end, the two issues will be touched on to deepen the keen understanding.

Theoretical analysis of static stiffness.[12]   Ito and coworkers conducted a series of investigations into the bolted joint and proposed the analytical expressions for the static joint stiffness of the bolted joint of type A under normal and tangential loading [35–37]. In the analysis, the bolted beam is replaced by the mathematical model, which is of the beam on the elastic supports or on the elastic foundation depending on the tightening and external loading conditions. In the model of elastic supports, the front edge of the base or rear edge of the beam must be considered one of the supports, whereas in the model of elastic foundation, the overlap condition of the pressure cone must be considered to rationally determine the spring constant. In a certain case, the spring constant must be doubled at the overlap area.

In short, the bending stiffness can be written as follows.

1. In case of normal loading

$$K_b = K_0/\{F(\lambda,\kappa) - G(\lambda,\kappa)[Q/P]\} \tag{7-15}$$

where K_b = bending stiffness of bolted beam.
 K_0 = bending stiffness of idealized beam, one end of which is firmly fixed $(= 3EI/L^3)$
 L = length of cantilever
 P = external bending load

$F(\lambda,\kappa)$ and $G(\lambda,\kappa)$ = nondimensional coefficients determined by connecting conditions such as area of joint surface, joint surface condition, diameter of connecting bolt. For a bolted joint, where connecting bolt nearest to loading point elongates, these coefficients are positive.

2. In case of tangential loading

$$K_{bh} = K_0/[1 + H(\lambda,\kappa)K_0] \tag{7-16}$$

[12]Pič, Iosilovich [S4], Schofield [S3], and Plock [3] conducted also the theoretical analysis of the static joint stiffness, and of these that of Plock is worth referring, because his assumptions are very close to the actual joint condition.
Pič, J., "Die Starrheit der Schraubenverbindung," *Konstruktion*, 1967, 19(1): 7–12.

where $K_0 = 3EI/L^3$ and $H(\lambda,\kappa)$ is the nondimensional coefficient determined by the connecting conditions and the function of the characteristic β_T.

$$H(\lambda,\kappa) = [-L/(2\beta_T{}^2 EI)]\{1/[1 - \cosh(2\beta_T l)]\}\,\{L\beta_T[\sin(2\beta_T l) + 2\sinh(2\beta_T l)]$$
$$- 2\cos(2\beta_T l)] - (1/\beta_T L)[\sinh(2\beta_T l) - \sin(2\beta_T l)]\}$$

where l = length of joint surface
$\beta_T = \sqrt[4]{k_{eq}/(4EI)}$ and $k_{eq} = bk_h$
k_h = tangential joint stiffness per unit area

These expressions are available not only for the bolted beam with line arrangement of the connecting bolt, but also for the multiple-bolted joint; however, there are certain differences in the detailed formulas of the nondimensional coefficients.

In these expressions, furthermore, we can consider the influences of the bending deflection, shear deflection, and additional deflection due to the bolt-hole and bolt head on the overall deflection of the bolted beam.

Figure 7-44 is a reproduction of available limits of the analytical expressions showing with the load-stiffness diagram proposed by Ito [39],

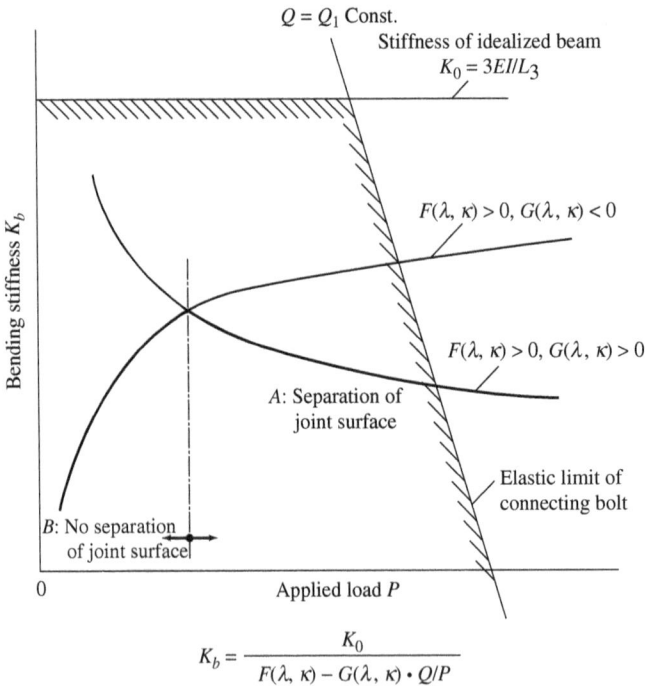

$$K_b = \frac{K_0}{F(\lambda,\kappa) - G(\lambda,\kappa) \cdot Q/P}$$

Figure 7-44 Available region of theoretical expression proposed by Ito et al.

in consideration of the elastic limit of the connecting bolt and the stiffness of the idealized bolted beam, i.e., monolithic beam. In short, the boundary conditions can be written as

$$\{G(\lambda,\kappa)/[1 - F(\lambda,\kappa)]\}Q \leqq P \leqq (1/4)[\varphi/\psi_i]\sigma_e\pi d^2 - [\xi_i/\psi_i]Q \quad (7\text{-}17)$$

where ψ_i/φ and ξ_i/φ = nondimensional functions, which indicate effects of length ratio and spring constant ratio of bolted beam on reaction forces

σ_e = elastic limit of bolt material

d = stem diameter of connecting bolt

suffix i = order number of supporting point. In case of upward loading, connecting bolt No. 3 in mathematical model is at issue, and thus $i = 3$

The regions A and B correspond with the bolted joints showing and not showing the joint separation under loading, respectively. As a matter of course, in the former, the stiffness of the bolted joint is under the control of that of flat joint, showing considerable nonlinearity to both the applied load and the tightening force. In the latter region, the stiffness of the connecting bolt itself governs the stiffness of the bolted joint.

Figure 7-45 shows a comparison between the theoretical and experimental values, and as can be seen, both values are qualitatively in good agreement, and Fig. 7-46 shows the effect of the resistance moment caused by the bolt head under upward bending loading.[13] In addition, the stiffness of the bolted beam decreases with the increase of the stiffness of the clamped beam itself, i.e., that of joint surroundings, resulting in the growing importance of the tightening force as the stiffness of the beam becomes larger. Obviously, the expression proposed by Ito and coworkers has fairly good applicability.[14]

---

[13]Shimizu et al. conducted experimental research into the effect of the bolt head on the joint stiffness in detail.

Shimizu, S., M. Ito, and R. Fukuda, "Influences of the Hexagon Headed Bolt Head on the Static Behavior of the Bolted Joint in Connecting," *J. of JSPE*, 1983, 49(2): 184–189.

[14]Although the spring constant of the connecting bolt should be calculated using that of Plock, as already stated, the spring constant was, in general, calculated by assuming that the bolt elongates at the thickness of the clamped beam. In contrast, the spring constant of the base was calculated by assuming the local deformation of a semifinite elastic body under concentrating force. These assumptions yield to a certain deterioration of the calculating accuracy. In this context, Kobayashi and Matsubayashi reported a noteworthy result: The meshing portion of the bolt thread with the threaded hole in the base has considerable effect on the stiffness of the bolted beam. The more underneath the meshing portion, the larger is the stiffness of the bolted beam.

Kobayashi, T., and T. Matsubayashi, "Considerations on the Improvement of the Stiffness of Bolted Joints in Machine Tools," *Trans. of JSME (C)*, 1986, 52(475): 1092–1096.

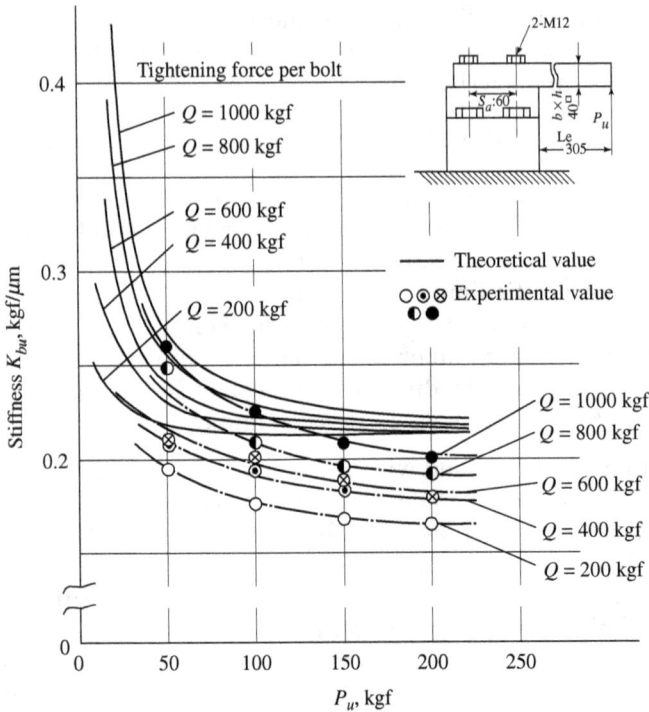

Figure 7-45 Comparison of theoretical and experimental values.

Interface pressure distribution.[15] From the academic research point of view, a dire necessity is to measure the topographical information, i.e., two-dimensional interface pressure distribution. In fact, an interface pressure distribution at certain cross section of a bolt-flange assembly, i.e., one-dimensional pressure distribution, leads us often to misunderstanding; however, such simplified measurement can, on the contrary, provide us with the valuable information when we conduct the engineering design.

Apart from works of Ito and coworkers, there were, in fact, no reports so far to ascertain experimentally even the contact pattern, i.e., qualitative interface pressure distribution, in the single bolt-flange assembly, when maintaining the joint surface as it is. In addition to those already shown in Figs. 7-14, and 7-17, therefore, some other interesting behavior

---

[15]Details of the ultrasonic waves method will be stated in App. 1 and some measured results have already been shown in the preceding sections, i.e., those related to the pressure cone and bolt pocket.

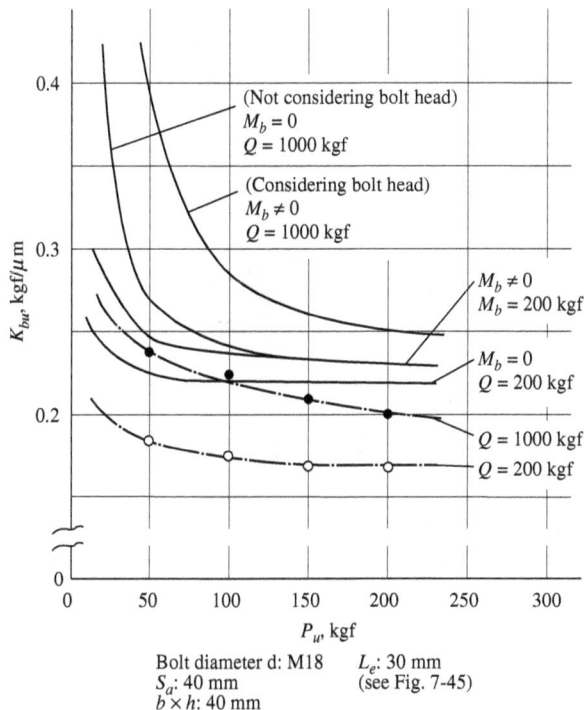

Figure 7-46   Effect of bolt head to increase bending stiffness.

is shown in Figs. 7-47 and 7-48. Summarizing all these measured results, the following marked observations can be pointed out.[16]

1. The interface pressure distribution depends largely upon both the flange material and the finishing quality of the joint surface, and also to some extent upon the flange thickness. Of these, we can anticipate the larger influence of the machining method of the joint surface within the area closer to the bolt-hole to the interface pressure.

2. The interface pressure distribution is in closer relation to the relationship between the joint stiffness and the stiffness of the joint surroundings. More specifically, the interface pressure distribution becomes more gently sloped as the flange material and joint surface become softer and rougher, respectively, because of lower joint stiffness. In due course, the interface pressure distribution approaches a more gently sloped curve with the increase of the flange thickness.

---

[16]The shape and size of the bolt head may affect the interface pressure distribution, and thus Shimizu conducted an interesting research into this subject.

Shimizu, S., "Relationships between the Pressure Distribution of the Bolt Head Bearing Surface and of the Joint Interface," *J. of JSPE*, 1983, 49(12): 1645–1651.

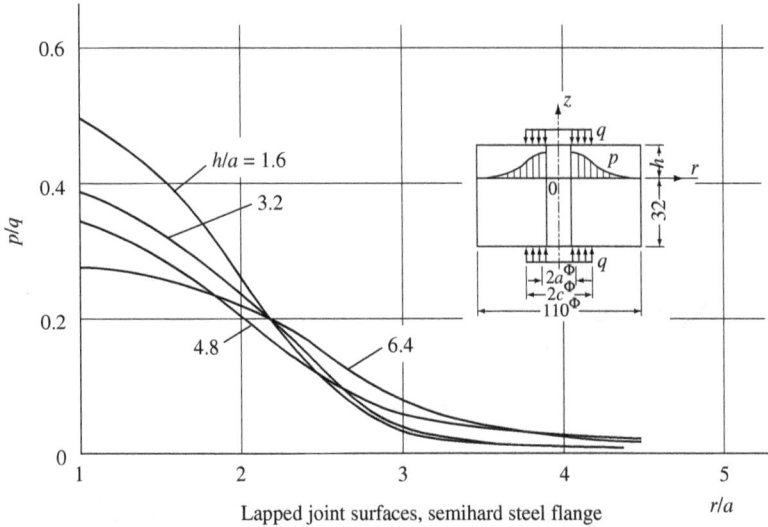

Lapped joint surfaces, semihard steel flange

Figure 7-47   Effects of flange thickness on interface pressure distribution.

7.1.5   Representative researches and their noteworthy achievements—dynamic behavior

There have been very few activities on the dynamic behavior of the bolted joint compared with those for static behavior. This trend may be attributed to the uncertainty of the damping capacity in the bolted joint together with the difficulty in the measurement of the damping capacity.[17] In due course, at issue is the estimation of the damping capacity, and thus a preliminary trial for the laminated beam has already been introduced in the preceding section. In retrospect, damping at the mating surface was, as already mentioned, investigated vigorously to unveil the macroscopic slip damping at the "Christmas tree (fir tree) joint" in the turbine;[18] however, such earlier research activities

---

[17]The measurement of the damping capacity is carried out using, e.g., the following excitation and displacement detection. In the utmost preferable case, the noncontact excitation and displacement detection are recommended.

- Impact excitation and detector of capacitance type.
- Electrohydraulic exciter and detector of eddy-current type.
- Electromagnetic exciter and detector of piezoelectric type.

To measure the frequency response, the exciter can apply the sinusoidal exciting force, which is superimposed onto the static preload.

[18]Goodman, L. E., and J. H. Klumpp, "Analysis of Slip Damping with Reference to Turbine-Blade Vibration," *Trans. of ASME*, September 1956, p. 241.

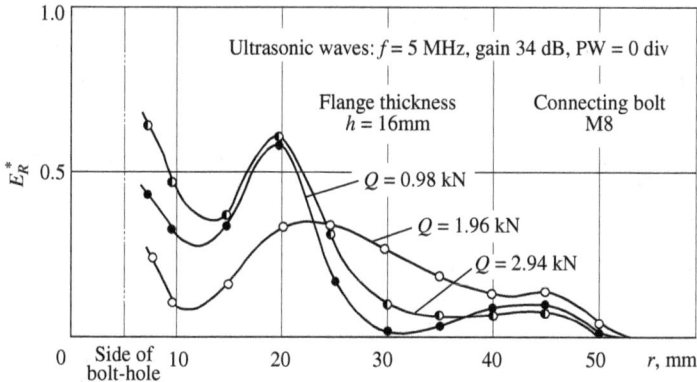

Figure 7-48 Measured pressure distribution for bolt-flange assembly with lapped wavy joint surface.

did not consider the characteristic feature of the bolted joint in the machine tool. In the bolted joint, the crucial problems are, as already mentioned, how to deal with the microslip of less than a few micrometers and with the displacement dependence of tangential force ratio in the state of microslip.

Aiming at finally the estimation of the damping capacity, Groth [1], Weck and Petuelli [40], and Ito and coworkers[2, 41–43] conducted researches into the dynamic behavior of the bolted joint. These earlier works have clarified such general characteristics of the dynamic behavior of the bolted joint as follows.

1. When a machine tool structure shows larger damping, its static stiffness deteriorates considerably.

2. The damping capacity of a machine tool as a whole is from 0.05 to 0.2 in terms of logarithmic damping decrement. These values are 4 to 10 times larger than the internal material damping of the steel or cast iron.

3. The damping capacity of the bolted joint is more largely dependent upon the tightening force, as shown in Fig. 7-6 which also shows, in a certain case, the peak at a certain tightening force, as already demonstrated in Fig. 7-7. In general, the larger the tightening force, the lower the damping capacity.

4. The damping capacity and natural frequency are maximum at a certain value of h/H, where h and H are the thicknesses of the flange and base, respectively.

5. The damping capacity is dependent on the vibration amplitude, and this amplitude dependence is subjected to the machining method

and roughness of the joint surface, and to the interfacial layer. In general, damping increases with increasing vibration amplitude, the same as the material damping in the cast iron (see Table 9-2 and related materials).[19]

6. The eigenfrequency (natural frequency) in the first vibration mode is not so far from that of an equivalent solid. In general, the eigenfrequency increases with the tightening force.

7. As can be readily seen from the damping mechanism, the vibration mode has considerable effect on damping. In fact, there is no damping when the joint is in node under the vibration.

In the following, some of the behavior mentioned above is detailed.

Static preload component of exciting force. When the tightening force is lower, damping of the bolted joint reduces with increasing static preload, whereas the static preload has no effect on the damping capacity when the tightening force is higher. It is furthermore said that the exciting force has, in general, no effect on the damping capacity,

Effects of machining method and surface roughness of joint. Figure 7-49 shows a relationship between the logarithmic damping decrement and the tightening force when the machining method is varied. Admitting the difficulty in suggesting the general rule, it is at least said that the damping capacity of the bolted joint reduces with improving the quality of the joint surface. In addition, the machined lay orientation has a large effect on the damping capacity.

Effects of interface layer. On the basis of the knowledge obtained from the earlier works, there are three cases in connection with the behavior of damping, when the interfacial layer is applied to the joint.

1. The damping capacity is nearly equal to that of the interfacial layer.

2. In addition to damping of the interfacial layer, the damping derived from the microslip at the dry joint contributes considerably to the damping capacity.

3. The damping capacity does not change by the interfacial layer at all.

---

[19]In the case of the solid interfacial layer, there are no apparent effect of the vibration amplitude, whereas in the case of the fluid interfacial layer, the vibration amplitude shows certain effect on damping. In the latter case, the oil viscosity is one of the leading factors for controlling the effect of vibration amplitude.

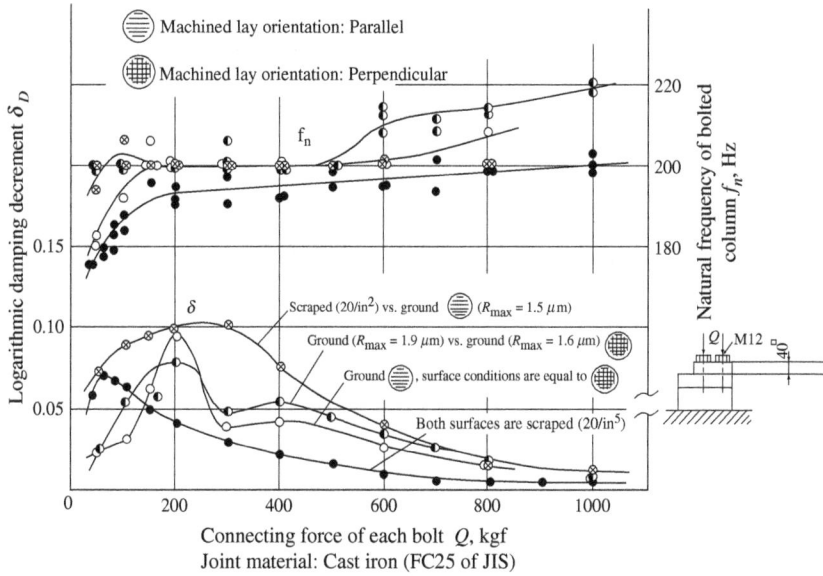

Figure 7-49 Effects of machining methods on damping capacity and natural frequency.

More importantly, it is a myth that the damping capacity of the bolted joint always increases by applying the oil or plastics to the joint surface. This is a very interesting observation, and Fig. 7-50 shows such results [43], and as clearly shown, the machine oiled joint shows lower damping than the dry joint. These imply the importance of the viscosity and penetrating ability of the fluid interface layer. In fact, the lower the viscosity, the larger the damping capacity.

7.1.6 Representative researches and their noteworthy achievements—thermal behavior

From the academic point of view, the thermal contact resistance has been already clarified to a large extent; however, its application to practical problems is far from completion. For example, Fontenot conducted a series of basic researches into the loosening phenomena of the bolted and riveted joints, intending to apply the due knowledge to the practical problems in the space vehicle [44]. Such a loosening phenomenon is caused by the temperature difference between the day and the night. In fact, there remains something to be seen in the application procedure, and the same story may be admitted in the case of the machine tool joint.

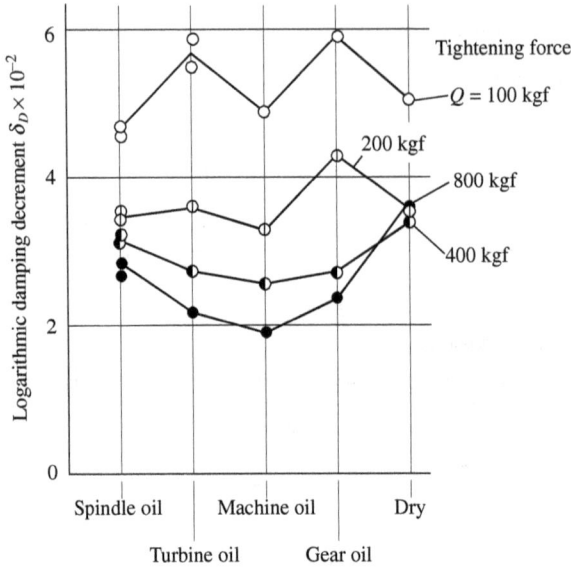

(Material: SS41B)

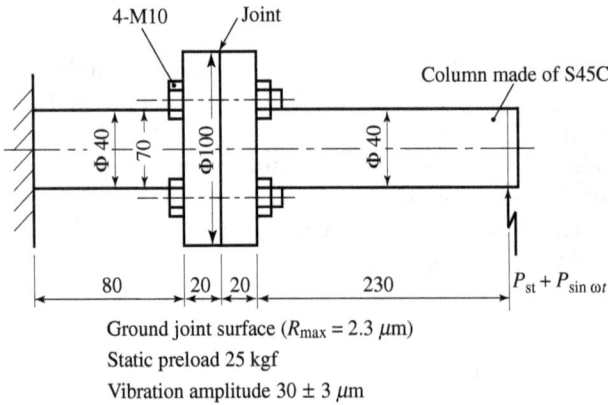

Ground joint surface ($R_{max} = 2.3\ \mu m$)
Static preload 25 kgf
Vibration amplitude $30 \pm 3\ \mu m$

Figure 7-50   Effects of interface layers on damping capacity.

As already stated in Chap. 6, the thermal behavior of the single flat joint has been unveiled to a large extent: however, there have been very few researches into the thermal behavior of the bolted joint. As a result, even in the very late 1990s, Fukuoka and Xu [45] conducted a series of researches. A root cause of difficulties lies in the shortage of knowledge about the unstable change of the interface pressure distribution, which is core in the concept of the closed-loop effect as already mentioned in

Chap. 6. For the ease of understanding, the closed-loop effect in the bolt-flange assembly will be detailed in the following.

1. By the tightening force of the clamping bolt, an interface pressure distribution can be given first, and then it changes by the external loading.

2. The thermal contact resistance is given in accordance with the interface pressure distribution, and it changes by the thermal loading, resulting in a *temperature distribution* across the whole bolt-flange assembly.

3. In accordance with the dynamic boundary conditions and temperature distribution, the bolt-flange assembly shows certain deformation duly including the thermal expansion of the connecting bolt.

4. The deformation of the bolt-flange assembly induces a new interface pressure distribution.

A primary concern is thus how long the closed-loop effect can be continued or how many times it can be repeated. Itoh et al. conducted a research into this subject using the ultrasonic waves method and Cu-Co thermocouples to measure simultaneously the contact pattern and temperature distribution [46]. Figure 7-51 shows the typical changes of the contact pattern and temperature difference at the joint when the steady-state thermal load is applied. In short, Itoh et al. suggested that the closed-loop effect appears not to repeat too often. In addition, they reported some interesting observations as follows.

Temperature distribution along axial direction

1. The thermal contact resistance increases with the distance in the r direction, e.g., smaller and larger around the center and skirt of the bolt-flange assembly, respectively, and also decreases with the flange thickness.

2. The distribution of the thermal contact resistance is in good correspondence with the interface pressure distribution.

3. In certain joints, the gradients of the temperature in the upper and lower flanges differ from, especially in the case of thinner flange. This phenomenon reveals the presence of a radial heat flow, which is directed from the circumference to the center of the flange, resulting in the reduction of the heat flux in the axial direction.

Interface pressure distribution

1. In the case of thinner flange, there are no changes of the interface pressure distribution. As a result, it is recommended that the ratio

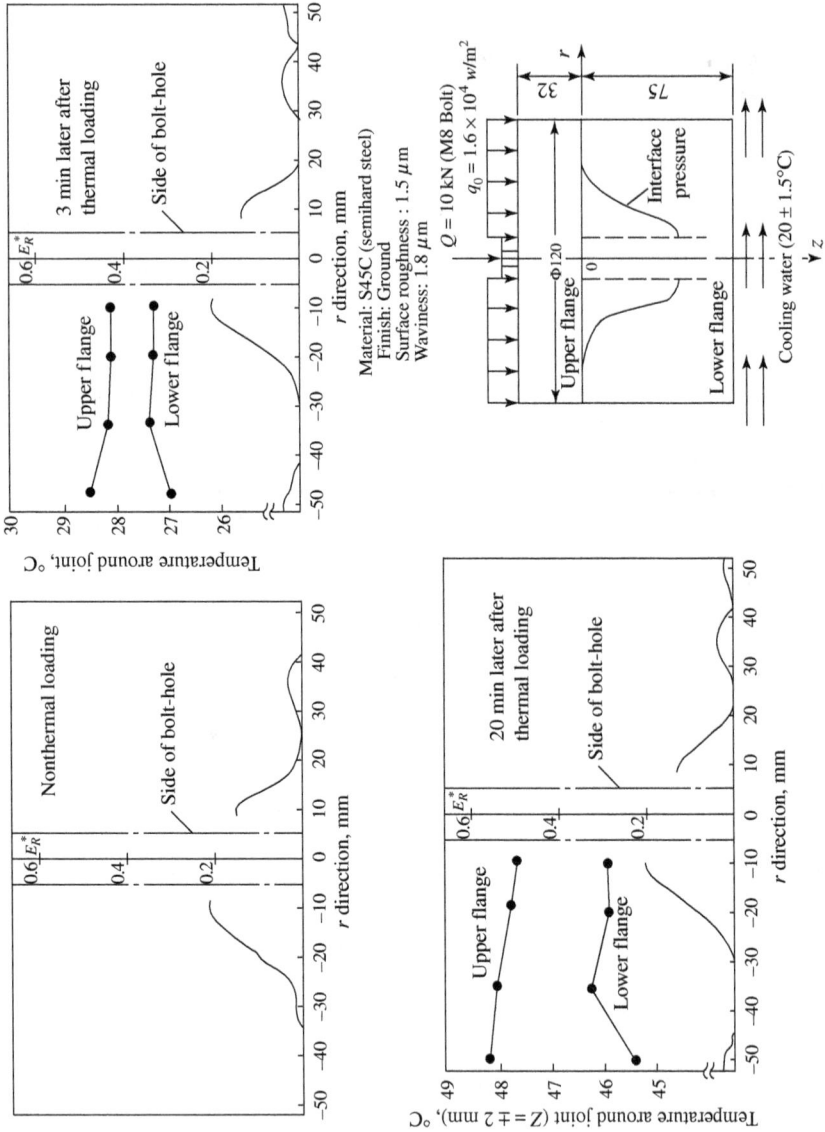

Figure 7-51 Changes of interface pressure distribution when applying steady-state thermal load.

h/d (h is flange thickness, d is bolt diameter) be lower than 2 to have the stable bolted joint for thermal loading.

2. In the case of thicker flange, the additional interface pressure appears at the outer joint surface by thermal loading. In addition, the thicker flange shows a slight decrease of the interface pressure around the center[20] and considerable elongation of the connecting bolt.

7.2 Foundation

The foundation is one of the most important joints in machine tools, especially in large-size machine tools; and the static, dynamic, and thermal behavior of a machine tool as a whole is governed by the behavior of the foundation to a various and large extent. This is because the foundation determines the boundary conditions of a machine tool and, as can be readily seen, the thermal deformation is changed considerably by the boundary condition. Figure 7-52 shows the effects of the installation method, i.e., boundary condition, for the spindlestock on the temperature distribution [47], and the discontinuity in the temperature distribution is obvious when the heat insulating effect is larger than that of usual installation method. As another example, it has been widely known that the deflection of a long and relatively flexible bed subjected to the traveling load and cutting force is derived from the deflection of the leveling block, i.e., one of the boundary conditions.

Despite its great importance, the foundation has not been investigated vigorously because of its structural complexity. In fact, the foundation of leveling block type consists of several joints, as already shown in Fig. 5-8, i.e., those between the leveling block and sheet plate, leveling block and machine base, and sheet plate and grout. The most distinguishing feature of the foundation from that of other machine tool joints is that there are the metal-to-metal and metal-to-grout or metal-to-concrete contacts together with the leveling or anchor bolt. In addition, the foundation can be typified by several variants, e.g., common foundation across whole workshop, independent concrete block (foundation), common or independent steel plate on workshop floor. In consequence, the characteristic features are very different from those of other joints, although the foundation shows similar behavior to those of the bolted joint to some extent. From the viewpoint of machine tool joints, the joint between the concrete base

---

[20]In the case of shocklike thermal loading, the bolt-flange assembly with thicker flange loses considerably the effect of the interface pressure over nearly all joint surfaces, and gradually recovers the interface pressure with the lapse of time, finally showing a pressure distribution similar to that of the initial stage.

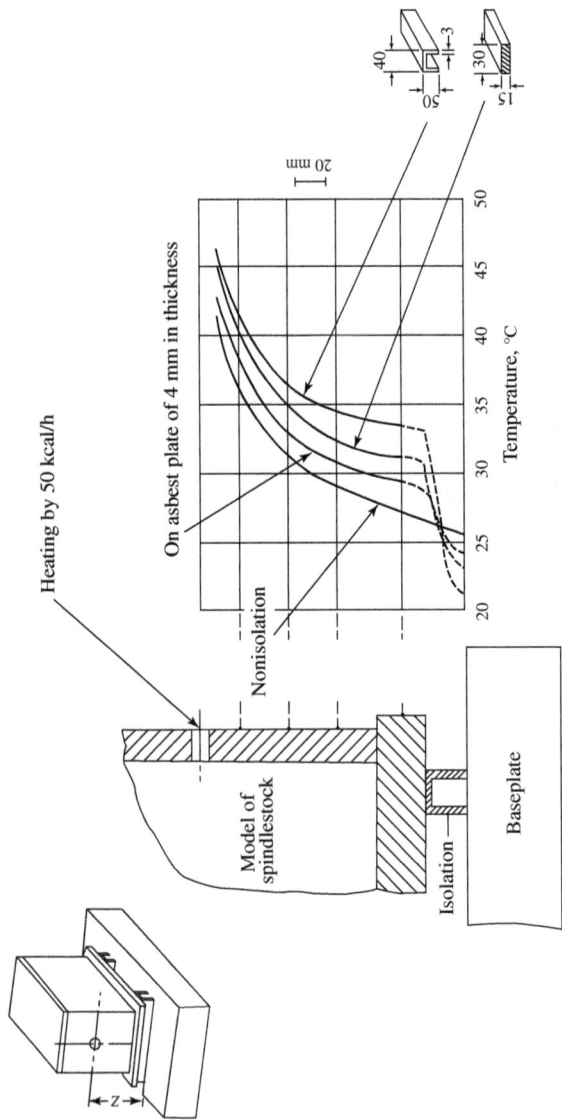

Note: After 240 min of heating

Figure 7-52 Discontinuity in temperature distribution caused by isolation methods (by Opitz and Schunck).

and the soil is also one of the objectives; however, the major character-istics of such a joint cannot be clarified without using the knowledge of civil engineering.

There are the two major types of the foundation: one is of direct type and the other is of leveling block type. In both types, the con-crete base plays an important role as the joint surroundings, and at present it has not yet been clarified how much the concrete base itself contributes to the stiffness of a foundation. To understand the foundation, a primary concern is knowledge about the natures of the soil and concrete block. In this regard, Eastwood [48], Kaminskaya [49], and others have conducted the due investigation, especially put-ting main stress on the deformation calculation, i.e., determination of the depth, width, and length of the concrete base. In short, to determine the suitable depth of the concrete base, the following fac-tors should be considered.

1. The stiffness of the concrete base is largely dependent on the soil prop-erties, for instance, the waterproof, creeping properties and sensitiv-ity for vibration. As a result, the concrete base has a time-dependence characteristic and needs a long time up to its stabilization together with its own time dependence in the base deflection. Figure 7-53 shows the time- dependence of the bed deflection for a planer with table of 4 m length [49, 50]. The bed deflection increases gradually

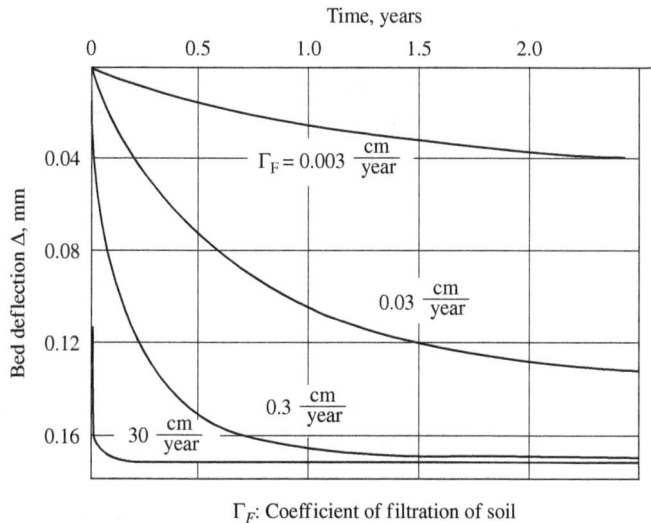

Figure 7-53   Time dependence of machine bed deflection (by Kaminskaya).

TABLE 7-3   Values of E_0, υ, and Γ_F for Different Types of Soils (by Kaminskaya)

Soil	E_0, kgf/cm^2	ν	Γ_F cm/yr
Sand	200–2000 (250–500)	0.25–0.30	$3 \times 10^7 - 3 \times 10^3$
Sandy loam	100–500 (150–350)	0.28–0.35	$3 \times 10^5 - 3 \times 10$
Loamy	50–1000 (100–300)	0.33–0.37	$3 \times 10^3 - 3 \times 10^{-1}$
Clay	25–5000 (50–250)	0.38–0.45	$3 \times 10 - 3 \times 10^{-3}$

or rapidly depending on the coefficient of filtration Γ_F, where the filtration means that the water included within the soil is squeezed out. The large and small values of Γ_F correspond to the soils consisting of the coarse-grained sand and clay, respectively (see Table 7-3).

2. The time-dependent damping of the concrete base is approximately evaluated by the hydrodynamic stress theory, because the soil includes considerable water.

3. The ultimate load of soil. In general, the required value is more than 5 tonf/m^2 [51].

4. The settlement of concrete base. In the case of clays, (a) elastic compression, (b) plastic deformation, and (c) consolidation are issues.

5. Movement of the ground caused by the moisture-content change.

Reportedly, the kind, number, and supporting points of the machine tool have furthermore considerable effect on the torsional deformation of the bed. In this context, Poláček reported the importance of the supporting point through a model testing for the milling machine of bed type [52], while moving the heavy work on the table from one to another critical ends of the table stroke. Figure 7-54 shows the effects of the supporting point and machine bed structure on the relative deflection between the main spindle and the work, where δ_X and δ_Y are the relative deflections in longitudinal and cross directions of the table, respectively. As can be readily seen, the relative deflection depends largely on the allocation of the supporting point of the machine bed and to some extent on the bed structure. An interesting behavior can be observed especially in the case of three-point supporting.

In addition, the bed with closed structural configuration as shown in Fig. 7-54(h) is in relatively small deflection compared with that of open structural configuration. In this case, the machine bed can be regarded

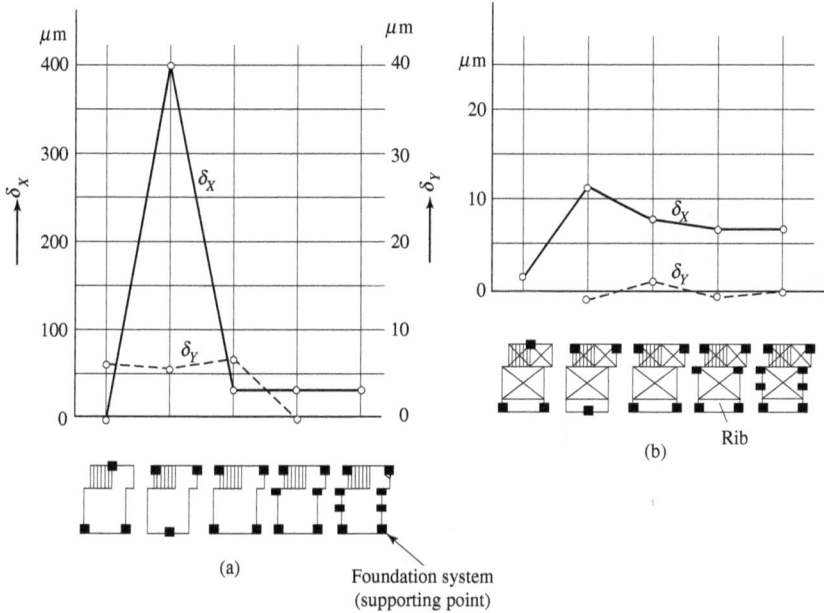

(a)

Foundation system
(supporting point)

(b)

Rib

Note: In all cases, the relative displacement in vertical direction Z is negligible.

Figure 7-54 Effects of foundation system and bed construction on relative displacements between main spindle and workpiece: (a) Open-type bed and (b) closed-type bed (by Poláček).

as one of the joint surroundings, and is reinforced by the stiffening ribs and bottom plate. The stiffer the joint surroundings, the larger the joint stiffness.

Obviously, the foundation has another important function, i.e., to maintain the accurate alignment of the machine base or bed, which is mandatory to obtain the allowable machining accuracy. In this regard, for instance, the idea of the leveling block of servo type was proposed by Hailer [53]. In this leveling block consisting of a hydraulic cylinder, the alignment can be automatically compensated by the servomechanism, and it is always constant, even when the load acted on the base or bed changes to some extent.

In fact, there have not been active researches and engineering developments with decreasing use of the large-size machine tool; however, some notable contrivances have been carried out, and these can verify the importance of the foundation. In fact, Fig. 7-55 shows some variants of the foundation, and Fig. 7-56 shows the compact connector [54, 55], which can be used instead of the leveling block. Within the compact connector, that of Gemex GmbH & Co. KG was patented in 1974 (No. 2304132).

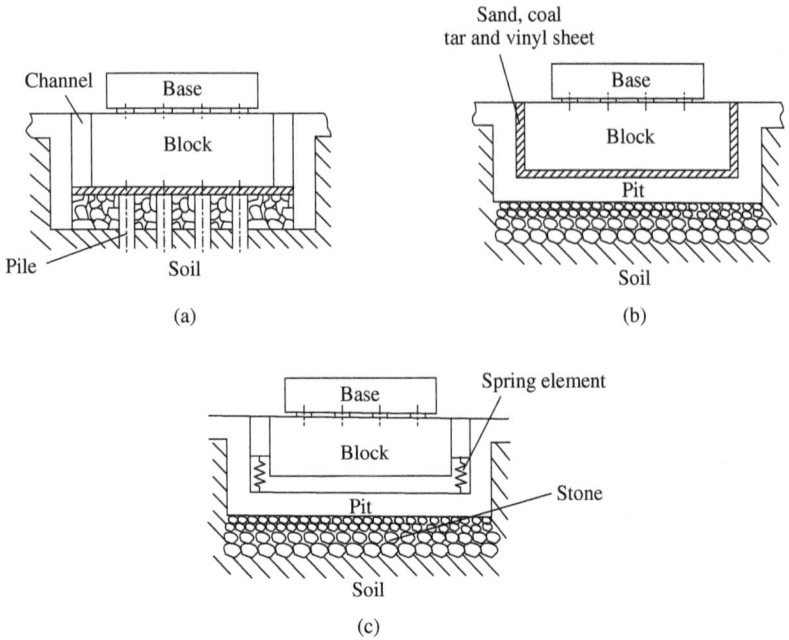

Figure 7-55   Variants of foundation system: (a) Foundation with antivibration channel, (b) two-layered foundation of stationary type, and (c) two-layered foundation of suspension type.

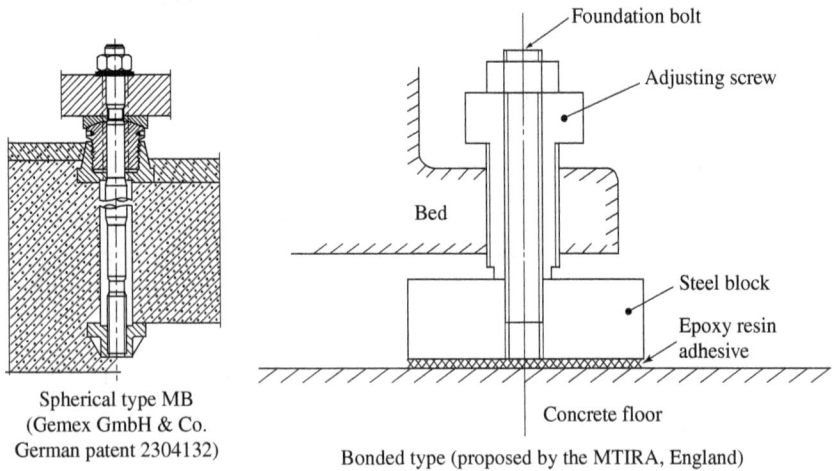

Figure 7-56   Some variants of leveling block.

7.2.1 Engineering calculation for foundation

Although there are a considerable number of variants, as shown above, within a foundation context, the primary concerns of the engineering calculation are how to determine the depth of the concrete base, including the supporting force of the pile in certain cases, and to calculate the stiffness of the leveling block. In general, a mathematical model for the base of large-size machine tool is the elastic beam or plate on the elastic foundation.

Depth of concrete base. On the basis of decaying settling, Kaminskaya [49] investigated how to determine the depth of the concrete base and necessary intervals for conducting the realignment of a machine. In the sphere of civil engineering, the foundation settling means the stabilization of vertical displacement of the concrete base, which is derived from the load transmitted from the concrete base to the soil. The foundation settling is thus in closer relation to the compaction of the soil and the duration reaching to its stabilized condition, i.e., full settlement, after passing a long time from the installation of the machine. The actual factors for full foundation settling are (1) applied load and its type, (2) dimensions of the concrete base and its type, and (3) compressibility factor of the soil.

For the rate of the settlement, we must furthermore consider (4) the permeability factor and (5) the creep factor of the soil. As a result, the time-dependence in the settlement is very important, because the non-steady change during the settlement induces unfavorable deformation of the base or bed.

In the determination of the depth of concrete base, an available mathematical model is that of a beam on an elastic foundation together with assumption of the direct proportionality between the soil displacement and the reaction. In addition, the time-dependence of the modulus of soil should be considered. The model is as same as that for the flat joint with local deformation (see Chap. 6), and Kaminskaya [49] proposed an expression to determine the modulus of the soil K_{so} as follows.

$$K_{so} = (\pi/2 \ln 4\xi_\alpha)[E_0/b(1 - v^2)][1/(1 - e^{-N})]$$
$$N = \pi^2 C_v t/4H_p$$
$$C_v \approx K_1 E_0/W_V \tag{7-18}$$

where E_0 = modulus of total deformation of soil
  v = coefficient of transverse deformation of soil (Poisson's ratio of soil)
  ξ_α = L/b (ratio of length L to width b of concrete base)
  H_p = depth of soil layer
  T = time
  Γ_F = filtration factor
  W_V = 0.001 kgf/cm^3 (volumetric weight of water)

Table 7-3 summarizes the values of E_0, ν, and Γ_F for different types of soil, where the figures in parentheses are those of closely related values. For the actual engineering calculation, however, it is recommended that a test with full-size be carried out to determine these values.

In addition, Kaminskaya pointed out that the bed or base deformation of the machine tool should be calculated for the load far exceeding the uniform distribution load, which is caused by the dead weight of the structural body component. In actual cases, the depth of the concrete base is (a) 0.07 to 0.15 L for the planer and planomiller and (b) 0.08 to 0.1 L for the lathe. According to a report of Naxous Union Co., the required stiffness of the concrete base is at least 5000 kgf/μm for the weight of machine, carrying work, and concrete base itself.

Although the concrete has undesirable properties, such as high sensitivity to temperature and humidity changes, which cause a considerable movement and setting shrinkage, concrete is a very popular material for the foundation. In the case of heavy machine tool, its concrete base is as much as 5 m deep, and in consequence the temperature distributions in the machine base and concrete base differ greatly from each other, when the temperature fluctuates. This causes the large thermal deformation of the base guideways, resulting in the deterioration of the guiding accuracy. To reduce such an influence, Innocenti Co., one of the leading machine tool manufacturers, used a foundation base of honeycomb type, through which the air blown by a fan was flowed.

Supporting force of pile. In the case of very poor ground, the concrete block should be laid on the pile; however, the pile does not reach to a base rock in nearly all cases. The concrete base must be thus supported by the frictional force between the outer surface of the pile and the soil. The supporting force can be given by the following expression [51].

$$P = f * \pi l [(d_1 + d_2)/2] \qquad (7\text{-}19)$$

where P = supporting force
  f^* = supporting force per unit area determined by friction between soil and pile
  d_1 and d_2 = diameters of pile at both ends
  l = penetrating length of pile

Table 7-4 shows data of the supporting force determined by the friction. Importantly, the allowable magnitude for the long-term load-carrying capacity of the soil must be specified in designing the supporting force of the pile. For example, such capacities of the hard rock bed, tight

TABLE 7-4   Supporting Force by Friction

Soil	Pile	Supporting force by friction ton/m^2
Clay	Concrete pile with rough surface	2.5
	Wooden pile with rough surface	2.0
	Iron plate with rivets	1.5
Sand and sand with pebbles	Concrete pile with rough surface	3.5
	Wooden pile with rough surface	3.0
	Iron plate with rivets	2.0

gravel, and sandy clay are 400, 60, and 30 tonf/m^2, respectively, and in general, the capacity of 5 tonf/m^2 can be recommended in consideration of the safety rate in the design. In addition, there have not been any reports on the modulus of the soil with piles.

7.2.2   Stiffness of leveling block

The leveling block is the utmost representative within the foundation system, and Faingauz [56] conducted very interesting research using model testing. Figure 7-57 shows the test rig, in which the wedge shoe

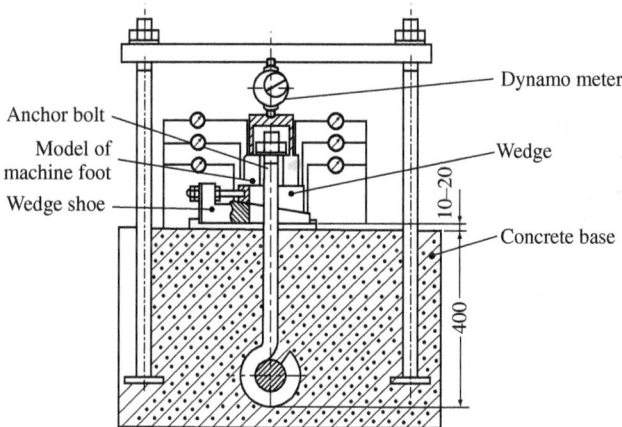

Figure 7-57   Test rig for model of leveling block (by Faingauz).

For a support clamped with a
30 mm diameter foundation bolt
with a force of 6 tons

Ditto to 2
(with a force of 3 tons)

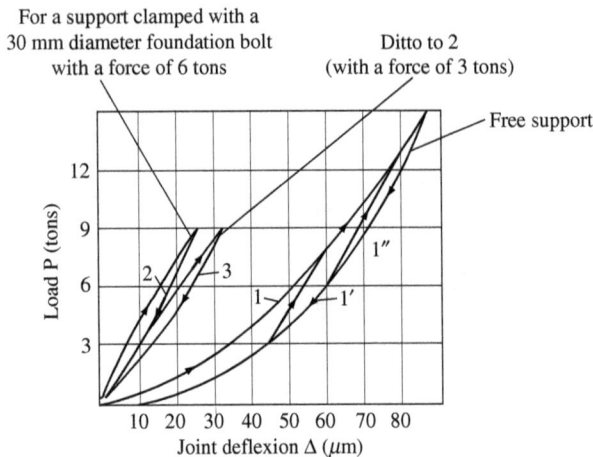

Note: Leveling block is grouted-in with a fluid mix.

Figure 7-58   Load-joint deflection curves of leveling block (by
Faingauz).

was held in place by the grout applied to the concrete base. Faingauz
investigated the supporting stiffness of the leveling block, i.e., effects of
the wedge shoe, grout, curing time of grout, anchor bolt, and its tightening force on the joint stiffness.

As can be readily seen, the tightening force of the anchor bolt
has a considerable effect on the total stiffness of the leveling block.
Figure 7-58 shows the external load-joint deflection curves for several
tightening conditions, where the curve is similar to that observed in
the flat joint under lower normal loading. As compared with the free
support, i.e., that without tightening force, the stiffness of the clamped
support is larger and increases with the tightening force, approaching the stabilized condition, when the tightening force is more than
6 tons, i.e., the interface pressure between the shoe body and the
grout is 25 kgf/cm^2. Table 7-5 shows some average values of the stiffness measured when the tightening and grout conditions are varied,
where the scatter in measurement is ±(20–25)% at the tightening
force of 3 tons, because of noncentral loading in part and irregular
deformation of the support. In addition, it is noticeable that the maximum support stiffness can be achieved 25 to 30 days after grouting
in the wedge shoe.

In the leveling block system, the supporting stiffness k_L can be
written as

$$1/k_L = (1/k_{m1}) + (1/k_{m2}) + (1/k_g)$$

TABLE 7-5   Stiffness of Leveling Block (by Faingauz)

Bolt diameter, mm	Type of grout mix	Time after grouting in shoe, days	Clamping force of bolt, tonf	Stiffness of leveling block with load of (kgf/μm)(1/cm^2)	
				3 tonf	More than 9 tonf
—	F	7	—	0.6	2.3
—	F	30	—	0.9	2.8
	S	30		0.4	1.9
25	F	7	3	2.4	
	F	30		2.7	
25	F	7	6	2.6	
	F	30		3.1	
30	F	30		3.7	
	S	30		2.7	
25	F	7	9	3.1	
25	F	30		3.2	
30	F	30		3.7	
30	F	30	12	3.7	
	S			3.25	

F: fluid, S: stiff.

where k_{m1} = stiffness between foot and shoe wedge, which can be estimated using knowledge about flat joint

k_{m2} = stiffness between wedge and shoe body

k_g = stiffness between shoe body and concrete base with grout

In consequence, the primary concerns are k_{m2} and k_g to clarify the characteristic features of the leveling block, whereas the stiffness of the anchor bolt must be taken into consideration when the machine base is loaded upward.

Figure 7-59 shows the change of the stiffness k_{m2} and k_g with varying interface pressure, where the joint areas for k_{m2} and k_g are 115 and 270 cm^2, respectively. The stiffness k_{m2} increases with the interface pressure, similar to that of flat joint. In contrast, the stiffness k_g varies considerably depending on the grout curing time and fluidity of concrete mixture, i.e., grout condition. The necessity is thus to ensure the reliable cohesion of the metal to the concrete, and duly the bottom surface of the shoe body must be cleaned of rust and then wetted with the water. More specifically, the stiffness k_g is in satisfactory condition when grouting in with a fluid mixture, which can fill all the uneven parts across the whole joint surface. As a result, we can expect the formation of the dense monolithic layer. This action of the fluid mixture can be interpreted as

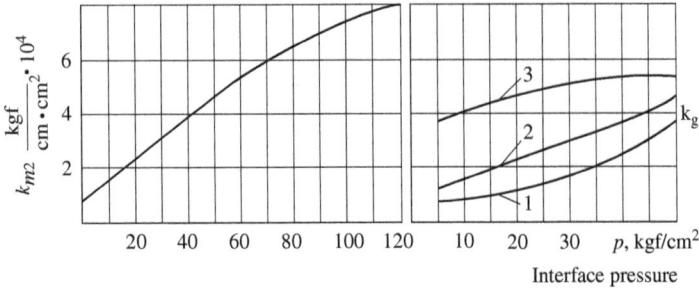

Notes: 1. Stiffness of joint between the shoe and the base was
measured, when grouted in with a stiff mix (curve 1),
seven days after grouting in with a fluid mix (curve 2)
and after 30 days (curve 3).

2. Composition of the concrete base is one volume of
Portland cement to three of sand.

Figure 7-59   Values of k_{m2} and k_g (by Faingauz).

that of adhesive in the bonded joint (see Chap. 9), whereas the
stiff mixture produces a porous layer after curing, resulting in insufficient adhesion over the joint surface.

Importantly, Kaminskaya summarized a generalized formula for calculating the stiffness of the leveling block ranging from the leveling
blocks with and without tightening bolt to the leveling block with tightening bolt of split holding-down type [57]. Apart from the contribution
of the tightening bolt, the stiffness k_L of the leveling block can be written as

$$k_L = 1/(\Sigma C_{oi} + C_{of})$$

$$C_{oi} = C_i/A_i \tag{7-20}$$

where C_{oi} = compliance of ith butt joint in leveling block, e.g., those
between machine foot and wedge, and wedge and wedge
shoe

A_i = area of contact in ith butt joint, cm^2

$C_{of} = C_{oc}/A_{oc}$, compliance between wedge shoe and concrete
base. For not grouting, $C_{oc} = (10-30) \times 10^{-4}\,\text{cm}^3/\text{kgf}$, and
for grouting C_{of} is mainly determined by deformation of
concrete foundation

A_{oc} = area of supporting surface of wedge shoe

C_i = coefficient of contact compliance of ith butt joint,
cm^3/kgf, given by Fig. 7-60. As can be readily seen,
C_i indicates a stiffness distribution diagram within
leveling block

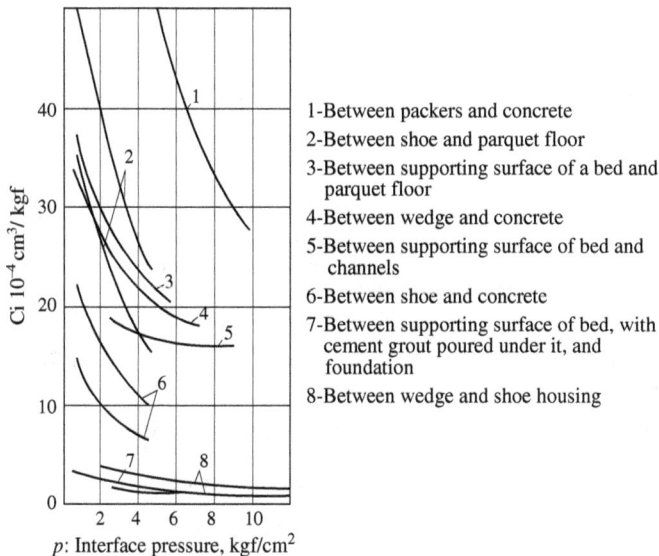

Figure 7-60   Diagram to determine coefficient C_i (by Kaminskaya).

When the leveling block is used, furthermore, the tightening force of the bolt should be within 500 kgf/cm^2 to avoid the undesirable plastic deformation of the concrete base.

To this end, other activities not mentioned above will be introduced to deepen the understanding of what was underway in the foundation.[21] Although we need more sophisticated foundation with the growing importance of higher-accuracy and higher-speed machining, there have not been any relevant activities on the foundation since the 1980s.

---

[21]We can, without any difficulties, enumerate the following materials.

Jìrek, B., "Foundations and Levelling Pads in Heavy Machine Tools," in S. A. Tobias and F. Koenigsberger (eds.), *Proc. of 6th Int. MTDR Conf.*, Pergamon,1966, pp. 123–138.

Brogden, T. H. N., "The Stiffness of Machine Tool Foundations," Research Report No. 33 of MTIRA, 1970.

Redchenko, A. G., "Installing Heavy Machine Tools," *Machines and Tooling*, 1971, 42(6): 9–10.

Hoshi, T., "Parameters of Mounting and Foundation Affecting the Structural Dynamics of Machine Tools," *Annals of CIRP*, 1973, 22(1): 129–130.

McGoldrick, P. F., and B. S. Baghshahi, "A Technique for the Determination of the Depth of Concrete Required for a Machine Tool Foundation," in J. M. Alexander (ed.), *18th Int. MTDR Conf.*, Macmillan, 1978, pp. 539–543.

References

1. Groth, W. H., "Die Dämpfung in verspannten Fugen und Arbeitsführungen von Werkzeugmaschinen," Dr.-Ing. Dissertation, Januar 1972, RWTH Aachen (Rheinisch-Westfälischen Technischen Hochschule Aachen).
2. Ito, Y., and M. Masuko, "Experimental Study on the Optimum Interface Pressure on a Bolted Joint Considering the Damping Capacity," in F. Koenigsberger and S. A. Tobias (eds.), *Proc. of 12th Int. MTDR Conf.*, Macmillan, 1972, pp. 97–105.
3. Plock, R., "Untersuchung und Berechnung des elastostatischen Verhaltens von ebenen Mehrschraubenverbindungen," Dr.-Ing. Dissertation, Mai 1972, RWTH Aachen. (Quick note: Plock, R., "Steifigkeitsuntersuchungen an Schraubenverbindungen," *Industrie-Anzeiger*, 1971, 93(82): 2041–2045.)
4. Plock, R., "Die Übergangssteifigkeit von Schraubenverbindungen," *Industrie-Anzeiger*, 30 März 1971, 93(27): 571–575.
5. Ito, Y., J. Toyoda, and S. Nagata, "Interface Pressure Distribution in a Bolt-Flange Assembly," *Trans. of ASME, J. of Mech. Des.*, April 1979, 101: 330–337.
6. Ito, Y., "A Contribution to the Effective Range of the Preload on a Bolted Joint," in S. A. Tobias and F. Koenigsberger (eds.), *Proc. of 14th MTDR Conf.*, Macmillan, 1974, pp. 503–507.
7. Fernlund, I., "Druckverteilung zwischen Dichtflächen an verschraubten Flanschen," *Konstruktion*, 1970, 22(6): 218–224.
8. Gould, H. H., and B. B. Mikic, "Areas of Contact and Pressure Distribution in Bolted Joints," *Trans. of ASME, J. of Eng. for Ind.*, Aug. 1972, pp. 864–870.
9. Itoh, S., Y. Murakami, and Y. Ito, "Engineering Calculation Method on the Spring Constant of Bolt-Flange Assembly," *Trans. of JSME (C)*, 1985, 51(467): 1816–1822.
10. Tsutsumi, M., A. Miyakawa, and Y. Ito, "Topographical Representation of Interface Pressure Distribution in a Multiple Bolt-Flange Assembly — Measurement by Means of Ultrasonic Waves," *Design Engineering Conference and Show*, April 1981, 81-DE-7, ASME.
11. Itoh, S., Y. Ito, and T. Saito, "Interface Pressure Distribution in Single Bolt-Flange Assembly — Development of a Measuring Equipment for Two Dimensional Interface Pressure Distribution and a Few Measured Results," *Trans. of JSME (C)*, 1984, 50(458): 1816–1824.
12. Itoh, S., Y. Murakami, and Y. Ito, "Interface Pressure Distribution of Bolt-Flange Assembly under Complex Loading Condition," *Trans. of JSME (C)*, 1985, 51(469): 2414–2418.
13. Bradley, T. L., T. J. Lardner, and B. B. Mikic, "Bolted Joint Interface Pressure for Thermal Contact Resistance," *Trans. of ASME, J. of Appl. Mech.*, June 1971, pp. 542–545.
14. Thompson, J. C., et al., "The Interface Boundary Conditions for Bolted Flanged Connections," *Trans. of ASME, J. of Pressure Vessel Technol.*, Nov. 1976, p. 277.
15. Birger, I. A., "Determining the Yield of Clamped Components in Threaded Connections," *Russian Eng. J.*, 1961, 41(5): 35–38.
16. Mitsunaga, K., "On Stress Distribution in Clamped Components of Threaded Connections," *Trans. of JSME*, 1965, 31(231): 1750–1757.
17. Shibahara, M., and J. Oda, "On Spring Constant of Clamped Components in Bolted Joint," *J. of JSME*, 1969, 72(611): 1611–1619.
18. Shibahara, M., and J. Oda, "On Spring Constant of Clamped Components in Multiple-Bolted Joint," *Trans. of JSME*, 1971, 37(297): 1033–1040.
19. Motosh, N., "Determination of Joint Stiffness in Bolted Connections," *Trans. of ASME, J. of Engg. for Ind.*, August 1976, pp. 858–861.
20. Tsutsumi, M., Y. Ito, and M. Masuko, "Deformation Mechanism of Bolted Joint in Machine Tools," *Trans. of JSME*, 1978, 44(386): 3612–3621.
21. Connolly, R., and R. H. Thornley, "Determining the Normal Stiffness of Joint Faces," *Trans. of ASME, J. of Engg. for Ind.*, Feb. 1968, pp. 97–106.
22. Opitz, H., and J. Bielefeld, "Modellversuche an Werkzeugmaschinenelementen," Forschungsberichte des Landes Nordrhein-Westfalen, 1960, NR. 900, Westdeutscher Verlag.

23. Schlosser, E., "Der Einfluß ebener verschraubter Fugen auf das statische Verhalten von Werkzeugmaschinengestellen," *Werkstattstechnik und Maschinenbau*, 1957, 47(1): 35–47.
24. Opitz, H., and R. Noppen, "A Finite Element Program System and Its Application for Machine Tool Structural Analysis," in S. A. Tobias and F. Koenigsberger (eds.), *Proc. of 13th Int. MTDR Conf.*, Macmillan, 1973, pp. 55–60.
25. Thornley, R H., "The Effect of Flange and Bolt Pocket Designs upon the Stiffness of the Joint and Deformation of the Flange," *Int. J. Mach. Tool Des. Res.*, 1971, 11: 109–120.
26. Ito, Y., S. Itoh, and S. Endo, "Effects of Bolt Pocket Configuration on Joint Stiffness and Interface Pressure Distribution," *Annals of CIRP*, 1988, 37(1): 351–354.
27. Schlosser, E., "Feinmessung elstostatischer Formänderungen an ebenen verschraubten Fugen von Werkzeugmaschinen-Versuchsgestellen," *Werkstattstechnik und Maschinenbau*, 1957, 47(2): 81–88.
28. Ito, Y., and M. Masuko, "Effect of Number and Arrangement of Bolts on a Normal Bending Stiffness of Bolted Joint," *Trans. of JSME*, 1971, 37(296): 817–825.
29. Ito, Y., M. Koizumi, and M. Masuko, "One Proposal to the Computing Procedure of CAD Considering a Bolted Joint — Study on the CAD for Machine Tool Structures, Part 2," *Trans. of JSME*, 1977, 43(367): 1123–1131.
30. For example, M. Masuko, Y. Ito, and N. Urushiyama, " Experimentelle Untersuchung der Statischen Biegesteifigkeit von Verschraubten Fugen an Werkzeugmaschinen," *Trans. of JSME*, 1968, 34(262): 1159–1167.
31. Ito, Y., and M. Masuko, "Experimental Study on the Optimum Interface Pressure on a Bolted Joint Considering the Damping Capacity," in F. Koenigsberger and S. A. Tobias (eds.), *Proc. of 12th Int. MTDR Conf.*, Macmillan, 1972, pp. 97–105.
32. Tsutsumi, M., Y. Ito, and M. Masuko, "Dynamic Behaviour of the Bolted Joint in Machine Tool," *J. of JSPE*, 1977, 43(1): 105–111.
33. Ockert, D., "Zur Dämpfung am einfach geteilten Biegestab," *Maschinenmarkt*, Oktober 1961, pp. 39–49.
34. Masuko, M., Y. Ito, and K. Yoshida, "Theoretical Analysis for a Damping Ratio of a Jointed Cantibeam," *Trans. of JSME (Part 3)*, 1973, 39(317): 382–392.
35. Ito, Y., and M. Masuko, "Untersuchung über die statische Biegesteifigkeit von verschraubten Fugen an Werkzeugmaschinen (1)," *Trans. of JSME*, 1968, 34(266): 1789–1797.
36. Ito, Y., and M. Masuko, "Untersuchung über die statische Biegesteifigkeit von verschraubten Fugen an Werkzeugmaschinen (2)," *Trans. of JSME*, 1968, 34(266): 1798–1804.
37. Ito, Y., and M. Masuko, "Study on the Horizontal Bending Stiffness of Bolted Joint," *Trans. of JSME*, 1970, 36(292): 2143–2154.
38. Thornley, R H., et al., "The Effect of Surface Topography upon the Static Stiffness of Machine Tool Joints," *Int. J. Mach. Tool Des. Res.*, 1965, 5(1/2): 57–74.
39. Ito, Y., "Study on the Static Bending Stiffness of Bolted Joint in Machine Tools," Dr.-Eng. Thesis of Tokyo Institute of Technology, October 1971.
40. Weck, M., and G. Petuelli, "Steifigkeits- und Dämpfungskennwerte verschraubter Fügestellen," *Konstruktion*, 1981, 33(6): 241–245.
41. Ito, Y., and M. Masuko, "Study on the Damping Capacity of Bolted Joints — Effects of the Joint Surfaces Condition," *Trans. of JSME*, 1974, 40(335): 2058–2065.
42. Tsutsumi, M., Y. Ito, and M. Masuko, "Dynamic Behaviour of the Bolted Joint in Machine Tool—In the Case of Dry Joints," *J. of JSPE*, 1977, 43(1): 105–111.
43. Tsutsumi, M., Y. Ito, and M. Masuko, "Dynamic Behaviour of the Bolted Joint in Machine Tools — The Effect of Lubricant," *J. of JSPE*, 1977, 43(5): pp. 567–572.
44. Fontenot, J. E., Jr., "The Thermal Conductance of Bolted Joints," Doctoral dissertation of Louisiana State University, May 1968.
45. Fukuoka, T., and Q. T. Xu, "Evaluations of Thermal Contact Resistance in an Atmospheric Environment," *Trans. of JSME (A)*, 1999, 65(630): 248–253.
46. Itoh, S., Y. Shiina, and Y. Ito, "Behavior of Interface Pressure Distribution in a Single Bolt-Flange Assembly Subjected to Heat Flux," *Trans. of ASME, J. of Engg. for Ind.*, May 1992, 114: 231–236.

47. Opitz, H., and J. Schunck, "Untersuchung über den Einfluß thermisch bedingter Verformungen auf die Arbeitsgenauigkeit von Werkzeugmaschinen," Forschungsberichte des landes Nordrhein-Westfalen, 1966, Nr. 1781, Westdeutscher Verlag.
48. Eastwood, E., "Machine Tool Foundation," Research Report of MTIRA, April 1963, No. 1.
49. Kaminskaya, V. V., "Determining Foundation Depth for Large Tools," *Machines and Tooling*, 1967, 38(12): 5–9.
50. Kaminskaya, V. V., "Calculation and Research on Machine Tool Structures and Foundation," in S. A. Tobias and F. Koenigsberger (eds.), *Proc. of 8th Int. MTDR Conf.*, Pergamon, 1968, pp. 139–161.
51. Ishige, S., "Foundation for Machine Tools," *Hitachi Hyoron*, 1964, 46(9): 1546–1553.
52. Poláček, M., "Vorausbestimmung der optimalen Auslegung des Rahmens von Werkzeugmaschinen mit Hilfe von Versuchsmodellmaschinen," *Maschinenmarkt*, 1965, 71(37): 37–43.
53. Hailer, J., "Die Selbsttätige Ausrichtung von Werkzeugmaschinen," *Maschinenmarkt*, Nov. 1962, no. 88, pp. 40–47.
54. Burdekin, M., Z. J. Huang, and S. Hinduja, "Predicting the Influence of the Foundations on the Accuracy of a Large Machine Tool," in B. J. Davies (ed.), *26th Int. MTDR Conf.*, Macmillan, 1986, pp. 227–237.
55. Raue, K., "Schwingungskontrollierte Maschinenlagerung," *Werkstatt und Betrieb*, 1973, 106(10): 799–804.
56. Faingauz, V. M., "Stiffness of Wedge Supports for Installing Machine Tools," *Machines and Tooling*, 1970, 41(5): 9–11.
57. Kaminskaya, V. V., "Combined Design of Beds and Foundations," *Machines and Tooling*, 1971, 42(11): 19–25.

Supplement 1: Firsthand View for Researches in Engineering Design in Consideration of Joints

Figure 7-S1 depicts a firsthand view of the research into the engineering calculation and computation for the structural characteristics in consideration of the joint. As can be seen, up to the 1980s, there were a considerable number of researches; however, with the advent of powerful software, such researches become useless rapidly.

From these earlier researches, some valuable suggestions can be obtained such as follows.

1. As exemplified by Back et al., the joint can be replaced by the spring element or beam element. In the practical case, there are no apparent differences between the computed results with spring and beam elements.

2. The constant of spring element can be given by the expression of Ostrovskii, although it is capable of taking only the normal joint stiffness into consideration. In contrast, the beam element can handle the normal, torsional, flexural, and shear stiffness of the joint.

3. In the computation, the interface pressure distribution and joint deflection are to be determined in full consideration of the deformation of the joint surroundings. As a result, the iterative method should

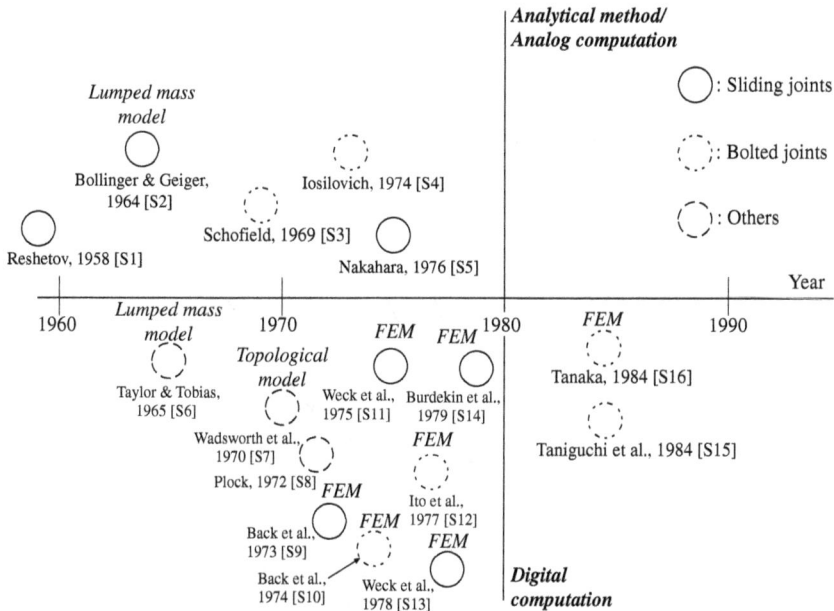

Analytical method/ Analog computation

○ : Sliding joints

⊙ : Bolted joints

◯ : Others

Lumped mass model

Bollinger & Geiger, 1964 [S2]

Iosilovich, 1974 [S4]

Schofield, 1969 [S3]

Reshetov, 1958 [S1]

Nakahara, 1976 [S5]

Year

1960   1970   1980   1990

Lumped mass model

Taylor & Tobias, 1965 [S6]

Topological model

Wadsworth et al., 1970 [S7]

Plock, 1972 [S8]

Back et al., 1973 [S9]

Back et al., 1974 [S10]

Weck et al., 1975 [S11]

Weck et al., 1978 [S13]

FEM   *FEM*

Burdekin et al., 1979 [S14]

FEM

FEM

Ito et al., 1977 [S12]

FEM

FEM

Tanaka, 1984 [S16]

Taniguchi et al., 1984 [S15]

Digital computation

Note: Number in square bracket indicates reference paper listed in final part of Chap. 7.

Figure 7-S1 Firsthand view for researches into and proposals to structural design in consideration of joints.

be employed. In the iterative method, furthermore, the cross section of the spring element can be varied stepwise, or, in certain cases, the modulus of elasticity of the spring element may be varied.

To understand the engineering calculation, a procedure proposed by Plock [S8] will be stated in the following by taking the static characteristics of a multiple-bolted joint as an example.

STEP 1: Determination of mathematical model

STEP 2: Equilibrium of loads acted on joint surface and estimation of local interface pressure

STEP 3: Determination of spring characteristics of single bolt-flange assembly

STEP 4: Determination of load-deformation diagram of bolt-flange assembly

STEP 5: Calculation of joint stiffness and deflection at cutting point

In retrospect, Weck et al. developed a program named FINDYN, which was capable of simulating the dynamic behavior of the machine tool

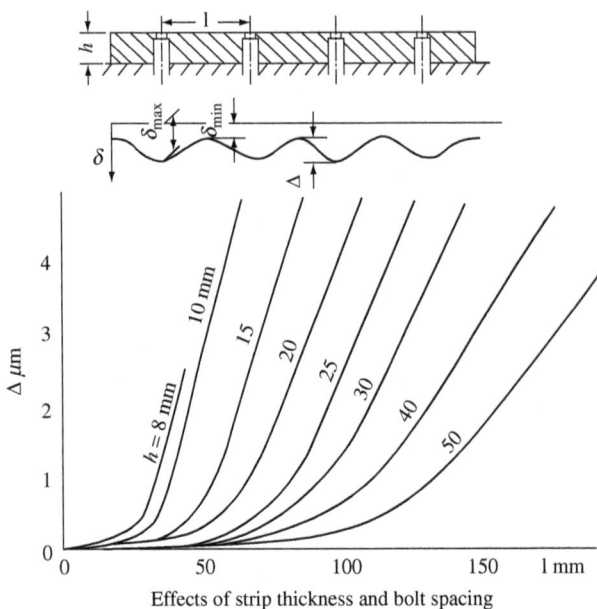

Effects of strip thickness and bolt spacing

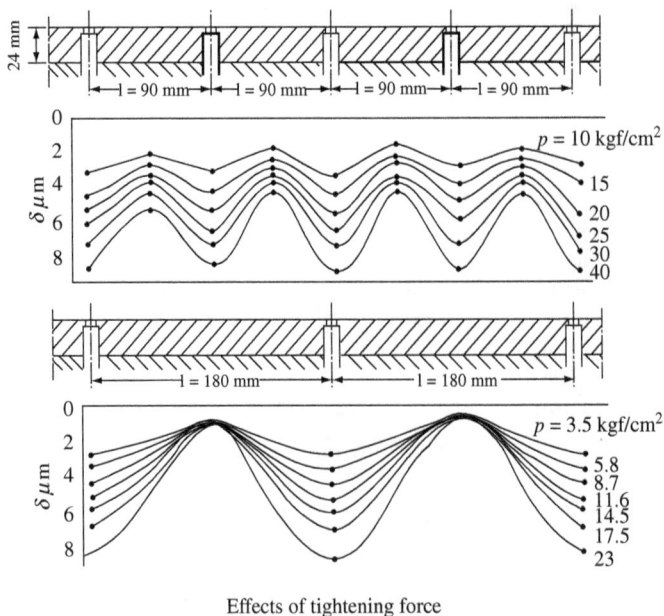

Effects of tightening force

Figure 7-S2 Determination of preferable bolt spacing in hardened strip bolted on bed slideway (by Levina).

structure with joints. In this program, the joint was replaced by the spring-dashpot coupling; also the damping matrix of joints can be incorporated within the structural matrix, where damping of either stiffness-proportional or velocity-proportional type can be considered. In the simulation, the constants in the spring-dashpot model were first determined to match the computing value with the experimental one by using the simplified joint.

Supplement 2: Influences of Joints on Positioning and Assembly Accuracy

As already described in Chap. 5, another primary concern in the machine tool joint is how to enhance the positioning accuracy and assembly accuracy in the structural body complex. For example, in the former case, the locating accuracy of the stacked blanks mounted on the arbor is at issue when the preparatory work is performed in the hobbing machine [S17]. In fact, the latter case is one of the representatives within the bolted joint, and a typical example is the hardened strip bolted onto the base or bed slideway. Reportedly, the bolt spacing has a larger effect on the waviness of the bolted strip [S18]. The thinner the strip and larger the tightening force, the larger the waviness, such as shown in Fig. 7-S2. In general, regrinding is required after bolting the hardened strip in the production and repair.

Supplement References

S1. Atscherkan, N. S., "Werkzeugmaschinen, Band 1," S. 269. 1958, VEB Verlag Technik.

S2. Bollinger, J. G., and G. Geiger, "Analysis of the Static and Dynamic Behaviour of Lathe Spindles," *Int. J. of Mach. Tool Des. and Res.*, 1964, 3(4): 193–209.

S3. Schofield, R E., "Schraubenverbindungen im Werkzeugmaschinenbau," *Maschinenmarkt*, 1969, 75(35): 736–740.

S4. Iosilovich, G. B., "Calculation for Joints with Circular Contacting Flanges, under the Action of Tensile Loads," *Russian Engg J.*, 1974, 54(6): 24–26.

S5. Nakahara, T., T. Endo, and Y. Ito, "Analysis for a Local Deformation of Two Flat Surfaces in Contact," *J. of JSLE*, 1976, 21(11): 764–771.

S6. Taylor, S., and S. A. Tobias, "Lumped-Constants Method for the Prediction of the Vibration Characteristics of Machine Tool Structures," in S. A. Tobias and F. Koenigsberger (eds.), *Proc. of 5th Int. MTDR Conf.*, Pergamon, 1965, pp. 37–52.

S7. Wadsworth, R., A. Cowley, and J. Tlusty, "Theoretische und experimentelle dynamische Analyse einer Horizontalbohr- und –fräsmaschine," *fertigung*, 1970, 70(4): 121–130.

S8. Plock, R., "Untersuchung und Berechnung des elastostatischen Verhaltens von ebenen Mehrschraubenverbindungen," Dr. Dissertation des RWTH Aachen, 1972.

S9. Back, N., M. Burdekin, and A. Cowley, "Pressure Distribution and Deformations of Machined Components in Contact," *Int. J. Mech. Sci.*, 1973, 15: 993–1010.

S10. Back, N., M. Burdekin, and A. Cowley, "Analysis of Machine Tool Joints by the Finite Element Method," in S. A. Tobias and F. Koenigsberger (eds.), *Proc. of 14th Int. MTDR Conf.*, Macmillan, 1974, pp. 529–537.

S11. Weck, M., et al., "Anwendung der Methode Finiter Elemente bei der Analyse des dynamischen Verhaltens gedämpfter Werkzeugmaschinenstrukturen," *Annals of CIRP,* 1975, 24(1): 303.

S12. Ito, Y., M. Koizumi, and M. Masuko, "One Proposal to the Computing Procedure of CAD Considering a Bolted Joint," *Trans. of JSME*, 1977, 43(367): 1123–1131.

S13. Weck, M., et al., "Finite Elemente bei der Analyse des dynamischen Verhaltens gedämpfter Werkzeugmaschinenstrukturen," *fertigung,* 1978, 78(1): 15–19.

S14. Burdekin, M., N. Back, and A. Cowley, "Analysis of the Local Deformations in Machine Joints," *J. Mech. Eng. Sci.,* 1979, 21(1): 25–32.

S15. Taniguchi, A., M. Tsutsumi, and Y. Ito, "Treatment of Contact Stiffness in Structural Analysis—1st Report, Mathematical Model of Contact Stiffness and Its Applications," *Bull. of JSME,* 1984, 27(225): 601–607.

S16. Tanaka, M., "An Application of FEM to Threaded Components—Part 4," *Trans. of JSME (C),* 1984, 50(456): 1502–1511.

S17. Zakharov, V. A., "How Deformation of Flange Affects Locating-Face Positions during Assembly," *Machines and Tooling,* 1973, 44(5): 21–24.

S18. Levina, Z. M., "Research on the Static Stiffness of Joints in Machine Tools," in S. A. Tobias and F. Koenigsberger (eds.), *Proc. of 8th MTDR Conf.,* Pergamon, 1968, pp. 737–758.

8

Design Guides, Practices, and Firsthand View of Engineering Developments—Sliding Joints

There are many kinds of sliding joint in machine tools, as shown in Fig. 8-1, and of these the utmost representatives are the guideway and main bearing. As already stated in Chap. 5, the guideway can be classified into three types, i.e., hydrodynamic (slideway), hydrostatic, and rolling guideways, depending upon their lubrication mechanism and kind of intermediate. In retrospect, Black reported such a comparison for the general characteristics of guideways as shown in Table 8-1 [1]. In this context, Poláček and Vavra carried out some measurements of the dynamic behavior using the feed unit with various types of guideway and trapezoid screw-nut or recirculating ball screw-nut driving systems, shown in Fig. 8-2 [2]. Figure 8-3 shows the resonance amplitude and frequency in the feed direction when the table was subjected to the constant exciting force generated by the electrodynamic exciter, its exciting frequency ranging from 25 to 250 Hz. As can be readily seen from Fig. 8-3, there are no essential differences in the dynamic behavior apart from the slideway. In the slideway, the dynamic behavior is similar to that of other guideways when the sliding velocity is more than 300 mm/min, although it differs completely when the sliding velocity is less than 300 mm/min.

In the past, the slideway and hydrostatic guideway were in the leading positions in the structural design of machine tools; however, nowadays the rolling guideway is major especially in the case of the conventional NC turning machine and MC, although the hydrostatic guideway has been employed in the case of the large-size and ultra-precision machine tools.

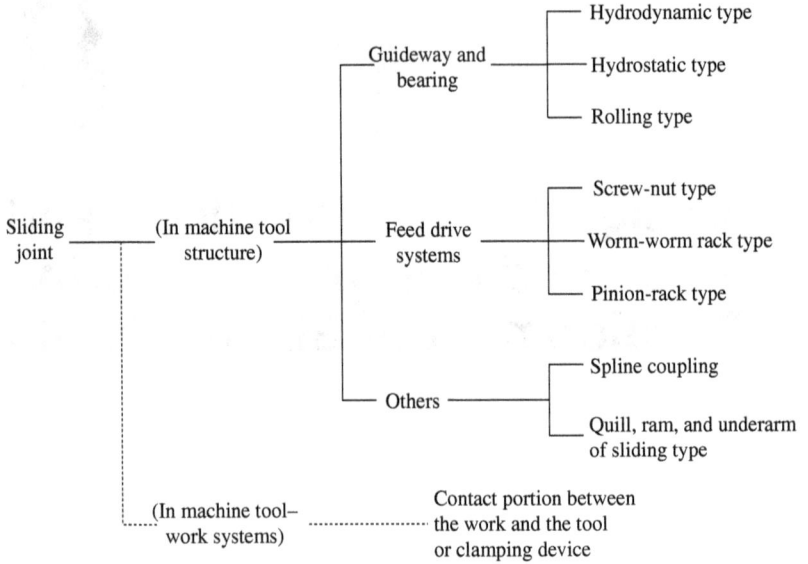

Figure 8-1 Classification of sliding joint.

TABLE 8-1   **General Characteristics of Guideways (by Black)**

Kinds of guideway	Representative characteristics			
	Coefficient of friction	Load capacity	Possibility of arising stick-slip	Others
Slideways	High	Very high	Yes	Simplicity of design Inadequate lubrication may cause wear
Hydrostatic guideways	Extremely low	Very high	No	Special design needed Piping a problem
Flat-type ball ways	Extremely low	Medium	No	Available off-the-shelf Compactness and low profile
Round-type ball ways (ball bushings)	Low	Light and medium	No	Available off-the-shelf Ease of Installation Simplicity of design Compactness
Flat-type roller ways (way bearings of U.S. patent No. 3003828)	Extremely low	Heavy	No	Available off-the-shelf Extremely compact Low profile
Round-type roller ways	Low	Heavy	No	Available off-the-shelf Ease of Installation Simplicity of design

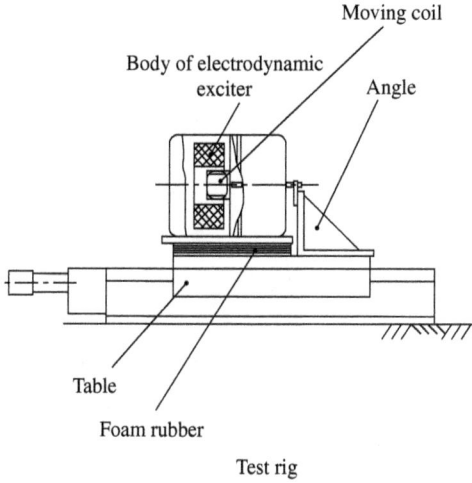

Test rig

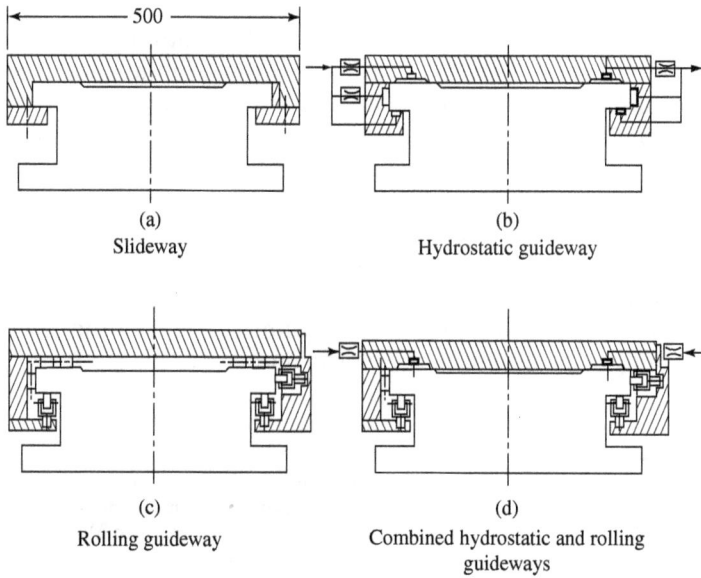

Type of guideway examined

Note: Lubrication oil: (1) For sliding guideway 2.5°E, 4.7°E, and 5.8°E/50°C
(2) For hydrostatic guideway 2.5°E/50°C

Figure 8-2 Schematic view of test rig used by Poláček and Vavra.

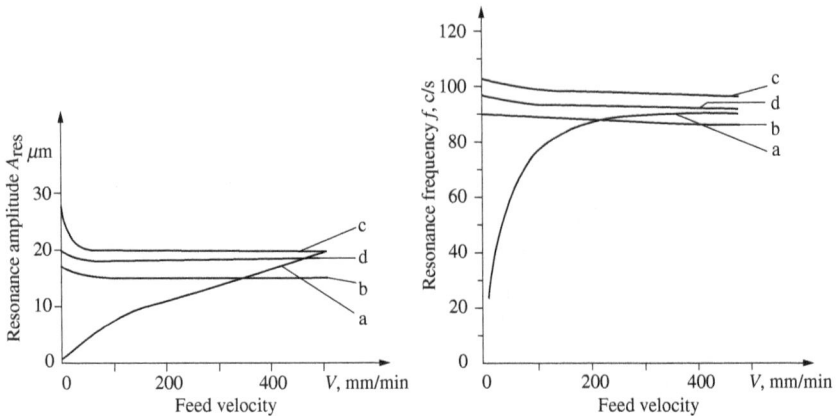

Note: Symbols a–d correspond with those shown in Fig. 8-2.

Figure 8-3   Effects of feed velocity on resonance characteristics of various guideways.

One of the reasons is that the slideway has certain limitations in its allowable traveling speed. As widely accepted, the maximum allowable speed is less than 20 m/min in general.[1] Table 8-2 summarizes the representative technological subjects for the guideway of the NC machine tool in the 1990s, and we can see various subjects explicitly and implicitly related to the joint.

Importantly, the hydrostatic guideway and rolling bearing have established their own realms as already stated in Chap. 5, and thus the major focus in this chapter is the slideway.[2,3] In this context, we must be aware that a considerable number of variants of the sliding joint have been employed to enhance the performance of the guideway.

---

[1]Some manufacturers tried to overcome the barrier in 2005. For instance, Kitamura Machinery has contrived an innovative remedy by considering the compatibility among the contact points of scraping in any 1 in^2 of bearing area, bonded Turcite-to-Meehanite cast iron combination, zigzag arrangement of connecting bolts for keep plate and kind of lubricant. As a result, the allowable speed is 50 m/min maximum while machining, and the machining accuracy is guaranteed to be better than 1 μm.

[2]Within the rolling bearing context, primary concerns in the engineering design are the dimensional specifications, bearing life, load-carrying capacity, and static stiffness, whereas at issue is the damping capacity in the machine tool design. In addition, there have been less design data for the bearing stiffness in consideration of the effects of bearing surroundings and for the linear rolling guideway. Thus, this chapter will touch on some knowledge about such machine tool–oriented design data and concerns within the rolling bearing and linear rolling guideway.

[3]Within the bearing context, the magnetic type is also on market; however, there are very few applications to the machine tool. Thus, in this chapter, the magnetic bearing will not be discussed.

TABLE 8-2   Technological Subjects for Guideway of NC Machine Tools in the 1990s

Guiding method and structural configuration of guideway	Guiding method and structural configuration for better accuracy and higher rigidity: Ex. Larger damping capacity for traveling direction in rolling guideway
	Structural optimization for hybrid guideway
	Thermal deformation of slideway
	Enhancement of slideway characteristics—oil groove pattern and lubrication method/new machining method replaceable of scraping/realization of uniform interface pressure distribution
Guiding accuracy	Change of table posture: Ex. Floating by lubricant/table posture change at movement transition
	Realization of better traveling accuracy: Ex. Gib configuration and adjustment/suppression of pitching and yawing in vertical guideway
Guideway materials	Prevention of galling Evaluation method of Turcite bonded guideway Application of ceramic guideway
Others	Total comparative evaluation among hydrodynamic, hydrostatic, and rolling guideways
	Estimation method for life and aging
	Development of innovative wiper

For example, one is a combination of sliding joint and linear rolling guideway of package type, i.e., linear roller guide, as shown in Fig. 8-4,[4] and another is a combination of the linear guide and ball screw as exemplified by that of NSK make in the late 1990s. In addition, we can observe a further variant in the machine-tool-work system, which is called the *cutting stiffness*. As widely recognized, the cutting stiffness is in closer relation to the self-excited chatter, which was first investigated in the 1940s and even at present is not completely solved. In addition, we must solve many new chatter problems, which have arisen with the advance of the machining technology as well as the advent of new materials. In consideration of such an essential feature of chatter problem, we call it *one of the oldest, but the newest problems within the machine tool design.*

8.1 Slideways

As widely recognized, the slideway can be characterized by its interfacing conditions, e.g., solid contact, boundary and mixed lubrication, and also the full lubrication, depending upon the velocity of the slider, e.g., table, cross slide, and carriage. Figure 8-5 is a reproduction of

---

[4]The combination of the sliding joint with the linear guide was also employed in the past. For example, such a guideway can be observed in the cross rail guideway for the milling head of the planomiller.

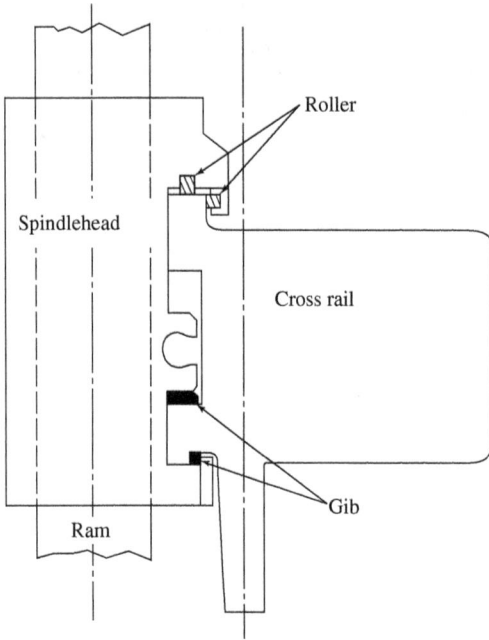

Figure 8-4 Guideway of hybrid type in five-face processing machine (type MCR-B II-HP, 1996, courtesy of Okuma).

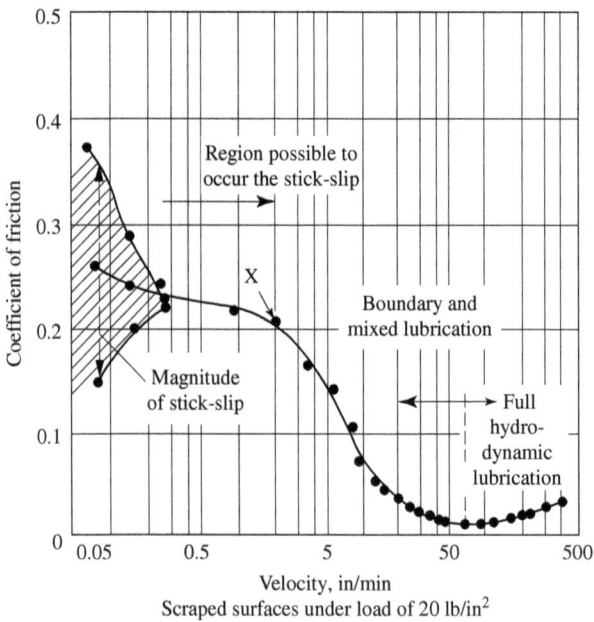

Figure 8-5 General lubrication characteristics of slideway (by PERA).

the general lubrication behavior publicized by the PERA else-where. Obviously, the stiffness of the slideway is completely sub-jected to that of lubricant, i.e., oil film stiffness, in the full lubricating condition.[5,6]

As can be readily seen from Fig. 8-5, the sliding velocity of the trav-eling component is one of the leading design factors; however, there have been very few researches that can provide the designer with the valuable design data. In fact, we can appreciate those of Bell, Corbach, Groth, Higashimoto, and Furukawa;[7] however, these researches were carried out under the lower sliding velocity of less than 3000 mm/min. A problem in the year 2000 was to investigate the static and dynamic behavior of the slideway under a higher sliding velocity of 20,000 mm/min. Despite such a constraint, some valuable findings will be stated in the following.

Bell and Burdekin [3, 4] reported the damping characteristics and normal joint stiffness of the full-size plain slideway with polar additive lubricant such as shown in Fig. 8-6. The experiment was carried out using a test rig similar to that of Poláček and Vavra, and can be char-acterized by investigating various combinations of the slideway mate-rial. As shown in Fig. 8-6, the materials chosen are cast iron (scraped), plastics, i.e., Ferobestos and Tufnol (thermosetting laminated plastics with cotton fabric base)[8] and nitride steel with hardness of 650 V.p.n.

---

[5]The stiffness of oil film in the hydrodynamic and hydrostatic guideways can be calculated by the lubrication theory. In this regard, see the book of Pinkus and Sternlicht.

Pinkus, O., and B. Sternlicht, *Theory of Hydrodynamic Lubrication*, McGraw-Hill, 1961.

[6]For the hydrodynamic lubrication, the basic theory is the generalized Reynolds equation. In contrast, to analyze the behavior of the hydrostatic guideway, the follow-ing three expressions should be simultaneously solved.

1. Generalized Reynolds equation
2. Continuous equation for flow
3. Load equilibrium equation

In short, the most characteristic feature of the hydrostatic guideway is its higher damp-ing capacity together with ensuring considerable high static stiffness. This is not the rule of the general behavior of the machine tool joint.

[7]In relation to those of Eisel and Corbach, refer to Chap. 6, where the sliding velocity is 1300 mm/min maximum. In contrast, Higashimoto investigated the table with the slid-ing speed up to 4000 mm/min and publicized observations similar to those of Bell and Burdekin; however, at issue is an assumption in which the slideway was replaced with a vibrating system with one-degree of freedom.

Higashimoto, A., et al., "On a Method for Measuring of the Damping Capacity of Machine-Tool Table Guide Ways," *J. of JSPE*, 1975, 41(12): 1134–1140.

[8]The sliding surface of the plastics bonded onto the cast iron backing strip was fine-planed.

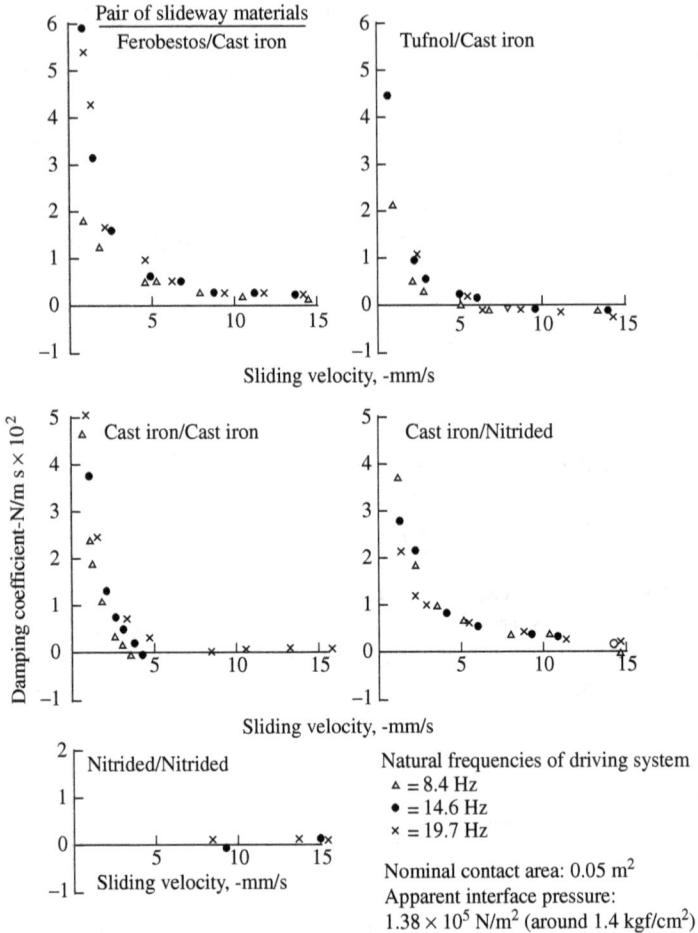

Figure 8-6 Damping characteristics of slideway with polar additive lubricant Tonna (courtesy of Bell).

(ground). In addition, the damping coefficient was mathematically determined from the damping ratio, which was measured from the decay free oscillation, and also was compensated inherent damping of the test rig. In short, damping decreases with the sliding velocity apart from the nitride steel-to-nitride steel slideway. In addition, Bell and Burdekin imply that the lubricant viscosity does not play the important role in damping as reported elsewhere.

Figure 8-7 reproduces the result reported by Groth [5], in which the relative damping capacity means the ratio of both damping capacities between the still and the operating conditions of the column or headstock.

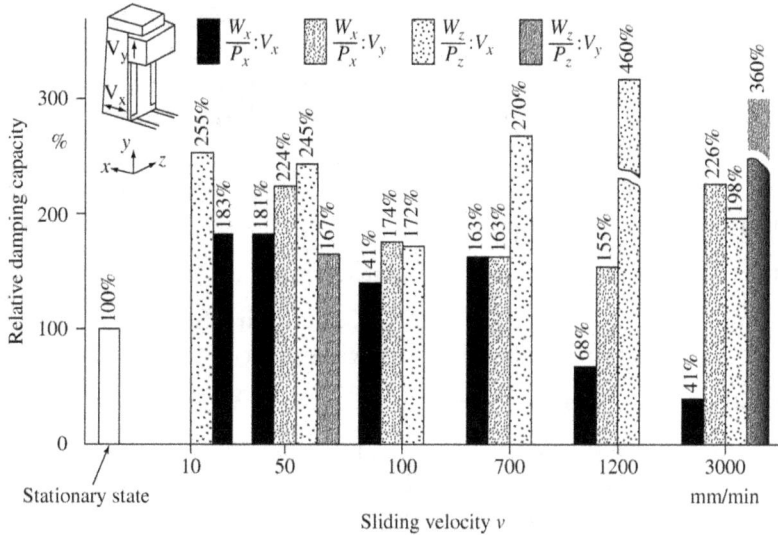

Note: The finishing method of the slideway is not stated.

Figure 8-7   Relative damping capacity under bending vibration mode—milling machine of open and column traveling type (courtesy of Groth).

In addition, the symbol W_z/P_z; v_x means the relative damping capacity while traveling to the X direction under the excitation to the Z direction. From Fig. 8-7, we can observe the interesting relationships between the relative damping capacity and the sliding velocities v_x and v_y, when the column and headstock are in bending and rocking vibration modes, respectively.

1. The damping capacity of the slideway in operating condition is, in nearly all cases, considerably larger than that in sill stand.

2. There is a directional orientation effect of the exciting force on the damping capacity.

In referring to these earlier researches, it can be said that the slideway can benefit under low speed range by showing higher damping capacity, although there are some possibilities of the stick-slip phenomenon occurring.

   In addition to these researches, Furukawa and Moronuki proposed an estimation method for the machining accuracy derived from the slideway deflection [6]. Although their method is, in principle, the same as that of Kaminskaya [7] and Levina [8, refer to Supplement], we can appraise its value as follows.

1. Clarification of constants in Ostrovskii's expression for the fluoric resin (Turcite)-to-cast iron joint, which is widely employed in the conventional NC turning machine and MC (refer to Sec. 8.1.1). In fact, λ (μm) $= 0.29p^{0.55}$ (kPa), resulting in four times larger deflection than that of cast iron-to-cast iron joint.

2. Clarification of the influencing rates of pitching, yawing, and rolling on the machining error.

Figure 8-8 shows such an error distribution diagram where

ε_x, ε_y, and ε_z = machining errors in X, Y, and Z directions (reference of ε_y is the driving point)

x_0 and z_0 = parallel displacements of table to X and Z directions

In accordance with the estimation calculation for Fig. 8-8, the machining error becomes minimum at $L_x/L_z \fallingdotseq 3$, and the larger the value of L_x/L_z, the larger the influence of yawing.

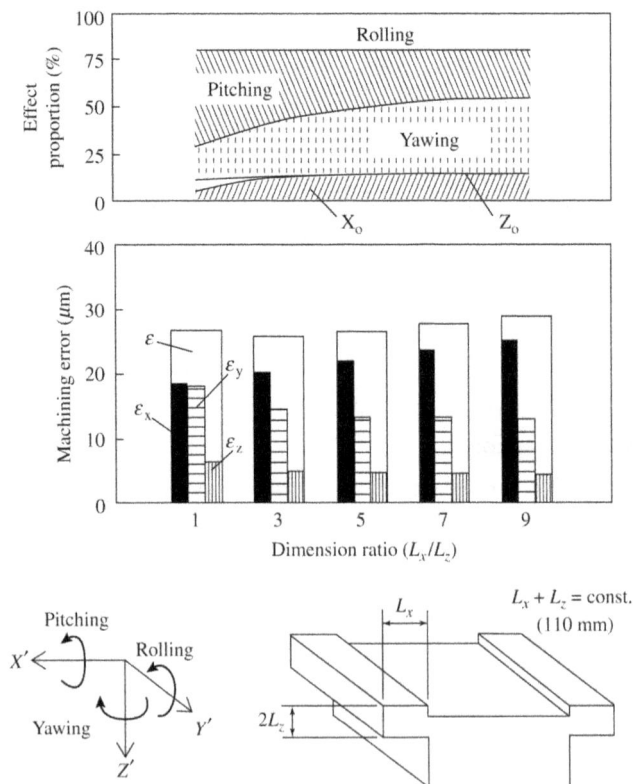

Figure 8-8 Changes of machining error with dimensional ratio, and influencing rate of pitching, yawing, and rolling (courtesy of Furukawa).

Within this context, another issue is to which point of the table the driving force must be acted. The widely accepted design rule is to allocate the driving point in vertical and horizontal planes as close as possible to the main slideway and centerline of both slideways, although no academic research has been conducted in detail, but the rule is only based on long-standing experience. More specifically, the driving point in the horizontal plane must be allocated in consideration of the leverage of the frictional forces acting on both slideways. To verify such an empirical design rule, Tsutsumi et al. conducted a basic research into the relationships between the slideway stiffness and the driving point [9]. Figure 8-9 shows some valuable data, and as can be readily seen, the driving point has considerable influence on the slideway stiffness

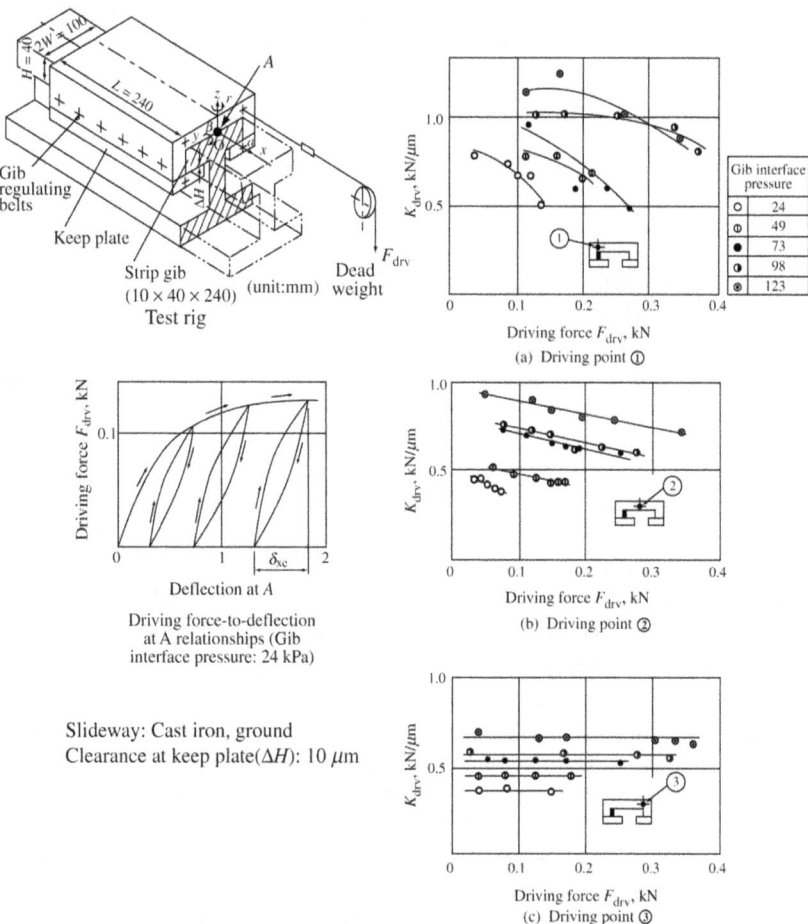

Figure 8-9 Effects of driving point on stiffness of slideway K_{drv}.

especially in relation to the gib allocation. In Fig. 8-9, we must mind the following.

1. The interface pressure at the gib is calculated by dividing the total clamping force of the gib adjusting bolts by the gib area.
2. The sliding stiffness K_{drv} is given by F_{drv}/δ_{xe} where δ_{xe} is the elastic deflection in the traveling direction measured at the center of the table. In fact, the slideway deflection shows the hysteresis loop to be the same as that observed in the flat joint under tangential loading, provided that the driving force is less than the starting frictional force.

It is furthermore interesting that the stable and constant stiffness can be obtained when the driving force is applied at the opposite point of the gib. In contrast, the higher stiffness can be observed when the driving force is applied to the gib side; however, the stiffness itself is unstable.

8.1.1 Design knowledge—slideway materials

In discussing the performance of the slideway, primary concerns are the interface pressure and its distribution, sliding velocity, and a pair of slideway materials. In addition, the less-frictional force and higher ability for wear resistance are leading design factors. In this context, the MTIRA of Macclesfield, United Kingdom, often published valuable research reports related to the friction and wear of the slideway. Within a slideway context, the joint stiffness must be in leverage at least with the friction and wear characteristics, and thus the slideway material becomes the focus.

In general, the slideway is made of gray cast iron, hardened cast iron, hardened steel, and plastics by choosing a preferable combination from these materials for bed and table slideways. In the choice of the material combination, the leading criterion is to give a solution that can simultaneously realize the preferable wear resistance with less friction, better linearity of traveling, rotational motion with higher accuracy, and higher joint stiffness. Although this is a typical ill-defined problem, further materials and processing methods have been contrived, and nowadays the applicable materials for slideway can be summarized as shown in Fig. 8-10. In fact, to accept the leading criteria, there are various measures such as the hardened slideway, fastening the flat bar made of hardened steel onto the bed slideway, fusion of special alloy materials with higher wear resistance onto the slideway, and so on. Within a sliding joint context, that of hard-facing type using ceramics, self-fluxing alloy melt-deposited or self-fluxing alloy melt-injected is of great importance, because it has high potential to provide a desirable

Materials for slideway

(Matrix itself) —
- Gray cast iron
- Alloy cast iron
- Meehanite cast iron

(Bonding) —
- Bronze
- Ceramics
- Hardened steel
- Asbestos-based (e.g., ferobestos LA3)
- Fabric-based (e.g., bakelite/tufnol/end-grain laminate)

(Bonding & coating with impregnated backing) —
- PTFE (e.g., ferobestos X389/Tufnel 2F-3-PTFE/sintered bronze/turcite)

(Self-fluxing powder flame-sprayed) —
- (E.g., 18-8 stainless steel/high Cr stainless steel/molybdenum/ceramics)

Others —
- Cast iron with Cr coating
- Epoxy resin coating

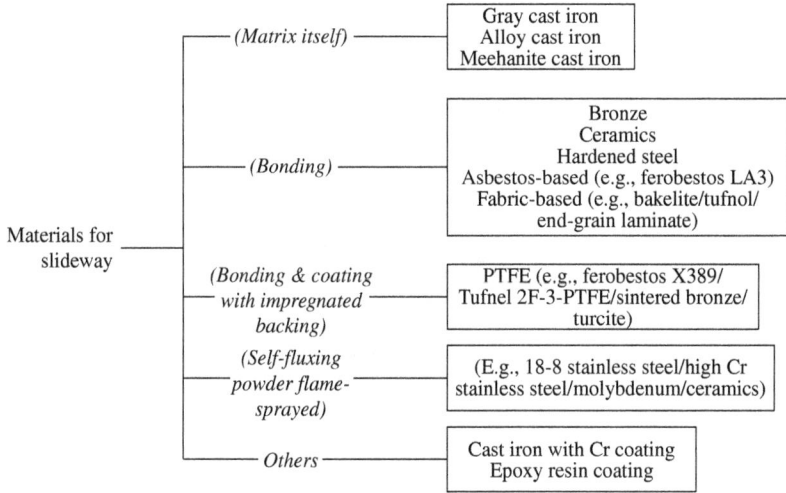

Notes 1 In certain cases, PTFE is combined with lead or graphite.
2 Self-fluxing powder flame-sprayed is, in general, employed to cast iron or steel.

Figure 8-10   Materials applicable to slideway.

solution for the ill-defined problem. One such hard-facing type is shown in Fig. 8-11, and as can be seen, its joint characteristics are very complicated. For example, the melt alloy is injected on the surface, which is to be machined to have a comblike form to ensure higher stick strength of melt alloy, simultaneously giving certain benefits on the joint stiffness.[9]

Importantly, the composite material with lower friction was dominant in the year 2000 and beyond in the production of the conventional NC turning machine and MC. More specifically, the commercially named Turcite, Rulon, Moglice, and SKC are often employed on the slideway, although these were already on market in the late 1960s and early 1970s. Turcite is made of Teflon-impregnated sheet-type material and first prevailed in the United States, whereas Moglice and SKC are made of epoxy-based replication materials containing molybdenum disulfide (MoS_2) and graphite and are widely employed by European manufacturers [10].

It envisages that Turcite has a unique characteristic in friction; i.e., the coefficient of friction falls as interface pressure increases and flattens to 0.04 at 0.7 MPa. Unfortunately, the interface pressure within

---

[9]The *self-fluxing alloy* is on the market under the commercial name Colmonoy or Metacolly (Cr-based alloy).

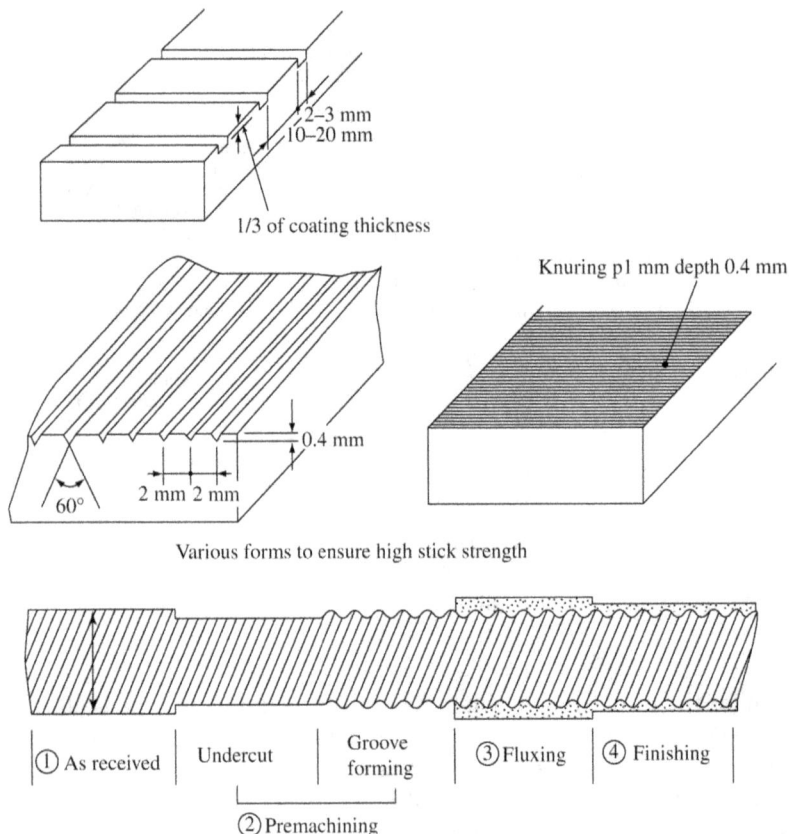

2–3 mm
10–20 mm

1/3 of coating thickness

Knuring p1 mm depth 0.4 mm

0.4 mm

60°   2 mm   2 mm

Various forms to ensure high stick strength

① As received | Undercut | Groove forming | ③ Fluxing | ④ Finishing

② Premachining

Figures are the process sequence number.

Figure 8-11   Producing process of slideway of hard-facing type.

nearly all slideways is less than 0.2 MPa. Table 8-3 shows some values of the coefficient of friction for representative slideway materials. In contrast, Moglice can be very accurately molded, resulting in the duplicate as a copy of the slideway reference. In consequence, the slideway produced has nearly 100% bearing contact, which is not preferable from the lubrication point of view, and thus the slideway must be light-milled to produce tightly crisscrossed oil-groove pattern to obtain 60% bearing surface. As can be readily seen, the joint characteristic could become very complicated, and to facilitate understanding, Fig. 8-12 shows the shape and dimensions of the bed slideway prior to molding the Moglice and also the finished surface [10]. In general, the coating thickness of Moglice is 1.5 to 2 mm, and the opposite table slideway is finished with surface roughness of 2 to 6 μm.

TABLE 8-3 Static Coefficient of Friction for Slideway Materials in Lubrication (courtesy of Devitt and of SME)

Materials	Static coefficient (lubricated)
Cast iron	0.21
Bronze	0.16
Phenolic	0.18
Moglice P-500	0.06
Turcite	0.04

In the choice of the slideway material, we must furthermore consider what is suitable for the finishing method of the sliding surface. In many respects, scraping and grinding have been widely employed, and scraping demonstrates substantial importance and maintains the old-fashioned, technician-based workshop technology. In scraping, a common standard is 10 to 15 points in any 1 in^2 of bearing area. When seeking the

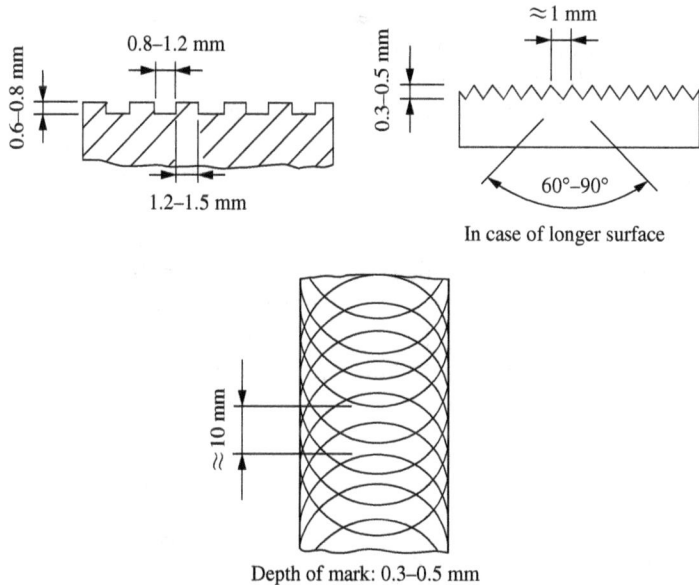

In case of longer surface

Depth of mark: 0.3–0.5 mm

Figure 8-12 Preparatory machining of bed slideway and finished surface with crisscrossed mark (courtesy of Devitt and of SME).

absolute best quality for the slideway, the designer employs the scraping surface with 30 to 40 contact points; however, such a slideway is very costly.[10]

8.1.2   Design knowledge—keep plate and gib configurations

In the structural design of the slideway, a root cause of difficulties lies in the complete accommodation of the *narrow guide principle* together with maintaining the smooth movement of the traveling unit. As a result, a suitable clearance should be provided to the sliding portion, simultaneously having the larger static stiffness and damping. In the design procedure, furthermore, it is necessary to consider the limited space allowable for the clearance regulating elements, such the keep plate and gib.

In general, the gib can facilitate the better realization of the narrow guide in the slideway by regulating the clearance between the base and the table slideways. In nearly all cases, both the gib and the keep plate thicknesses are insufficient from the stiffness point of view. As is widely known, the taper gib is very common because of the ease of its handling; however, according to the report of Zelentsov [11], the actual contact area on the sliding surface in most machine tools is only 15% to 20% of the nominal area, resulting in the deterioration of the joint stiffness. This is caused by the sinusoidal wavelike deformation of the taper gib, which is the two or three half-wave bends depending upon the length, thickness, and taper of the gib. In short, there are three contact points on a sliding surface and duly two contact points on an opposite surface; and thus from the old days, any external loads must not be directed to the gib branch of the slideway, because under such a loading condition, the static stiffness of the slideway reduces considerably.

Figure 8-13 shows a notched taper gib to improve the contact condition, and on the basis of similar idea, Spieth-Maschinenelemente KG contrived a new gib with the function of clearance regulation and adjustment such as shown in Fig. 8-14 [12]. These gibs can be improved in part the shortcomings of the taper gib, e.g., necessity of scraping to

---

[10]Even in the year 2000 and beyond, scraping is one of the important finishing methods of the slideway, and primary concern is the scraped mark. Because of the nature of scraping, the scraped mark is, in the ideal case, of dimple with steep digging in and gently flattening out (dimple with long tail-off), where the depths are between 1.5 and 3.0 μm for better finishing and between 3.5 and 5.0 μm for ordinary finishing, respectively

Tohoku, K., and S. Nishimoto, *Workshop Technology Series—Handwork Finishing*, Sangyo Tosho, 1961, pp. 47–76.

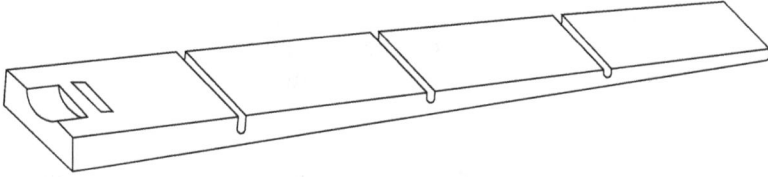

Figure 8-13 Notched gib for cross slide of vertical milling machine.

have the better contact, and that of Spieth-Maschinenelemente benefits
to form the lubrication film, because the small inclination can be produced
by the regulation. In addition, the joint surroundings are liable to deform
compared with those in other joints, and there is a high possibility of
the appearance of additional interface pressure and deformation caused
by the deformation of joint surroundings.

Notwithstanding their very importance, there have been very few
research activities on the keep plate and gib, and thus their design was,
in general, carried out on the basis of empirical knowledge. For example,
in the beginning of the 1960s, Levina and Ostrovskii [13] conducted some
notable investigations into the effects of keep plate, gib configuration,

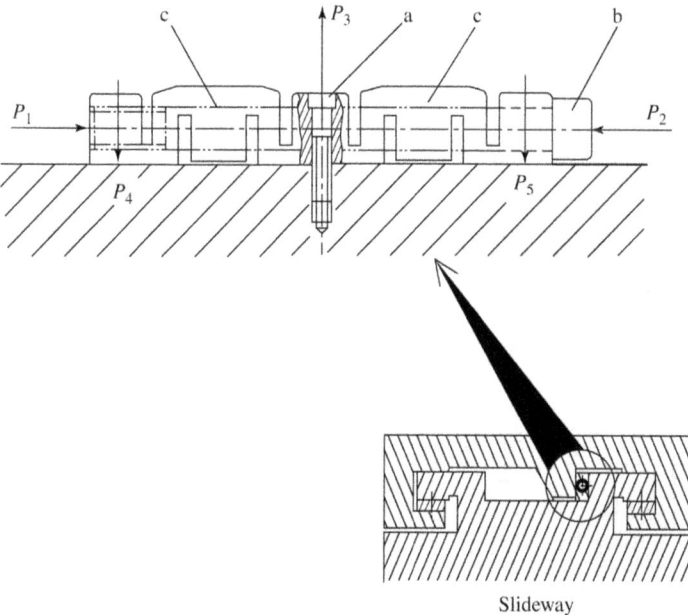

Slideway

Figure 8-14 A new gib with clearance regulating functionality (by Spieth-
Maschinenelemente KG, courtesy of Carl Hanser).

and dovetail on the stiffness. Although the FEM analysis can facilitate, at present, the structural design for the keep plate and gib, the design knowledge obtained from the earlier work is often valuable and thus will be quickly stated in the following.

Rectangular keep plate.   The keep plate is in the cantilever configuration with the clamping bolt (type A in Chap. 7) and thus shows relatively low stiffness. Figure 8-15 reproduces some measured results by Levina and Ostrovskii, and obviously the deflection of the keep plate can be calculated using the mathematical model of the elastic beam on elastic foundation such as shown in Fig. 8-16.

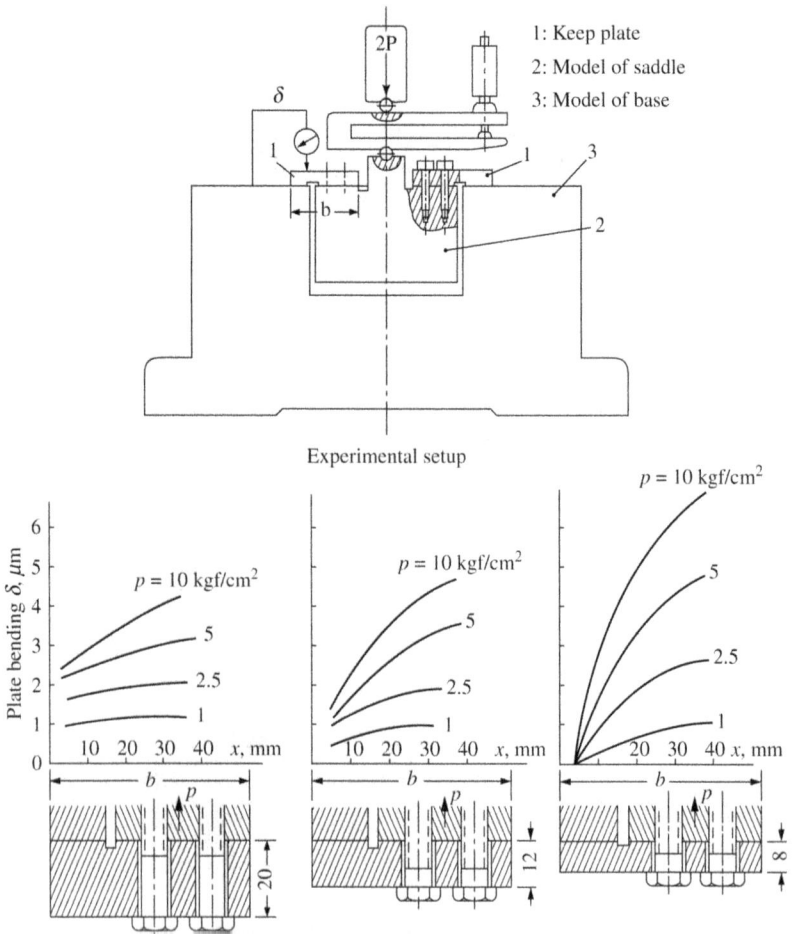

Figure 8-15   Elastic deflections of keep plates (by Levina and Ostrovskii).

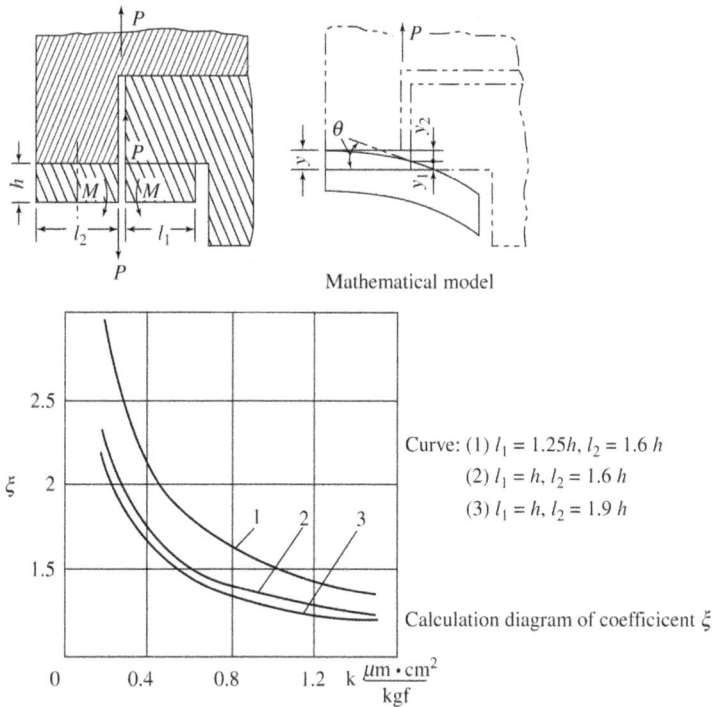

Mathematical model

Curve: (1) $l_1 = 1.25h$, $l_2 = 1.6h$
       (2) $l_1 = h$, $l_2 = 1.6h$
       (3) $l_1 = h$, $l_2 = 1.9h$

Calculation diagram of coefficicent ξ

Figure 8-16 Mathematical model of keep plate and calculation diagram of coefficient ξ (by Levina and Ostrovskii).

The deflection of the keep plate per unit length in its cross-sectional direction can be written as

$$y = y_1 + y_2 = pC_{k1}\xi$$

$$\xi = 4\lambda_1\{U_1 - \gamma^*V_1 + \psi^{*3}[U_2 + (\gamma^*V_2)/\psi^*]\}$$

$$\gamma^* = (V_1/2W_1)\{[1 - \psi^{*2}(V_2/V_1)]/[1 + \psi^*(W_2/W_1)]\}$$

$$\psi^* = \sqrt[4]{[C_{k2}/C_{k1}]}$$

where p = mean pressure per unit length P/l_1 [(kgf/cm) · (1/cm)]
C_{k1} and C_{k2} = compliances corresponding to sections l_1 and l_2,
                 respectively [μm · (cm^2/kgf)]
$\lambda_i = 0.51_i{}^4\sqrt{10^4}/(4EIC_{ki})$ $i = 1,2$
E = Young's modulus, kgf/cm^2
I = second moment of cross section, cm^4
U_1, V_1, and W_1 = coefficients depending on value λ_1
U_2, V_2, and W_2 = coefficients depending on value λ_2

In this engineering calculation, the following assumptions are employed.

1. A section of the beam with length l_2 is clamped, and its joint stiffness is larger than that of section l_1, which is in free condition.

2. The clamping force is sufficient to prevent the opening of the joint surface over the section l_2, when the external load is applied.

In the expression mentioned above, the coefficient ξ is introduced for the ease of calculation, and a calculation diagram is shown in Fig. 8-16, when $C_{k2} = 0.05\ \mu m \cdot (cm^2/kgf)$. In the engineering design, the thickness of the keep plate must be greater than the width of the abutment lip, but not too thick.

To this end, the dovetail configuration will be touched on. In short, the following data are recommended.

1. The ratio a/h should be, at least, 1.0 minimum.

2. The ratio a/h should be from 1.3 to 1.5 minimum in heavier loading.

Here a is the length between the side face of slide and the tip of V-shaped abutment lip of the base, and h is the depth of the dovetail.

Gib configuration. In short, at issues in the gib design are as follows.

1. Gib deformation

2. Regulating mechanism of the clearance between the gib and the slideway in consideration of joint surroundings

Figure 8-17 shows some representative gib configurations, which can be classified into the three basic types depending upon the kind of gib and

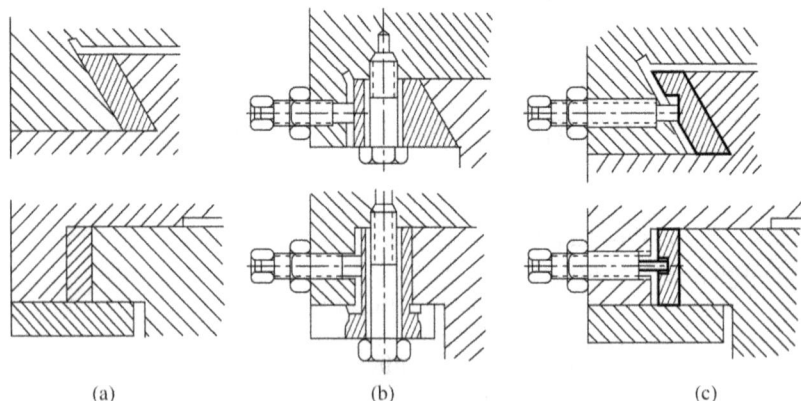

(a)   (b)   (c)

Figure 8-17 Representative gib configurations: (a) Taper gib, (b) strip gib of fixed type, and (c) strip gib.

allocation of the adjusting screw. As can be readily seen, the adjusting screw is another leading element to form the gib structural configuration. In other words, the gib is one of the variants of the bolted joint, and can be characterized by its amphibious feature, i.e., combination of stationary and sliding joints within its functionality.

Figure 8-18 shows the relationships between the gib configuration and the stiffness of slideway, where the stiffness is expressed by the swiveling angle of the traveling unit under the moment. The slideway with trapezoidal strip gib lightly clamped after the regulation is relatively rigid and shows around 2 to 2.5 times larger stiffness than that with taper gib. These experimental results verify the lower rigidity of the taper gib or adjusting screw, resulting in the nonuniform distribution

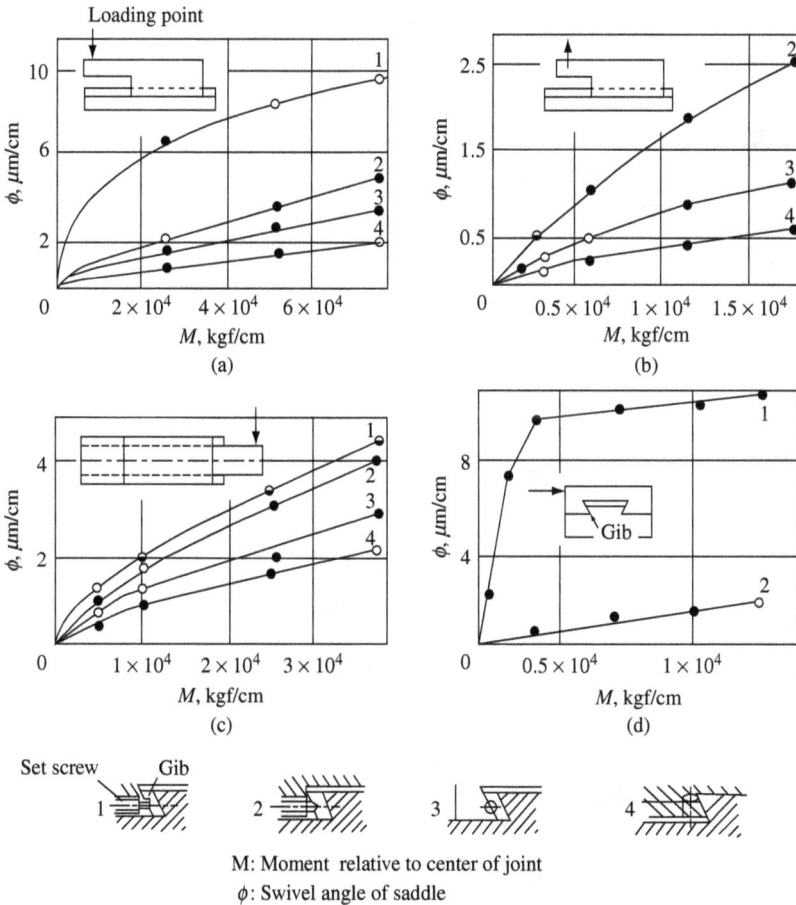

M: Moment relative to center of joint

ϕ: Swivel angle of saddle

Figure 8-18 Joint stiffness of dovetail slideways with various gib configurations (by Levina and Ostrovskiĭ).

of the interface pressure between the gib and the slideway. In the case of the flat strip gib, furthermore, the allocation distance of the adjusting screw along the gib length is of great importance and can be calculated using a mathematical model, which consists of an infinite length beam on elastic foundation. Theoretically, it is desirable that the $l_p/h = 3$ to 4, where l_p and h are the distance between both the adjusting screws and gib thickness, respectively; however, this condition is very difficult to realize in the practical design.

In general, the taper gib has been widely employed because of its ease of handling together ensuring relatively high stiffness. A primary concern in the taper gib is the clearance regulating and adjusting mechanism; and, e.g., Fig. 8-19 reproduces the ENIMS standard of U.S.S.R. in relation to the taper gib, where the taper should be in proportion to the length of the gib or to the length of the component to which the gib is attached, i.e., taper 1/50 for the length up to 500 mm, 1/75 for 500 to

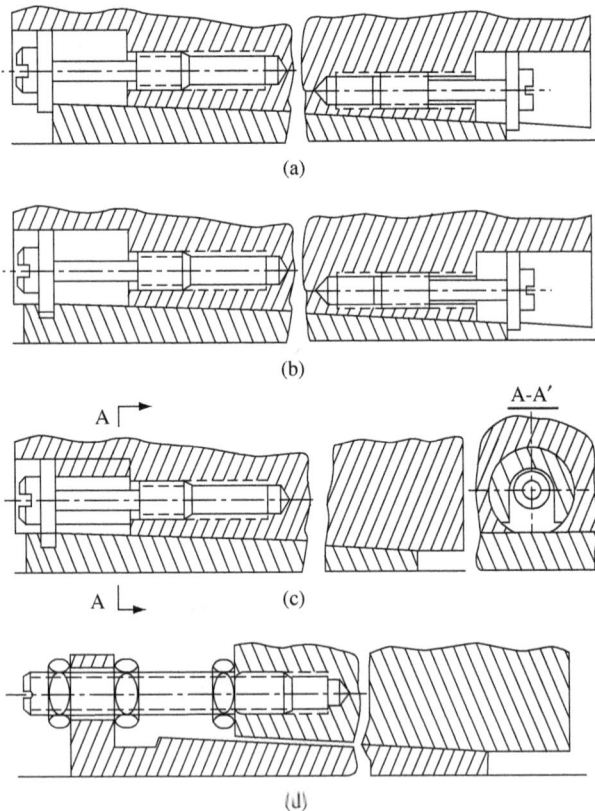

(a)

(b)

(c)

(d)

Figure 8-19   Various configurations of taper gib.

750 mm, and 1/100 for above 750 mm. These recommended values are determined in consideration of the bent of the gib in the working positions caused by both the residual stress and the longitudinal deflection when adjusting screws are at work in both ends of the gib. In addition, two taper gibs are generally arranged end-to-end in the slide when the length is more than 120 mm [14].

8.2   Linear Rolling Guideways (Linear Guide and Rolling Guideways)

In both the linear rolling guideway, i.e., linear guide, and the rolling guideway, a ball- or roller–to–flat surface contact is of basic form, and as is well known, Hertz' theory has been applied to analyze such a contact problem. In Hertz' theory, the idealized surface is objective; however, we cannot produce such an idealized surface, as can be readily seen from Chap. 6. In fact, the maximum pressure is lower and the interface pressure distribution shows a long tail-off, in the actual ball–flat surface contact compared with those in the idealized contact.

Admitting such a limitation, nearly all rolling bearing manufacturers rely on Hertz' theory even in the year 2000, and this induces certain shortages of design data for practices. Within this context, we must also be aware that the static and dynamic behavior of the linear rolling guideway is especially dependent upon the differing load per each rolling element, i.e., uneven supporting load distribution across the whole guideway and the uneven interface pressure distribution within a rolling element. These unfavorable loading conditions are caused by the flatness and linearity deviations of the guideway, elastic deformation of the joint surroundings, and machining errors of the components consisting of the rolling guideway. In the past, the elastic deflection of the rolling guideway was of the same magnitude as the deviations of the rolling body and guideway from their ideal shapes. More specifically, primary concerns were the diameter deviation within rolling elements, angle deviation of guideway, and conelike shape of the roller, resulting in the local deformation of the table and carriage.[11]

In the linear rolling guideway, the leading characteristic attributes are the load-carrying capacity, stiffness, accuracy, sensitivity to motion, uniformity of motion, resistance to motion, and damping properties; and the linear rolling guideway may be furthermore characterized by the

---

[11]Levina and Reshetov investigated the influence of manufacturing errors on the rigidity of the linear rolling guideway, and their report is worth referring to in discussing the research history of the linear rolling guideway.

Levina, Z. M., and D. N. Reshetov, "Rigidity Calculations and Investigations of Antifriction Slideways," *Machines and Tooling*, 1961, 32(11): 7–17.

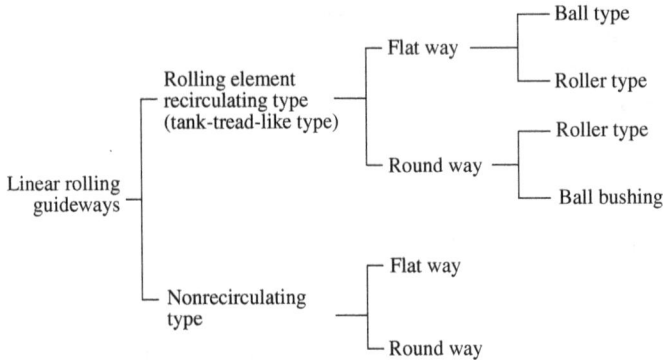

Note: Flat way includes those of V-form and convex races. In certain cases, the race consists of a combination of hardened small bars or wires.

Figure 8-20   Classification of linear rolling guideways.

lubrication mechanism, i.e., higher possibility of elastohydrodynamic lubrication than the rolling bearing.[12]

Design data of rolling guideways.   The linear rolling guideway can be classified as shown in Fig. 8-20, and of these the tank-tread-like configuration has been evolved as a package type to ease the practical use, i.e., linear guide (rolling guideway of circulating type). The linear guide has prevailed since the 1980s, especially when producing the conventional NC turning machine and MC, and this is one of the reasons why such kinds of machine tools of Japanese make have been very powerful in the world market. Importantly, the rolling guideway of noncirculating type has a beneficial factor, i.e., realization of higher guiding accuracy and load-carrying capacity by the selective combination method of rolling element. In addition, the noncirculating type has the benefit of relaxing the elastic deflection and flatness deviation of the table guideway using the cylindrical roller of hollow type as exemplified by the surface grinder of portal type of Okuma make (type GSA in the 1960s). Figure 8-21 shows the utmost representative application of the noncirculating flat type to the CNC jig grinder of Moore Co. make (type G-18CP around 1985). In this case, the base slideway is made of hardened steel and finished by grinding, which is bolted on the scraped surface of the base body, whereas the cross slide slideway is made of cast iron and finished by scraping.

---

[12]For example, see Napel W, E, and R, Bosma, "The Influence of Surface Roughness on the Capacitive Measurement of Film Thickness in Elasto-hydrodynamic Contacts," *Proc. of Inst. Mech. Engrs.*, 1971, 185(37/71): 635.

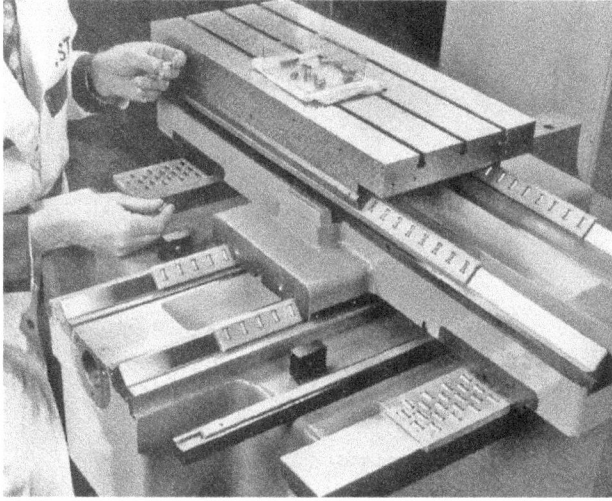

Figure 8-21   Flat linear roller guide of noncirculating type—CNC jig grinder (type G-18CP, late 1980s, courtesy of Moore Co.).

By contrast to the flat way, the round way is not so common; however, the bar guide is often very useful in the production of certain kinds of the machine tool. A disadvantage of round way is attributed to insufficient contact area between the rolling element and the guide compared with that of flat way. Figure 8-22 is one of the attempts to overcome this shortcoming [15].

With the growing employment of the linear guide beyond our estimation, there has been rapid development of the dimensional and performance specifications, and in the beginning of the 2000s, we have such a linear roller guide of NSK make as shown in Fig. 8-23. From this, it is very easy to understand what is the joint within the linear roller guide, although its

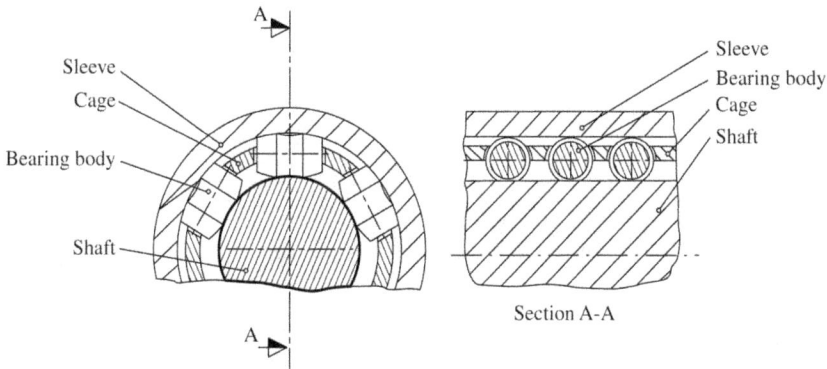

Figure 8-22   Linear roller guide for round way—roller with concave track (by Renker).

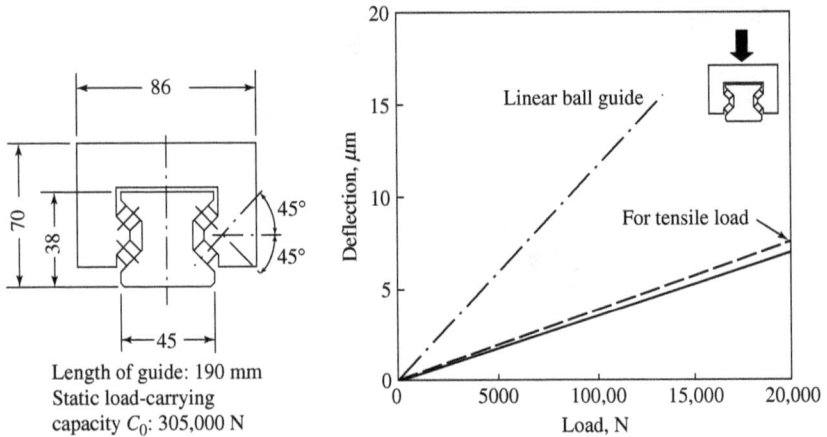

Length of guide: 190 mm
Static load-carrying
capacity C_0: 305,000 N

Note: Dimensional specifications of ball guide are similar to those of roller guide.

Figure 8-23   Normal stiffness of linear roller guide (courtesy of NSK).

structural configuration and functionality are essentially not changed. However, the performance has been improved considerably since the 1960s, because of the amazing advancement of related design and production technologies. For example, the stiffness of the linear roller guides was around 200 kgf/μm in the 1960s and is 300 kgf/μm in that of NSK.

As can be seen, the static stiffness of the linear guide can be obtained from the manufacturer without any difficulties. In general, the roller linear guide is stiffer than the ball linear guide, and the influencing factors of the roller guideway at present are the number and arrangement of rollers, nonuniformity of diameter of rolling elements, preload being applied to rolling elements, and load distribution on rolling elements. It is furthermore said that the misalignment caused by the machining error of the bed guideway and poor adjustment may be flattened and minimized, resulting in the better straightness of the table than one-fifth of that of the finished guideway.

Despite the various beneficial features of the linear guide, at issue even in the year 2000 is how to estimate its damping capacity. The linear guide has surely lower damping capacity than the slideway, although there are, dare to say, very few reliable data.[13] Within this context, Hallowes and Bell conducted research into the effects of the squeeze film device on the dynamic behavior of the linear guide [16]. The experiment was carried out using a model of guideway, where the static and

---

[13]We can understand the lower damping capacity of the linear guide through a product if IPM make, which has a damping block between both the linear guides.

Kieser, D., "Führungsprinzipien," *Industrie-Anzeiger*, 1991, 43: 120–124.

dynamic loads were applied by the dead weight (50 kg mass), and electrohydraulic exciter, respectively. In addition, the squeeze film block is of rectangular pad form with a working gap of 25, 50, or 150 μm, and the roller guideway consists of four Tychoway units (see Fig. 5-9): a unit has two roller tracks and contacts with hardened flat guideway of 76 mm width by 14 rollers (roller diameter 5 mm, roller length 5 mm). In this case, the static stiffness and natural frequency of the roller linear guide are 90 kgf/μm and 195 Hz, respectively, and the dynamic stiffness increases with reducing the working gap. More specifically, the effects of the squeeze film can be obviously observed as follows.

1. The dynamic stiffness is around 15 kgf/μm at the resonance frequency of the table-roller system, and it recovers up to 350 kgf/μm, when the oil viscosity and working gap are 0.6 poise (P) and 25 μm, respectively.

2. The dynamic stiffness is 120, 210, and 350 kgf/μm by varying the working gap at 125, 50, and 25 μm.

In retrospect, the linear rolling guideway was on the market and duly investigated even in the 1960s, when the linear guide did not yet prevail. In addition, the primary concern was the influence of the manufacturing error on the characteristics on that occasion. For example, Levina publicized a valuable report to assist the fundamental understanding of the linear rolling guideway in 1965 [17]. Figure 8-24 shows

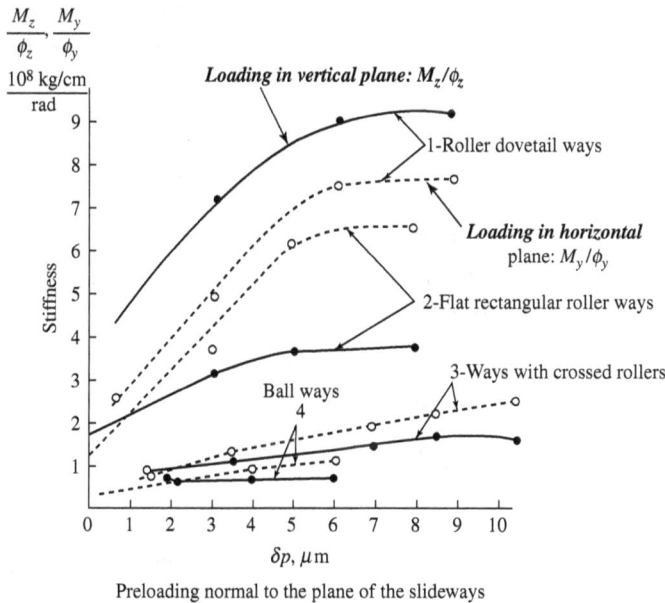

Figure 8-24   Relationships between stiffness of linear rolling guideways and preloading (by Levina).

the relationships between the stiffness under the moment and the pre-load, one of the data in her report. As can be readily seen, with increasing preload, the stiffness increases sharply in the initial stage, and then remains almost constant after reaching a specific value. Importantly, these data are applicable to units of low weight (100 to 300 kg). In fact, she conducted a comparative research into the characteristics of the linear rolling guideway using the test rig, where all the slides and beds were the same size and in most cases with identical dimensions. In addition to Fig. 8-24, she reported the following interesting characteristics.

1. When the external load is applied at the center or off-center of the linear rolling guideway, the relationships between the elastic deformation and the load are nearly linear.

2. The stiffness of ball guideway is around one-half of that of roller guideway.

3. The stiffness of crossed-roller guideway is approximately 50% lower than that of roller guideway with the same number of rollers.

Although the linear guide of forerunning type was already contrived as exemplified by the Rotax type, such a linear guide was not in mature on that occasion, and for further understanding, Fig. 8-25 shows such traditional rolling guideways [18].

Reportedly, the preload plays very important roles in the linear rolling guideway, and in general, the stiffness increases considerably with the preload. Hajdu verified also this fact in the case of the cross-roller chain type, and it is very interesting that the stiffness of the roller guideway of cross chain type increases with the external load [19].

8.3   Main Spindle-Bearing Systems

The main spindle system can be, in many respects, replaced with the mathematical model consisting of the elastic beam on two or three

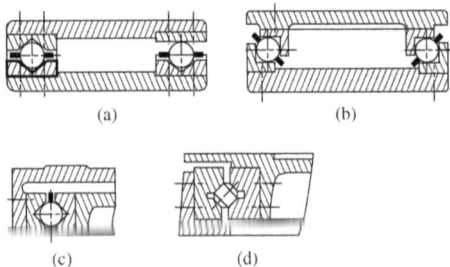

(a)   (b)

(c)   (d)

Figure 8-25   Various linear rolling guideways of noncirculating type (by permission, courtesy of Carl Hanser).

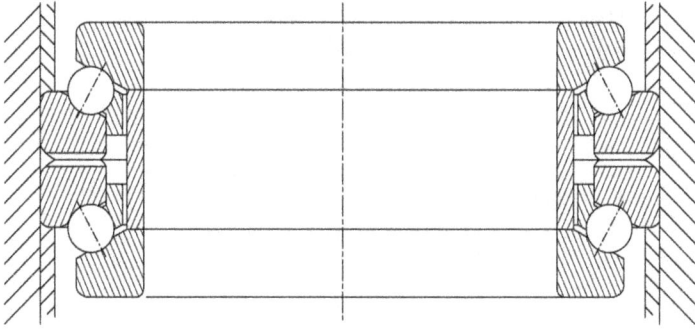

Figure 8-26 Various joints within a rolling bearing—in case of angular contact thrust ball type.

restraints (supporting points). The restraint corresponds with the main bearing, and its model is duly a couple of spring and dashpot.[14,15]

The leading characteristic attributes of the rolling bearing are the same as those for the linear rolling guideway; however, the primary concern is duly the static stiffness, i.e., spring constant, and damping capacity, i.e., coefficient of dashpot, when we carry out the engineering calculation or computation of the behavior of the main spindle system. Intuitively, even the spring constant of the bearing in still stand is very difficult to estimate correctly, because the bearing has a considerable number of stationary and rolling joints within it such as shown in Fig. 8-26. For example, Weck and Ophey proposed a mathematical model for the bearing itself, as shown in Fig. 8-27, by not taking the mass of bearing into consideration, because it is negligible compared with the masses of the shaft and bearing housing [20]. In addition, they suggested that the expressions to calculate the static stiffness so far

---

[14]There have been myriad researches and engineering developments to estimate the characteristics of the main spindle-bearing system. In all cases, the basic expression is *partial differential equation* for loading along the beam. Although the basic expression is the same, the engineering computations in practice differ from one another depending upon the available tools. In retrospect, the lumped mass model was dominant, and the basic expression was converted to *finite difference equations* in the 1960s. In due course, the characteristics were simulated using the analog computer.

Bollinger, J. G., and G. Geiger, "Analysis of the Static and Dynamic Behavior of Lathe Spindles," *Int. J. Mach. Tool Des. Res.*, 1964, 4: 193–209.

Heinrich, I., "Das dynamische Verhalten des Systems Hauptspindel—Lagerung einer Werkzeugmaschine," *Industrie-Anzeiger*, 1967, 89(6): 25–28.

[15]For an engineering calculation for static stiffness of the main spindle system, see that of Opitz et al.

Opitz, H., et al., "Untersuchung an Werkzeugmaschinenspindeln, Wälzlagern und hydrostatischen Lagerungen," *Forschungsberichte des landes Nordrhein-Westfalen*, Nr. 1331, 1964, West deutscher Verlag.

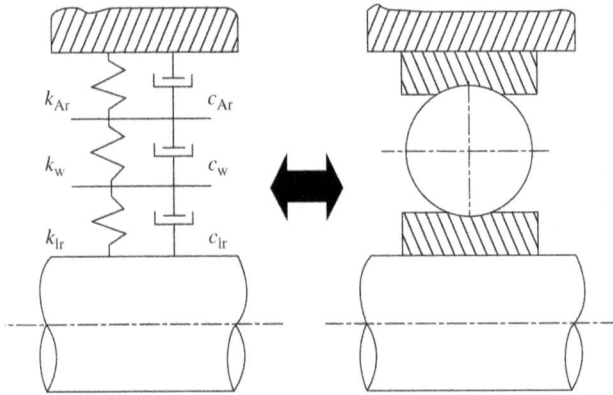

Figure 8-27 A mathematical model of rolling bearing (by Weck, courtesy of Industrie-Anzeiger).

proposed are well applicable. In contrast, there are no expressions to estimate the damping capacity.

More specifically, the joint stiffness within the bearing is in closer relation to the load-carrying capacity, and thus some marked remedies have been suggested by the SKF [21]. In fact, the SKF proposed the rollers with logarithmic curve to realize the uniform interface pressure distribution at the race and also with toroidal curve to increase the roller length by paying the special attention to the compactness of bearing. These remedies may improve the contact condition of the roller to the race, resulting in an increase of joint stiffness. In addition, the behavior of the main spindle system can be facilitated by the preloading mechanism of the rolling bearing, and the bearing nut was very popular up to the 1990s. With the growing rotational speed, the stepped sleeve has been employed especially to reduce the unbalance, and following to it, the KMT-C nut was newly developed by the SKF. These new elements became gradually dominant, and Fig. 8-28 is a qualitative comparison of the performance of these two elements.

Importantly, the stiffness of the bearing itself has been investigated in nearly all cases; however, the joints between the outer ring and the housing, and between the inner ring and the shaft, must be considered. In accordance with the report of Levina and Kotlyarenko [22], the elastic deflection in such joints is relatively small compared with that caused by the races, although it depends upon the fitting tolerance. In details, they investigated the deflection of the single-row and double-row cylindrical roller bearings (inner and outer diameters being 50 and 80 mm) as well as single-row ball bearing (inner and outer diameters being 40 and 80 mm) by varying the fitting tolerance. According to their report, the elastic deflections in bearing/shaft seating and in a housing are up

KMT-C nut   Stepped sleeve

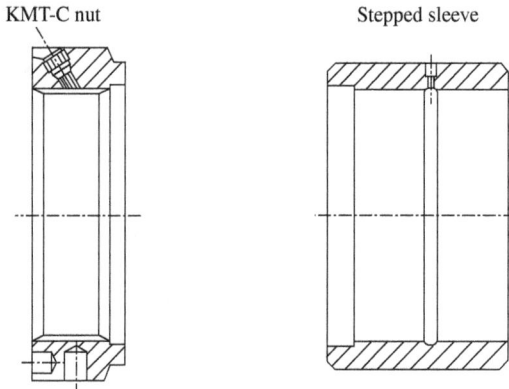

	KMT-C	Stepped sleeve
Assembly accuracy	5	5
Allowable axial load	5	5
High speed	4	5
Antishock load	4	5
Unbalance	4	5
Ease of maintenance	5	3

5: Best, 4: Good, 3: Having certain problem

Figure 8-28   Comparison of stepped sleeve and KMT-C nut (by Japan SKF, 1990s)

to 20% of the total elastic deflection of the bearing system under large loading. This magnitude cannot be disregarded when we discuss the joint stiffness of the bearing system.[16]

8.3.1 Static stiffness of rolling bearing

The rolling bearing is one of the very popular machine elements and very widely employed in various machines and equipment, which are, in nearly all cases, designed on the *principle of allowable stress*. This

---

[16]Notwithstanding its importance, we have no reliable and valuable database regarding the effects of the bearing housing on the bearing stiffness. At present, Koyama reviewed an example, and furthermore Podshchekoldin and Golubovskii investigated the effect of the radial clearance at seating surface on the vibration amplitude of the bearing. Of these, an interesting behavior is the directional orientation effect of the bearing housing, when the bearing housing is not a symmetric configuration.

Koyama, T., "Load Distribution and Fatigue Life of Rolling Bearings Considering Deformation of Mounting Structure around Bearings," *J. of JSLE*, 1979, 24(7): 424.

Podshchekoldin, M. I., and V. I. Golubovskii, "Elastic Oscillations in Anti-friction Bearings Assemblies," *Machines and Tooling*, 1968, 39(7): 16.

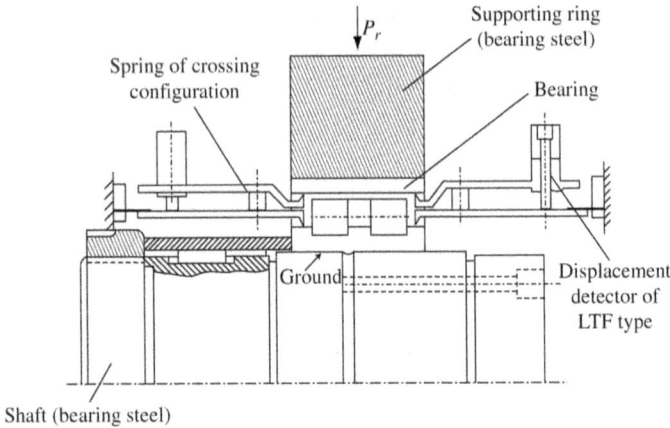

Figure 8-29   Experimental setup for measurement of bearing stiffness (by Günther).

means the important characteristics of the bearing are load-carrying capacity, rotating accuracy, and durability. As a result, there is an acute shortage of knowledge about the static stiffness. In the desirable case, the stiffness of the rolling bearing can be calculated theoretically in full consideration of the surface roughness at joints within the bearing.

In retrospect, Günther carried out a series of investigations, aimed at the establishment of a new calculation method for the cylindrical roller bearing of double-row type (NN30 and NNU 49 Series) and with both the positive and negative fitting tolerances [23, 24]. Figure 8-29 is a schematic view of the experimental setup, which can be characterized by its sophisticated features such as follows.

1. Consideration of manufacturing errors of bearing, such the out-of-roundness of race and roller

2. Consideration of the directional orientation of manufacturing errors in the bearing

Figure 8-30 is one of the measured results with respect to the elastic deflection to radial load, and on the basis of this knowledge, Günther proposed an expression by modifying that of Lundberg-Stribeck as follows.

$$\text{Radial stiffness } K_r = \{P_r^{0.1}/C_r^{0.9}\}\{[(n_i n_z)^{0.9} l_w^{0.8}]/0.54\}$$

where P_r = radial load
l_w = roller length
n_i = number of rows
n_z = number of rollers per row
C_r = load distribution factor (see Fig. 8-31)

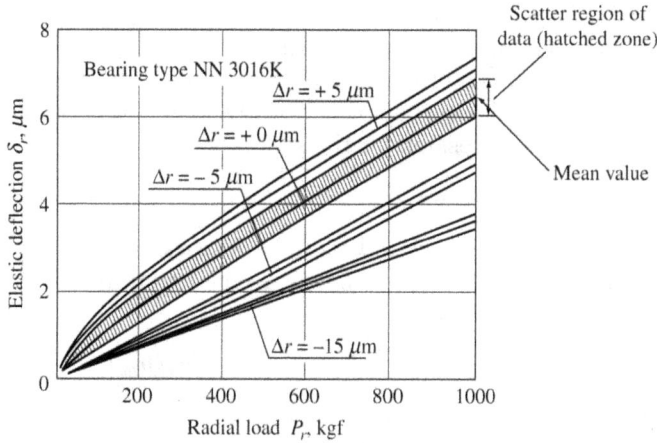

Δr: Fitting tolerance between the bore of inner ring and the shaft.

Figure 8-30 Measured values of bearing stiffness (by Günther).

The load distribution factor C_r is accommodated in order to consider the variation of the number of rollers being loaded, which is in dependence on the fitting tolerance.

In accordance with Lundberg-Stribeck, the coefficient is 0.6; however, to be compatible with the experimental values, Günther modified it to 0.54. This difference could be derived from the manufacturing errors and

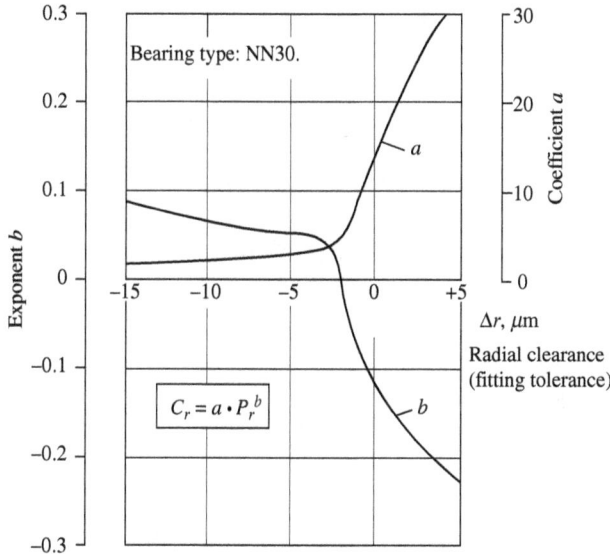

Figure 8-31 Calculation diagram of load distribution factor C_r (by Günther).

joint deflection within a bearing. Needless to say, the stiffness calculated by Günther is in good agreement with the measured value.

A similar attempt was made by El-Sayed for the radial stiffness of the deep-groove ball bearing [25]. On the basis of Hertz' contact theory, he derived first a theoretical expression for the stiffness K_r as

$$K_r = C_f P_r^{1/3}$$

where P_r and C_f are the radial load and stiffness factors, respectively. The stiffness factor is dependent on the bearing type and dimensions, and furthermore can be written as

$$C_f = \{531.6[(2.4d_o + d_i)/(d_o - d_i)]^{2/3}\{(M_i/m_i)C_i^{1/3} + (M_o/m_o)C_o^{1/3}\}^{-1}$$

where
$$d_o = \text{outer diameter of bearing, mm}$$
$$d_i = \text{bore diameter of bearing, mm}$$
$$M/m = \text{constant}$$
$$C_i = 2/r + 1/r_{ri} - 1/r_i$$
$$C_o = 2/r - 1/r_{ro} - 1/r_o$$
$$r = \text{ball radius, mm}$$
$$r_{ri} \text{ and } r_{ro} = \text{radii of raceways in rolling plane, mm}$$
$$r_{pi} \text{ and } r_{po} = \text{radii of raceways in perpendicular plane, mm}$$
subscripts i and o = inner and outer radii, respectively

Figure 8-32 shows some comparisons of theoretical and experimental stiffness after compensating the experimental values by subtracting

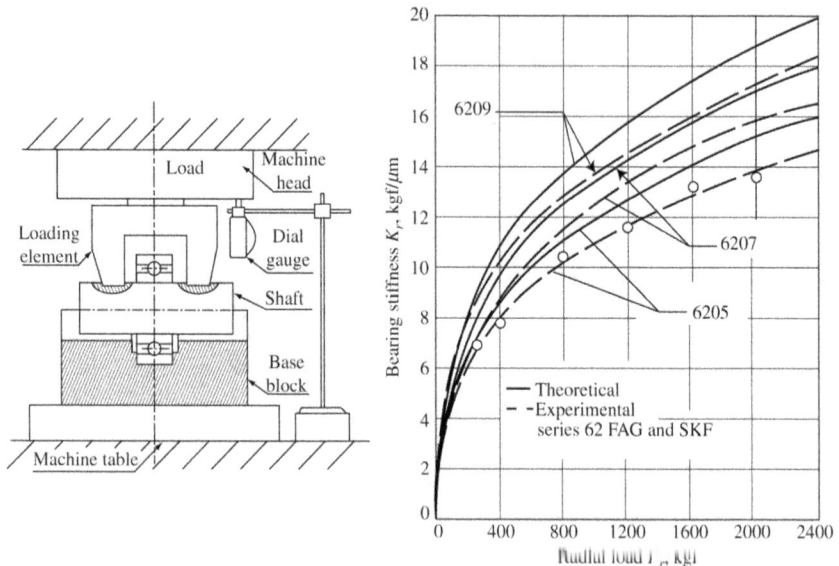

Figure 8-32 Comparison of theoretical and measured bearing stiffness (by El-Sayed).

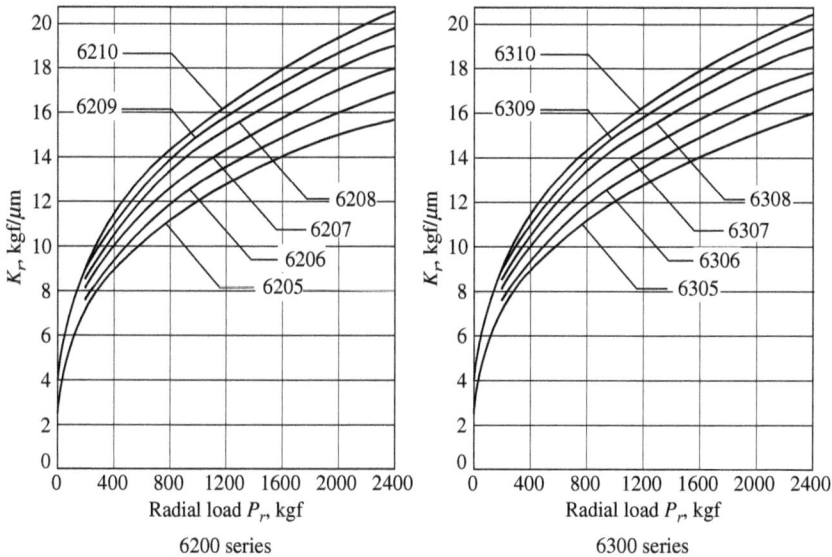

Figure 8-33 Calculation diagram for radial stiffness of ball bearings (by El-Sayed).

the corresponding shaft deflection. As in the work conducted by Günther, there are certain differences between both values, and El-Sayed considered that such differences may be caused by (1) the initial clearance and (2) approximation of the dimensional specifications of the bearing. However, it appears that the difference is, in part, due to the joint surface. For the convenience of the design, Fig. 8-33 shows the calculation diagrams for the stiffness of ball bearings.

In the main spindle, a combination of the rolling bearings for radial loading and for thrust is very common, and a facing problem is the estimation of the axial stiffness. In this case, a root cause of difficulties lies in a considerable number of joints such as already shown in Fig. 8-26. As verified by Borshchevskii et al. [26], the expression of Ostrovskii is applicable to such a bearing unit together with considering the manufacturing errors, e.g., nonparallelism of ball tracks, size differences, and out-of-roundness of balls. In short, the computed axial stiffness is 9770 kgf/mm, whereas the experimental stiffness is 10,810 kgf/mm in the case of a combination of two angular contact ball bearings (55 mm × 90 mm × 18 mm) and a thrust ball bearing (60 mm × 85 mm × 17 mm) under preloading of 100 kgf. Figure 8-34 is a calculation diagram for axial stiffness of the thrust ball bearing. In addition, Fig. 8-35 shows the axial stiffness of various ball bearings, i.e., deep-groove, angular contact, and thrust types, and roller thrust bearing. In short, the axial stiffness increases with the contact angle, whereas the maximum allowable speed decreases.

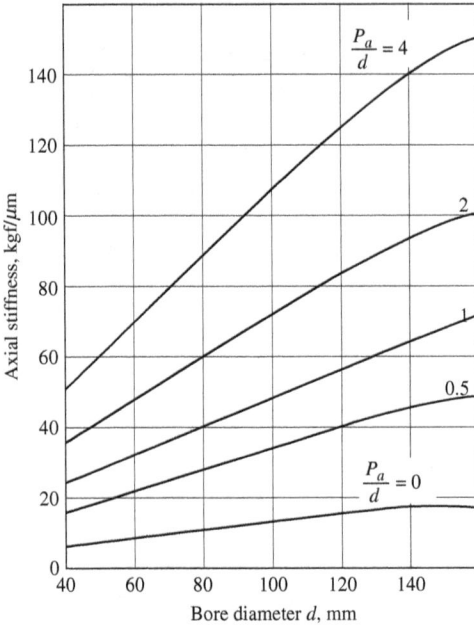

Pa: Preload (kgf)

Bearing type 8112: Bore diameter 60 mm
Outer diameter 85 mm
Width 17 mm

Figure 8-34 Axial stiffness of thrust ball bearing (by Borshchevskii et al.).

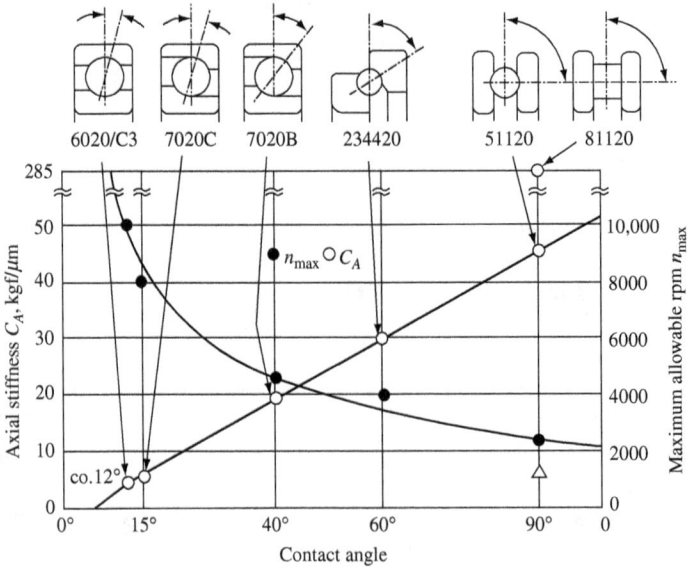

Figure 8-35 Stiffness and maximum allowable speed of ball and roller bearings (by Kunkel).

8.3.2   Dynamic stiffness and damping capacity of rolling bearing

Even in the case of static stiffness, we have not had the reliable database systematically publicized and available for the practical design, although the rolling bearing manufacturers endeavor to provide the customer with necessary data on demand. In due course, we face more difficulties in the case of the dynamic stiffness especially when the bearing is rotated.

Walford and Stone measured the radial dynamic stiffness of the ball bearing of angular type while rotating and lubricating by the gravity feed system [27]. In the measurement, a pair of identical bearings of 60 mm bore was used to realize the axial preloading in a usual bearing arrangement and vibrated with the electrodynamic exciter. Figure 8-36 shows

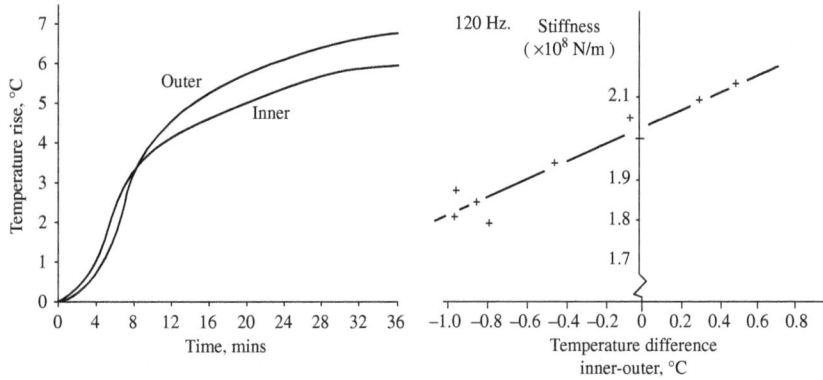

(a) Graph of bearing temperature against time after start at 2000 rpm

(b) Effect of raceway temperature on stiffness

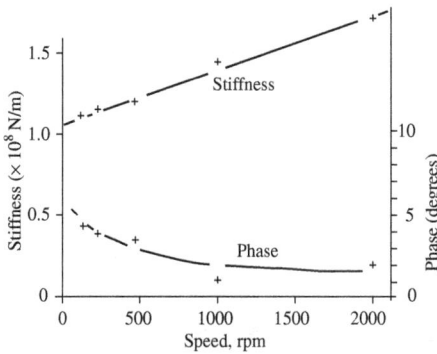

(c) Effect of speed on stiffness
Excitation: frequency 120 Hz
Force amplitude 283 N

Figure 8-36   Changes for dynamic stiffness of ball bearing with rotational speed and running temperature (by Walford and Stone, courtesy of I MechE).

some measured results, and as can be seen, the dynamic stiffness increases and decreases with the rotational speed and force amplitude, respectively. Importantly, the most marked behavior is that the dynamic stiffness is in larger dependence upon the temperature of inner and outer races. This implies that the dynamic stiffness becomes lower as the operating time is longer.

To estimate the dynamic characteristics of the main spindle-bearing system, at issue is to have the data for the damping capacity of the bearing itself while rotating. Although the engineer has well recognized such an importance, it is, in general, very difficult to carry out the theoretical calculation correctly or to measure accurately the damping capacity. In short, a dire necessity is, at least, to produce the damping distribution diagram for the shaft-bearing-bearing housing system.[17]

Reportedly, damping in the rolling bearing may be caused by the following factors.

1. Deformations of rolling elements and rings

2. Deformation in ring fits

3. Friction of rolling elements against rings especially arising out of the change of inclination angle of the shaft

4. Friction of rolling elements against the cage

5. In the tapered roller bearing, the friction between the roller ends and the ribs of inner ring, resulting in the appearance of peculiar behavior

Having in mind such difficulties, Elsermans et al. and Tsutsumi et al. conducted the valuable researches into the damping capacity of the rolling bearing while rotating.[18] In short, Elsermans et al. measured the damping capacity of the single-row tapered roller bearing while its inner ring is rotating [28], where the inner and outer diameters of the bearing are 60 and 110 mm, respectively. In contrast, Tsutsumi et al. measured the damping capacity of various rolling bearings while rotating the outer ring [29]. To carry out an accurate measurement, they used a

---

[17]Reshetov and Levina measured the damping capacity of the bearing in still stand. Following the work of Reshetov and Levina, Peters reported a measured result of the damping capacity for the main spindle of engine lathe, showing the importance of the preload. In fact, the damping capacity is maximum at certain preload.

Reshetov, D. N., and Z. M. Levina, "Damping of Oscillations in the Couplings of Components of Machines," *Vestnik Mashinostroyeniya*, 1956, 12: 3 (translated into English by PERA).

Peters, J., "Damping in Machine Tool Construction," in S. A. Tobias and F. Koenigsberger (eds.), *Proc. of 6th Int. MTDR Conf.*, Pergamon, 1966, pp. 23–36.

[18]Itumlı and Palıgauz measured the damping capacity of the main spindle of boring and milling machine in 1967 (see Chap. 5).

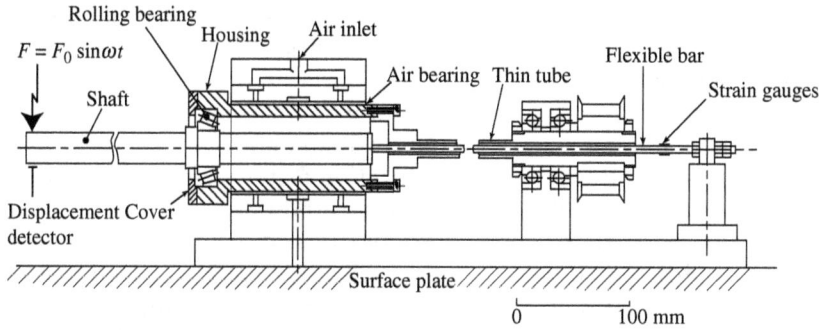

Figure 8-37   Experimental setup to measure damping of bearing while rotating (courtesy of Tsutsumi).

simple model such as shown, e.g., that of Tsutsumi, in Fig. 8-37. In both experimental setups, they were very keen to eliminate the disturbance from the driving system and to avoid the unfavorable disturbance from surroundings. For instance, in Tsutsumi et al., the shaft is fixed using a flexible bar, the bearing housing is supported by the air bearing to minimize the influence of damping of surroundings, and the housing is driven by the DC motor through the gear-toothed belt and thin tube. In addition, the bearing and its surroundings are air-cooled, and oil bath lubrication is used, so that the temperature of the shaft-bearing system is constant.[19]

It is very interesting that Elsermans et al. suggested the three kinds of damping in the tapered roller bearing, i.e., radial, axial, and clamping damping. Of these, clamping damping is dominant, and thus they measured it when applying the grease as a lubricant and varying the preload and rotational speed, where clamping damping (Nm·s/rad) is said to be the energy dissipation due to the angular oscillating motion between the outer and inner rings of a bearing. The clamping damping capacity increases and decreases with the preload and rotational speed, respectively, and shows a steep increase at the lower preload range. In short, the magnitudes of the clamping damping capacity are, e.g., as follows, when the axial preload is varied between 2500 and 7600 N.

1. In the rotational speed of 220 rpm, clamping damping ranges from 42 to 95 (N · m × s/rad) in the order of the axial preload.

2. In the rotational speed of 800 rpm, clamping damping ranges from 24 to 75 (N · m × s/rad).

---

[19]As pointed out by Walford and Stone [27], the bearing temperature has a considerable effect on damping of the bearing.

In the experiment, clamping damping was measured with the impact excitation and curve-fitting method. In addition, the outer and inner rings were press-fitted.

Figures 8-38 and 8-39 show the effects of the rotational speed and preload on the damping ratio reported by Tsutsumi et al. The damping ratio was measured from the magnification factor in second vibration mode. As can be seen from these results, the tapered roller bearing shows obviously different characteristics from other kinds as well as larger damping than others. More specifically, damping of the tapered roller bearing is strongly dependent upon the preload and rotational speed, simultaneously showing the maximum value at certain rotational speed. It can be deduced that such characteristic behavior is attributed to the structural configuration of the tapered roller bearing. In fact, the larger damping capacity may be facilitated with the contacts between the cone back face rib of inner ring and the larger end of roller. In addition, Tsutsumi et al. reported the following observations.

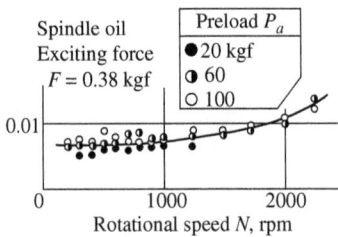

(a) Deep-groove ball bearing

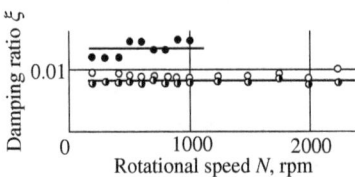

(b) Angular contact ball bearing A

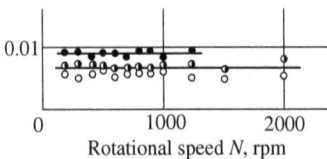

(c) Angular contact ball bearing C

(d) Tapered roller bearing

Deep groove ball bearing	No. 6206
Angular contact ball bearing C	No. 7206A
Angular contact ball bearing C	No. 7206C
Tapered roller bearing	No. 30206

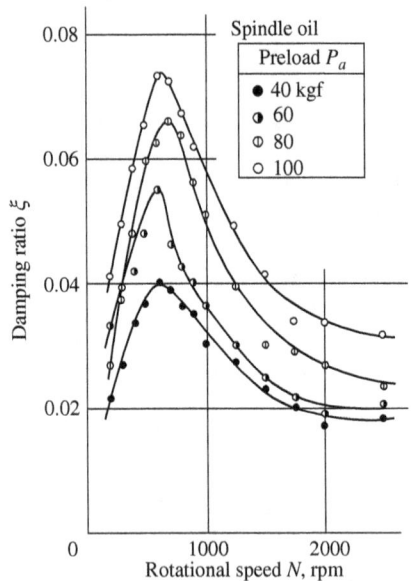

Figure 8-38 Relationships between rotational speed N and damping ratio ξ (courtesy of Tsutsumi et al.).

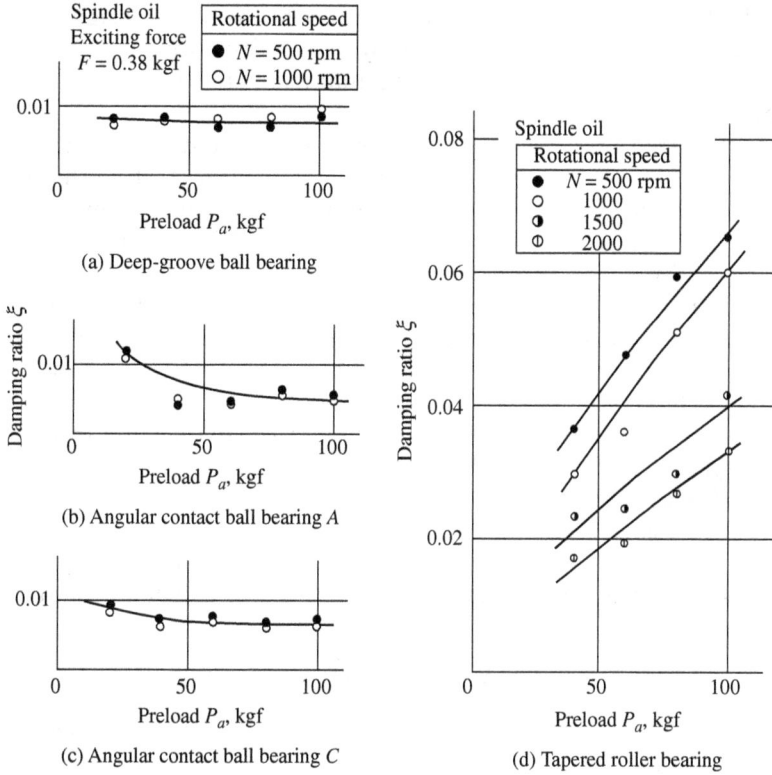

Figure 8-39 Relationships between preload P_A and damping ratio ξ (courtesy of Tsutsumi).

1. Apart from the cylindrical roller bearing, the bearings of other types do not show any vibration amplitude dependence.

2. The lubricant plays certain roles to increase the damping capacity of the tapered roller bearing, although it is difficult to find an obvious trend.

To this end, we must discuss the fitting tolerance of the bearing and the temperature rise in the main spindle-bearing system.

In fact, a crux is how to determine rationally the fitting tolerance between the outer race and the housing, and between the shaft and the inner race. The designer has been accustomed to choosing the preferable fitting tolerance on the basis of the standard without any doubts. A questioned point is that the fitting tolerance is actually regulated in the static condition by measuring the related dimensions of each part. Something uncertain remains thus in the fitting tolerance when the machine tool is at work. More specifically, a dire necessity is belatedly to clarify the difference in the fitting tolerance between the design stage (static fitting tolerance) and the operating condition (dynamic fitting tolerance), and the first work on the dynamic fitting tolerance is credited to Inaba in 1995 (refer to App. 1).

Inaba reported that the interface pressure varies obviously with the running time, showing a peak value immediately after starting the run and gradually decreasing its magnitude with the running time. This behavior can be interpreted as obvious evidence to demonstrate a considerable difference in fitting tolerance between the still and running conditions, resulting in considerable changes of the joint stiffness. More importantly, we must envisage that a dire necessity is to reconsider the preferable fitting tolerance so as to enhance the performance of the machine tool [30, 31].

Another crux is how to estimate the temperature in the main spindle-bearing system, and because of its importance, there are a considerable number of related researches and engineering developments. For example, Jędrzejewski et al. recommend the following expressions to estimate the frictional moment in the bearing by modifying a mathematical model of Snare-Palmgren, which is basically necessary to calculate the temperature rise [32].

$$M = M_0 + M_1$$

where $M_0 = 10^{-7} f_0 (\nu\, n)^\sigma d_m^{\,3}$, N · mm
$M_1 = f_1 \Gamma d_m$, N · mm
σ = 2/3 for jet lubrication and 1/3 for grease lubrication
ν = dynamic viscosity at running temperature, mm²/s
n = rotational speed of bearing, rpm
d_m = pitch circle diameter of bearing, mm
Γ = equivalent load, N
C_{M0} = coefficient depending on kind and lubricant of bearing
C_{M1} = coefficient depending on kind and loading of bearing

Table 8-4 shows the values of f_0 and expressions for Γ.

8.4   Sliding Joints of Special Types

As shown already in Table 8-1, the machine tool has the screw-and-nut, double-pinion, worm-worm rack, and pinion-rack driving systems, and of these, ball screw-and-nut driving is one of the variants of screw-and-nut driving and has become very popular nowadays. Another variant is a fixation of the screw end using the thread-nut connection. In addition, the machine tool of a certain kind has a sliding joint of special type, which facilitates the machine with special functionality. For example, the utmost representative is the spindle complex in the horizontal boring

TABLE 8-4   Coefficient f_0 and Equivalent Loads Γ (courtesy of Jędrzejewski)

Kinds of bearing	f_0		Γ F_r : Radial load F_a : Axial load
	$\sigma = 2/3$	$\sigma = 1/3$	
Cylindrical roller two-row NNU 49***K NN30***K	1.0 1.5	10 12	F_r F_r
Cylindrical roller single-row	1.3	12	F_r
Angular contact ball $\alpha = 15°$ $\alpha = 25°$ $\alpha = 40°$	0.5 0.7 1.0	8 10 13	$3.2\,F_a - 0.1\,F_r$ $1.80\,F_a - 0.1\,F_r$ $F_a - 0.1\,F_r$
Angular contact thrust ball 2344***2347***	1.5	14	$0.4\,F_a - 0.1\,F_r$
Tapered roller	1.8	18	Around $9\,F_a$
Thrust ball	0.8	9	$(2–3)F_a - 0.1\,F_r$

and milling machine, where the boring spindle placed inside the hollow milling spindle can travel axially, while its rotating. In other words, the objective is two-layered spindle configuration.

8.4.1   Screw-and-nut feed driving systems

Shuvalov et al. [33] conducted a research into damping of the screw-and-nut transmission and supporting bearings of the feed screw. Although not showing the experimental setup, they evaluated the damping capacity of the screw-and-nut transmission of friction, ball, and hydrostatic types using the area of hysteresis loop or damped free vibration. Figure 8-40 shows the energy loss factor ψ, i.e., the ratio of dissipated to potential energies per cycle, of the screw-and-nut transmission of sliding type, when we vary the vibration frequency f and preinterface pressure p_i between the screw and nut flanks. As can be seen, the lubricant has large effects on the loss factor, and the loss factor decreases with the preinterface pressure. In the experiment, unfavorable damping due to the test rig was subtracted from the measured value by estimating it from the equivalent solid transmission system. To deepen the understanding, Table 8-5 summarizes the damping characteristics of the ball screw-and-nut transmission, where the loss factor is independent of the vibration amplitude, vibration frequency, and interface pressure and shows no effects of lubricants.

Importantly, the data shown in Fig. 8-40 and Table 8-5 include those of supporting bearings at both ends of the screw, and thus Table 8-6

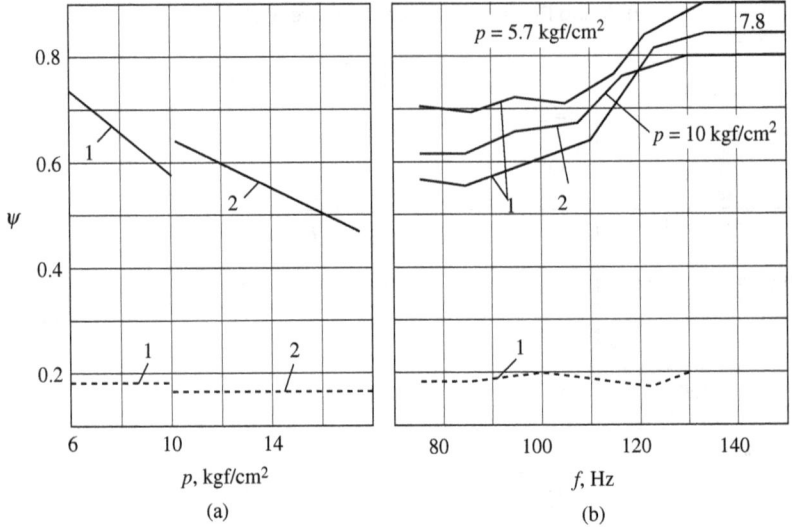

Curve 1: Screw dimension $70^\phi \times 12$ mm
Curve 2: Screw dimension $50^\phi \times 8$ mm

Full lines: With lubricant (industrial 45 oil)
Broken lines: Without lubricant

Figure 8-40 Energy Loss Factor of Screw-and-Nut Transmission (by Shuvalov et al.).

TABLE 8-5 Energy Loss Factor of Ball Screw-and-Nut Transmission (by Shuvalov et al.)

Dimensions of ball screw-and-nut transmission, mm	Axial preload, μm	Presence of lubricant	ψ when f (Hz) is:				
			80	100	120	140	160
$\phi 50 \times 8$	20	Yes	0.35	0.36	0.34	0.33	0.32
		No	0.35	0.35	0.34	0.32	0.32
	40	Yes	0.35	0.34	0.33	0.32	0.31
		No	0.35	0.34	0.32	0.32	0.31
$\phi 70 \times 10$	20	Yes	0.35	0.35	0.34	0.34	0.33
		No	0.35	0.34	0.32	0.33	0.32
	40	Yes	0.34	0.35	0.34	0.33	0.33
		No	0.34	0.32	0.33	0.33	0.32

TABLE 8-6   Damping Capacity of Supporting Bearings in Screw-and-Nut Transmission (by Shuvalov et al.)

Type of bearing	Diameter of screw, mm	Presence of lubricant	ψ at f (Hz)			
			80	100	120	140
Two angular ball + one radial ball	ϕ 50	Yes	0.22	0.22	0.22	0.21
		No	0.19	0.19	0.19	0.18
	ϕ 70	Yes	0.22	0.22	0.22	0.21
		No	0.19	0.19	0.19	0.18
Two thrust ball+ one radial ball+ one radial ball	ϕ 50	Yes	0.29	0.29	0.29	0.28
		No	0.23	0.23	0.23	0.23
	ϕ 70	Yes	0.3	0.3	0.3	0.3
		No	0.24	0.24	0.24	0.23

Note: Axial preload 100 kgf

shows the damping capacity of supporting bearings, whose average value is between 0.2 and 0.3. From Table 8-6, furthermore, we can observe that damping of the thrust bearing unit is higher than that of the angular ball unit.

On the basis of these experimental results, Shuvalov et al. proposed an empirical expression for the damping capacity of the screw-and-nut transmission of sliding type as follows.

$$\psi = C_d l \sqrt[3]{p}$$

where C_d is a factor depending upon the kind of lubricant and p is the interface pressure in kgf/cm^2.

8.4.2   Boring spindle of traveling type

In general, primary concern is the two-layered spindle as mentioned above; however, in the horizontal boring and milling machine with face plate-integrated type, the three-layered spindle must be considered, although only the boring spindle can travel. The machine with face plate is now in resurgence after modernizing to meet the new machining requirements of the present, those for system machine as mentioned in Part 1. From the structural design point of view, a root cause of difficulties lies in the determination of the clearance between the boring and the milling spindles in full consideration of the leverage of smoother moving and high stiffness. In addition, the boring spindle is in the cantilever configuration and under rotational cutting force in certain boring work.

Figure 8-41 shows a schematic view of main spindle of the horizontal boring and milling machine and also a mathematical model to calculate the static behavior of the two-layered spindle proposed by Bykhovskii [34]. Intuitively, the model is acceptable; however, at issue is how to quantify the nonlinear characteristics of the clearance

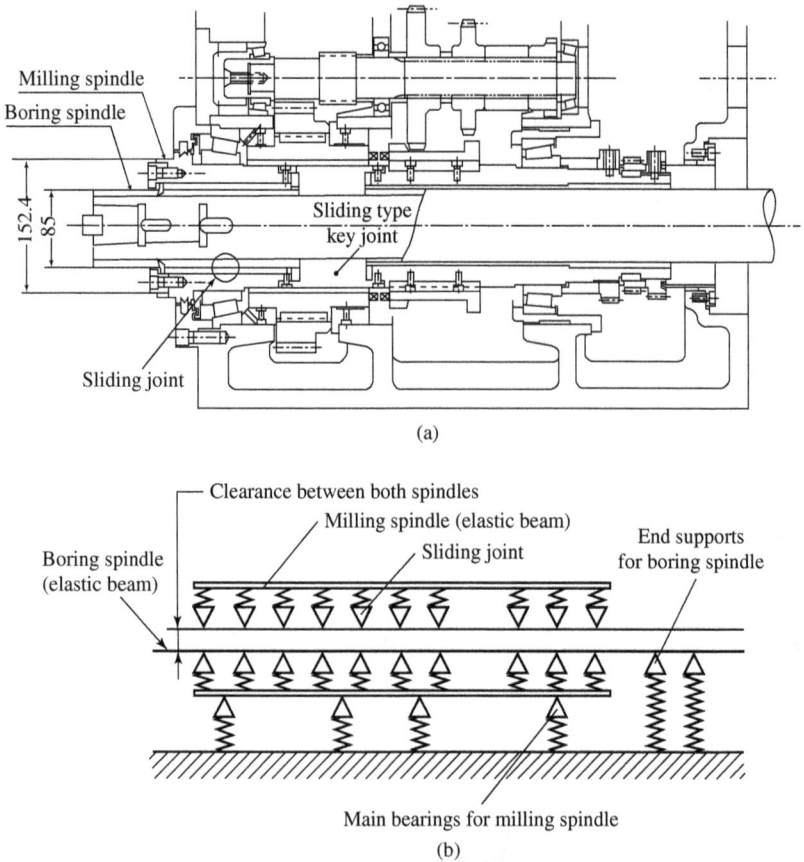

(a)

(b)

Figure 8-41   Multiple-layered main spindle of traveling type and its mathematical model: (a) Main spindle of boring and milling machine and (b) mathematical model for two-layered spindle (by Bykhovskii).

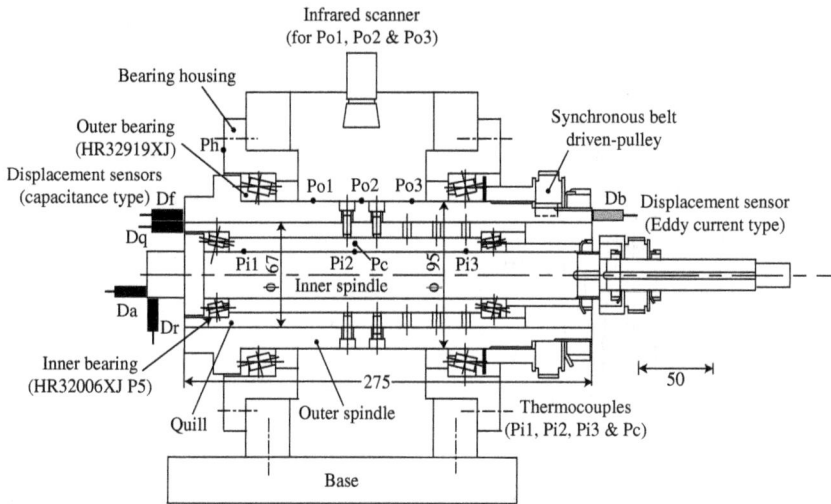

Figure 8-42   Schematic view of experimental setup for an innovative two-layered spindle.

with special respect to the joint stiffness. Even in the year 2000, the stiffness for the cylindrical joint with clearance was not clarified as yet.

Figure 8-42 shows the test rig employed in a recent research into the two-layered spindle. In consideration of the growing importance of the multiple-layered spindle as a system machine, Inaba et al. have proposed a new configuration, in which the boring spindle is supported with the rolling bearing [35]. This configuration can be characterized by the differing higher rotational speed between the outer and inner spindles, and this benefits to suppress the thermal displacement of the main spindle as follows.

1. The temperature rise of the inner spindle can be minimized at certain rotational speed of the outer spindle, while rotating the inner spindle.

2. A cooling system for the outer bearing is very effective, when the outer spindle is rotated.

In the horizontal boring and milling machine of large-size, furthermore, the underarm and ram are considered as to be a variant of the main spindle. In this context, a further variant is that of traveling ram in the vertical turning machine as shown in Fig. 8-43.

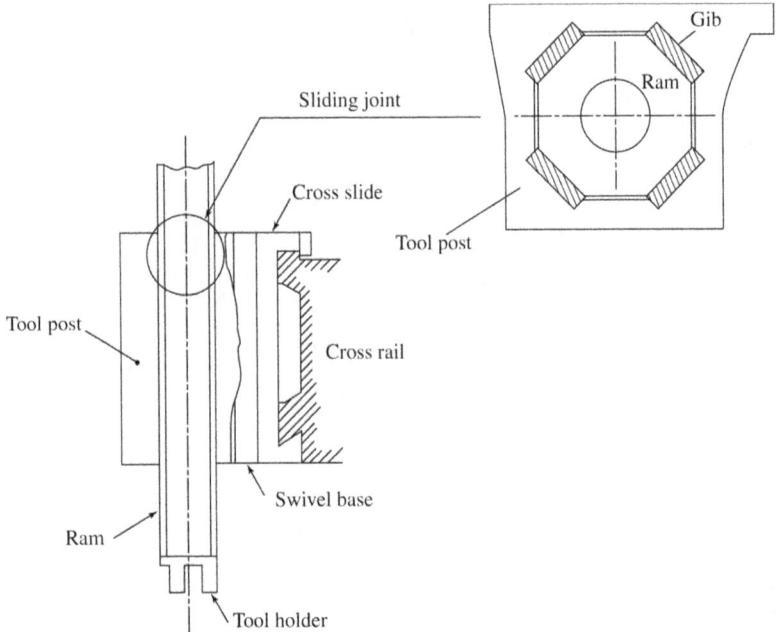

Figure 8-43   Traveling ram of vertical turning machine.

References

1. Black, T. W., "Machine Tool Way Bearings: A Brief Guide to Their Selection," *Machinery*, June 1966, pp. 106–113.
2. Poláček, M., and Y. Vavra, "The Influence of Different Types of Guideways on the Static and Dynamic Behaviour of Feed Drives," in S. A. Tobias and F. Koenigsberger (eds.), *Proc. of 8th Int. MTDR Conf.*, Pergamon, 1968, pp. 1127–1138.
3. Bell, R., and M. Burdekin, "The Friction Damping of Plain Slideways for Small Fluctuations of the Velocity of Sliding," in S. A. Tobias and F. Koenigsberger (eds.), *Proc. of 8th Int. MTDR Conf.*, Pergamon, 1968, pp. 1107–1125.
4. Bell, R., and M. Burdekin, "The Influence of Slideway Materials and Lubricants on the Dynamic Characteristics of Plain Slideways," *Proc. of Inst. Mech. Engrs.*, 1971, C79/71: 128–134.
5. Groth, H., "Die Dämpfung in verspannten Fugen und Arbeitsführungen von Werkzeugmaschinen," Dissertation der RWTH Aachen 1972.
6. Furukawa, Y., and N. Moronuki, "Contact Deformation of Machine Tool Slideway and Its Effect on Machining Accuracy," *Trans. of JSME (C)*, 1987, 53(485): 228–234.
7. Kaminskaya, V. V., et al., "Bodies and Body Components of Metal Cutting Machine Tools," *Mashgiz* (translated by PERA), 1960.
8. Levina, Z. M., "Calculation of Contact Deformation in Slideways," *Machines and Tooling*, 1965, 36(1): 8–17.
9. Tsutsumi, M., K. Hanaguri, and Y. Ito, "Influence of the Position of Driving Force Application on the Characteristics of Slideways," *J. of JSPE*, 1981, 47(6): 663–668.
10. Devitt, A. J., "Sliding-way Design Primer," *Manuf. Engg.*, 1998, 120(2): 68–72.
11. Zelentsov, V. V., "Deformation of Taper Gibs for Machine-Tool Slides," *Machine & Tooling*, 1966, 37(10): 19–22.
12. Metzger, H., "Spieleinstellung bei Flachführungen," *Werkstatt und Betrieb*, 1976, 109(1): 23–24.

13. Levina, Z. M., and V. I. Ostrovskii, "Influence of Gib Deformations on Pressure Distribution and Rigidity of Slideways," *Machines and Tooling*, 1963, 34(9): 10–15.
14. Burkov, V. A., "Taper Gib Slide-Adjusting Mechanisms," *Machines and Tooling*, 1969, 40(1): 11–15.
15. Renker, H., "A New Type of Linear Roller Bearing," in S. A. Tobias and F. Koenigsberger (eds.), *Proc. of the 10th Int. MTDR Conf.*, Pergamon, 1970, pp. 469–474.
16. Hallowes, J. G. M., and R. Bell, "The Dynamic Stiffness of Antifriction Roller Guideways," in S. A. Tobias and F. Koenigsberger (eds.), *Proc. of the 13th Int. MTDR Conf.*, Macmillan, 1973, pp. 107–112.
17. Levina, Z. M., "Main Operating Characteristics of Anti-Friction Slideways," *Machines and Tooling*, 1965, 36(7): 10–15.
18. Bankmann, G.., "Wälzführungen in der Feinwerktechnik," *Werkstatt und Betrieb*, 1976, 109(4): 203–206.
19. Hajdu, G.., "The Influence of the Characteristics of Machine Tool Guideways concerning the Dynamic Behaviour of Machine Tool Slides," in F. Koenigsberger and S. A. Tobias (eds.), *Proc. of the 14th Int. MTDR Conf.*, Macmillan, 1974, pp. 473–478.
20. Weck, M., and L. Ophey, "Experimentelle Ermittlung der Steifigkeit und Dämpfung radial belasteter Wälzlager," *Industrie-Anzeiger*, 1981, 103(79): 32–35.
21. Catalog of Japan SKF, 1990s.
22. Levina, Z. M., and L. B. Kotlyarenko, "Elastic Displacements in Anti-Friction Bearing Seatings," *Machines and Tooling*, 1971, 42(11): 39–41.
23. Günther, D., "Untersuchung der Federung von Hauptspindel-Largerungen in Werkzeugmashinen," *Industrie-Anzeiger*, 1965, 87(78): 319–326.
24. Opitz, H., et al., "The Study of the Deflection of Rolling Bearings for Machine Tool Spindles," in S. A. Tobias and F. Koenigsberger (eds.), *Proc. of 6th Int. MTDR Conf.*, Pergamon, 1966, pp. 257–269.
25. El-Sayed, H. R., "Stiffness of Deep-Groove Ball Bearings," *Wear*, 1980, 63: 89–94.
26. Borshchevskii, V. M., et al., "Axial Stiffness of Spindles Supported by Thrust Ball-Bearings," *Machines and Tooling*, 1974, 44(7): 18–20.
27. Walford, T. L. H., and B. J. Stone, "The Measurement of the Radial Stiffness of Rolling Element Bearings under Oscillating Conditions," *J. Mech. Engg. Sci.*, 1980, 22(4): 175–181.
28. Elsermans, M., M. Hongerloot, and R. Snoeys, "Damping in Taper Roller Bearings," in F. Koenigsberger and S. A. Tobias (eds.), *Proc. of 16th Int. MTDR Conf.*, Macmillan, 1976, pp. 201–206.
29. Tsutsumi, M., N. Nabeta, and N. Nishiwaki, "Damping in Single-Row Rolling Bearings," *Proc. of the 4th ICPE*, , JSPE and JSTP, Tokyo,1980, pp. 374–379.
30. Inaba, C., et al., "In-Process Measurement of Contact Pattern Variations in Rotating Main Bearing of Machine Tools Using Ultrasonic Waves Method," *Trans. of JSME (C)*, 1995, 61(588): 3375–3381.
31. Inaba, C., et al., "Estimation of Contact Pressure between Bearing and Bearing Housing by Means of Ultrasonic Waves," *Trans. of JSME (C)*, 2000, 66(645): 1674–1680.
32. Jędrzejewski, J., W. Kwaśny, and J. Potrykus, "Beurteilung der Berechnungsmethoden für die Bestimmung der Energieverluste in Wälzlagern," *Schmierungstechnik*, 1989, 20(8): 243–244.
33. Shuvalov, V. Yu, Z. M. Levina, and D. N. Reshetov, "Damping Longitudinal-Vibrations in Screw-and-Nut Transmission," *Machines and Tooling*, 1973, 44(4): 6–11.
34. Bykhovskii, A. N., "Stiffness of the Spindle Assembly for Horizontal Boring Mills," *Machines and Tooling*, 1973, 44(9): 13–15.
35. Inaba, C., et al., "Remedies for Thermal Deformation of Two-Layered Spindle for System Machine Tools," *Trans. of JSME (C)*, 2000, 66(648): 2864–2870.

Supplement: Deflection and Interface Pressure Distribution of Slideway

Figure 8-S1 is a schematic view of the carriage of the engine lathe, and for it the engineering calculation will be stated in the following. In this case, the carriage can be regarded as a stiff body, i.e., the stiffness of joint

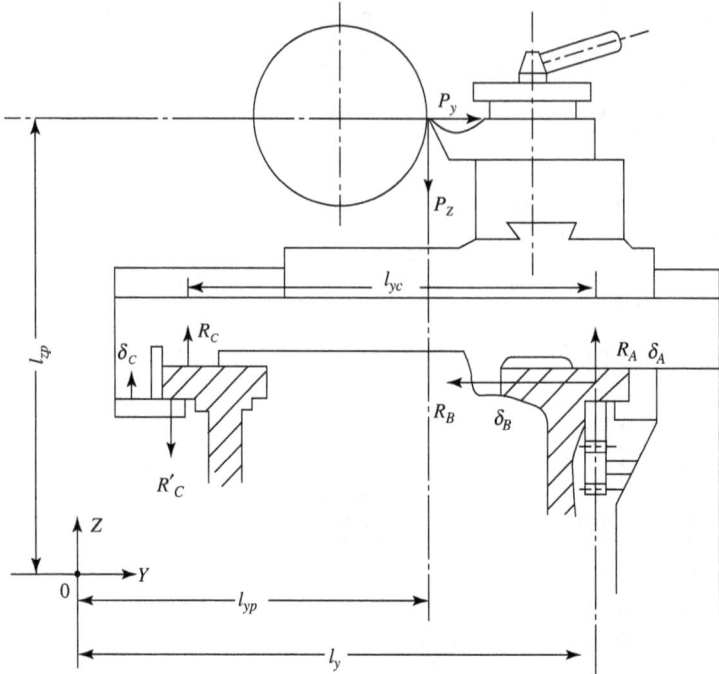

Figure 8-S1   Carriage of engine lathe.

surroundings is definitely larger than the joint stiffness, and thus the following three expressions are applicable.

1. The joint deflection λ, as already stated in Chap. 6, can be written as

$$\lambda = C_s p_n$$

where p_n is the mean interface pressure and C_s is given by Table 8-S1 [8].

2. The compliance of gib C_g can be given by modifying the value C_s in consideration of the weakness of gib.

$$C_g = \xi_g C_s$$

where ξ_g is 1.5 to 2.5. ξ_g must be smaller when the gib is longer and p_{ng} (average interface pressure between both gib adjusting screws) is less than 3 kgf/cm^2, whereas it must be larger when the gib is shorter and p_{ng} is 10 to 15 kgf/cm^2. In the case of the gibs widely employed, ξ_g is as shown in Fig. 8-S2.

TABLE 8-S1   Values of C_s

Mean interface pressure p_m, kgf/cm^2	Width of slideway mm	C_s $\mu m \cdot cm^2/kgf$
	< 50	0.5–0.7
	< 100	1.0
< 3.0	< 200	2.0–2.5
	< 300	3.2
	< 400	4.0
> 3.0–4.0	Reduce to 40%–50% of values for p_m, < 3.0	
Increase to 50%–70% of values for p_m, < 3.0 when local deflection appears Increase to 30%–40% of values for p_m, < 3.0 in case of vertical slideway		

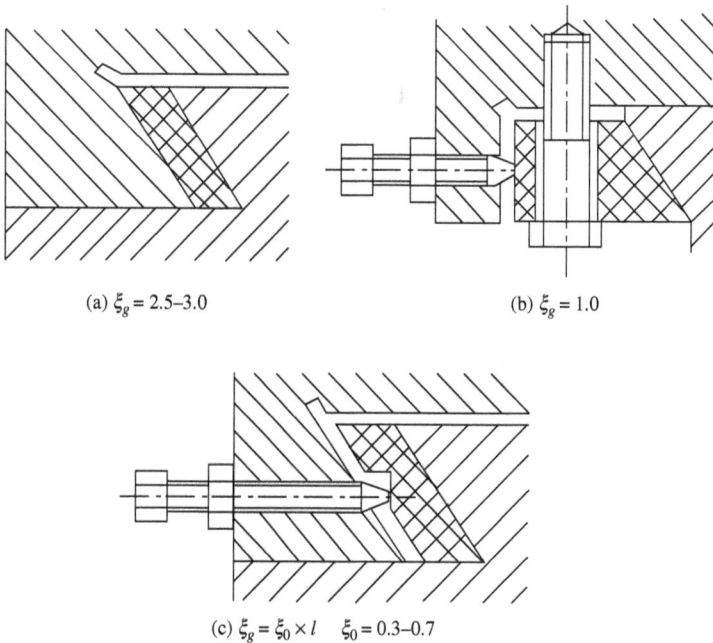

(a) $\xi_g = 2.5$–3.0

(b) $\xi_g = 1.0$

(c) $\xi_g = \xi_0 \times l$ $\xi_0 = 0.3$–0.7

Figure 8-S2   Values of ξ_g.

3. The compliance of keep plate C_k must be modified the same as that of gib.

$$C_k = \xi_k C_s$$

where ξ_k is, in general, 1.5 to 2.5 when the average interface pressure at keep plate p_{nk} is smaller than 10 to15 kgf/cm^2, and can be given by Fig. 8-S3.

For the details of ξ_g and ξ_k, refer to Levina and Ostrovskii [13].

Elastic deflection and interface pressure in *Y-Z* plane of slideway. In consideration of the equilibrium of forces and moments, the reaction forces R_A, R_B and $R_C(R_{C'})$ can be obtained and duly the inclination angle of the carriage yields to

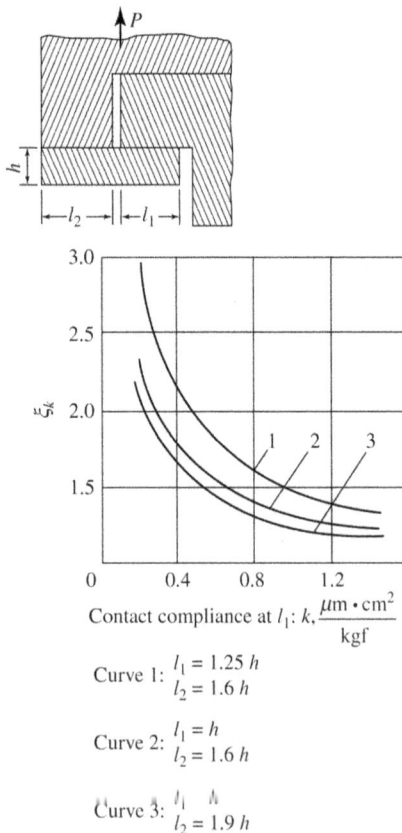

Contact compliance at l_1: $k, \dfrac{\mu m \cdot cm^2}{kgf}$

Curve 1: $\begin{aligned} l_1 &= 1.25\,h \\ l_2 &= 1.6\,h \end{aligned}$

Curve 2: $\begin{aligned} l_1 &= h \\ l_2 &= 1.6\,h \end{aligned}$

Curve 3: $\begin{aligned} l_1 &= h \\ l_2 &= 1.9\,h \end{aligned}$

Figure 8-S3   Values of ξ_k

1. $\varphi_x = C_s(p_{nA} - p_{nC})/l_{yc}$,   when $p_{nA} > 0$ and $p_{nC} > 0$ ($p_{nA} > p_{nC}$)

    $p_{nA} = R_A/l_a L_G$ $p_{nC} = R_C/l_c L_G$

where l_a, l_c, and l_b are the widths of front and rear slideways, and also the abutment lip.

2. $\varphi_x' = C_s(p_{nA} + \xi_k p_{nC'})/l_{y'c}$,   when $p_{nA} > 0$ and $p_{nC'} > 0$

    When we consider the clearance Δ at the slideway,

$$\varphi_{x0} = \varphi_x + \Delta/l_{y'c}$$

In addition, the displacement of the origin O to Y and Z axes can be given by

$$\delta_{y0} = C_s p_{nB} \quad \delta_{z0} = C_s(p_{nA} + p_{nC})/2 \quad \text{or} \quad \delta_{z0} = C_s(\xi_k p_{nC'} - p_{nA})/2$$

Therefore, the deflection of the arbitrary point within the Y-Z plane yields to

$$\delta_y = \delta_{y0} \pm \varphi_x z \qquad \delta z = \delta_{z0} \pm \varphi_x y$$

In due course, the displacement at the edge of the cutting tool and maximum interface pressure in the rear slideway can be written as

Deflection of cutting point to Y direction $\delta_{yp} = \delta_{y0} + \varphi_x l_{zp}$

Deflection of cutting point to Z direction $\delta_{zp} = \delta_{z0} - \varphi_x l_{yp}$

$$p_{nCmax} = p_{nC} + \varphi_x l_c/2C_s$$

Elastic deflection and interface pressure in X-Z plane of slideway. The moment distributions M_{Ay}, M_{Cy} at both slideways can be obtained, provided that the inclination angle of the saddle is the same at both slideways. Figure 8-S4 is a calculation diagram for M_{Ay} and M_{Cy} (for the details of the calculation, see Atscherkan [S1]). After M_{Ay} and M_{Cy} have been obtained,, the maximum interface pressure p_{nmax} can be determined using the calculation diagram shown in Fig. 8-S5. In Fig. 8-S5, p_n is the average interface pressure in the longitudinal direction of the saddle, and the horizontal axis corresponds with the ratio of moment to concentric load at each slideway.

Assuming that the moment M_y acting on both slideways can be divided into two components and the magnitude of each component is in proportion to the width of each slideway, the inclination angle φ_y of the saddle can be written as

$$\varphi_y = C_s C_\varphi\{12\,M_y/[(l_a + l_c)L_G^3]\}$$

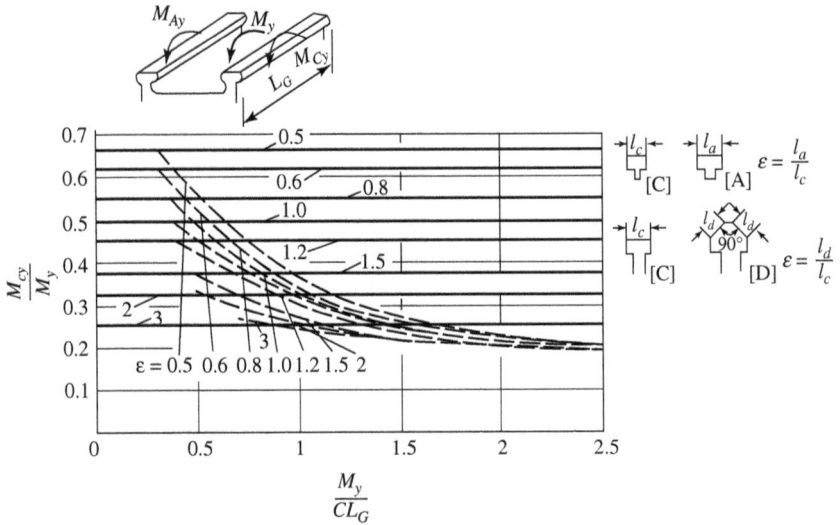

Solid line: Keep plate is on work
Broken line: Keep plate is not on work

Figure 8-S4   Calculation diagram for M_{Ay} and M_{Cy}.

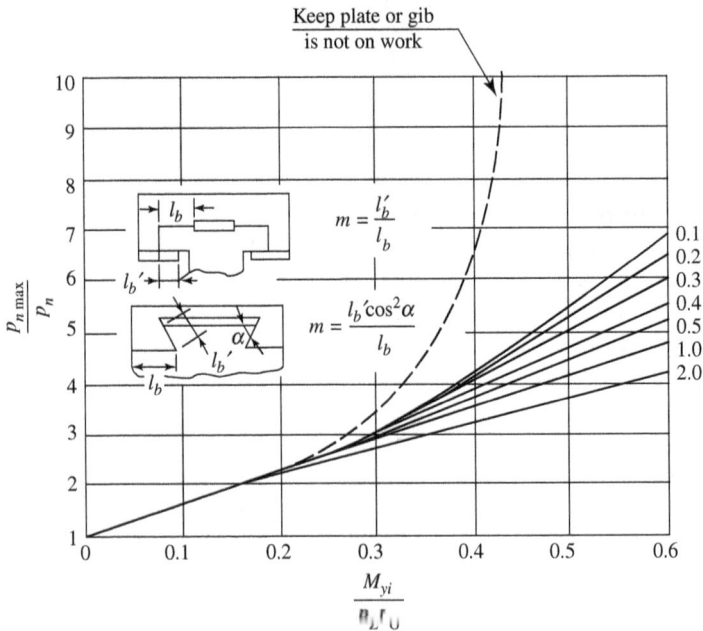

Figure 8-S5   Calculation diagram for maximum interface pressure.

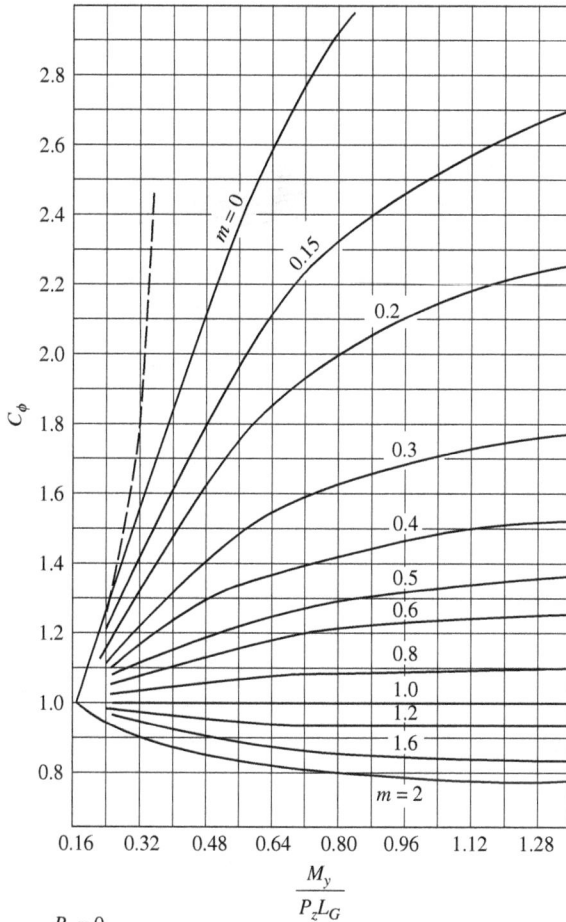

Figure 8-S6   Calculation diagram for C_φ.

$P_z = 0$							
m	0.1	0.2	0.3	0.4	0.5	0.6	0.7
C_ϕ	4.3	2.5	2.0	1.67	1.45	1.32	1.2

where C_φ is a coefficient to consider the noninterface pressure area across the whole slideway length, and can be obtained from Fig. 8-S6, and H is the length of saddle wing.

Within an engineering calculation context, the following three expressions can be finally obtained.

1. The displacement of the cutting edge:

$$\delta_{xp} = \varphi_y l_{zp} = C_s C_\varphi (12 M_y l_{zp})/[(l_a + l_c)L_G^3]$$

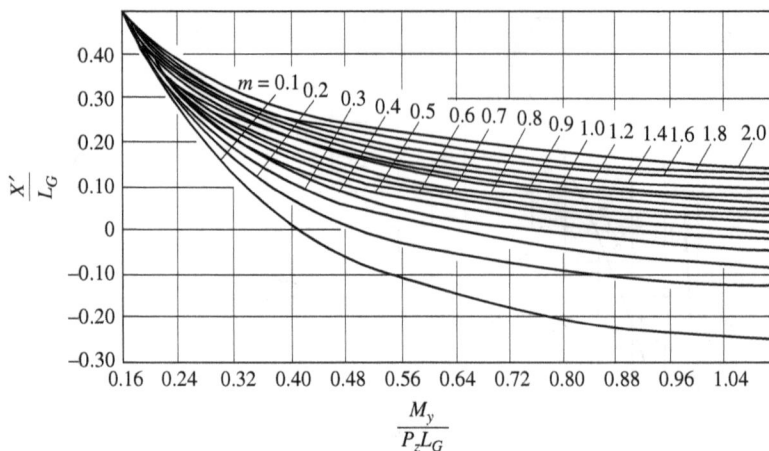

Figure 8-S7   Calculation diagram for X'/H.

2. The deflection along Z axis at distance x from the center of saddle:

$$\delta_z = \varphi_y(x + x') = C_s C_\varphi [12 M_y/(l_a + l_c) L_G^2][x/L_G + x'/L_G]$$

where x' is the distance between the center of saddle and neutral axis of the interface pressure distribution, and can be determined by Fig. 8-S7.

3. The interface pressure distribution:

$$p(x) = C_\varphi \{12 M_y/[(l_a + l_c) L_G^2]\}\{x/L_G + x'/L_G\}$$

Supplement Reference

S1. Atscherkan, N. S., "Werkzeugmaschinen Band 1," *VEB Verlag Technik Berlin,* 1958, pp. 269–289.

Rudimentary Engineering Knowledge about Other Joints

As already shown in Fig. 5-5, the machine tool has a considerable number of the joints ranging from the bolted joint for the assembly of structural body components, through guideway and main bearing, to Curvic coupling of turret head and taper connection. When we change the viewpoint from the structural design to the machine element, there are various joints, e.g., key fixation of shaft and gear, clutch plate, retaining ring to fix machine elements and gear-spline shaft connection. It is thus vital to reconsider to what extent the machine tool joint must be included from the viewpoint of the modular design. Intuitively, we have, at present, no answer to this question, and thus this chapter will touch on some joints that are considered to be important from both the structural design and the utilization technology of the machine tool. In the case of the utilization technology, one of the important issues is the joint between the main spindle and the auxiliary devices, such as the chuck, cutting tool, jig, and fixture, as is well known from the old days and systematically classified, e.g., by Gunsser [1]. In short, the objectives of this chapter are the joints for the light-weighted structure, i.e., the welded joint and bonded joint and furthermore the taper connection and chucking.[1] For further convenience, Fig. 9-1 summarizes the research and engineering design map for the rest of the joints.

---

[1] In this chapter, some of the figures and tables are in the form of engineering design data sheets.

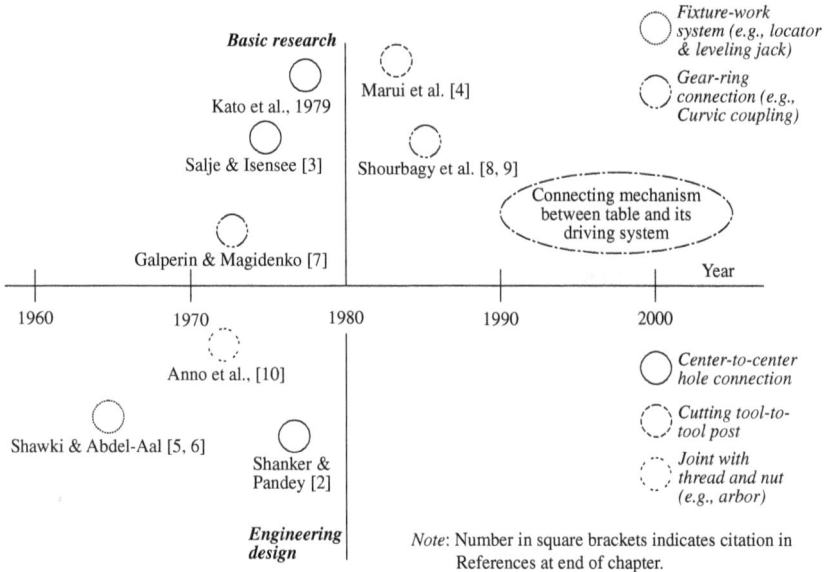

Figure 9-1   Firsthand view of research and engineering development in other joints.

9.1   Joints for Light-Weighted Structures

As widely understood, the machine tool is required to have simultaneously the larger static stiffness per unit weight and higher damping capacity. In due course, the light-weighted structure has, in principle, a higher possibility of being the machine tool structure to respond to such requirements. On the basis of long-standing experience, we can summarize the threefold remedies to be the desirable structure in reality such as shown in Fig. 9-2, i.e., those from the structural configuration, structural materials, and jointing method of the structural body component. More specifically, primary concerns in the light-weighted structure are the steel welded and bonded structures together with consideration of the sandwich and panel plates as the raw material for the structures. Typically, both the welded and bonded joints are special cases in the machine tool joint. In short, they are not combined to the structural body component in principle, but can fabricate the structural component itself. In addition, at issue is the structural configuration with the larger stiffness per unit weight, e.g., how to allocate and arrange the stiffening rib, connecting rib, double wall, and partition with aperture, in the expectation of increasing the damping capacity in certain cases. In this context, much of the knowledge for cast structure so far accumulated can facilitate the design of the light-weighted structure.

Importantly, from the viewpoint of the structural design principle, the steel bonded structure is very desirable, as can be seen from Fig. 9-2;

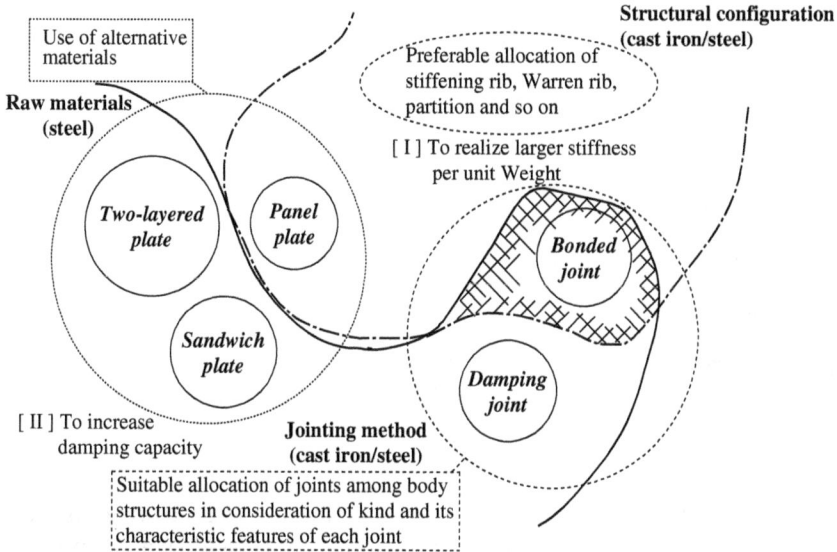

Figure 9-2 General guides for light-weighted structural design—various measures to increase static stiffness per unit weight and damping capacity.

however, for reasons of reliability and durability, the steel welded structure has prevailed.

9.1.1 Welded joint

The welded structure was one of the driving forces to carry out the research and development for the machine tool joint as already described in Chap. 5. On that occasion, the welded structure showed higher damping than that of cast structure; however, after then nearly all engineers believed that the welded structure has lower damping capacity than the cast structure. Obviously, this myth is derived from the considerable difference of the material damping between the structural carbon steel and the cast iron, both of which are the fundamental structural materials in the machine tool. In Table 9-1, some of the measured material damping capacities are summarized, and as can be seen, the damping capacity of cast iron is around 10 times larger than that of steel [11].

In contrast, Fig. 9-3 shows a comparison for the dynamic behavior between the welded and cast columns of full-size in a horizontal boring and milling machine of floor type. In this case, the rest of the structural body components are identical, and importantly there are no obvious differences between the columns. This is, as widely accepted now, due to the joint in a whole structure together, expecting the increase of damping by welding. In addition, we must be aware of the following.

TABLE 9-1   Specific Damping Capacities of
Representative Materials (by Stansfield)

Material	Specific damping capacity, %
Mild steel	1.5
3% Ni-Cr steel	0.8
Austenitic stainless steel	1.8
13% Cr stainless steel	3.8
Cast iron	10.6
70–30 brass	0.35
Gunmetal	0.9
Nickel aluminium bronze	0.1
Zn-Zr-Th magnesium alloy	12.5
Sonoston*	30.0

*Composition of Sonoston: Cu 25–50%, Ni 0.5–3.5%,
Al 2.5–6% Fe less than 5%.

1. Because of the differing material properties, the welded structure can be designed to have the thinner thickness for the side wall compared with the cast structure, and can thus accommodate the higher internal stress. This results in the larger material damping, as shown also in Table 9-2. Reportedly, material damping can be determined by (1) frequency of cyclic stress, (2) heat treatment and grain size of material, (3) circumference temperature, and (4) stress history. For example, the material damping of carbon steel is extremely large under shear stressing than tension-compression stressing.[2]

2. As suggested by Bobek et al., the welded joint shows the effect of *disturbance of shear stress flow* (die Störung des Schubspannungsflußes), which results in the increase of damping [12].

In general, the logarithmic damping decrement of the structural material is between 0.005 and 0.01, whereas the damping capacity of the joint ranges from 0.1 to 0.2, showing absolutely larger damping than the material damping itself. At present, the machine tool designer understands very well that there is no serious difference in the damping capacity between the welded and cast structures.

---

[2]There are many definitions for the damping capacity, and they are interrelated.

$$Q_D = \pi/\delta_D = 1/(2D) = 2\pi/s_D = 1/\eta$$

where Q_D = magnification factor
δ_D = logarithmic decrement
D = damping ratio
s_D = specific damping capacity
η = loss factor

	Natural frequency, Hz						Logarithmic damping decrement 10^{-2}					
	X direction/bending		Y direction/bending		Torsion		X direction/bending		Y direction/bending		Torsion	
	1st mode	2nd mode	1st	2nd	1st	2nd	1st	2nd	1st	2nd	1st	2nd
With cast column	24	140	14	140	73	265	4.6	4.1	9.6	4.6	4.6	4.4
With welded column	20	103	12	122	53	275	8.2	5.0	4.6	6.4	4.4	1.4

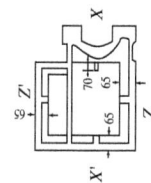

Diameter of boring spindle: 130 mm

Welded structure
Wall thickness   19 mm
Rib thickness   19 mm
Height of column   4880 mm

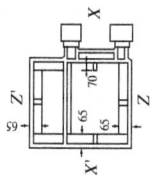

Cast structure
Wall thickness   30 mm
Rib thickness   20 or 25 mm
Height of column   3680 mm

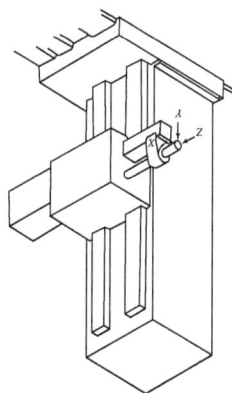

Cross-sectional configurations of columns

Figure 9-3 Comparison between cast and welded columns for horizontal boring and milling machine of floor type—experimental results of Yasui of MERL of MITI in 1971 (within the activities of JSME).

TABLE 9-2   Relationships between Internal Stress and Material Damping

Advantageous aspects:

Increase of material damping, which is derived from, e.g., thinner wall thickness, resulting in higher stress within the body structure. In general, the material damping increases with the stress being applied.
In addition, the steel welded or bonded light-weighted structure can easily form the boxlike configuration, i.e., those with closed cross-sectional configuration.

M. Kronenberg, P. Maker, and E. Dix, "Practical Design Techniques for Controlling Vibration in Welded Machines," *Machine Design*, July 12, 1956, pp. 103–109.
Yorgiadis, A., "Damping Capacity of Materials," *Product Eng.*, Nov. 1954, pp. 164–170.

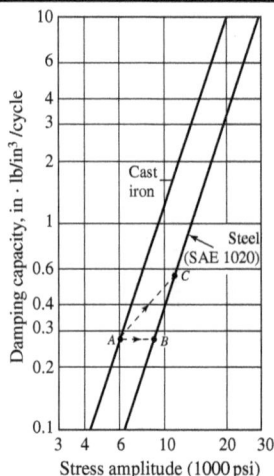

Disadvantageous aspects:

The structure is liable to show the cross-sectional distortion, membrane vibration, drum effect, and so on.

Necessity of providing the body structure with the desirable rib, stiffener, partition, and so on at suitable location to reduce such unfavorable deformation, maintaining larger static stiffness per unit weight	If the stress within the wall of cast structure is at point A, the stress within the wall of steel welded structure is at point C when both structures have the same stiffness, resulting in the welded structure showing about two times larger damping.

In consideration of those evidences mentioned above, the engineering design for the welded joint aims eventually at the increase of damping derived from welding together with positively using the advantageous features of the welded structure. In short, the stiffer the welded structure, the more difficult is to have larger damping as a rule of engineering design of the structure. For example, the intermittent joint shows considerable deterioration of the static stiffness, but larger damping compared with the joint of continuous welding.

Table 9-3 exemplifies the benefits and defects of the welded structure in general compared with the cast structure. It is, however, envisaged now that there are no differences in the static, dynamic, and thermal performances between the welded and cast structures. In fact, the employment of the welded structure relies on whether the manufacturing cost including the purchasing cost of the raw material can be reduced, and also on the market state of the foundry. In contrast, it is very difficult to estimate the economic benefits of the welded structure, because the manufacturing cost depends upon the manufacturing volume, available facilities, complexity of structure, level of worker's wedges, and so on.

Within a structure design context, the welded structure has been prevailed with growing production volume of the NC machine tool of

TABLE 9-3   Benefits and Defects of Welded Structure Compared with Cast Structure

Benefits	Defects
Suitable for structure by mass manufacturing or one-off manufacturing	Unsuitable for structure by batch manufacturing
Wide flexibility of structural design and redesign, because of not necessary wooden pattern	Liability of occuring local deformation or cross-sectional distortion of structure High possibility of local membrane vibration, i.e., "drum effect"
Increase of productivity, because no necessities in adjustments of interrelation are among pattern making, casting, and machining	Certain necessity of carrying out residual stress relief annealing
Capability of producing light-weighted structure with higher static and dynamic stiffness	
Simplicity of repair or refabrication of structure	
High possibility of providing high damping capacity using proper structural configuration and innovative welded joints	

small- and medium-sized, although it was only applied to the large-size machine tool in the past. In addition, the modular design often becomes applied to the NC machine tool and can in turn be facilitated by the welded structure, because of its ease of handling. As can be readily seen from Fig. 9-2, the engineering design data for the welded structure are threefold, i.e., (1) larger stiffness per unit weight, (2) larger damping capacity, and (3) use of new and innovative raw material. Some information regarding these will be quickly noted in the following.

Larger stiffness per unit weight. In principle, Young's modulus of steel is about two times larger than that of cast iron, and thus the wall thickness of the welded structure can be at least reduced up to one-half of the cast structure, when the designer intends to obtain the same static stiffness. In addition, there is no necessity to provide, e.g., the flange, seat, and boss, in the fabrication of the structural body component from the raw materials. As a result, the welded structure can be considered one of the desirable light-weighted structures; and to enhance the beneficial feature of the welded structure, the designer should duly aim at the realization of the structure with larger stiffness per unit weight.

In this regard, the engineering design knowledge for the cast structure is available to a greater extent, provided that the defect shown in

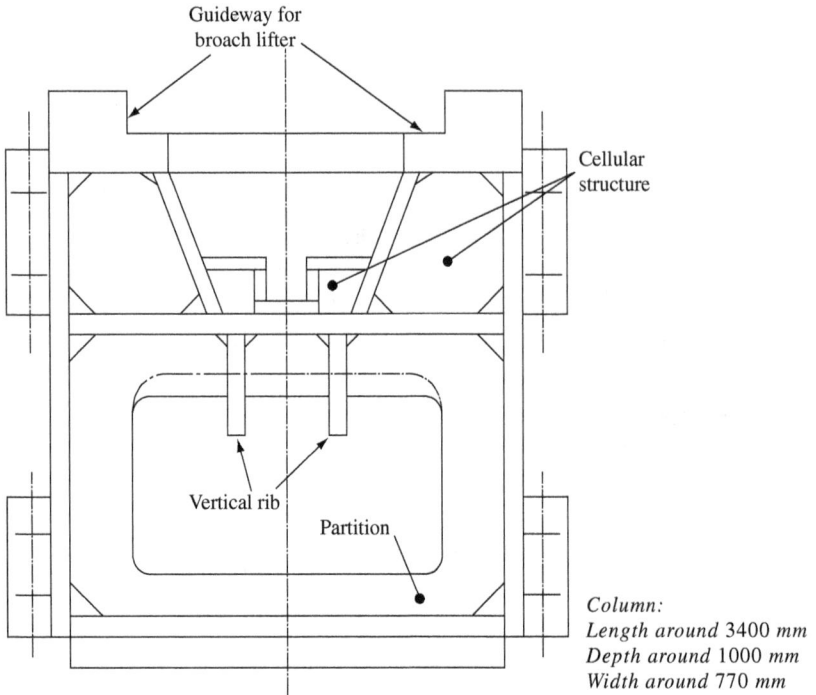

Figure 9-4 Steel welded column of vertical broaching machine (type NUV, from 1960s to 1990s, courtesy of Fujikoshi).

Table 9-3 is overcome. In fact, the local membrane vibration of thin plate, i.e., drum effect [13], appears often especially in the welded structure under dynamic loading with relatively high frequency. Figure 9-4 shows a representative of the welded structure, i.e., column of broaching machine. The basic structure of this machine was designed in the 1960s and up to 2000 has been maintained, although some minor improvements have been carried out. As can be seen, the machine consists of box and cellular structure, and of the stiffening partition and vertical rib, resulting in the typical welded structure. More specifically, the engineering design for welded structure was established to a larger and various extent between the 1960s and 1970s, and from it the following design principle can be recommended.

1. Employment of closed box and cellular configuration. The welded structure can be characterized by having these structural entities. In the case of cast structure, we face fatal hindrances to use these structural entities, because of difficulties in wooden pattern making and fettling.

Ferrite resin: Aggregate is ferrite powder produced from the wasted disposal water of factory. Utmost benefit is its good machinability (so far used as a bearing housing)

Ceramics plate
bonded on guideway

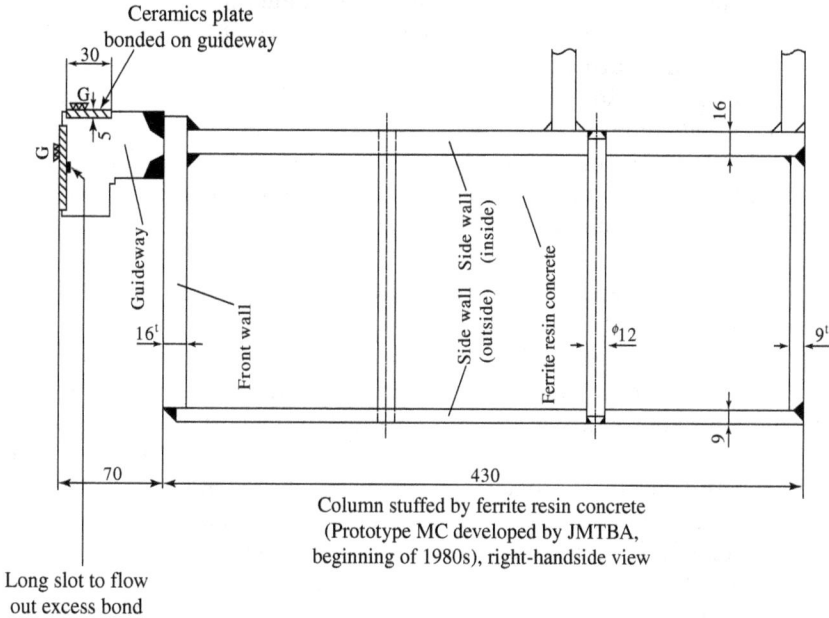

Column stuffed by ferrite resin concrete
(Prototype MC developed by JMTBA,
beginning of 1980s), right-handside view

Long slot to flow
out excess bond

Ito, Y., "Research and development activities to enhance market competitiveness of products in Japanese machine tool industry." in Rasmussen, L., F. Rauner (eds). *Industrial Culture and Production—Understanding Competitiveness.* Springer-verlag, London, 1996, pp. 107–133.

Figure 9-5   Applications of polymer concrete to MC column.

2. Employment of rib, partition, and double wall. These remedies are very common in the case of the cast structure.[3] However, a further consideration is required to prevent the cross-sectional distortion by relieving residual stress with aging. Figure 9-5 shows a portal column of MC with double wall, and the noteworthy feature is the use of thinner outer side and rear walls to absorb unfavorable distortion. In contrast, the column slideway and inner side wall are thicker to ensure better guiding accuracy.

---

[3]There are a handful of noteworthy reports. For example, that of Loewenfeld depicts the correlation between the torsional stiffness and rib arrangement in the square and oblong plates.

Loewenfeld, K., "Die Gestaltung von Platten," *Konstruktion,* 1957, 9(5): 180–187.

In retrospect, the welded structure was used first by directly replacing the cast iron with the steel, i.e., with the dimensional specifications and structural configuration being the same in both the welded and cast structures. As can be imagined, the welded structure has been improved to accommodate its beneficial features to a larger extent, resulting in the maturation of engineering design. Within this context, Bobek et al. [12], Kopitsyn [14], and Frank [15] already arranged and systematized some noteworthy data for the engineering design on the basis of the achievements of the past. Reportedly, their works are very informative and valuable even now, and available for the engineering design at present such as already exemplified in Fig. 9-4.

Larger damping capacity.   In the fabrication of the structural body component, various welded joints are used in consideration of their beneficial aspects, and in general, we use the continuous welded joints of fillet type to ensure satisfactory static stiffness. Obviously, the continuous welded joint shows larger static stiffness, but lower damping capacity, and thus at issue is to realize simultaneously both the higher static stiffness and damping capacity. On the basis of our long-standing experience, there are two remedies to leverage such discrepant requirements, i.e., applications of (1) stress concentration and (2) shear effect. The shear effect means the energy dissipation caused by microslip at the welded joint.

Figure 9-6(a) and (b) shows the shapes and dimensions of various continuous and intermittent welded joints of butt type with V-shape groove, and their measured logarithmic damping decrements, respectively [16]. The measurement was carried out by comparing those of equivalent solids made of steel and cast iron, and with the cantilever configuration in bending free vibration, in which the effective cantilever length was 360 mm. As can be readily seen, additional damping due to welding is not so large in the case of butt welding, and in addition, damping increases with the vibration amplitude. Table 9-4 shows furthermore the static stiffness, eigenfrequency, and logarithmic damping decrement, when the vibration amplitude is 100 μm in full scale. Reportedly, in the butt type, the static stiffness of the welded joint increases with the seam length, whereas the damping capacity decreases in the reverse way. These marked observations can be attributed to the stress concentration due to the notch effect at the welded joint. In fact, the stress concentration can be obviously observed at the welded joint as shown in Fig. 9-7, which was measured for one of butt welding shown in Fig. 9-6, when downward bending is applied. Obviously, the static and dynamic behavior of the welded joint is largely dependent upon the arrangement and allocation of the joint itself as well as the operation technique for welding.

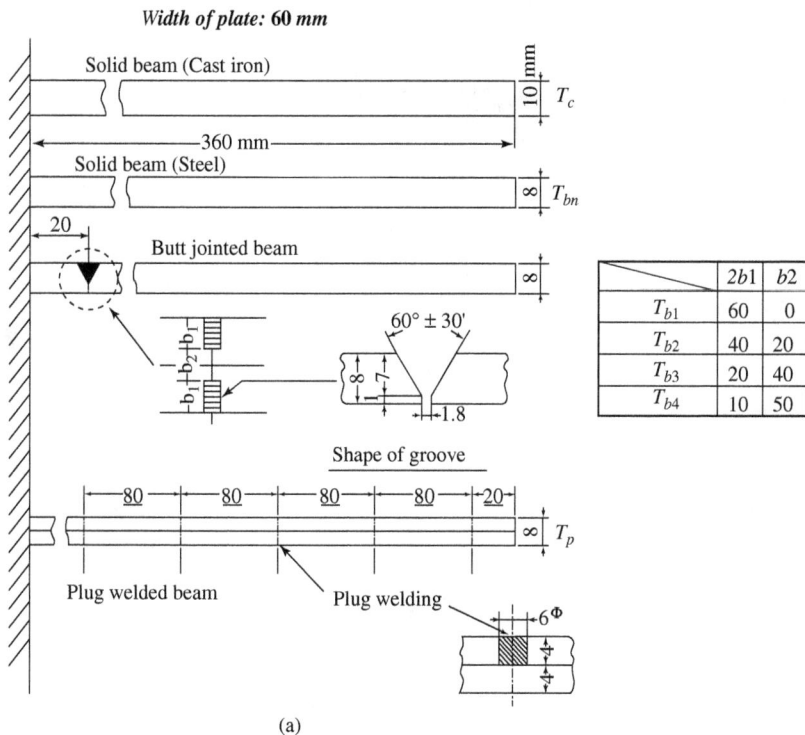

Width of plate: **60 mm**

Solid beam (Cast iron) — 10 mm — T_c

— 360 mm —

Solid beam (Steel) — 8 — T_{bn}

20

Butt jointed beam — 8

b_1 b_2 b_1

$60° \pm 30'$

8
7

1.8

Shape of groove

	2b1	b2
T_{b1}	60	0
T_{b2}	40	20
T_{b3}	20	40
T_{b4}	10	50

— 80 — 80 — 80 — 80 — 20 — 8 — T_p

Plug welded beam

Plug welding

6 Φ

(a)

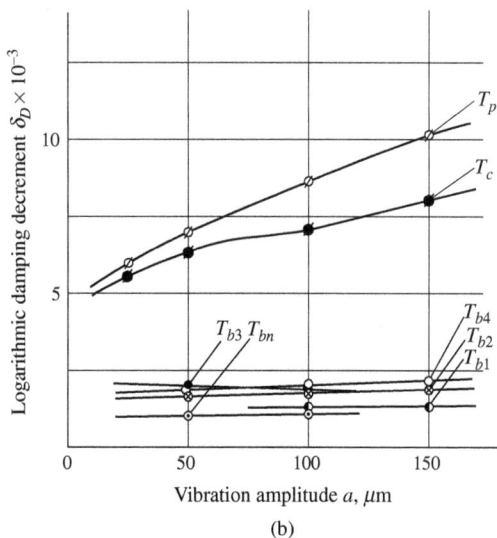

Figure 9-6 Damping capacity for various welded joints: (a) Shapes and dimensions of various welded joints and (b) effects of vibration amplitudes on damping capacity.

TABLE 9-4   Static Stiffness and Damping Capacity of Welded Plates in Cantilever Configuration

Testpieces	Logarithmic damping decrement $\delta_D \times 10^{-3}$	Eigenfrequency in 1st vibration mode f_n, Hz	Static stiffness k_x, kgf/mm
T_c	7.2	46	3.8
T_{bn}	1.3	48	3.8
T_{b1}	1.3	48	3.8
T_{b2}	1.8	47	3.6
T_{b3}	1.9	43	3.1
T_{b4}	2.1	39	2.6
T_p	8.6	46	3.4

Note: The vibration amplitude is 100 μm maximum.

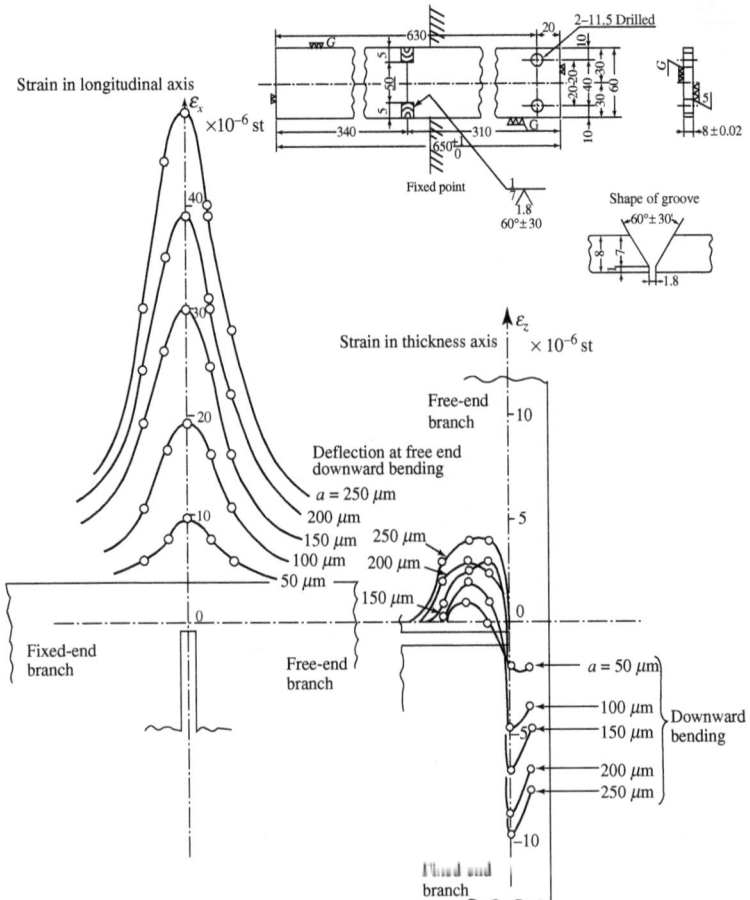

Figure 9-7   Stress concentration at intermittent welded joint.

Although damping of the welded joint increases by the stress concentration, a marked increase in damping can be expected when the shear effect functions as exemplified in the case of plug welding shown already in Fig. 9-6 and Table 9-4. More specifically, the welded joint of plug type shows larger damping than material damping of cast iron without deteriorating the static stiffness. In addition, the free decay vibration for the cantilever with plug welding behaves like that having viscous damping, which is very similar to that of bolted joint. In consideration of the welding pattern, plug welding has no apparent stress concentration, and in fact the simple flat joint with plug welding shows obvious microslip in its shear load-displacement curve as reported elsewhere. The effective area of the interface pressure may be considered as to be within the plug diameter, and thus an underlying hypothesis is that the microslip at plug welding differs from that of the bolted joint. As already stated in Chap. 7, the microslip at the bolted joint is caused by the elastic and plastic deformations of the seizure points within the joint surface. In consequence, Anno et al. implied that the microslip in plug welding is derived from the mediate layer between the welded deposit and the parent material, such as shown in Fig. 9-8, i.e., coarse grain in recrystallization zone.

As can be readily seen, larger damping at the welded joint can be facilitated by the shear effect, and thus there have been various trials to use the shear effect practically. In this regard, a facing necessity is to establish an engineering calculation for damping capacity, and thus at burning issue is how to estimate the interface pressure at the welded joint.

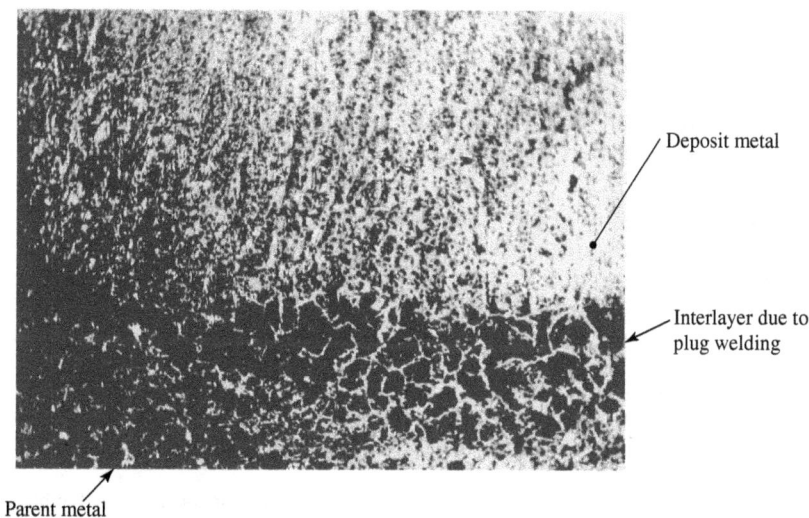

Figure 9-8 Cause of microslip at plug welding.

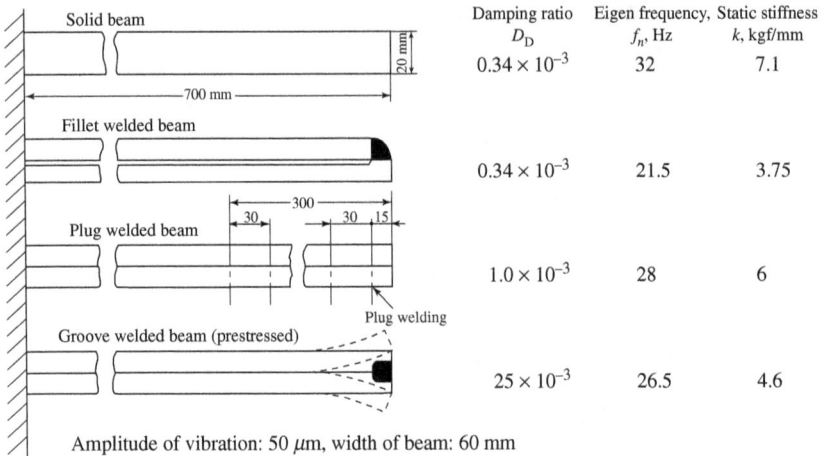

	Damping ratio D_D	Eigen frequency, f_n, Hz	Static stiffness k, kgf/mm
Solid beam	0.34×10^{-3}	32	7.1
Fillet welded beam	0.34×10^{-3}	21.5	3.75
Plug welded beam	1.0×10^{-3}	28	6
Groove welded beam (prestressed)	25×10^{-3}	26.5	4.6

Amplitude of vibration: 50 μm, width of beam: 60 mm

Figure 9-9 Larger damping in laminated and welded beam when using shear effect (by Bobek et al.).

To get some ideas on this subject, those of Ockert and Ito are very useful. They tried to calculate the damping capacity of the laminated plate of cantilever type (refer to the bolted joint in Chap. 7).[4]

Damping joint. In the book by Bobek, a comparison of damping capacity for laminated beam is quickly stated such as shown in Fig. 9-9, and larger damping can be obtained when the shear effect is applied adequately. Importantly, Kronenberg et al. contrived a marked welded joint called the *damping joint*, a basic configuration of which is shown in Fig. 9-10 [17]. In short, the damping joint utilizes effectively both the shrinkage stress caused by fillet welding and the frictional energy loss at filed flats. Consequently, the V- or U-shaped rib should be arranged so that welding can cause the compression stress at the filed flats. In due course, Kronenberg et al. applied the damping joint to the internal grinder of Bryant make by simultaneously using the plug, intermittent and continuous welding adequately and employing the closed-box section or cellular configuration to ensure higher rigidity. As a result, it was reported that the grinder could be free from disturbing vibration at least up to 180,000 rpm.

---

[4]In this context, refer to the following.

Eisele, F., and H. Drumm, "Steifigkeit und Dämpfung geschweißter Bauelemente," *Maschinenmarkt*, 1959, 2: 19–22.

Katzenschwanz, N., "Dynamische Stabilität geschweißter Konstruktionen im Hinblick auf die Erfordernisse im Werkzeugmaschinenbau," *Maschinenmarkt*, 1961, 79: 29–39.

Example of damping joints

For structure made of Al alloy

No limitation in thickness

Flat

For structure made of steel

Flat

$t_2 \gg t_1$

Continuous welds---1, 4, 9, 13, through18
Intermittent welds--2, 8, 11, 20
Plug welds---5, 12, 19
Damping joins--- ⊕
Partial damping joints--- ●

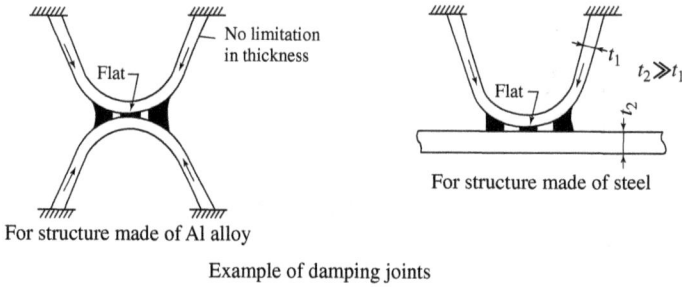

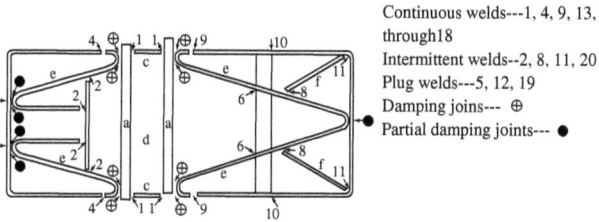

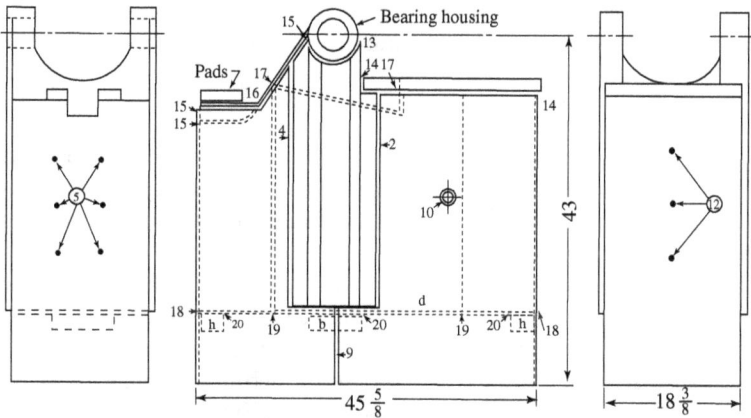

Practical application of damping joints to internal grinder

Bearing housing

Pads

$45\frac{5}{8}$

$18\frac{3}{8}$

Figure 9-10   Damping joint proposed by Kronenberg et al.

More specifically, the following three rules must be considered in the design of the damping joint.

1. The joint surface is under the prestressed condition. For instance, the two plates in prebent configuration are assembled so that the curvature of each plate is in opposite direction, and then are fillet-welded at their ends, after two compulsory plates are joined.

2. The joint surface under prestress should be as wide as possible.

3. The prestress should be as high as possible, provided that the microslip can be accommodated.

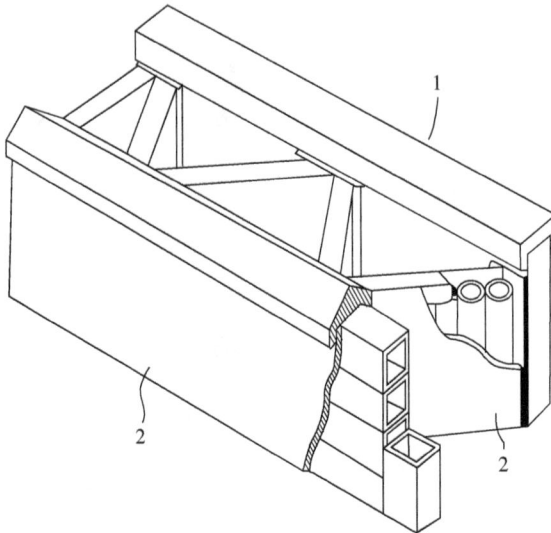

Application to lathe bed

Figure 9-11   A panel structure with temperature control function.

In addition, it is worth suggesting that damping of the welded joint caused by the microslip and the compression stress due to the shrinkage at the welded joint results in the frictional loss energy of viscous type, but not the dry type.

In the welded structure, as already shown in Fig. 9-2, what is the raw material is another crucial problem. Admitting that there have been many examples of using steel tube to realize the higher torsional stiffness and to reduce the manufacturing cost, the panel structure may have a certain potential as the raw material for welded structure. Figure 9-11 reproduces thus a proposal of Usi and Sakata (Japanese patent no. 1384337, June 26, 1962), although the structure is, in principle, of bonded type. In this panel structure, the tube can facilitate either cooling or heating fluid to flow, resulting in the structure with temperature control function.

In addition, we must be aware of further new materials called *porous material*[5] and Iso TRUSS^{TM} Grid Structure. In the former, we can

---

[5]In accordance with an achievement in prototype, the bending deformation by self-weight of 6.6 tons is 14 μm in the cross rail (5900 × 1400 × 940 mm) made of the sandwich plate with porous metal (AlMg1Si0.6: density 0.5 g/cm^3). In contrast, that of steel welded structure (self-weight: 6.3 tons) is 34 μm. Obviously, at issue is the welding technology to connect the porous material with the steel parts.

Neugebauer, R., T. Hipke, and S. Ihlenfeldt, "Hochdynamische Werkzeugmaschinen-strukturen und-komponenten," *ZwF,* 2001, 96(9): 445–450.

observe its typical application to the parallel link machine, where the cross rail must be light the same as the link mechanism, resulting in, e.g., traveling speed of 40 m/min with an acceleration of 10 m/s^2. In addition, such a cross rail shows the amazing increase in damping. In contrast, the latter is of composite material and nonstuffed gabionlike configuration. The Elektornik-Entwicklung has been claimed to apply this material to its product, i.e., five-axis high-speed processing center, type HSM-MODAL, around 2003. In short, the graphite/epoxy Iso Truss of 5-tow/6-node shows the following properties, when its geometric properties are 2.90 cm^2 in nominal area, 31.78 cm in average length, 8.43 cm in average diameter, and 17.9 cm^4 in torsion constant.

1. Approximate moduli: 25.5 in simple tension, 35.3 GPa in simple compression, and 527 MPa in simple torsion.

2. Approximate toughness: 0.37 MPa in simple compression and 0.26 in simple torsion.

To this end, Fig. 9-12 shows a representative column fabricated by welding. The column is for the boring machine with 8 or 10 inch spindle diameter of Giddings Lewis make and can be characterized by positively utilizing the beneficial features of the welded structure together

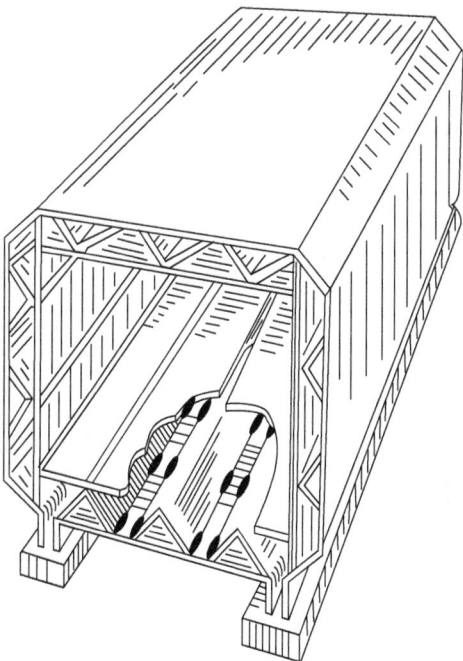

Figure 9-12   A representative welded column using panel structure, U.S. Patent No. 2789480 (Giddings Lewis make).

with realizing the panel structure. In fact, we can observe the following interesting features.

1. Use of box stiffeners with rolled triangular or bent profile as a rib to increase the rigidity of plate itself, high local wall and cross-sectional profile rigidities, although employing relatively thin wall thickness

2. In welding the box stiffeners, use of intermittent welded joint, resulting in higher damping

3. Use of double wall configuration to realize the maximum rigidity per unit weight

Table 9-5 is furthermore a proposal for the design database of welded structures.

9.1.2 Bonded joint

Although there have been a handful of marked trials for applying the steel bonded structure to the machine tool, the bonded joint is far from being a commonly available method for producing the machine tool structure. In fact, the bonded joint has lower reliability with special respect to the bonding strength and shows the deterioration of certain properties under the machining environment. Apart from the slideway bonded the hardened steel strip on, and also the bed bonded on the concrete base [18], the bonded structure has been employed, e.g., at the connection of the grip to the shift lever and in the manufacture of the gang gear, which plays no important roles in the structural stiffness.

TABLE 9-5 A Proposal of Design Database for Welded Structure

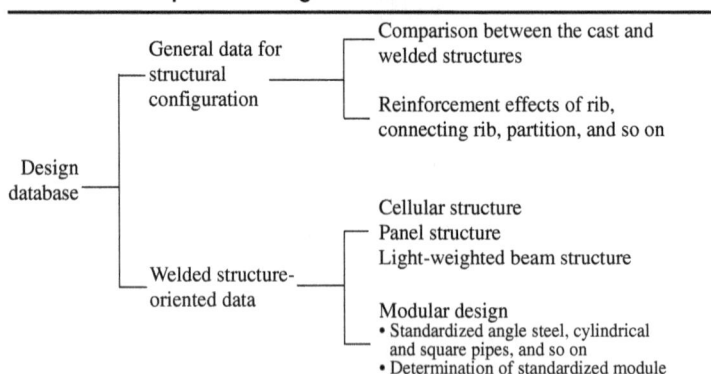

Design database
- General data for structural configuration
  - Comparison between the cast and welded structures
  - Reinforcement effects of rib, connecting rib, partition, and so on
- Welded structure-oriented data
  - Cellular structure
  - Panel structure
  - Light-weighted beam structure
  - Modular design
    - Standardized angle steel, cylindrical and square pipes, and so on
    - Determination of standardized module

At issue are the following.
1. Determination of standardized modules including widely available standardized raw materials
2. Leverage between the increase of stiffness and the productivity. The productivity becomes lower with providing more ribs and partitions to increase stiffness.

TABLE 9-6   Advantages and Disadvantages of Bonded Strip Slideway

Advantages	Disadvantages	Remedies to disadvantages
Wider availability of strip made of suitable antifriction materials with both required hardness and close tolerances

In certain cases, no machining requirements for slideway up to closer tolerances

Ease of repair of worn slideway | Necessity of machining of bonded strip after bonding, to produce better reference surface with less flatness deviation | Further machining after bonding not necessary
Supply of enough glue at bonding surface and squeezing excess glue from bonding surfaces using proper channels on bed slideway
Using a part of strip as slideway
Adhesive injected between strip and bed slideway through grooves located on top surface of bed slideway |
| No necessity of employing alloyed cast iron to improve quality of slideway | Large difference of temperature between strip and bed slideway | *Providing suitable tension to strip*
Strip slideway having less thermal susceptibility: Use of adhesive mingled portland cement as filler after sand blasting bed slideway |

In the bonded strip slideway, we can observe some prominent benefits, e.g., flatness deviation being equal to that obtained by scraping or grinding, and disadvantages as shown in Table 9-6. Concerning these disadvantages, Scharmann, one of the leading German machine tool manufacturers, has suggested some remedies, as shown also in Table 9-6; however, a serious problem is the rise of temperature of the strip more than that of the structural body component while running the machine tool. This is due to the low thermal conductivity of the adhesive, and thus a remedy is to provide the strip with suitable tension at its end by using some contrived devices. In addition, Dolgov and Nizhnik suggested that the length of the strip to be bonded must be limited to within a certain value [19].

In short, there are two-pronged applications of the bonded joint in consideration of its beneficial features. One is the high damping capacity in reality together with improving the fixing reliability, and the other is to realize the sure fixation of both parts with ease of handling. Obviously, the former is the target in the case of the machine tool joint, and in fact the most interesting trial using the bonded joint is credited to Lamb and Al-Timimi in 1978 [20]. They produced a full-sized milling machine as a prototype, and its noteworthy feature is the desirable use of the standardized joint, by which the steel plates can be connected with one another with epoxy resin bond. In addition, they proposed the design philosophy and some considerations for fabricating the bonded structure on the basis of both the experiment and the prototype production.

In the case of bonded structure, it is desirable to use positively the standard angles, channels, and I-shaped beams the same as in the welded structure. We should thus pay further special attention to the applicability of these standard members to the structure without showing the weakness at the joint under bending. In consequence, one standardized joint is to be reality such as a *double containment joint*, i.e., "plug-in-like joint," in full consideration of the perfect supply of bonding adhesives into the joint.

Importantly, the beneficial features of the bonded structure can be summarized as follows.

1. Higher possibilities to produce the light-weighted structure with higher damping capacity
2. Improvement of flatness deviation produced by machining, resulting in the improvement of the assembly accuracy and ease of assembly together with increasing both the static joint stiffness and the damping capacity in certain cases

More specifically, the bonded joint can be characterized by the following factors, which differentiate the bonded joint from the welded joint, although both are very effective in the production of the light-weighted structure.

1. The bonded structure shows lower stiffness and larger damping capacity than the welded structure. The bonded structure should thus be reinforced by such stiffening rib, button dowel with slot, spigot location, and so on.
2. In general, it is difficult to fabricate the welded structure consisting of different materials. In contrast, the bonded joint is applicable to the structure made of different materials.
3. For the welded structure, the stress relieving is of great importance; however, the bonded structure does not require such a treatment while producing, resulting in the reduction of the production cost. This is the utmost beneficial feature in manufacturing the large-sized machine tool.

To understand the beneficial characteristics of the bonded joint, Chowdhury et al. compared the stability chart for the chatter vibration of the milling machines, where their overarms are made of cast, welded, or bonded type; however, to accommodate advantageous features of each type, the cross sectional views of the overarm are not of the same structural similarity [21]. Given such a difference, the milling machine with bonded overarm shows better resistance for the chatter vibration, which is attributed to the higher damping capacity of the bonding agent of cold curing type, i.e., epoxy resin. The experiment was carried out using the milling cutter of 4 in diameter and test piece of tapered strip type, the basic idea of which was proposed by the MTIRA of the United Kingdom to measure the stability limit of the chatter vibration. In fact, the limiting widths of cut in the stability chart are 210, 265, and 312×10^{-3} inch for cast, welded, and bonded types, respectively.

In the bonded joint, the shear strength is also of great importance, and the behavior of the bonded joint changes considerably depending upon the polymerization time, i.e., aging time, heating time of the adhesive, and the materials of joint surroundings.

Within a basic research context, Thornley and Lees [22] investigated the static and dynamic characteristics of various bonded joints in detail and clarified an optimum bonding technique. In their research, the experiment was carried out with the bonded joint under normal loading, which was of square plate with 232 cm² apparent contact area and made of mild steel. As the interface bonding material, four types of epoxy resin adhesives curing at room temperature were employed, and these adhesives were applied to the joint surfaces machined by shaping (R_{CLA}: 87 to 107 μm), peripheral grinding (R_{CLA}: 10 to 32 μm) with the machined lay orientation being at a constant 90° throughout the experiment, and scraping (minimum of 21 contact points per any one square inch). In addition, Thornley and Lees expected a smoothing effect of the bonded joint, and thus the flatness deviation, i.e., peak-to-valley depth of the long-term surface waviness, of the bonded joint was varied from about 10 to 60 μm. In the case of the sandwich plate under bending, the damping capacity is apparently high while high bending stiffness is maintained. This is attributed to the fact that the external load cannot be supported by the adhesive, but by the metallic plates. Referring to this point, Thornley and Lees provided the adhesives to the joint surface as a filling medium for all the valleys within the flatness deviation, and thus the bonded joint has a considerable region of metal-to-metal contact.

As can be readily seen, the deflection of the dry joint is in quadratic relation with the interface pressure, whereas that of bonded joint is in hyperbolic relation, i.e., in more linear relation with the interface pressure, when the uniform static preload is 2840 kN/m² maximum. More specifically, the noteworthy behavior of the bonded joint can be summarized as follows. In this context, we must be furthermore aware that these characteristic features of the bonded joint are derived from the adhesive and cannot be observed in other machine tool joints, although there is a larger influence of microscopic air bubbles in the adhesive, as pointed out by Thornley and Lees.

1. The joint stiffness can approach that of equivalent solid, when the layer thickness of an adhesive film is kept to the same order of the magnitude of long-term surface waviness involved.

2. The increasing rate of the joint stiffness depends upon the machining method and quality of the surface finish, showing less improvement for a joint, which has greater stiffness by nature. In certain cases, the bonded joint shows only the elastic deflection after first loading.

3. In both ground and scraped bonded joints, the dynamic stiffness increases and decreases steeply with the static preload and excitation frequency, respectively.

4. In general, the type of adhesive appears to have little or no influence on the static stiffness.

Figure 9-13 reproduces the dynamic stiffness of the bonded joint examined by Dekoninck [23, 24]. In his experiment, the joint was subjected to

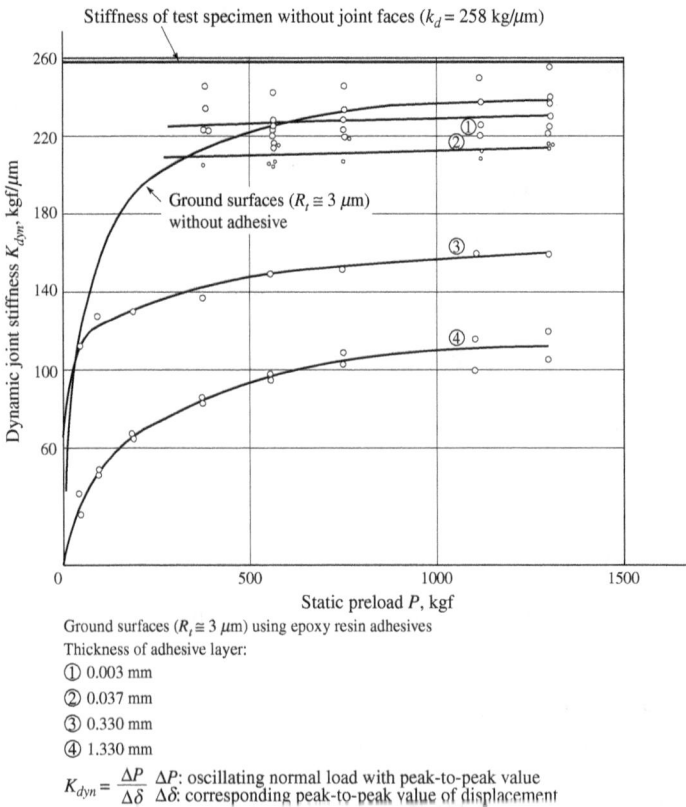

Stiffness of test specimen without joint faces (k_d = 258 kg/μm)

Ground surfaces ($R_t \cong 3\ \mu$m) without adhesive

Ground surfaces ($R_t \cong 3\ \mu$m) using epoxy resin adhesives
Thickness of adhesive layer:
① 0.003 mm
② 0.037 mm
③ 0.330 mm
④ 1.330 mm

$K_{dyn} = \dfrac{\Delta P}{\Delta \delta}$ ΔP: oscillating normal load with peak-to-peak value $\Delta \delta$: corresponding peak-to-peak value of displacement

Figure 9-13 Dynamic stiffness of bonded joints under normal loading (by Dekoninck).

an oscillating normal or tangential load, and we can summarize the further characteristic features of the bonded joint as follows together with considering the achievement of Thornley and Lees.

5. The dynamic joint stiffness increases with increasing static preload, and depends also considerably on the interfacial layer. In the case of normal loading, the high-quality adhesive has a very favorable effect on the joint stiffness, when special attention is paid to the thickness of the adhesive layer.

6. Under tangential loading, the larger damping capacity can be obtained when the relatively large load is applied, provided that the adhesive layer is only filled up to the valley.

In short, the higher dynamic stiffness of the bonded joint results from the high static joint stiffness, which may derived from the improvement of flatness of the joint by supplying the adhesive layer.

To this end, it is emphasized that the research and technology development for the bonded joint have not been vigorous since the 1980s as shown in a firsthand view of activities in Fig. 9-14, although the bonded joint has multifarious advantages. This unacceptable situation may be, as suggested beforehand, attributed to the lower reliability and short durability of the bonded joint.

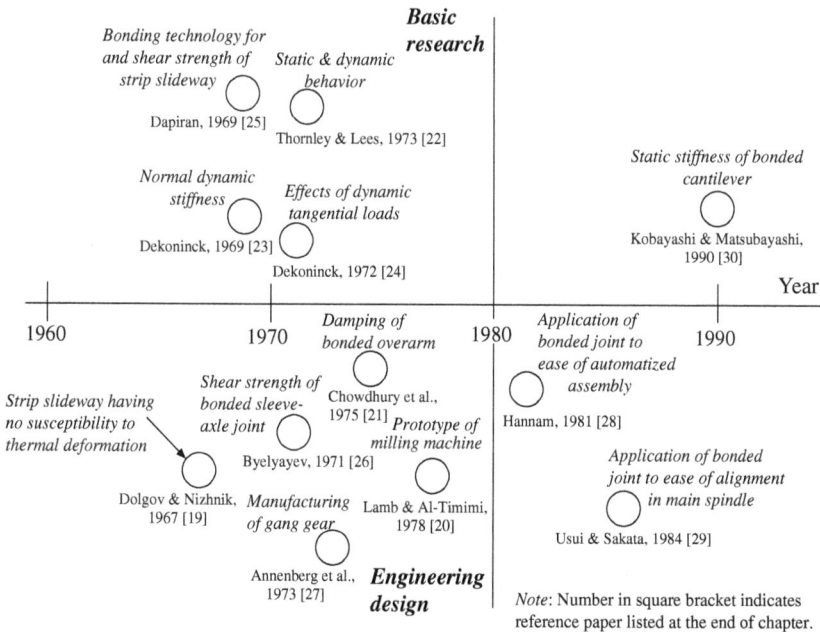

Figure 9-14 Firsthand view of research and engineering development in bonded joint.

9.2   Taper Connection

The typical examples of the taper connection are between the nose hole in the main spindle and the cutting tool in the case of MC, and also between the spindle nose and the chuck in the case of NC turning machine. These taper connections are in closer relation to the setting accuracy of the cutting tool and chatter stability of a machine-tool–work system, because they could be regarded as the weaker portion within the system, resulting in, e.g., the deterioration of the antichatter vibration capability.

As a reflection of the long developing history of the machine tool, there are various kinds of taper, as shown in Table 9-7, and they are of two categories. One is the self-holding type, and the other is the self-releasing type, depending upon the magnitude of the taper angle. In the self-holding type, the taper angle is between 2° and 3°, and the holding rigidity is, by nature, subjected to the frictional force between the mating surfaces. As a result, the taper of self-holding type shows the lower holding rigidity, when it has no driving key, cotter, or slot-to-tongue connection. In contrast, the taper angle of the self-releasing type is larger than 16°, and the additional axial force is, in general, required to ensure sufficient holding rigidity. In addition, the self-releasing type has a guide key on the front face of the spindle nose.

Although there are myriad kinds of tapers, at present the primary concerns are the metric, Morse, and National tapers, because these are widely employed in the conventional NC turning machine and MC. In addition, the Jacobs taper has been commonly employed in the drilling machine of bench type, and this taper can be characterized by its accommodation configuration to the tapered nose in the main spindle.

Within an engineering design context, the deflection of the taper connection should be estimated first, and Levina proposed an engineering calculation as follows [31, 32].

Figure 9-15 shows a taper connection in a quill-grinding spindle system and its mathematical model, respectively. The mathematical

TABLE 9-7   Kinds of Tapers

Self-holding taper	Metric taper— American standard taper, self-holding type
	Morse taper
	Brown & Sharpe taper
	Jarno taper— Reed taper (Short Jarno taper)
	Jacobs taper
	Others
Self-releasing taper	National taper— American standard taper, steep type

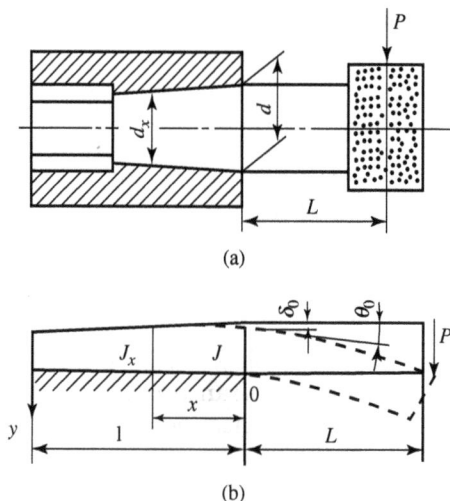

(a)

(b)

Figure 9-15 A quill-grinding spindle system and its mathematical model: (a) Actual structure and (b) mathematical model (by Levina).

model is an elastic beam on an elastic foundation in the theory of elasticity, and duly the following expressions can be drawn.

$$d_x = d\,[1 - (2\alpha_T/d)x]$$

$$EI_x = E(\pi\,d_x^4/64) \qquad (9\text{-}1)$$

$$b_x = 0.5\pi\,d_x$$

where d_x = diameter of tapered shank at x
  D = maximum diameter of tapered shank
  $2\alpha_T$ = taper angle
  E = Young's modulus
  b_x = equivalent width of beam

In consideration of the complete contact across the whole taper, it can be assumed that Levina's expression shown in Table 6-1 is available, and thus the differential equation for the deflection of a grinding spindle can be written as

$$d^2/dx^2\,[EI_x(d^2y/dx^2)] + b_x y/C = 0 \qquad (9\text{-}2)$$

where y is the bending deflection of a grinding spindle. At the loading point, the deflection δ can be written as

$$\delta = PL^3/(3EI) + \delta_0 + \theta_0 L \qquad (9\text{-}3)$$

where P = radial load, kgf
  L = protruded length, cm
  $I = \pi d^4/64$, cm^4
$\delta_0 + \theta_0 L$ = deflection of spindle derived from taper connection

TABLE 9-8   Values for A_1, A_2, and A_3 (by Levina)

	βl										
	Morse taper					7/24 taper					
	2	3	4	5	6	1.5	2.0	2.5	3.0	3.5	4.0
A_1	1.23	1.10	1.06	1.05	1.04	2.34	2.16	1.65	1.45	1.34	1.30
A_2	1.2	1.08	1.04	1.03	1.03	2.06	1.94	1.64	1.48	1.36	1.34
A_3	1.02	1.015	1.01	1.0	1.0	1.70	1.35	1.17	1.07	1.05	1.04

In short, the deflection at the entrance of tapered hole $\delta_0 = 2M\beta^2 C(A_1/b) + 2P\beta C(A_2/b)$ μm, and the inclination angle at the entrance of the tapered hole $\theta_0 = 4M\beta^2 C(A_3/b) + 2P\beta C(A_1/b)$ (μm/cm). In these equations,

$$M = PL$$

$$\beta = \sqrt[4]{b10^4/(4EIC)} \ 1/\text{cm}$$

$$b = \pi d/2 \ \text{cm}$$

A_1, A_2, and A_3 = compensating coefficients for various diameters of tapers (see Table 9-8)

Levina has furthermore recommended such values for C as shown in Table 9-9. According to the research of Martinez et al., however, the average interface pressures measured are as follows [33].

1. For Morse taper No. 5, between 0.06 and 1.0 kgf/cm^2
2. For National taper No. 40, between 0.5 and 8.0 kgf/cm^2

TABLE 9-9   Compliance C for Taper Connection (recommended by Levina)

Unit: μm · cm^2/kgf

Kinds of taper	Interface pressure, kgf/cm^2	
	Less than 35	70–150
Morse taper		
Nos. 2 & 3	0.06	
No. 4	0.04	0.015–0.020
Nos. 5 & 6	0.03	
7/24 taper Nos. 35, 40 & 50	0.02	—

Note 1: Ratio of bluing to apparent areas is 80%.
    2: Surface finish is Class 9 in U.S.S.R. Standard.

The measurement was carried out using the engine lathe and milling machine under nonloading conditions, and the measured values were compensated by predicting the maximum cutting force and by assuming the full contact across the whole taper connection.

In relation to the torsional angle of the taper connection φ_T ($\mu m/cm$), Levina recommended the following expression.

$$\varphi_T = [(4k_\tau \cos\alpha_T)/\pi \, ld_m{}^3]M_T \qquad (9\text{-}4)$$

where M_T = torque, kgf · cm

k_τ = tangential joint stiffness, $\mu m \cdot cm^2/kgf$, where k_τ ranges from 0.04 (p = 26 kgf/cm^2) to 0.02 (p = 78 kgf/cm^2) depending upon the interface pressure p

l = connection length along shank axis, cm

d_m = mean diameter in taper connection, cm

α_T = taper half angle

As can be readily seen, those expressions of Levina can be accommodated for the engineering design, because of its ease of use; however, there are no proposals for the engineering calculation of the dynamic and thermal deflection.

Figure 9-16 shows a firsthand view of the research into and engineering development for the taper connection, and from these earlier

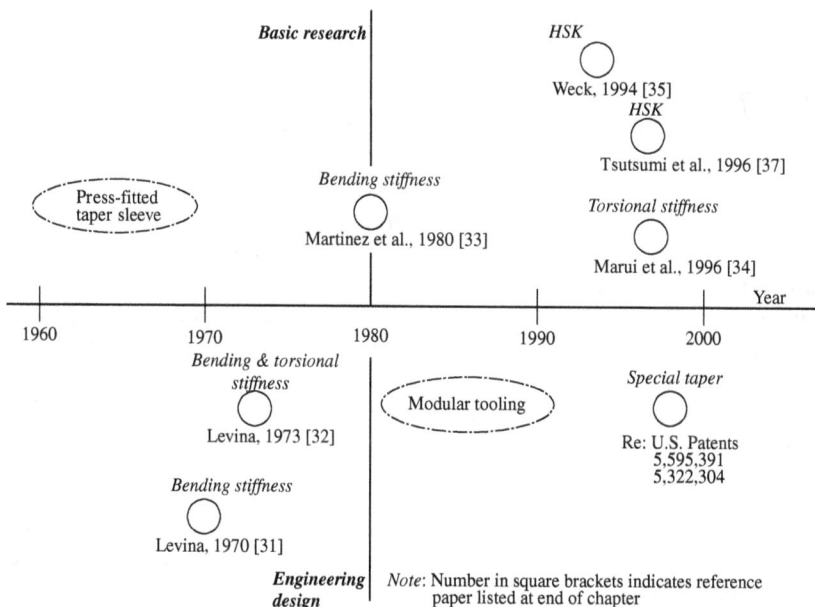

Figure 9-16 Firsthand view of research and engineering development in taper connection.

works, the general behavior of the taper connection under bending can be summarized as follows.

1. The deflection of tapered shank connected with the tapered hole is nearly in linear relation to the applied bending load, even when the interface pressure is lower. This is the very special case as the machine tool joint. In general, the deflection-load relation of the machine tool joint shows, without exception, nonlinearity.
2. The joint deflection is in linear relation to the interface pressure.
3. The interface pressure, applied load, and shank length have fewer effects on the stiffness of the tapered shank.
4. The diameter of the barrel (quill with tapered hole), i.e., joint surroundings, has no obvious effects on the stiffness of the tapered shank.

In many respects, the taper connection shows the typical behavior of the joint of closed type (see Chap. 5). Reportedly, the deterioration of the joint stiffness is less than 10%, as shown in Fig. 9-17, where the stiffness of the tapered shank is, in certain cases, equal to that of equivalent solid. Obviously, the joint stiffness of the taper connection depends also on the larger-end diameter of the tapered hole and the ratio of bluing to apparent contact area. In fact, from the old days, an engineering rule is to maintain the tight fit at the entrance of the tapered hole. Figure 9-18 reproduces a result reported by Martinez, where he compared the shanks with Morse and National tapers by controlling the fitting force, both of which have nearly equal larger-end diameter of tapered hole and are tight-fitted at the entrance of the tapered hole. As can be seen, we cannot observe the apparent difference between both the taper connections, and Levina et al. reported the same result [32].

In contrast, the torsional behavior of the taper connection has not been clarified yet, although Levina proposed an expression for the engineering calculation. Recently, Marui et al. investigated the static and dynamic behavior of the Morse taper connection and reported that the Morse taper connection under torsion shows very similar behavior to that of the flat joint under tangential loading. In addition, the taper connection shows the free decayed vibration curve with amplitude of stick-slip-like variation, which is the same behavior observed in the bolted joint [34].

More importantly, with the higher rotational speed of the main spindle in the MC, there arises, at present, a handful of such serious problems as follows.

1. The larger-end diameter of the tapered hole is expanded by the centrifugal force, resulting in the additional axial withdrawal of the tapered shank. In accordance with the experiences of Toyoda Iron Works and Makino Milling Machine in the mid-1990s, the axial withdrawal in the case of NT 50 is 40 μm at 30,000 rpm in spindle speed.

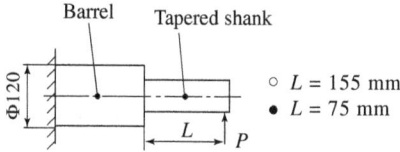

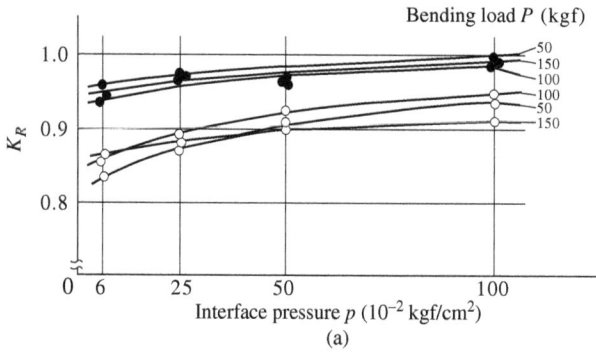

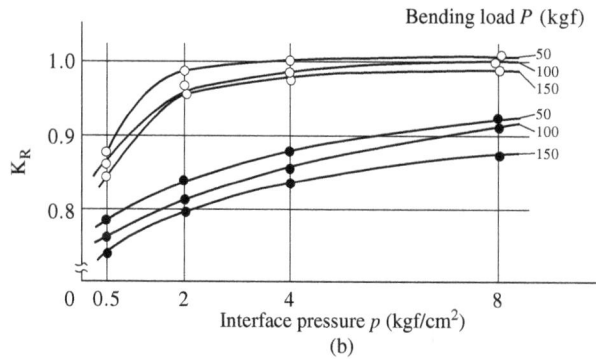

Figure 9-17 Stiffness ratio K_R in taper connection: (a) MT No. 5 and (b) NT No. 40.

2. The fretting corrosion is very liable to occur by nature, because of involving the microvibration with higher frequency while rotating the spindle.

3. There is unfavorable vibration due to the unbalance of the Belleville washer.

To solve these problems, there have been multifarious trials, e.g., FMT (flange mount tooling, see Fig. 9-19), KM (Kennametal) tooling, HSK,

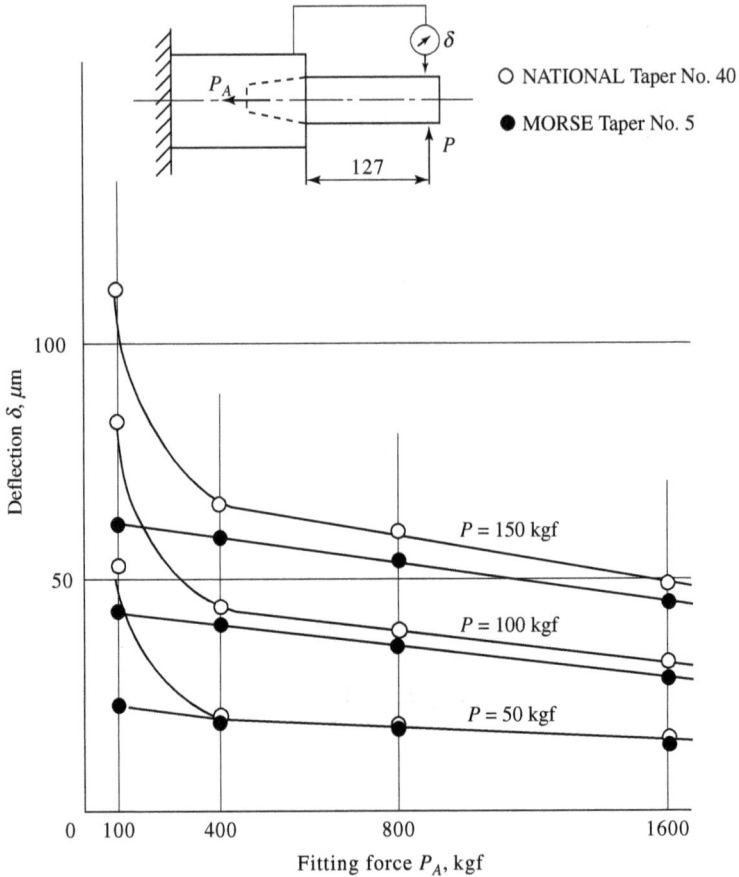

Figure 9-18 Relationships between deflection and fitting force for both types of tapers.

and HSK with taper sleeve. The FMT system is of short taper type, and its reproducibility of positioning accuracy is better than 3 μm. For accurate positioning, the taper nose of 10 mm length can deflect elastically. In addition, there were other trials to modify the standard 7/24 tool holder using the intermediate balls such U.S. Patent Nos. 5,595,391 and 5,322,304. In short, such trials aim at the "sure-holding by the simultaneous taper and end surface connection," and the leading trend is duly to employ the HSK [35].

The HSK has the two types, i.e., double cylindrical hollow tool shank and taper hollow shank. In the beginning of 2000, the HSK of taper type is dominant and has already been enacted as a DIN-Standard in 1993, because the double cylinder type has certain difficulties in its manufacturing. Figure 9-20 shows a DIN Standard and also a typical application

Main spindle

Taper: 20°

M115 p3

φ100

Notches:3

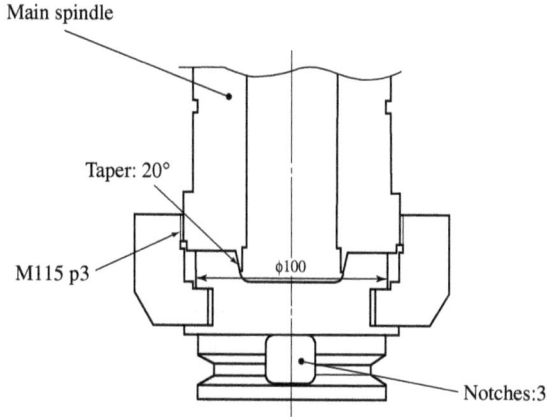

Figure 9-19 Flange mounting tooling developed in the beginning of the 1990s (courtesy of Kuroda Precision Industries).

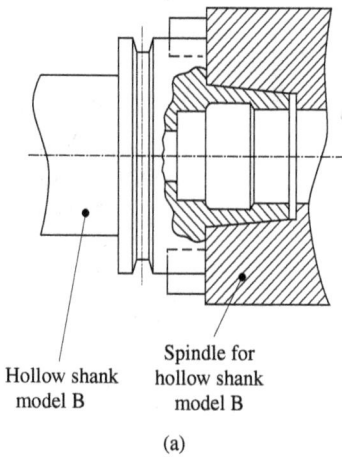

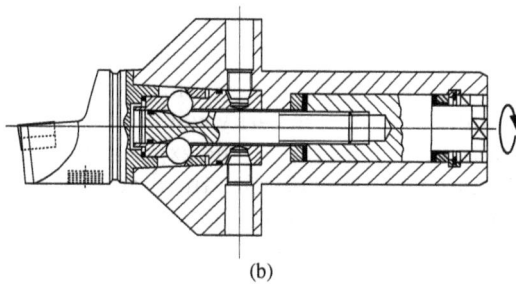

Hollow shank
model B

Spindle for
hollow shank
model B

(a)

(b)

Figure 9-20 HSK-DIN Standard and an application to cutting edge module: (a) Hollow shank model B in DIN 69893 and (b) cutting edge module (by Pegels, courtesy of Carl Hanser).

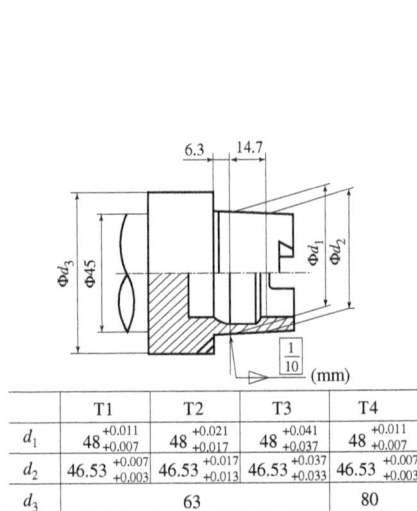

Dimensions of HSK tool shanks

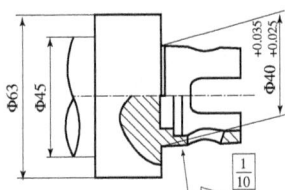

	T1	T2	T3	T4
d_1	$48 {}^{+0.011}_{+0.007}$	$48 {}^{+0.021}_{+0.017}$	$48 {}^{+0.041}_{+0.037}$	$48 {}^{+0.011}_{+0.007}$
d_2	$46.53 {}^{+0.007}_{+0.003}$	$46.53 {}^{+0.017}_{+0.013}$	$46.53 {}^{+0.037}_{+0.033}$	$46.53 {}^{+0.007}_{+0.003}$
d_3		63		80

Dimensions of KM tool shank (KM6350)

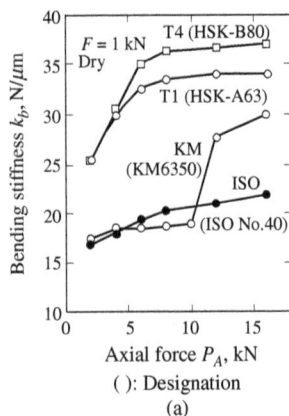

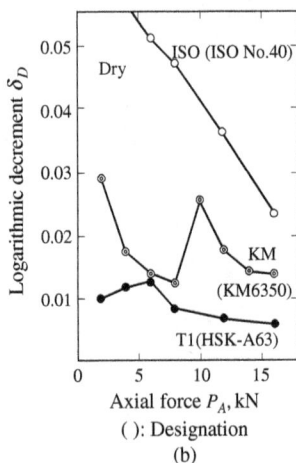

Figure 9-21 Static stiffness and damping capacity of HSK: (a) Comparison of bending stiffness of tools and (b) comparison of damping (courtesy of Tsutsumi).

of the HSK to the cutting edge module [36]. Figure 9-21 shows the static stiffness and damping capacity of the HSK under bending load. These results were reported by Tsutsumi et al. [37] to verify whether the HSK can solve the shortcomings of the National taper. The experiment was carried out using the spindle nose—tool model, with the diameter and length of cantilever (tool model) being 45 and 160 mm, respectively, and by comparing the HSK, that of Kennametal, and National taper (ISO No. 40), where the reference diameter of ISO No. 40 taper is close to that of HSK-A63 taper.

As can be seen, the HSK is superior to the National taper in bending stiffness, but has considerably lower damping capacity as a rule of the machine tool joint. It is interesting that the flange face of the HSK plays a very important role in its behavior, and this is attributed to the structural configuration of the HSK. In fact, the flange and taper play the roles of load-bearing surface and positioning, respectively. In both the traditional taper shank (BT type) and the HSK, the pull stud is either the collet or ball type.

9.3 Chucking

The holding device of the work or cutting tool is in general called the attachment of a machine tool, but as not literally shown, its extreme importance is notable in machining. As is well known, there are a handful of representative holding devices, such as the chuck, mandrel, center, faceplate, driving plate, work driver, and indexing table. Of these, the utmost representative is chucking, and Fig. 9-22 shows a typical machine-chuck-work-tool system. Importantly, chucking is a representative of the semiclosed joint described in Chap. 5.

In accordance with fulfilling various chucking requirements, many types of chuck have been developed and are in practical use, and the leading ones are the jaw and collet chucks. These chucks can be classified into several variants depending upon the number of jaws and the jaw moving mechanism. In general, the three-jaw and four-jaw chucks are self-centering and independent moving types, respectively. In addition, the three-jaw chuck can be classified into the scroll, lever, and wedge types, depending upon the jaw moving mechanism.

In discussing the chucking stiffness, at issue is, needless to say, the contact surface between the jaw and the work; however, we must be

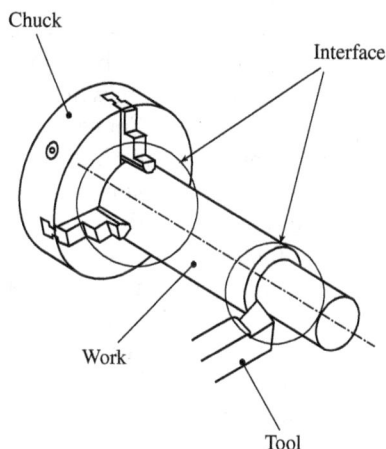

Figure 9-22 A machine-chuck-work-tool system.

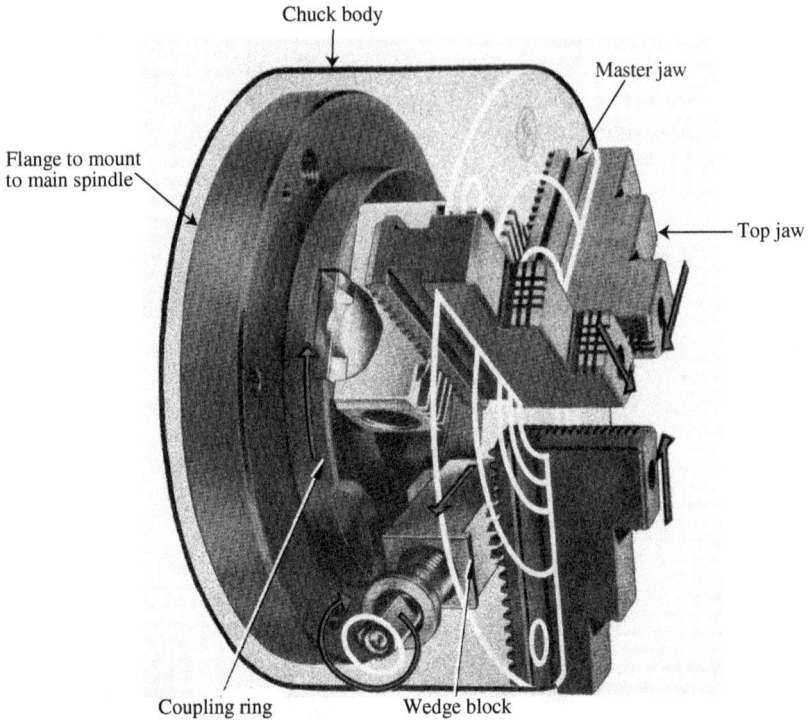

Figure 9-23   A sectional view of three-jaw wedge chuck.

aware of the implicit influence of the moving mechanism and traveling method of the jaw on the chucking stiffness. Figure 9-23 shows the overall sectional view of the three-jaw wedge chuck, and as can be seen, the chuck itself has many mating surfaces, which may function, more or less, as the influencing factor for the reduction of chucking stiffness. In this context, Semekhin reported that the stiffness of the three-jaw chuck with self-centering type is between 0.6 and 0.8 kgf/μm, whereas the lathe as a whole shows 3 to 4 kgf/μm in its stiffness [38]. Importantly, Semekhin suggested that the stiffness of the three-jaw scroll chuck is dependent on the clearance between the jaw slot and the jaw tongue, and on the guiding tolerance of the scroll disk. More specifically, Ivashchenko suggested the predictable factors to influence the stiffness of the three-jaw chuck with self-centering mechanism such as shown in Table 9-10 [39]. In addition, Ivashchenko conducted a series of investigations into the chuck stiffness, especially considering the effect of inscribing the outer diameter of work on the arc of the jaw. As a result, the characteristic features of the chucking stiffness can be unveiled as follows.

TABLE 9-10   Predictable Factors to Influence on Chucking Stiffness (by Ivashchenko)

Geometric factors	Condition of clamping surface of jaw	Type of three-jaw chuck		
		Scroll	Wedge	Lever
Clearance between abutment or inverse abutment lip of jaw tongue and jaw slot	A	Strong direct effect		
	B	Direct effect		
Clearance between widths of jaw tongue and slot	A & B	Feeble indirect effect		
Difference of diameter between inscribing circle of jaws and work	A	Direct effect		
	B	No effects		
Straightness error of chucked portion of work	A & B	Direct effect		
Clearance in connection of bore of chuck body or inner diameter of scroll disk	A & B	Not clarified as yet		
Range of lever arm sizes	A & B	—	—	Not clarified as yet
Difference of inclination angles to wedge projections of coupling	A & B	—	Not clarified as yet	—
Displacement of plane of spiral relative to axis of bore in pinion helix	A & B	Not clarified as yet	—	—

Clamping surface of jaws: A, not bored before given operation to work size;
    B, bored with preliminary compression of thrust ring.

1. The chucking stiffness is largely dependent upon the contact condition between the clamping surface of jaws and the work, and it is preferable that the contact be under the uniform interface pressure distribution. In this case, the larger the chucking force, the higher the chucking stiffness.

2. The clearance related to the abutment lip at the jaw tongue has large influence on the chucking stiffness, whereas the clearance related to the jaw slot has no effect.

3. Within the self-centering chuck, the moving mechanism of the jaw has certain effects on the chucking stiffness. In fact, the scroll type is superior to the wedge and lever types.

4. In the case of the three-jaw scroll chuck, truing the jaw is very effective to ensure the uniform clamping condition across the whole jaw bearing surface. In detail, the inscribed circle of jaws must be prebored to coincide with the periphery of the work to be clamped.

Following these earlier works, there have been myriad technological developments, as shown in Fig. 9-24, but little research has been carried out into the chucking stiffness so far. In this context, Rahman and

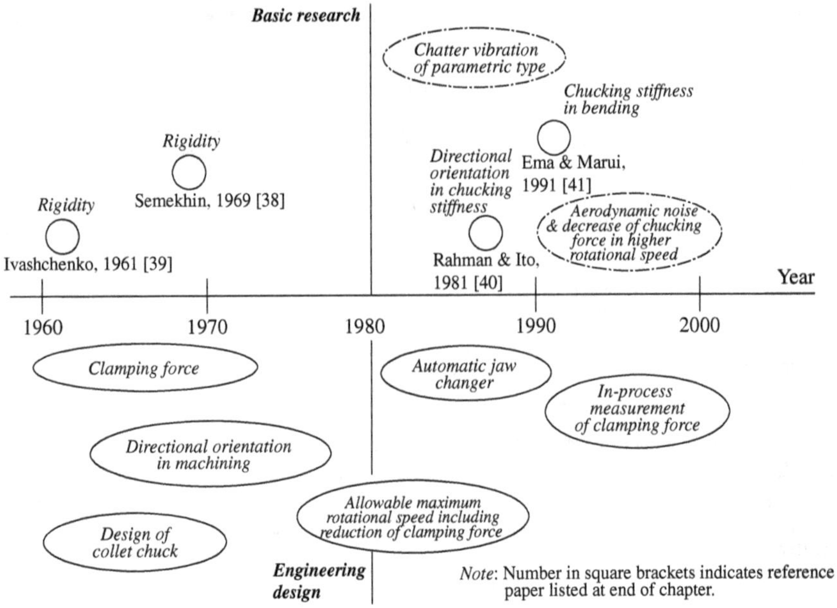

Figure 9-24   Firsthand view of research and engineering development in chucking.

Ito investigated the behavior of the stiffness and damping capacity in chucking in detail, and reported the following interesting observations

1. The same as other machine tool joints, the load-deflection curve of the chucked work consists of the elastic and residual components.

2. The residual displacement may be derived from the nonreversible displacement of the jaw in its slot and slipping of the work at the jaw.

In considering these characteristic features of chucking, it is very interesting to know what has relevance to chucking as a semiclosed joint. The answer is the directional orientation in chucking stiffness as obviously ascertained by Rahman and Ito [40], and Fig. 9-25 reproduces such a directional orientation, when the work diameter is varied under bending loading. In fact, the directional orientation is the stiffness variation of the chucked work at its every circumferential position, and can be apparently observed in the case of the three-jaw chuck. As a matter of course, the chucking stiffness is in peak around the jaw position and in the lowest between both jaws. In contrast to static loading, the directional orientation cannot be observed in the case of damping capacity.

Importantly, the directional orientation affects considerably the behavior of the chatter vibration in machining, as shown schematically the

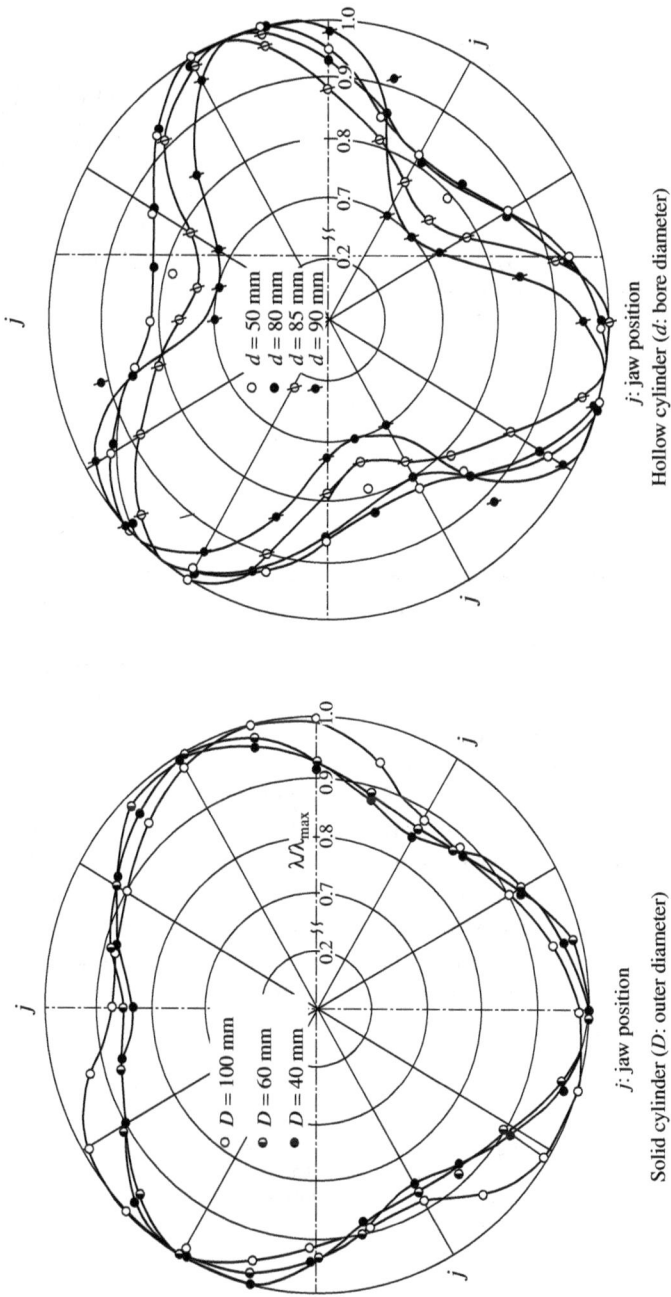

Figure 9-25 Directional orientation in chucking stiffness.

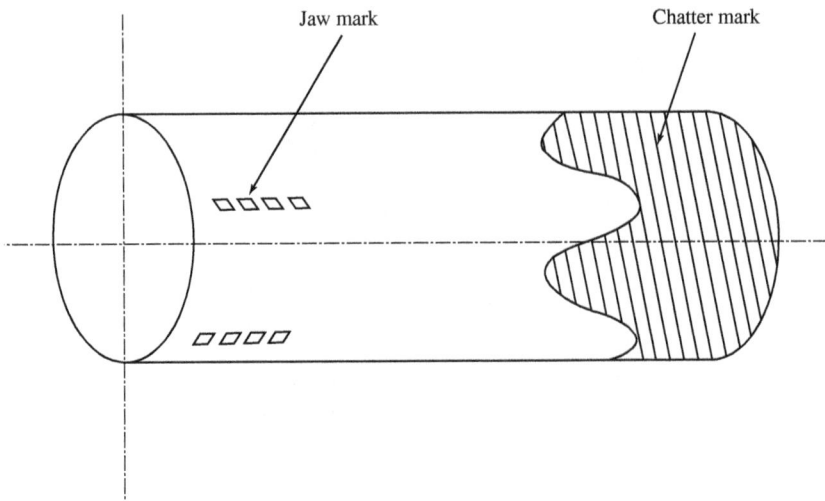

Figure 9-26   Effect of directional orientation on chatter mark.

chatter mark in Fig. 9-26. In the case of the work held by the three-jaw self-centering chuck, the chatter is of self-excited type mingling with the parametric vibration caused by the directional orientation of the stiffness. In addition, it appears that some of the stability charts so far reported involve uncertainties, if they determined the chatter threshold using the chatter mark.[6]

In chucking, furthermore, an ever-growing important subject is how to recover the reduction of the clamping force at the higher rotational speed. A widely known remedy is to use the counterbalance to compensate the centrifugal force acted especially on the top jaw. In contrast, the other remedy is to employ a new material, e.g., FRP (*fiber reinforced plastic*) to the top jaw. Obviously, the chuck with FRP top jaw shows less reduction of the radial rigidity compared with the steel top jaw, where the radial stiffness of the chucks with former and latter top jaws are 66 and 216 N/μm, respectively [42].

---

[6]As shown in Fig. 9-26, the chatter mark shows a wavelike pattern, which induces certain difficulty in the accurate determination of the chatter commencement, e.g., measurement of the length between the jaw end surface and the wavelike portion of chatter mark. In fact, we have no acceptable criterion for the chatter commencement yet, although some trials are being carried out.

Rahman, M, and Y. Ito, "A Method to Determine the Chatter Threshold," in S. A. Tobias and F. Koenigsberger (eds.), *Proc. of 13th Int. MTDR Conf*, Macmillan, 1979, pp. 191–196.

References

1. Gunsser, O., "Werkzeughalterung," *Werkstattstechnik,* 1959, 49(3): 153–160.
2. Shanker, A., and P. C. Pandey, "Comparative Performance of Turning Between Centres with Revolving and Dead Centres," in *Proc. of ICPE (Delhi),* 1977, 1: iii-202–iii-211.
3. Saljé, E., and U. Isensee, "Dynamisches Verhalten schlanker Werkstücke bei unterschiedlichen Einspannbedingungen in Spitzen," *ZwF,* 1976, 71(8): 340–343.
4. Marui, E., S. Ema, and S. Kato, "Relative Sliding Behavior and Damping Characteristic of Turning Tools," *J. of JSPE,* 1983, 49(10): 1404–1409.
5. Shawki, G. S. A., and M. M. Abdel-Aal, "Effect of Fixture Rigidity and Wear on Dimensional Accuracy," *Int. J. Mach. Tool Des. Res.,* 1965, 5: 183–202.
6. Shawki, G. S. A., and M. M. Abdel-Aal, "Rigidity Considerations in Fixture Design—Contact Rigidity at Locating Elements," *Int. J. Mach. Tool Des. Res.,* 1966, 6: 31–43.
7. Galperin, B. Ya, and S. B. Magidenko, "Optimum Turret-Head Clamping Force," *Machines and Tooling,* 1973, 44(11): 10–12.
8. Shourbagy, El Hazem, M. Tsutsumi, and Y. Ito,"Static Behavior of Turret Head with Curvic Coupling," in *13th NAMRC Proc.,* NAMRI of SME, 1985, pp. 277–282.
9. Shourbagy, El Hazem, et al., "A New Modular Tooling System of Curvic Coupling Type," in B. J. Davies (ed.), *Proc. of 26th Int. MTDR Conf.,* Macmillan,1986, pp. 261–267.
10. Anno, Y., et al., "Static Rigidity of Tool Considering Joint Surfaces—On the Case of Milling Arbor with Distance Collars," *Trans. of JSME,* 1970, 36(285): 862.
11. Stansfield, F. M., "Damping in Structural Materials," in The MTIRA "One-Day Conference on Damping in Machine Tool Structures," April 30, 1969.
12. Bobek, K., A. Heiß, and Fr. Schmidt, *Stahlleichtbau von Maschinen,* Springer-Verlag, 1955.
13. Oberst, H., "Entdröhnung von Stahlblechkonstruktionen," *Schiff und Hafen,* 1971, 23(4): 285–290.
14. Kopitsyn, V. I., "Housing Components of Machine Tools Fabricated by Welding," *Machines and Tooling,* 1961, 32(12): 4–9.
15. Frank, W., "Leichtbau von Werkzeugmaschinen—durch geschweißte Stahlkonstruktionen," *Werkstatt und Betrieb,* 1956, 89(6): 299–308.
16. Anno, Y., et al., "Study on the Damping Capacity of a Welded Structure," *Trans. of JSME,* 1970, 36(284): 663–672.
17. Kronenberg, M., P. Maker, and E. Dix, "Practical Design Techniques for Controlling Vibration in Welded Machines," *Machine Design,* July 12, 1956, pp. 103–109.
18. Grab, H., and P-H. Theimert, "Beton im Werkzeugmaschinenbau," *Werkstatt und Betrieb,* 1976, 109(4): 195–202.
19. Dolgov, K. P., and E. G. Nizhnik, "Design and Repair of Strip Slideways," *Machines and Tooling,* 1967, 38(3): 22–25.
20. Lamb, E. J., and K. Al-Timimi, "Design Concepts for Fabrication of Bonded Machine Tool Structures," in J. M. Alexander (ed.), *Proc. of 18th Int. MTDR Conf.,* Macmillan, 1978, pp. 561–567.
21. Chowdhury, M. I., M. M. Sadek, and S. A. Tobias, "The Dynamic Characteristics of Epoxy Resin Bonded Machine Tool Structures," in S. A. Tobias and F. Koenigsberger (eds.), *Proc. of 15th Int. MTDR Conf.,* Macmillan, 1975, pp. 237–243.
22. Thornley, R. H., and K. Lees, "Some Static and Dynamic Characteristics of Bonded, Machined Joint Faces," in S. A. Tobias and F. Koenigsberger (eds.), *Proc. of 13th Int. MTDR Conf.,* Macmillan, 1973, pp. 79–86.
23. Dekoninck, C., "Experimental Investigation of the Normal Dynamic Stiffness of Metal Joints," *Int. J. Mach. Tool Des. Res.,* 1969, 9: 279–292.
24. Dekoninck, C., "Deformation Properties of Metallic Contact Surface of Joints under the Influence of Dynamic Tangential Loads," *Int. J. Mach. Tool Des. Res.,* 1972, 12: 193–199.
25. Dapiran, A., "A New Method for Bonding Strip Slideways," in S. A. Tobias and F. Koenigsberger (eds.), *Proc. of 9th Int. MTDR Conf.,* Pergamon, 1969, pp. 897–905.
26. Byelyayev, G. S., "Strength of Bonded Sleeve-Axle Joints," *Machines and Tooling,* 1971, 42(7): 27.

27. Annenberg, E. A., "Adhesive-Bonded Cylindrical Gears," *Machines and Tooling*, 1973, 44(10): 40–41.
28. Hannam, R. G., "A Case Study on a Use of Anaerobic Adhesives in Machine Tool Assembly," in B. J. Davies (ed.), *Proc. of 22d Int. MTDR Conf.*, Macmillan,1981, pp. 109–117.
29. Usui, Y., and O. Sakata, "Assembling a Main Spindle of a Lathe with Adhesive Bonding," *Proc. of 5th Int. Conf. on Production Engineering*, JSPE, 1984, pp. 573–578.
30. Kobayashi, T., and T. Matsubayashi, "A Study on the Bending Rigidity of Adhesive Joints (The Case of Adhesive Lap Joints Subjected to a Lateral Bending Load)," *Trans. of JSME*, 1990, 56(531): 3148–3153.
31. Levina, Z. M., "Stiffness Calculations for Cylindrical and Taper Joints," *Machines and Tooling*, 1970, 41(3): 5–10.
32. Levina, Z. M., et al., "Taper-Connexion Stiffness," *Machines and Tooling*, 1973, 44(10): 21–26.
33. Martinez, J. M. P., Y. Saito, and Y. Ito, "A Stiffness of Taper Connection in a Main Spindle of Machine Tools," *J. of JSPE*, 1980, 46(2): 242–248.
34. Marui, E., et al., "Research on Joining Characteristics of Tapered Coupling Joint," *Trans. of JSME (C)*, 1996, 62(603): 4302–4308.
35. Weck, M., "New Interface Machine/Tool: Hollow Shank," *Annals of CIRP*, 1994, 43(1): 345–348.
36. Pegels, H., "Werkzeugtechnik für eine flexible, automatisierte Fertigung," *Werkstatt und Betrieb*, 1987, 120(10): 875–878.
37. Tsutsumi, M., et al., "Static and Dynamic Stiffness of 1/10 Tapered Joints for Automatic Changing," *Int. J. of JSPE*, 1996, 30(1): 23–28.
38. Semekhin, M. I., "Rigid Three-Jaw Self-centring Chuck," *Machines and Tooling*, 1969, 40(11): 28–29.
39. Ivashchenko, I. A., "Some Factors Affecting the Accuracy and Rigidity of Three-jawed Chucks," *Machines and Tooling*, 1961, 32(1): 31–33.
40. Rahman, M., and Y. Ito, "Some Necessary Considerations for the Dynamic Performance Test Proposed by the MTIRA," *Int. J. Mach. Tool Des. Res.*, 1981, 21(1): 1–10.
41. Ema, S., and E. Marui, "Gripping Characteristics of a Wedge-Type Power Chuck (Bending Load Dependence on Bending Stiffness)," *Trans. of JSME (C)*, 1991, 57(539): 2460–2465.
42. Spur, G., and U. Mette, "Clamping-Force Optimization Allows High-Speed Turning," *Production Engg.*, 1998, V(1): 55–58.

Measurement of Interface Pressure by Means of Ultrasonic Waves

The measurement of the pressure distribution at the jointed surfaces is very important to understanding the behavior of not only the machine tool, but also other machines, and thus a considerable number of methods have been proposed so far. For example, there have been (1) the pressure sensitive pin (reclamation pin) method mainly used in the plastic forming sphere, (2) the footprint method used in the bearing industry and academic sphere, and (3) the pressure sensitive paper method widely used in the industry. Figure A1-1 summarizes the measuring methods for the interface pressure and contact area so far contrived. However, the methods proposed up to now have certain disadvantages because they need some modification or reconfiguration of the joint surface in order to attach or settle down a measuring device, causing the pressure distribution to differ from that of the original state to some extent. To understand essential features of such measuring methods, Table A1-1 shows the quick notes for the pressure sensitive pin and paper methods, which are even now in the leading position.

In contrast, the UWM (*ultrasonic wave method*), i.e., authentic measurement of the interface pressure by means of ultrasonic waves, can facilitate the measurement of the interface pressure by maintaining the joint surface as its original states. In fact, the utmost representative characteristic feature of the UWM is nondestructive. Although the UWM is a very effective tool to measure the interface pressure, as already shown in the representative examples in Chaps. 7 and 8, a root cause of its difficulties lies in the quantification of the measured result. Thus at present, we can, in principle, use the UWM for the measurement

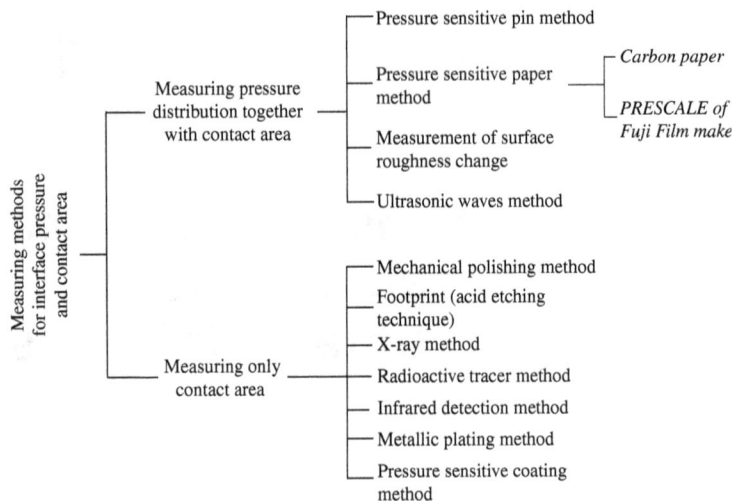

Figure A1-1   Measuring methods for interface pressure and contact area.

TABLE A1-1   Outlines of Pressure Sensitive Pin and Paper Methods

Measurement method	Measuring principle	Measuring accuracy	Available range	Advantages and disadvantages
Pressure sensitive pin method (contact pin technique, pressure pickup pin)	Displacement pickup caused by pressure (strain gage type is dominant)	Better than 10 %	Up to 10,000 kgf/cm²	Ease of measurement Need to calibrate load-displacement relationships beforehand Used widely, especially in spheres of press and forging dies Difficulties in pin arrangement To maintain high resolution, pin diameter must be 1 mm maximum Measuring accuracy depends on hardness of opposite surface
Pressure sensitive paper	Thin paper or sheet having sensitivity to change of pressure, e.g., carbon paper for duplicate typewriting In color type, microcapsules of smaller and larger diameters are destroyed by relatively higher and lower pressures	—	Up to 800 kgf/cm²	Ease of measurement and understanding by direct observation of changing color In repeated loading cycles, interface pressure for maximum loading can be measured In carbon type, joint must be ground or lapped In color type, calibration curve changes in dependence on temperature, humidity, and loading speed Capable of converting density change of color to interface pressure in color type developed by Fuji Film Co. (PRESCALE)

Note: The measuring accuracy is evaluated by comparing the integrated value obtained from the measured interface pressure distribution with the applied load.

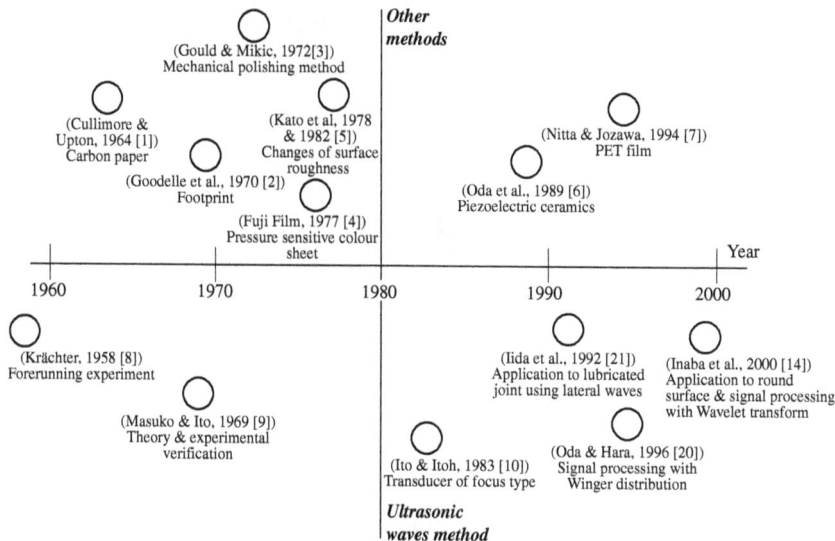

Other methods

(Gould & Mikic, 1972[3])
Mechanical polishing method

(Cullimore & Upton, 1964 [1])
Carbon paper

(Goodelle et al., 1970 [2])
Footprint

(Kato et al, 1978 & 1982 [5])
Changes of surface roughness

(Fuji Film, 1977 [4])
Pressure sensitive colour sheet

(Nitta & Jozawa, 1994 [7])
PET film

(Oda et al., 1989 [6])
Piezoelectric ceramics

Year

1960   1970   1980   1990   2000

(Krächter, 1958 [8])
Forerunning experiment

(Masuko & Ito, 1969 [9])
Theory & experimental verification

(Iida et al., 1992 [21])
Application to lubricated joint using lateral waves

(Inaba et al., 2000 [14])
Application to round surface & signal processing with Wavelet transform

(Ito & Itoh, 1983 [10])
Transducer of focus type

(Oda & Hara, 1996 [20])
Signal processing with Winger distribution

Ultrasonic waves method

Figure A1-2 Firsthand view for measuring methods of contact area, contact pattern, and interface pressure.

of the qualitative interface pressure and its distribution at the joint surface, i.e., relative contact intensity and contact pattern, across the whole joint surface.

Figure A1-2 is a firsthand view of the researches into the interface pressure measurement.

A1.1   Principle of Measurement and Its Verification

The concept of UWM is credited to Krächter in 1958 [8]. He tried to measure the contact pressure between two die blocks of an injection molding machine, and his proposal was patented (DP. Nr. 938273 in 1953). It is worth pointing out that Krächter of Mannesmann AG verified experimentally the validity of the proposed method in the mating surfaces with the higher interface pressure; however, he did not state the principle of the measurement from the theoretical aspect and the availability for the mating surfaces with lower interface pressure. In this context, the due achievements are credited to Masuko and Ito in 1969, and some quick notes will be stated below [9].

Principle of measurement.   There are, at present, two methods for the practical use depending on the type of ultrasonic transducer, i.e., beam and focus types. In both the beam and focus types, the measurement principle is the same, as will be stated in the following, although there

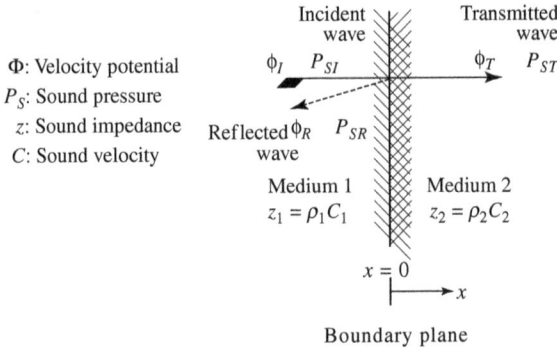

Figure A1-3 Reflection and transmission of plane waves at boundary plane.

are considerable differences in the measuring accuracies for the magnitude of interface pressure and directional resolution.

Now let us consider the plane waves incident perpendicular to a boundary plane, where the sound characteristics for both media differ from each other as shown in Fig. A1-3. In due course, the reflection and transmission of plane waves occur at the boundary plane. By assuming the sound pressure and particle velocity as to be not in time-intermittence and by taking the X coordinate as shown in Fig. A1-3, the amplitude ratio of the sound pressure on the boundary plane, i.e., the reflection rate of the sound pressure R_p, is given by

$$R_p = (\rho_2 C_2 - \rho_1 C_1)/(\rho_1 C_1 + \rho_2 C_2) \tag{A1-1}$$

where ρ = density of medium
C = sound velocity in medium
1 and 2 = subscripts corresponding to media 1 and 2, where plane waves are in progress

According to this characteristic, the ultrasonic waves are reflected at the boundary plane, and its sound pressure can be written as

$$P_{SR} = R_p \cdot P_{SI} \tag{A1-2}$$

where P_{SR} and P_{SI} are sound pressures of the reflected and incident waves, respectively.

The representative values of R_p for various materials are shown in Table A1-2, and in short the values of R_p are nearly nil and unit, when the waves are in progress within medium 1 made of iron and reach medium 2 made of iron and air, respectively. This means that the perfect transmission and nearly perfect reflection of waves occur at the

TABLE A1-2   Reflection Rate of
Sound Pressure R_p

Medium 2	R_p
	Around
Air	1.0
Oil	0.94
Water	0.88
Iron	0

Note: Medium 1 is made of iron, e.g., steel
or cast iron.

microscopic contact, i.e., iron-to-iron asperities contact, and at the iron-to-air contact. In contrast, the mating surface of the machined part is rough and wavy, even when the surface is finished by ultraprecision machining and its roughness is on the order of angstroms. Reportedly, the two surfaces in contact show the elastic and plastic deformation in their asperities, and the increase of contact area derived from such deformations is in proportion to the applied load, as widely recognized.

As a result, the quantity of transmitted waves increases, while that of reflected waves decreases in the dry flat joint with increasing normal applied load. Masuko and Ito analyzed the relationships between the normal load Q (interface pressure) and the sound pressure of the reflected waves in the dry flat joint made of iron, and gave the following expressions.

$$P_{SRm} = P_{SIm}[1 - (S + \Delta S)/S_0]$$

$$P_{SRm}/P_{SIm} = (1 - B_0) - B_2 Q$$

where P_{SIm} = peak of sound pressure of incident waves
P_{SRm} = peak of sound pressure of reflected waves
S_0 = area of idealized surface where ultrasonic waves are incident
S = area of initial contact
ΔS = increased area of contact by normal loading
B_0, B_2 = constants
B_0 = smaller than 1

Importantly, the relationships between the reflected sound pressure and the interface pressure can be depicted as shown in Fig. A1-4, where we can observe the nonlinearity and certain scatter caused by the following [9].

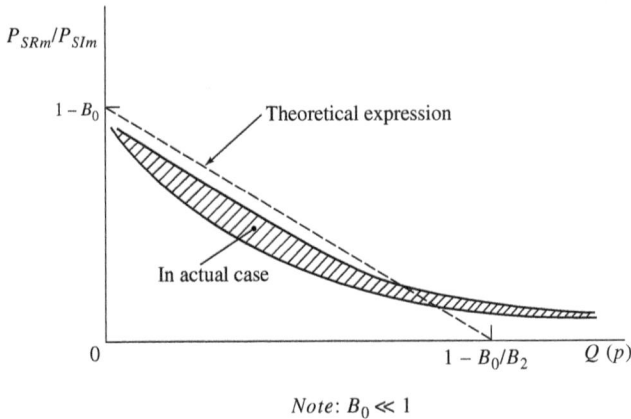

Note: $B_0 \ll 1$

Figure A1-4   Relationships between P_{SRm}/P_{SIm} and interface pressure p.

1. Even when the normal load increases extremely, the contact area with the air does not disappear.

2. For reasons of inhabiting random waviness and roughness at the contact surface, the new contact appears with increasing normal load.

3. There is damping of ultrasonic waves derived from the scatter of the waves at grain boundary and internal friction of the material.

When the ultrasonic flaw detector is used, the sound pressures of the incident and reflected waves coincide with the heights of the initial pulse and the echo on the CRT (*cathode ray tube*). It is thus recommended to use the ratio of h_e to h_{e0}, that is, the E_R value (echo ratio), to measure the interface pressure. In principle, the ratio of h_i to h_e must be used from the theoretical point of view; however, we can expect to avoid unfavorable noise by using the E_R value. In addition, $E_R{}^*$ is more suitable for the practical measurement than E_R, so that the measured value can be interpreted in accordance with the human sense, where $E_R{}^*$ is $1 - E_R$. For the ease of understanding, we can conceptualize the measuring principle and relationships between $E_R{}^*$ and the interface pressure as shown in Fig. A1-5. Within this context, Fig. A1-6 furthermore reproduced the relationships between the mean interface pressure and E_R value, which was reported by Masuko and Ito to verify the theoretical analysis. In this case, the measurement was carried out using the test rig, which was similar to that used by Krächter and shown together in Fig. A1-6. In fact, the UWM can measure the interface pressure accurately together with showing the good reproducibility when we repeat loading and unloading [9].

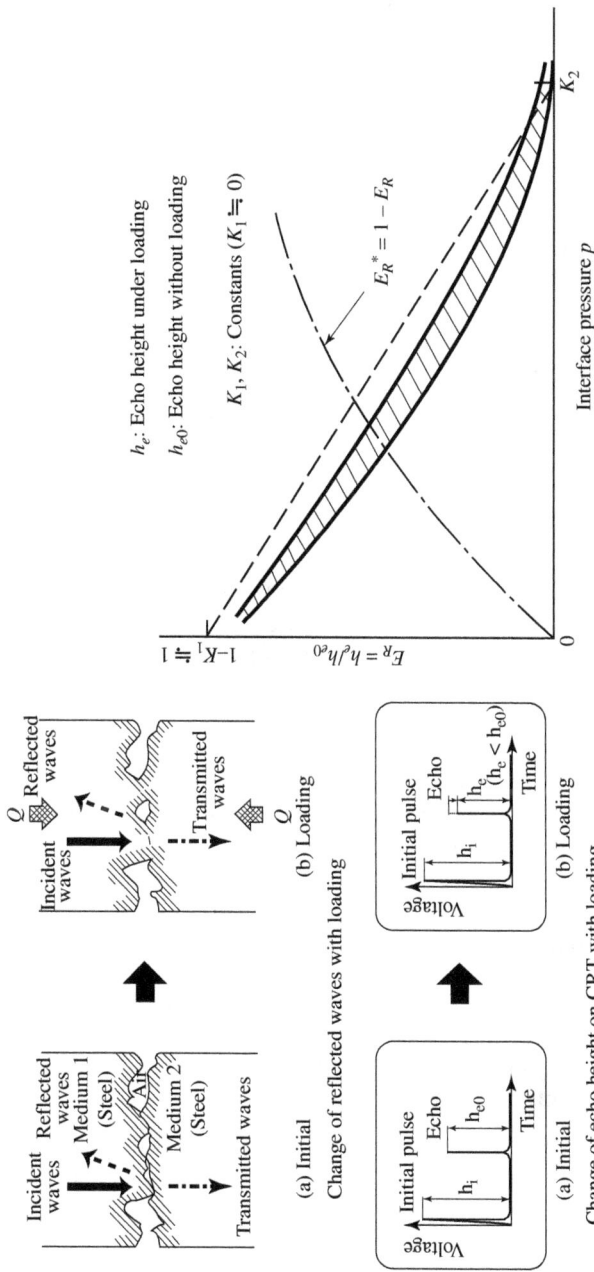

h_e: Echo height under loading

h_{e0}: Echo height without loading

K_1, K_2: Constants ($K_1 \fallingdotseq 0$)

$E_R^* = 1 - E_R$

Interface pressure p

$E_R = h_e / h_{e0}$

$1 - K_1 \fallingdotseq 1$

K_2

Incident waves **Reflected waves** Medium 1 (Steel) Air Medium 2 (Steel) **Transmitted waves**

(a) Initial

Change of reflected waves with loading

Q **Reflected waves** **Incident waves** **Transmitted waves** Q **Loading**

(b) Loading

Voltage Initial pulse Echo h_i h_{e0} Time

(a) Initial

Voltage Initial pulse Echo h_i h_e ($h_e < h_{e0}$) Time

(b) Loading

Change of echo height on CRT with loading

Figure A1-5 Schematic views for measuring principles and fundamental relationships between E_R^* and p.

461

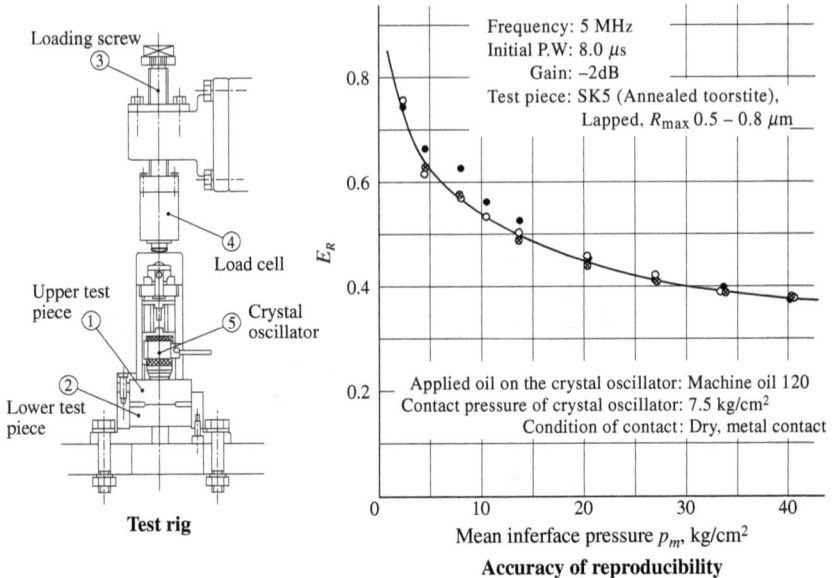

Frequency: 5 MHz
Initial P.W: 8.0 μs
Gain: –2dB
Test piece: SK5 (Annealed toorstite),
Lapped, R_{max} 0.5 – 0.8 μm

Applied oil on the crystal oscillator: Machine oil 120
Contact pressure of crystal oscillator: 7.5 kg/cm^2
Condition of contact: Dry, metal contact

Loading screw
③
Load cell
④
Upper test piece ①
Crystal oscillator ⑤
Lower test piece ②

Test rig

E_R

Mean inferface pressure p_m, kg/cm^2

Accuracy of reproducibility

Figure A1-6   Relationships between E_R and mean interface pressure.

In short, the UWM can be mostly characterized by the measurement of the change of the echo height on the CRT, i.e., use of voltage domain (sound pressure) together with maintaining the nondestructive aspect. In contrast, the ultrasonic waves have been applied, in wider scope, to the measurement of the reflected time of the incident waves, i.e., use of time domain, so far exemplified by the ultrasonic flaw detection and bolt tension measurement merchandised by Raymond Co. in the United States. As can be readily seen from the measuring principle, the UWM has higher potential to apply it to not only the machine tool joint but also other measuring objectives.

Admitting that the UWM has many beneficial features explicitly and implicitly through the considerable number of forerunning researches and engineering developments, the facing necessities are to develop some innovative techniques such as the following:

1. Focusing method of ultrasonic waves to measure the interface pressure at the smaller spot

2. Processing method of output signal to improve the measuring accuracy

3. Quantitative measuring method with higher accuracy: how to convert the $E_R{}^*$ value to the interface pressure in practical application

As already shown in Fig. A1-2, the UWMs using the transducer of focus type and with new signal processing by Wigner distribution have been proposed for subjects 1 and 2 mentioned above in the 1980s and

1990s, respectively; however, the quantified measurement by the UWM is far from completion, apart from the measurement in the university laboratory, because of the difficulty of calibration of E_R^* to the quantified value. As a result, the UWM is, in general, applicable to the measuring purpose of the qualitative interface pressure distribution, i.e., distribution of echo height E_R^*, which is called the contact pattern, as will be shown below in the cases of the piston ring and multiple-bolt-flange assembly. Importantly, the E_R^* is nearly proportional to the interface pressure, and thus the contact pattern can, in preferable case, be converted into the quantified interface pressure distribution with the assistance of the mathematical treatment.

Now let us consider how to concentrate the incident area of ultrasonic waves on the smaller spot. In the case of the conventional transducer, which has prevailed, the ultrasonic waves in a solid propagate like a beam near the transducer and are liable to diverge far from the transducer, as shown in Fig. A1-7. Of these, the diverging portion can be determined by the directional angle, which is in proportion to the wavelength and in inverse proportion to the diameter of the transducer. In the measurement by the conventional transducer with longitudinal waves, the diameter of the crystal oscillator is from 10 to 20 mm, and

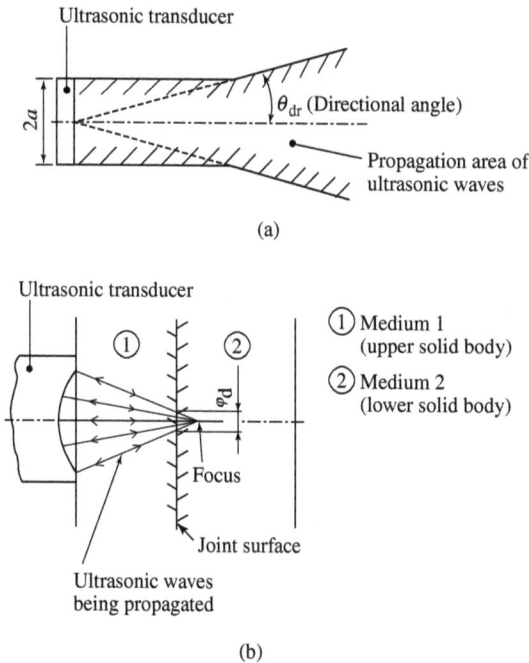

Figure A1-7 Types of ultrasonic transducers and propagation states of ultrasonic waves: (a) Conventional type and (b) focus type.

the corresponding oscillating frequency is from 10 to 3 MHz. In consequence, the incident area of ultrasonic waves becomes from 10 to 20 mm. More specifically, the mean interface pressure within a relatively large spot can be measured when the conventional transducer is used, and thus Ito and Itoh tried to use the transducer of focus type [10].

Although the measuring principle is the same in both beam and focus types, the transducer of focus type can control the incident diameter of ultrasonic waves at the joint surface, such as shown in Fig. A1-7. Obviously, the focus type has certain limitations in the measurement, which is derived from the short focus length. Figure A1-8 shows the comparison of the relationships between the E_R value and the interface pressure when varying the incident diameter of ultrasonic waves at the joint surface. Typically, as the incident diameter gets smaller, the measuring accuracy for the interface pressure itself deteriorates considerably; however, it may be expected to improve the measuring accuracy for the contact region, i.e., directional resolution accuracy, to some extent.

To clarify the beneficial feature of the focus type, some measurements were carried out for the contact pattern of press-fitting components. Figure A1-9 is a schematic view of the experimental setup. The shaft, both sides of which were cut off to form the flat planes with different widths, was press-fitted in cylinder 2, and the contact pattern of the shaft-to-cylinder connection was measured using the transducer of focus type 4 or conventional type 5. Both transducers were held in the holder 3 to maintain the constant contact condition, and the measuring points were arranged around the circumference of press fitting at 5 degrees intervals. Figure A1-10 shows some measured results, and as can be

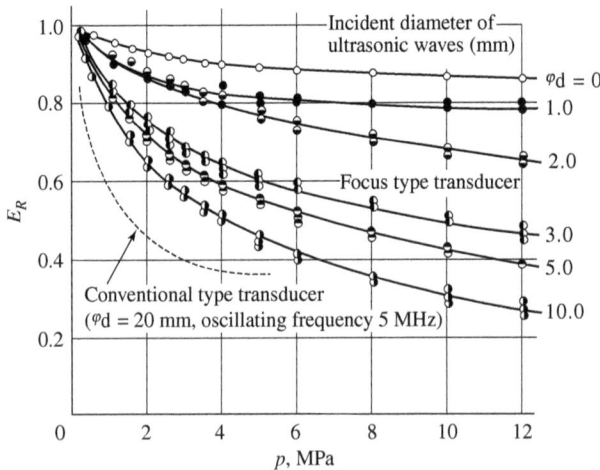

Figure A1-8   Effects of incident diameter d of ultrasonic waves on relationships between mean interface pressure p and value of E_R.

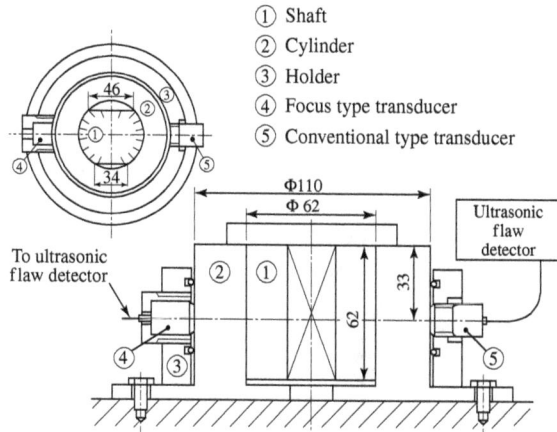

① Shaft
② Cylinder
③ Holder
④ Focus type transducer
⑤ Conventional type transducer

Figure A1-9   Experimental setup to measure contact pattern of press-fitting.

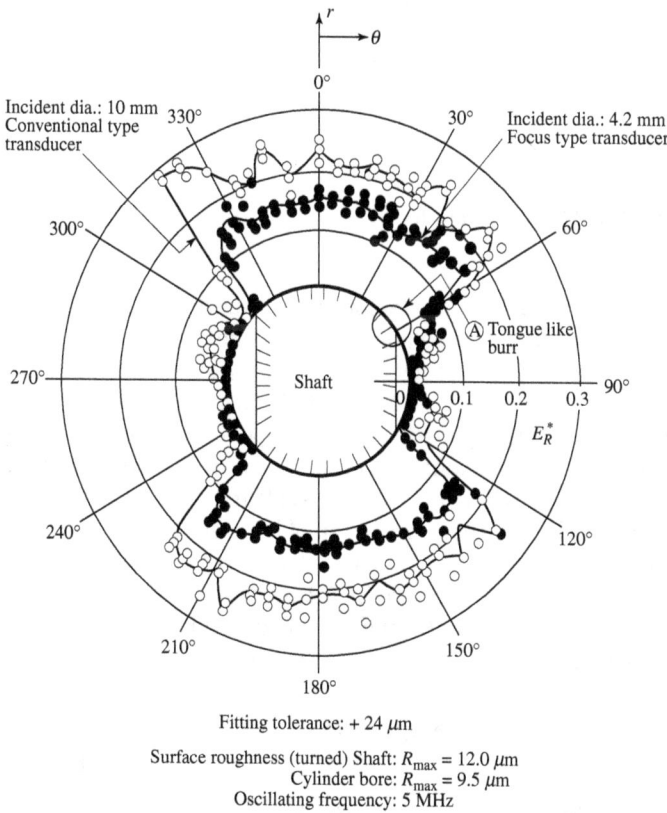

Fitting tolerance: + 24 μm

Surface roughness (turned) Shaft: R_{max} = 12.0 μm
Cylinder bore: R_{max} = 9.5 μm
Oscillating frequency: 5 MHz

Figure A1-10   Examples of measured contact patterns.

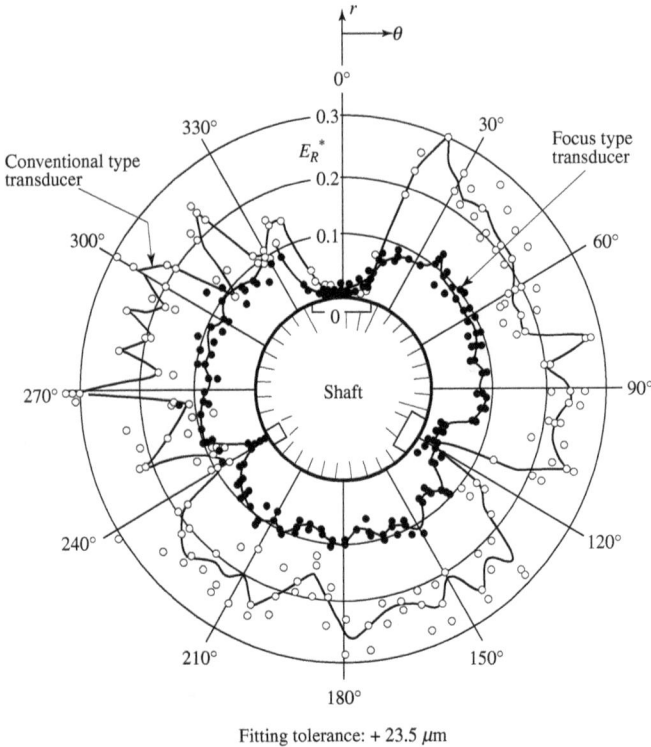

Fitting tolerance: + 23.5 μm

Figure A1-11   Availability of both the transducers.

seen, the contact pattern can be satisfactorily measured with both transducers. As demonstrated at the edge portion of the shaft, it is noteworthy
that the directional resolution accuracy is considerably improved when the
focus type is used. In addition, owing to the difference of the quantity of
waves detected, the measured contact patterns appear not to be identical
to each other. By keeping the incident area of the ultrasonic waves constant,
thus, the contact pattern for the fitting tolerance of 9 μm was also measured
using the focus type. Although the comparison of both measured results is
not shown, as can be readily seen, the intensity of contact coincides well with
that of fitting tolerance, i.e., the more tight-fitting, the larger the value of
E_R^*, together with showing sufficiently the distinct difference. Figure A1-11
shows furthermore another measured result for press-fitting.

A1.2   Some Applications and Perspectives in the Very Near Future

An application to machine tool joint.   In Chaps. 7 and 8, the two-dimensional
contact patterns were displayed from the engineering design point of

view. A representative measurement for the multiple-bolt-flange assembly will thus be shown below in full consideration that the three-dimensional contact pattern can facilitate the firsthand understanding of the essential features for the joint characteristics. In fact, there have been a considerable number of researches into a bolt-flange assembly, and within these researches the contact pattern at a certain cross section was examined. However, it may be considered unreasonable to presume the whole shape of the contact pattern from the result of one sectional contact pattern, especially when the joint has a flange of bay-type with multiple connecting bolts and the joint surface is very wavy.

Within this context, Tsutsumi et al. tried to measure the three-dimensional contact pattern and represented the results with topographical means [11]. The experimental setup is shown in Fig. A1-12, and the contact pattern between the bolted column and the baseplate was measured using the transducer, which was in contact at the bottom surface of the baseplate through the threaded hole in the rotary disk. The diameter and oscillating frequency of the transducer are 10 mm and 10 MHz, respectively.

Figures A1-13 and A1-14 show the effects of the flange thickness when the number of the connecting bolt is varied. We can first recognize the very importance of the topographical representation of the contact pattern compared with the cross-sectional representation, and then we can observe the interesting behavior of the multiple bolt-flange assembly as follows.

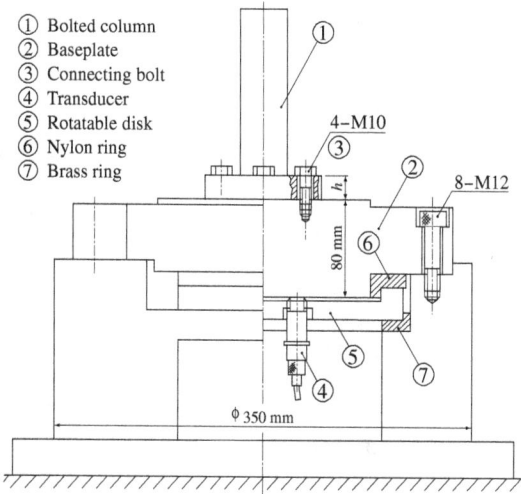

Figure A1-12 Experimental setup to measure three-dimensional contact pattern.

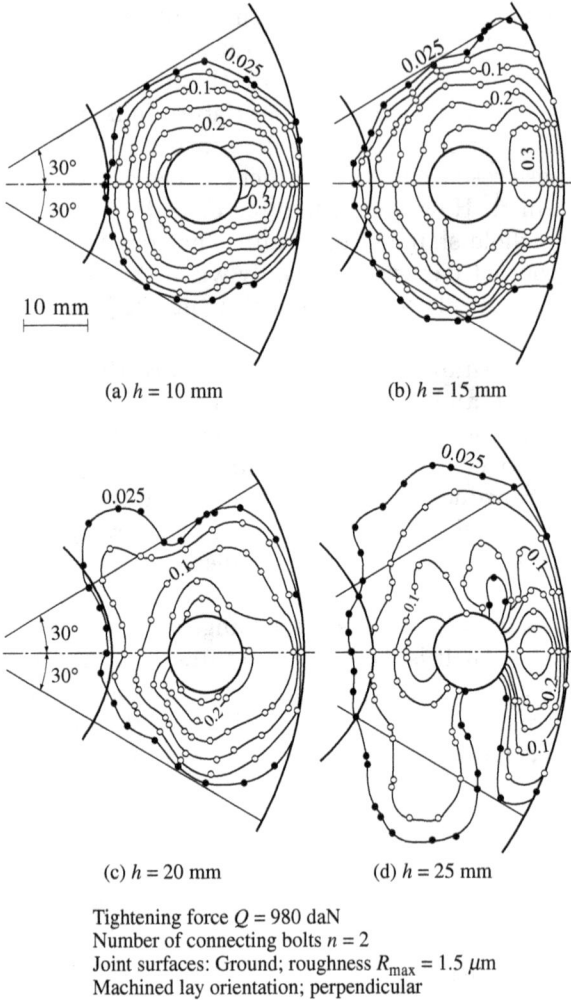

(a) h = 10 mm   (b) h = 15 mm

(c) h = 20 mm   (d) h = 25 mm

Tightening force Q = 980 daN
Number of connecting bolts n = 2
Joint surfaces: Ground; roughness R_{max} = 1.5 μm
Machined lay orientation; perpendicular

Figure A1-13   Interface pressure distribution in multiple-bolt-flange assembly.

1. The contact pattern depends on the flange thickness. In the case of the thin flange, the contact pattern is of concentric circlelike form, whereas in the thick flange, the contact pattern spreads in the circumferential direction, but does not spread so much in the radial direction. In short, the interface pressure is prone to concentrate on the weak portion in a bolted joint, resulting in the occurrence of local deformation. In other words, the flange portion of a bolted column shows the obvious pattern of furrowlike deformation especially in the case of thin flange, whose valley corresponds with the connecting bolt. These appear to be new findings, showing also the future potential of the UWM.

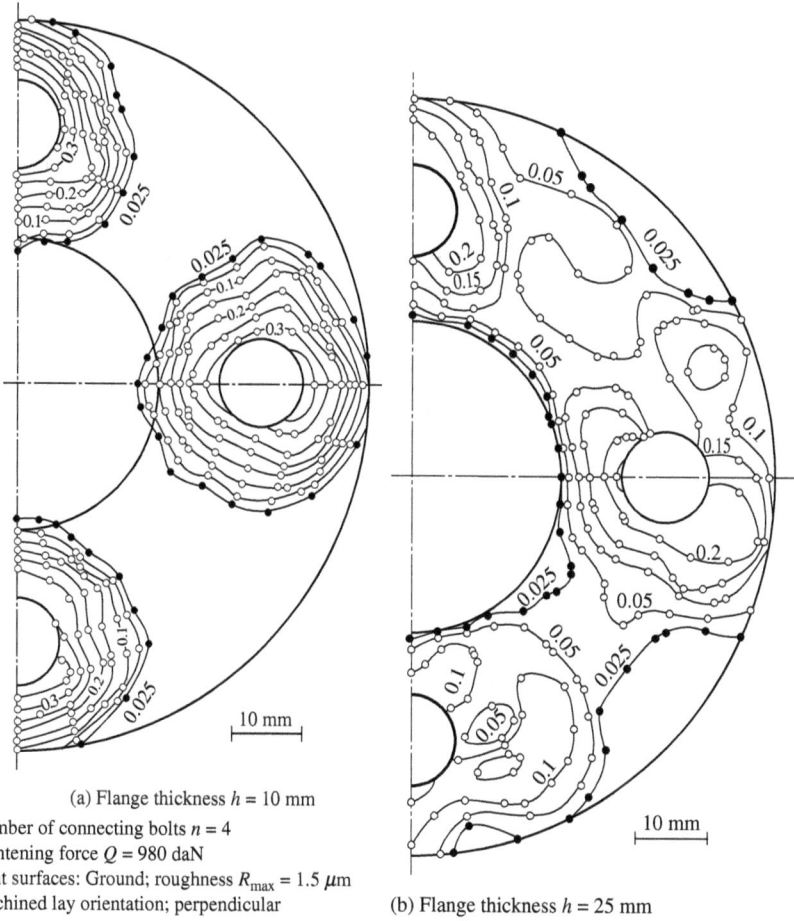

(a) Flange thickness h = 10 mm

Number of connecting bolts n = 4
Tightening force Q = 980 daN
Joint surfaces: Ground; roughness R_{max} = 1.5 μm
Machined lay orientation; perpendicular

(b) Flange thickness h = 25 mm

Figure A1-14   Interface pressure distribution in multiple-bolt-flange assembly.

2. The flatness deviation or waviness affects somewhat the contact pattern, resulting in the irregular distribution, and it becomes relevant when the flange thickness is thicker. In other words, the thinner flange is liable to deform largely, and as a result, the waviness pretends to have no effect on the contact pattern. The same behavior can be observed even in the single bolt-flange assembly shown in Fig. A1-15, where the irregular distribution is in good agreement with that of waviness across the whole joint surface.

Figure A1-16 shows the effects of the tightening force, and in short, it can be said that the tightening force has large effects on the magnitude of the interface pressure, keeping its working area constant.

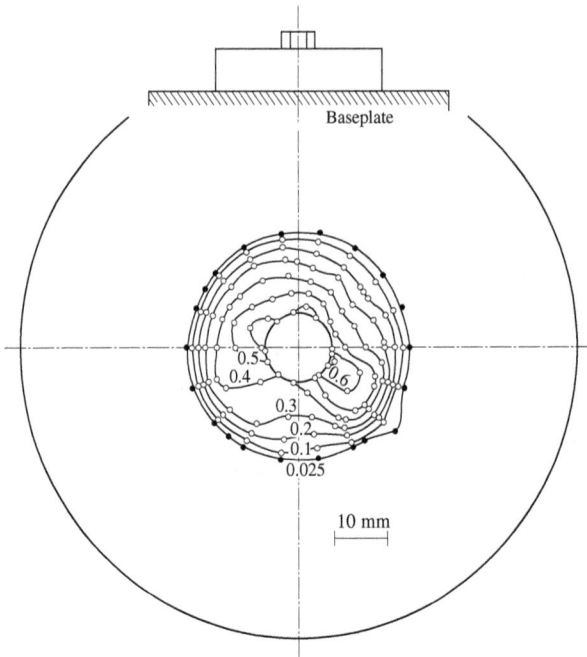

(a) $h = 10$ mm

Tightening force $Q = 980$ daN
Joint surfaces: Ground; roughness $R_{max} = 1.5$ μm
Machined lay orientation; perpendicular

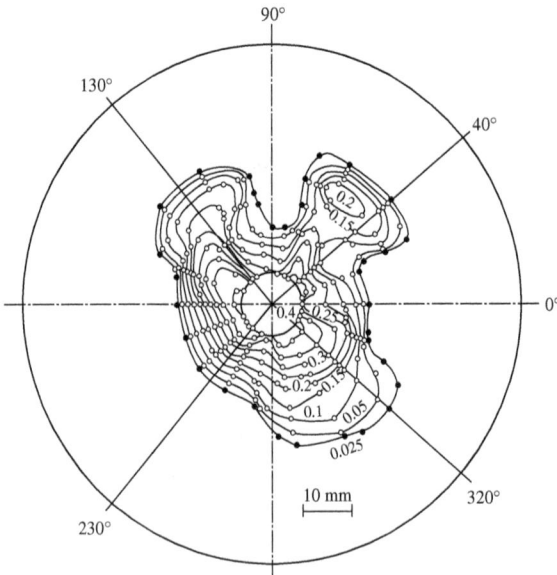

(b) $h = 25$ mm

Figure A1-15 Interface pressure distribution in single bolt-flange assembly.

Having the difficulty of the conversion in mind, the contact pattern was furthermore converted to the quantified interface pressure distribution using the calibration curve produced beforehand. Figure A1-17 is a reproduction of such a conversion, and the integrals of the interface pressures across the whole area surrounding each connecting bolt are within the differences of –25% to 10% compared with its tightening force.

As can be readily seen, the UWM is a very effective tool to measure the contact pattern, and the topographical representation of the contact pattern can facilitate the ease of understanding of the joint behavior to a large extent.

Some applications to other engineering spheres.   The UWM has not been widely broadcasted, but has been used by some professional people in

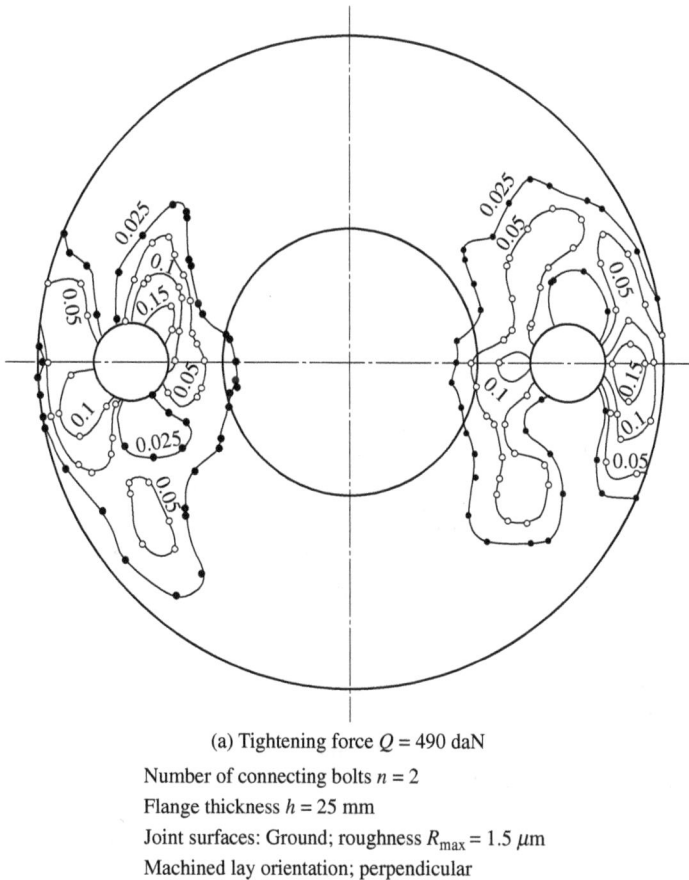

(a) Tightening force Q = 490 daN

Number of connecting bolts n = 2

Flange thickness h = 25 mm

Joint surfaces: Ground; roughness R_{max} = 1.5 μm

Machined lay orientation; perpendicular

Figure A1-16   Influence of tightening force on contact pattern.

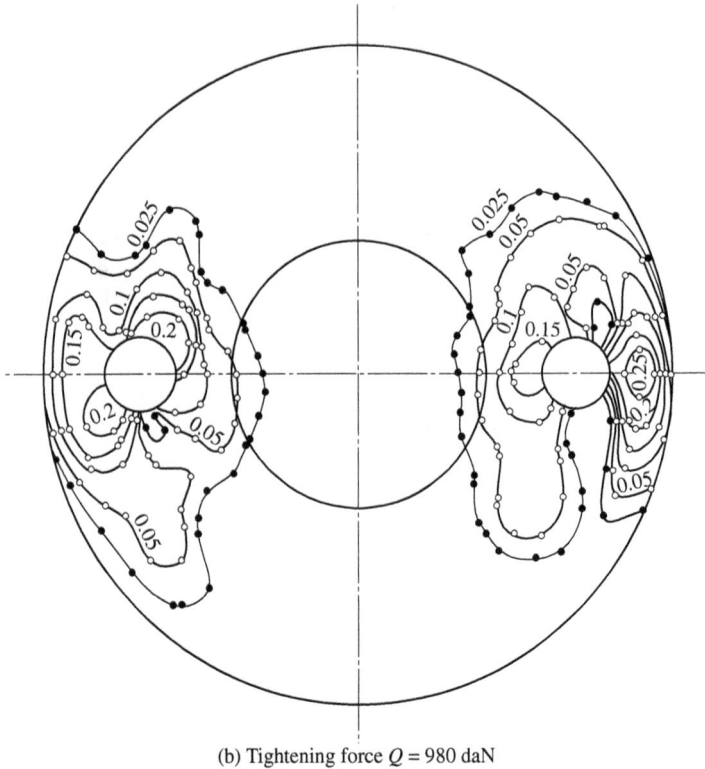

(b) Tightening force $Q = 980$ daN

Figure A1-16 (*Continued*)

accordance with their dire necessities up to the year 2000. Figure A1-18 is a firsthand view of such activities, and in the following, some representative applications of the UWM to other spheres will be stated.

In retrospect, the earlier applications are the measurements of the real contact area between the pin and disc in the sphere of the tribology [12], and also of the contact pattern between the cylinder bore and the piston ring in the internal combustion engine [13]. In the automobile, the sealing capability of the piston ring was at burning issue to improve the fuel consumption efficiency of the engine on that occasion. The measurement was carried out using lateral waves, which can be converted from longitudinal waves through the wedge located to the bore. Importantly, the contact pattern, i.e., echo changes in r-θ coordinate, can be obtained by traveling the transducer along the circumference of bore. Figure A1-19 is a reproduction of the measured contact patterns, and it appears similar to those already shown in Fig. A1-10. Although the sealing for the combustion pressure can be guaranteed, the contact pattern is not uniform, but shows obvious variation to some extent together

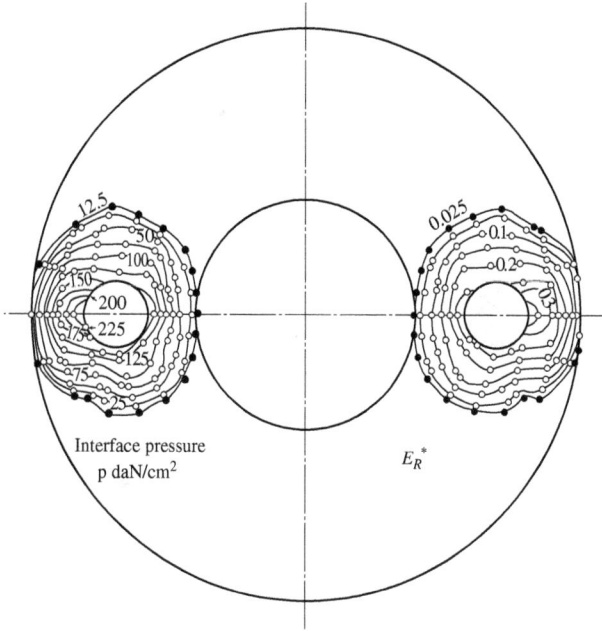

Number of connecting bolts $n = 2$
Tightening force $Q = 980$ daN
Flange thickness $h = 10$ mm
Joint surfaces: Ground; roughness $R_{max} = 1.5$ μm
Machined lay orientation; perpendicular

Figure A1-17   Qualitative and quantitative interface pressure distribution with topographical representation.

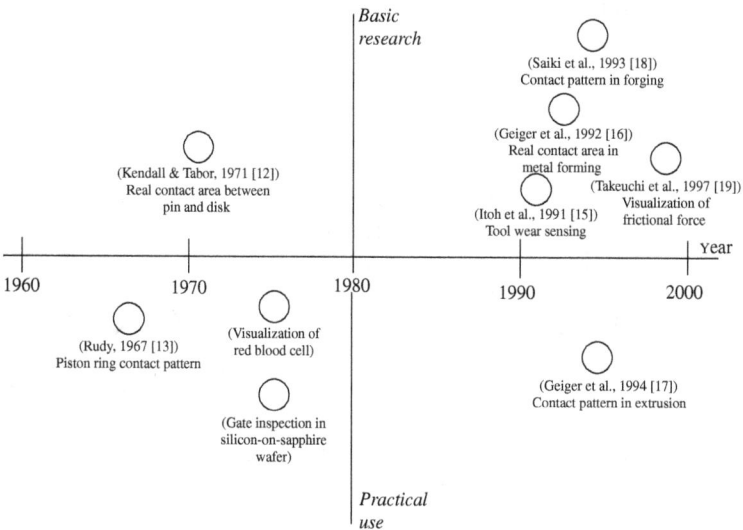

Figure A1-18   Firsthand view for applications of UWM to other industrial spheres.

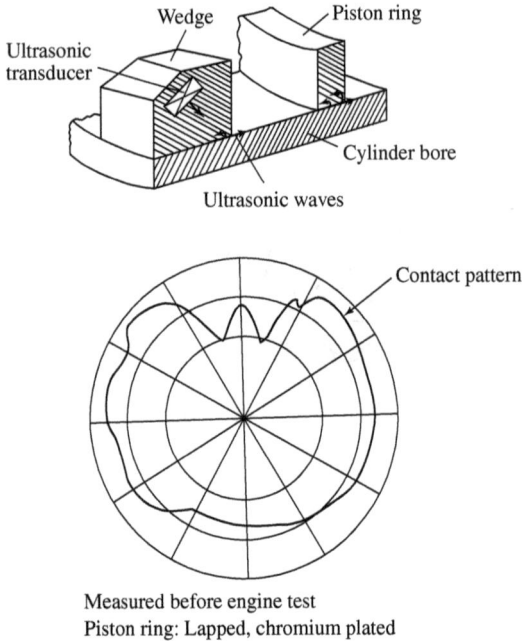

Figure A1-19   Contact pattern of piston ring to cylinder bore (by Rudy).

with implying that the wear of the piston ring is not uniform across the whole contact line. Through nearly all earlier research activities, the UWM using the conventional transducer and longitudinal waves has been dominant, and Inaba et al. have recently applied the focus type to measure the interface pressure between the outer ring of the bearing and the bore of the bearing housing [14]. In the cylindrical contact, the ultrasonic waves transmit more complicated state than in the flat contact, resulting in much inclusion of unfavorable reflection noise within the output signal. In this context, together with the contrivance for fixing the transducer at the round surface, reasonable signal detection can be facilitated with the Wavelet transform to identify the correct reflected waves from both the time and frequency domains. Figure A1-20 shows an example of signal processing, where the characteristic h_w is the absolute value of the transformed echo height. In addition, Inaba et al. conducted skillfully the conversion of h_w to the quantified interface pressure using the special comparator. Figure A1-21 is a schematic view of the spindle model, and with it the variation of the interface pressure at the operation was measured, as shown in Fig. A1-22. It is very interesting that the interface pressure increases immediately after starting the operation, which may be caused by the local expansion of the outer

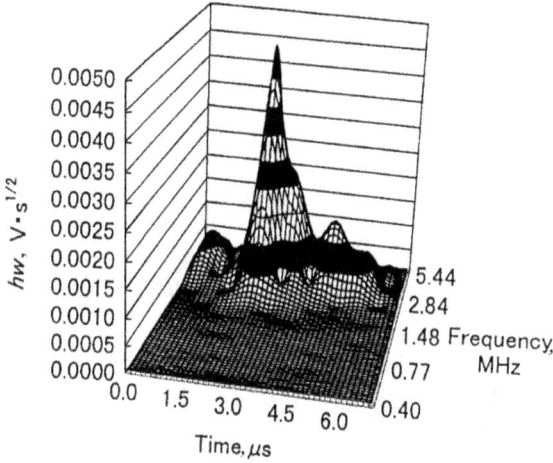

Figure A1-20 Wavelet transform of reflected waves.

ring of bearing due to the intensive temperature rise. In due course, the interface pressure reaches the lower pressure with the running time, resulting from the expansion of the whole spindle system. In addition, that of Inaba et al. implies a dire necessity to reconsider the validity of the design standard for the fitting tolerance. Intuitively, there remains

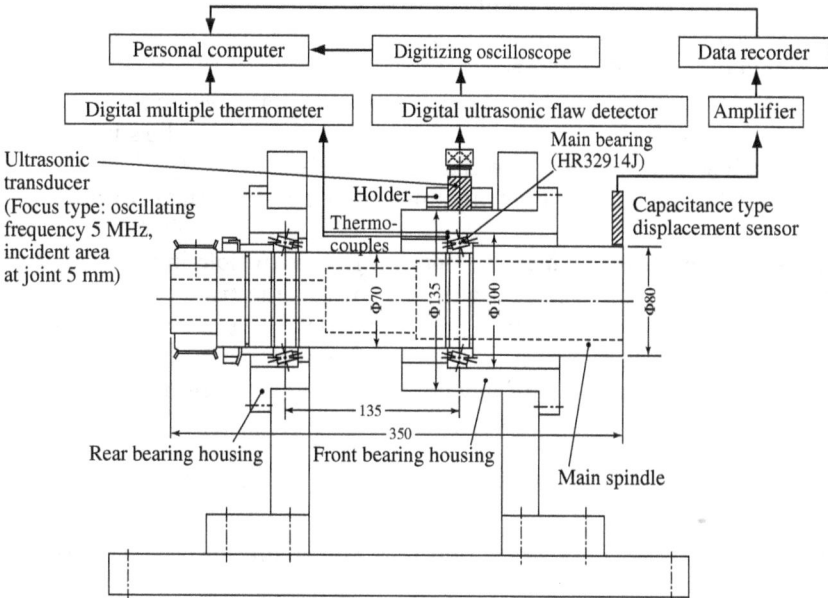

Figure A1-21 Schematic view of spindle model.

Fitting tolerance between the outer ring and the bearing housing: –6 μm
Surface roughnesses: Outer ring-Ground, 0.16 μm R_a
Inner surface of bearing housing-turned, 9.40 μm R_a
Preloading of bearing: 2000 N   Lubricant: Grease

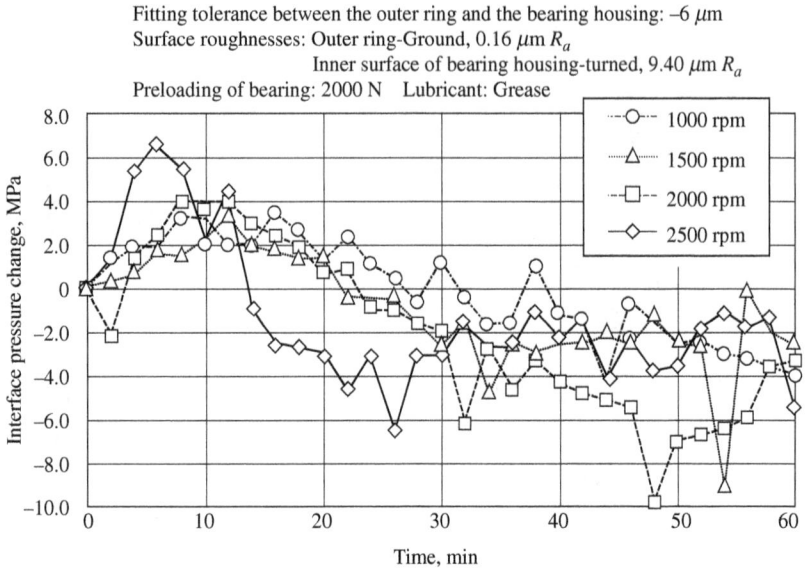

Figure A1-22   Variations of interface pressure with running time.

something to be seen whether the present design standard has been considered such a difference between the still stand and the operating condition, in the determination of the fitting tolerance standard.

Based on these applications, it can be suggested that at the beginning of the 2000s an industrial sector capable of effectively using the UWM might be that of die molding. For example, we may measure the contact pattern of the die base in injection and transferability of die surface texture with echo variation, as well as the interfacing quality of small mold parts in stamping die by the transmission loss or decaying rate of waves.

Recent improvements for measurement.   There is twofold improvement in the UWM. One is concerned with the ultrasonic transducer for non-contact transmission so far reported elsewhere, and the other is for output signal processing. At the beginning of the 2000s, the following new transducers are being developed; however, there are no reports related to their application to the UWM.

1. Microwaves transducer of focus type (available for dielectric materials: 30 to 300 GHz)

2. Air-coupled transducer for Lamb waves (available for plastics: 100 kHz to 23 MHz)

3. Laser-based type with amplitude of less than 1 nm.

In relation to signal processing, some marked improvements have been carried out by Oda and Hara [20] and by Inaba et al. [14]. Reportedly, the leading method so far used is based on the $E_R{}^*$ value; however, at issue is to enhance the interpretation of obtained output signal. Thus, to improve the reliability and to extend the applicability of the UWM, Oda and Hara have recently proposed another method using the Wigner distribution. In short, the interface pressure can be evaluated by the energy of the reflected echo WDE, which is represented with the integration of the due Wigner distribution in time and frequency domains.

In short,

$$\text{WDE} = \iint W_{xx}(t,f)\, dt\, df \qquad (\textstyle\int: -\infty - \infty)$$

where $W_{xx}(t,f)$ is the auto-Wigner distribution and can be given by

$$\int x(t + \tau/2)\, x^*\, (t - \tau/2)e^{-i2\pi ft}\, d\tau \qquad (\textstyle\int: -\infty - \infty)$$

where x^* = complex number of x
$\qquad x$ = reflected echo measured
$\qquad t$ = time
$\qquad f$ = frequency

In addition, they have newly proposed to use the countersunk head screw-like tube containing glycerin to focus the ultrasonic waves. Figure A1-23 shows an effect of the proposed method, and as can be seen, there is certain improvement to measure the qualitative interface pressure up to 0.6 MPa.

As mentioned above, the UWM is, in general, carried out with the longitudinal waves and duly results in the deterioration of the measuring accuracy derived from the interfacial layer, especially by the oil and

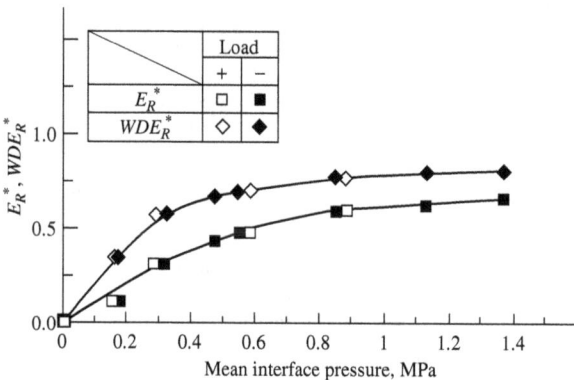

Figure A1-23   Correlation of $E_R{}^*$ and $WDE_R{}^*$ to interface pressure (courtesy of Hara).

grease. In the machine tool joint, the jointed surface is, in nearly all cases, very oily, and we have some difficulties to obtain the contact pattern and even the contact area. In due course, the necessity is to degrease the joint surface prior to the measurement; however, this is not acceptable from the engineering point of view. This is attributed to the characteristic feature of the longitudinal waves, which can propagate in both the solid and liquid.

In this context, it is furthermore worth pointing out that Iida et al. have tried to measure the contact area in the flat joint with oil [21]. They used the lateral waves, because only the longitudinal waves can transmit in the liquid. In other words, the jointed surface with the oil behaves as a dry joint when the lateral waves are incident. In due course, the echo height ratio E_R shows the obvious reduction with the intensity of contact.

References

1. Cullimore, M. S. G., and K. A. Upton, "The Distribution of Pressure between Two Flat Plates Bolted Together," *Int. J. Mech. Sci.*, 1964, 6: 13–25.
2. Goodelle, R. A., W. J. Derner, and L. E. Root, "A Practical Method for Determining Contact Stresses in Elastically Loaded Line Contacts," *ASLE Trans.*, 1970, 13: 269–277.
3. Gould, H. H., and B. B. Mikic, "Areas of Contact and Pressure Distribution in Bolted Joints," *Trans. of ASME, J. of Eng. for Industry*, Aug. 1972, pp. 864–870.
4. Catalog of Fuji Film Co., "PRESCALE," 1977.
5. Kato, S., K. Yamaguchi, and T. Kato, "Pressure Distribution Measurement for Metal-to-Metal Contact Surface Using Change of Surface Roughness," *Trans. of JSME*, 1982, 48(427): 408–417.
6. Oda, J., et al., "Development of Sensing System of Piezoelectric Ceramics Measuring Contact Pressure Distributions," *Trans. of JSME*, 1989, 55(513): 1230–1235.
7. Nitta, I., and T. Jozawa, "Measurements of Distributions of Contact Pressures between Ground Surfaces Using PET Films," *Trans. of JSME*, 1994, 60(579): 3970–3977.
8. Krächter, H., "Ein neuartiges Verfahren zur Messung von Druckkräften mit Ultraschall," *Werkstatt und Betrieb*, 1958, 91(5): 246–248.
9. Masuko, M., and Y. Ito, "Measurement of Contact Pressure by Means of Ultrasonic Waves," *Annals of CIRP*, 1969, XVII(3): 289–296.
10. Ito, Y., and S. Itoh, "Contact Pattern Measurement by Means of Ultrasonic Waves: Art of Present and Some Improvements of Its Performance," *Trans. of ASME, J. of Vib., Acoustics, Stress, and Reliability in Design*, 1983, 105: 237–241.
11. Tsutsumi, M., A. Miyakawa, and Y. Ito, "Topographical Representation of Interface Pressure Distribution in a Multiple Bolt-Flange Assembly—Measurement by Means of Ultrasonic Waves," Design Engineering Conference and Show, ASME, April 1981, 81-DE-7.
12. Kendall, K., and D. Tabor, "An Ultrasonic Study of the Area of Contact between Stationary and Sliding Surfaces," *Proc. Roy. Soc. Lond.*, 1971, 323A: 321–340.
13. Rudy, T., "An Ultrasonic Method of Measuring Piston Ring Bore—Contact Patterns," in *Proc. of Automotive Engineering Congress*, January 1967, No. 670027.
14. Inaba, C., et al., "Estimation of Contact Pressure between Bearing and Bearing Housing by Means of Ultrasonic Waves," *Trans. of JSME (C)*, 2000, 66(645): 1674–1680.
15. Itoh, S., et al., "Ultrasonic Waves Method for Tool Wear Sensing—In-Process and Built in Type," Proc. of Int. Mech. Engg. Conf. 1991, IE Australia, Sydney,
16. Geiger, M., U. Engel, and F. Vollertsen, "In-Situ Ultrasonic Measurement of the Real Contact Area in Bulk Metal Forming Processes," *Annals of CIRP*, 1992, 41(1): 255–258.

17. Geiger, M., U. Engel, and M. Pfestorf, "Ultrasonic Inspection of Stress State of Prestressed Extrusion Tools," *Production Engg.*, 1994, 1(2): 213–216.
18. Saiki, H., et al., "Estimation of Contact Conditions of Frictional Interface in Forging Processes Using Ultrasonic Examination," *Trans. of JSME (C)*, 1993, 59(562): 1934–1939.
19. Takeuchi, A., et al., "An Attempt at Observation of Local Friction Using Ultrasonic Support," *Trans. of JSME (C)*, 1997, 63(611): 2456–2463.
20. Oda, M., and T. Hara, "Contact Pressure Measurement Utilizing Time-Frequency Analysis of Ultrasonic Reflected Wave at the Rubber Surface in Sliding Contact," *Trans. of JSME (A)*, 1996, 62(598): 1425–1431.
21. Iida, T., et al., "A Preliminary Evaluation of Real Contact Area Using Ultrasonic Method," *Tribologist*, 1992, 37(1): 69–75.

2

Model Testing and Theory

From not only the enhancement of functionality and performance of a machine tool, but also the rationalization of the machining processes within the production morphology, it is desirable to estimate the static, dynamic, and thermal characteristics of the machine tool as a whole in the design stage. In the old days, dominant was the engineering calculation, which relied on the analytical method. There were thus many difficulties to evaluate exactly such characteristics at the design stage. With the advance of the CAD and CAE, e.g., digital engineering and rapid prototyping in recent years, however, we now have certain powerful tools to estimate roughly the characteristic features of the machine tool in the design stage.

Obviously, the machine tool design is, by nature, one of the ill-defined problems, and in consequence the computerized design has a certain limitation, although many machine tool designers are not aware of it. More specifically, the burning issues in the computerized design are the determination of the dynamic and thermal boundary conditions, and the prediction of the distribution frequency of the applied load during the machine life. Of these, the utmost difficulty is to determine the heat dissipation coefficient of the machine.

Although such problems remain to be solved, at present the computerized design is prevalent, and as a result, the model theory and testing are believed to be far afield from the design procedure. In retrospect, there were some methodologies to assist the designer, e.g., FOF analysis, deformation diagram, and model testing. Of these, model testing could again be of resurgence in consideration of the growing difficulty and complexity in the estimation of the thermal deformation, which is derived from the decoration with the panel and splash covers. Reportedly, the panel cover causes the thermal accumulation in one aspect and accelerates the thermal dissipation in another aspect. In addition, the advent of the machine tool of multiple-machining-function integrated type has accelerated the necessity of the model testing again, because of very complex loading during the machine life.

In fact, model testing is very effective, as has been verified through long-standing experience, although there is a severe constraint in the scale factor to obtain the reliable test result. In the following, thus, the basic knowledge will be quickly stated.

A2.1   Model Testing and Theory for Structural Body Component

There are three kinds of testing in the machine tool, i.e., analytical, assessment, and acceptance tests; and the analytical test is often carried out using the model. As a matter of course, model testing can facilitate the reduction of the research and engineering development cost by applying the test result obtained from the model, i.e., small specimen, to the full-scale structure after the result is interpreted in accordance with the model theory. In relation to the model testing, thus, the basic necessity is first to understand the basic model theory and testing for the monolithic structural body component. As can be readily seen, the structural body component with the joint shows more complicated behavior, and thus some further consideration must be given.

Within a model testing context, Saljé [1], Opitz and Bielefeld [2], and Thornley [3] made many contributions so far, and these earlier works can obviously show the characteristic and advantageous features of the model testing for the monolithic structural body component as follows.[1]

1. The model testing has higher accuracy and reliability, the error of which is on the order of a few percent to 10%, provided that the length scale factor is more than about 1/3.

2. The accuracy of model testing deteriorates amazingly, and the related error ranges from 20% to 30%, when it is applied to the structure with the joint.

---

[1]There are another two model tests: One is the photoelasticity method, and the other uses elastic materials. In the former, the model is made of photoelastic material, especially aiming at the investigation into the stress distribution. In the latter, the model is made of rubber or foam plastic, and the designer can use it behind the desk with ease of handling and visualization to understand, for instance, the local deformation at the aperture and abutment lip.

Shimamura, S., et al., "Photoelastic Analysis of Machine Tool Structure, Parts 1 and 2," Preprint of 11th Autumn Meeting of JSPME, October 1966

The MTIRA, "Annual Report 1973," p. 69.

Having the effectiveness of model testing in mind, it is vital to know how to compare or interpret the test result of the model with that of the parent structure, i.e., full-size structure. In this context, the model theory must be first stated.

To satisfy the perfect similarity of the model to the parent structure, the following three scale factors should be employed in the case of static and dynamic behavior.

1. Length scale factor λ_s to maintain the geometric similarity

2. Load scale factor κ_s to maintain the similarity of static loading condition

3. Time scale factor τ_s to maintain the similarity of the vibration period under dynamic loading

Figure A2-1 shows a beam supported at both ends. By taking this example, the model theory will be explained below, where the cross-sectional area A and second moment of area I can be written as

$$A = bh \quad \text{and} \quad A_m = b_m h_m$$

$$I = bh^3/12 \quad \text{and} \quad I_m = b_m h_m^3/12$$

By substituting the length scale factor, $A_m = \lambda_s^2 A$ and $I_m = \lambda_s^4 I$.

Under static bending loading, the differential equation for the deflection y of the beam is given by

$$d^2y/dx^2 = -M/(EI)$$

where E and M are Young's modulus and the applied moment, respectively.

$$d^2y_m/dx^2_m = (1/\lambda_s)\, d^2y/dx^2$$

In addition, using the representative length L and load W,

$$M = LW \quad \text{and} \quad M_m = L_m W,$$

then
$$M_m = \lambda_s L \kappa_s W = \lambda_s \kappa_s M$$

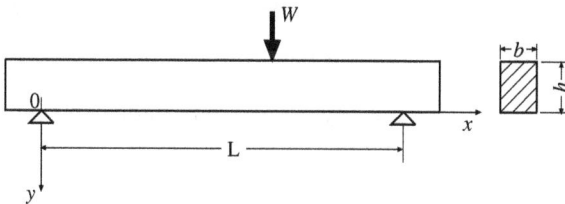

Note: Subscript m signifies model.

Figure A2-1   Beam simply supported at both its ends.

The differential equation for the model can be written as

$$\lambda_s \, d^2 y_m / dx^2{}_m = -M/(EI) \quad \text{and} \quad d^2 y_m / dx^2{}_m = -M_m/(E_m I_m)$$
$$= -(\lambda_s \kappa_s M)/(E_m \lambda_s{}^4 I)$$

Therefore,

$$\kappa_s/\lambda_s{}^2 = E_m/E$$

Providing that the model and parent structures are made of the same material, i.e., $E_m = E$, the expression yields to

$$\kappa_s = \lambda_s{}^2$$

Considering that the bending stiffness K is given by W/y, the model theory of the beam supported at both ends, and being applied to the static load, can be written as

$$K_m/K = (W_m/y_m)/(W/y) = \kappa_s/\lambda_s$$

Thus,

$$K_m = \lambda_s K$$

In the case of dynamic bending loading, the differential equation for the vibration of the beam is given by

$$EI \, d^4 y/dx^4 = -\mu d^2 y/dt^2$$

where the cross-sectional area of beam A is constant, and $\mu = A\gamma/g$ (g is acceleration of gravity, γ is specific weight). Thus

$$\gamma/E = (\gamma_m/E_m)(\lambda_s/\tau_s)^2$$

If the parent and model structures are made of the same material,

$$\lambda_s = \tau_s$$

The natural frequency of the beam supported at both ends is given by

$$f = (1/2\pi)(m\pi/l)^2 \sqrt{(EI)/(\gamma A)} \quad (m = 1, 2, 3, \dots)$$

The model theory for the natural frequency of the beam yields thus

$$f_m = (1/\lambda_s)f$$

In short, the model theories available for other cases are as shown in Fig. A2-2. Reportedly, the model theory is very simple, and the model

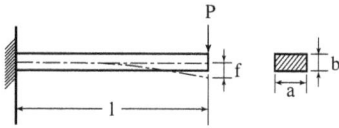

Length scale factor $l_1 : l_2 = \lambda_s$
Load scale factor $P_1 : P_2 = \kappa_s$
Time scale factor $T_1 : T_2 = \tau_s$

Loading	General model theory	The parent and model structures being made of the same material
Static	$\dfrac{\lambda^2_s}{\kappa_s} = \dfrac{E_2}{E_1}$	$\lambda^2_s = \kappa_s \longrightarrow K_1 = \lambda_s \cdot K_2$
Dynamic	$\dfrac{\tau^2_s}{\lambda^2_s} = \dfrac{\rho_1}{\rho_2} \cdot \dfrac{E_1}{E_2}$	$\tau_s = \lambda_s \longrightarrow f_1 = \lambda_s^{-1} \cdot f_2$

(a) Bending of a beam with cantilever configuration

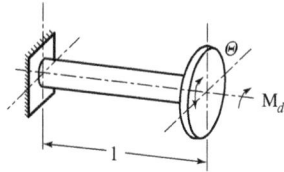

Loading	General model theory	The parent and model structures being made of the same material
Static	$\dfrac{\lambda^2_s}{\kappa_s} = \dfrac{G_2}{G_1}$	$\lambda^2_s = \kappa \longrightarrow K_{d1} = \lambda^3_s \cdot K_{d2}$
Dynamic	$\dfrac{\tau^2_s}{\lambda^2_s} = \dfrac{\rho_1}{\rho_2} \cdot \dfrac{G_1}{G_2}$	$\tau_s = \lambda_s \longrightarrow f_1 = \lambda_s^{-1} \cdot f_2$

(b) Torsion of a beam with cantilever configuration

K: stiffness, f: frequency, G: shear modulus of elasticity

Figure A2-2 Some examples of model theories.

testing is very easy to carry out, even when the materials of the parent structure and model differ. As a result, in model testing, the choice of material for the model is dependent only upon (1) the material and manufacturing costs, (2) the feasibility of manufacturing the model, and (3) the feasibility of conducting the model testing. It is, however, strongly recommended to employ the same material in the actual structure to avoid unnecessary deterioration of the testing accuracy. Tables A2-1 and A2-2 show the metallic and nonmetallic materials, which are considered suitable for model testing. Of these, the cast iron, steel, and plastics

TABLE A2-1   Metallic Materials Available for Model Testing and Their Properties

Materials	Young's modulus, $\times 10^6$ kgf/cm^2	Shear modulus of elasticity, $\times 10^6$ kgf/cm^2	Poisson's ratio
Mild steel	2.1	0.81	0.29
Cast iron	1.0	0.38	0.1–0.2
Brass	0.63	0.24	0.34
Copper	1.25	0.47	0.34
Aluminum	0.72	0.28	—

were, in general, employed, and thus Table A2-3 is a comparison of the advantages and disadvantages of these materials.

Through a series of researches conducted by Saljé, Opitz and Bielefeld, and Thornley, we can conclude that the model testing is applicable to the monolithic structural body component with accuracy better than 10%, provided that the length scale factor is larger than[2] 1/3. More specifically, we can state the following.

1. In the cases of static bending and torsion, the model testing is capable of being applied to the model, in which the length scale factor is larger than 1/3.

2. In the cases of bending and torsional vibrations, the model testing is also capable of being applied to clarify the resonance frequency and vibration modes.

Figure A2-3 shows a result reported by Thornley for the static torsional stiffness. In this case, the lathe beds examined are of steel-welded box type with and without connecting ribs (cross-sectional dimensions: 4×4 in^2 or 2×2 in^2, supporting length of large bed: around 23 in), where the connecting ribs are of Warren and Vierendeel types. The large and small beds can be regarded as the parent and model structures, respectively, and thus the length scale factor is 1/2. As can be readily seen from Fig. A2-3, the model testing shows very good results. In addition, Saljé stated on the basis of his experiment that the exponent of the length scale factor ranges from 2.7 to 2.9, resulting in 10% in the maximum difference between the experimental and theoretical scales.

---

[2]The accuracy of the model testing deteriorates in the case of dynamic loading, and the basic necessities to carry out the acceptable model testing are as follows.

1. To maintain the constant boundary conditions
2. To equip a special fixture for the model so that its support deflection does not affect the measured deflection
3. To eliminate the additional mass effect of the moving part of exciter

TABLE A2-2   Plastic Materials Available for Model Testing and Their Properties

Kind	Symbol	Specific gravity	Young's modulus, $\times 10^4$ kgf/cm^2	Tensile strength, kgf/cm^2
Methacryle	PMMA	1.17–1.20	2.46–3.52	562–773
Vinyl chloride	PVC	1.3 –1.45	2.46–4.22	352–633
Styrole	PS	1.04–1.07	2.81–3.52	351–633
Polyethylene	PE	0.94–0.97	0.56–1.05	218–387

Note: In general, the methacryle and hard-vinyl chloride plastics are employed in model testing.

A2.2   Model Testing in Consideration of Joints

As suggested from the past [2, 3], model testing shows certain accuracy deterioration when it is applied to the structure with the joint; however, there have been no attempts to challenge this subject except those of Ito, Hanks, and Schossig. Fortunately, each of these is related to the static stiffness, damping capacity, and thermal deformation, and thus it is, as shown below, very convenient to understand quickly what is the model theory and testing for the machine tool structure with the joint.

TABLE A2-3   Advantages and Disadvantages of Model Materials

Material	Advantages	Disadvantages
Cast iron	In most cases, the model and parent structures are capable of producing the same material, and thus it is easy to improve the accuracy of model testing, which is dependent of the properties of model materials. The manufacturing accuracy of model is considerably high, because machine tool manufacturers are familiar with using cast iron.	The model cost is relatively high due to the requirement of the wooden pattern to each model to be examined. The length scale factor is limited within 1/2 to1/3 minimum, because the available wall thickness of casting is 5 mm minimum. Young's modulus of cast iron changes with varying thickness, leading certain experimental errors.
Steel	The modification of a model is very easy with welding, featuring its extreme suitability for model testing. The material constants of steel do not change with its thickness. Steel is one of the most common materials in commercial use.	The technological differences between welding and casting should be considered when a parent structure is cast. The minimum thickness of steel plate is around 1 mm to avoid unnecessary welding distortion.
Plastics	The manufacture of model is easy due to the good machinability and feasibility of bonding. Large deformation is obtainable only by providing a relatively small applied load. The full fixed condition can be satisfied using a base made of cast iron or steel. The same as for steel, the modification of a model is very easy.	The elastic properties of materials depend considerably upon the circumferential temperature. The corresponding strain shows the time-dependent characteristics when the load is applied. Young's modulus shows the stress dependence. In dynamic testing, Young's and shear moduli should be corrected in consideration of the exciting frequency.

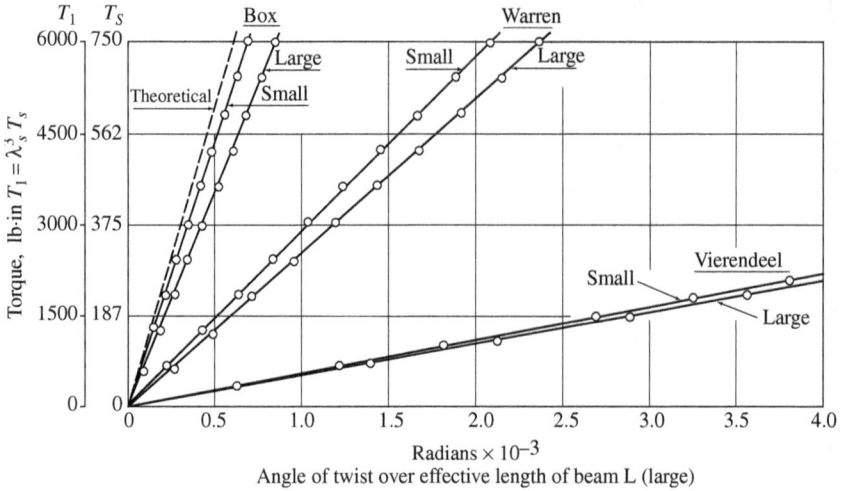

Figure A2-3   Model testing for lathe beds of boxlike type under torsion (by Thornley).

Static bending stiffness.   Following a series of basic researches, Ito and Masuko tried a research into the model theory for the bolted joint of A type [4]. On the basis of the expressions already shown in Chap. 7, they proposed and verified the model theory as follows.

For normal bending stiffness.   By considering the length and load scale factors to be the same as those for the monolithic structural body component, the due expressions yield

$$K_{om} = (E_m/E)\lambda_s K_o$$

$$F_m(\lambda,\kappa) = F(\lambda, \kappa) \qquad \text{and} \qquad G_m(\lambda, \kappa) = G(\lambda, \kappa)$$

where the suffix m designates the model. Thus

$$K_{bm} = (E_m/E)\lambda_s K_b$$

By carrying out the similar procedures mentioned above, e.g., the model theory for the elongation of the connecting bolt can be given, and in due course, we can conclude the following theory, provided that the materials of the parent and model are the same and $\lambda_s = \kappa_s^3$.

1. The static bending stiffness of the model is equal to the stiffness, which can be obtained by multiplying the stiffness of the parent structure by the length scale factor.

---

[3] In Ito and Masuko, the similarity to the deformation pattern of the bolted beam is not considered, but approved the similarity only at the loading and supporting points in the mathematical model.

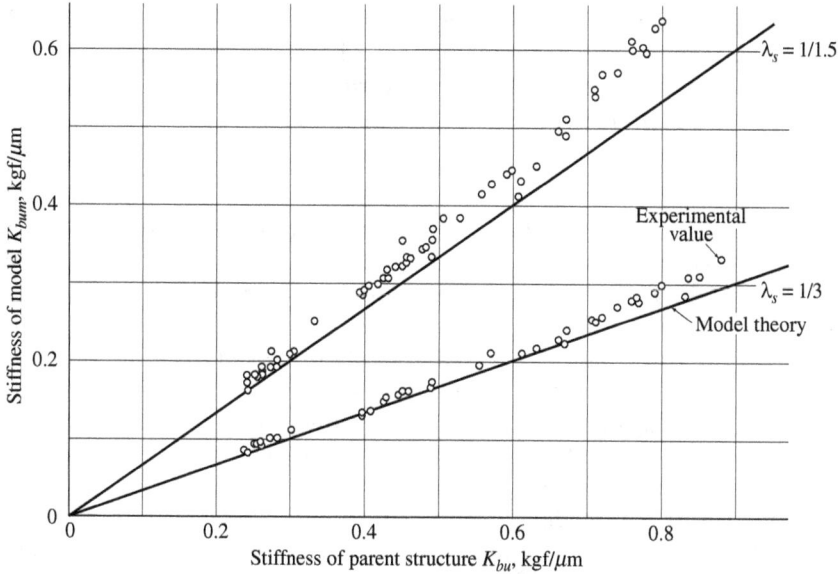

Figure A2-4 Model testing in consideration of joint: upward bending stiffness of bolted cantilever.

2. Notwithstanding the predetermining value of the scale factor, the elongation of the connecting bolt and the base compression in the model are equal to those of the parent structure.

Figure A2-4 shows one of the investigated results where the cantilever beam clamped with M18 bolts is regarded as the parent structure. From all the research results, the following interesting observations can be obtained.

1. For upward bending, the error of the model testing is around 10% maximum.

2. For downward bending, the error of the model testing is around 20% maximum.

3. The stiffness of the model is always larger than that of the parent.

For tangential bending stiffness.   In the case of tangential loading, $H_m (\lambda, \kappa) = [1/\lambda_s] H(\lambda, \kappa)$.

Thus, by manufacturing the parent and model structures with the same material and keeping their joint surfaces the same,

$$K_{bLm} = [\lambda_s K_o]/\{1 + [1/\lambda_s] H(\lambda, \kappa)\lambda_s K_o\} = \lambda_s K_{bL}$$

This model theory is as same as that for normal bending stiffness.

Assuming that the equivalence of both joint surfaces can be realized by manufacturing the flat joint surface with the same tangential stiffness, we can obtain the following condition, provided that the expression of Back et al. is available (see Table 6-8).

$$\lambda_s^2 = \kappa_s$$

Figure A2-5 shows an experimental result of the model testing. We can observe certain deterioration of the accuracy compared with that for normal bending, although we pay the special attention to the ease of the satisfaction of the boundary condition. In this case, the joint surfaces of both parent and model structures are duly produced to be the same.

In principle, the joint surface conditions, such as roughness, waviness, and machined lay orientation, should be reduced in accordance with the length scale factor in the model testing; however, this requirement is very difficult to fulfill. Thus, the joint surfaces are maintained the same in both parent and model structures, as mentioned above. As a result, it can be said that the friction may have relatively large effect on the accuracy of model testing for tangential loading.

Damping capacity of bolted joint. Hanks and Stephens [5] conducted a model testing for the damping capacity of a beam-joint assembly, in which the beam is bolted between the two angle brackets at the support, such as shown in Fig. A2-6. Because of aiming at the collection of

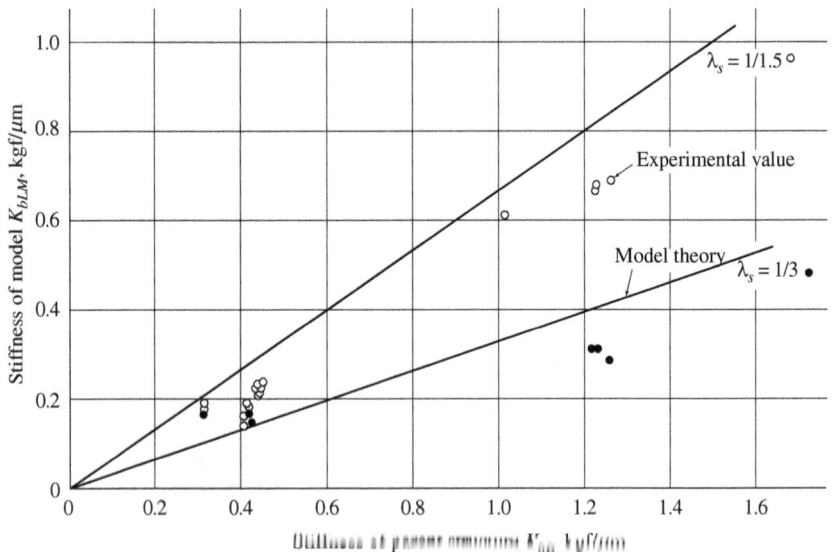

Figure A2-5 Model testing in consideration of joint : tangential stiffness of bolted cantilever.

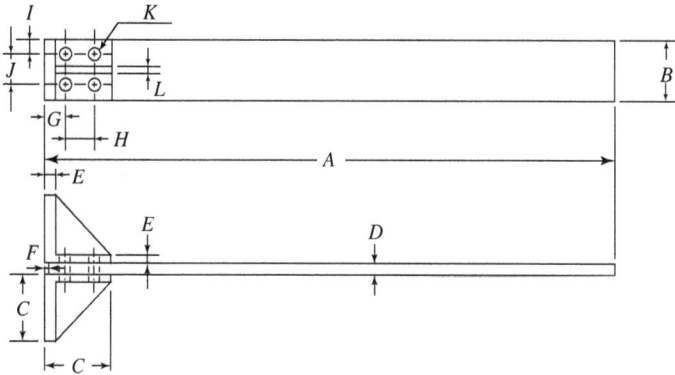

Symbol	Dimension, in	Symbol	Dimension, in
A	60.1 λ	G	2.19 λ
B	6.00 λ	H	3.13 λ
C	7.00 λ	I	1.69 λ
D	1.00 λ	J	2.63 λ
E	0.88 λ	K	1.13 λ
F	0.10 λ	L	0.63 λ

Scale factor λ: 1, 0.667, 0.333, 0.053

Figure A2-6 Bolted beam of cantilever type (by Hanks and Stephens).

engineering data for the space structure systems, both the beam and the bracket are made of 6061 aluminum alloy, and the joint surface is finished with 63 μin R_{rms}. In all the experiments, total damping in the first vibration mode was measured using a free decay vibration, under the control of the interface pressure between the beam and two brackets by the tightening torque of bolts.

Figure A2-7 is one of the measured logarithmic damping decrements, where y and h are the vibration amplitude at the beam tip and the beam thickness, respectively. As can be seen, the damping capacity decreases with the length scale factor up to 0.50 and flattens out beyond it. In addition, Hanks and Stephens investigated the effects of the interface pressure, vibration amplitude, and interfacial layer, e.g., viscoelastic film, upon the damping capacity. As exemplified by Fig. 2A-7, it is, at least, envisaged that the model theory for damping caused by the joint is not so simple as that for static stiffness, implying the further dire necessity of conducting the appropriate research.

Thermal deformation.   Importantly, Schossig tried once to establish the model theory and testing for the thermal deformation [6, see Fig. 5-23]. More specifically, he first analyzed the model theory after determining

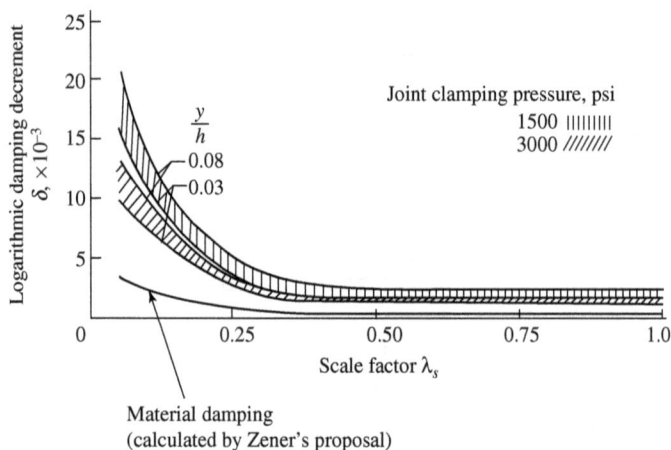

Figure A2-7   Model testing for damping capacity (by Hanks and Stephens).

several assumptions and in consideration of heat transfer through the joint surfaces. Following the analysis, the model testing was carried out using the steel welded headstocks, where the length scale factor is 0.667. The spindle displacements for both parent and model are in good agreement, although they do not show obviously to what extent the thermal contact resistance affects the accuracy of the model testing. On the basis of experiments to a various extent, Schossig reported that the model testing is available for the model having the length scale factor larger than 0.2.

References

1. Saljé, E., "Die Ähnlichkeitsmechanik ein Hilfsmittel für den Werkzeugmaschinen-Konstrukteur," in *Forschungsberichte Werkzeugmaschinenkonstruktion*, W. Girardet Verlag, 1955.
2. Opitz, H., and J. Bielefeld, "Modellversuche an Werkzeugmaschinen-elementen," *Forschungsberichte des Landes Nordrhein-Westfalen*, Nr. 900, 1960, Westdeutscher Verlag.
3. Thornley, R. H., "Models in Machine Tool Design," *J. of IPE*, 1961, 40(8): 520–541.
4. Ito, Y., and M. Masuko, "Influence of Bolted Joint on the Model Testing of Machine Tool Construction," *Trans. of JSME*, 1970, 36(284): 649–654.
5. Hanks, B. R., and D. C. Stephens, "The Mechanisms and Scaling of Damping in a Practical Structural Joint," 36th Shock and Vibration Symposium, October 1966, Los Angeles, California.
6. Schossig, H.-P., "Modellversuche über Wärmedeformationen an Werkzeugmaschinen-Bauteilen," *Fertigungstechnik umd Betrieb*, 1969, 19(2): 428–434.

Index